Studies in Applied Philosophy, Epistemology and Rational Ethics

Volume 31

About this Series

Studies in Applied Philosophy, Epistemology and Rational Ethics (SAPERE) publishes new developments and advances in all the fields of philosophy, epistemology, and ethics, bringing them together with a cluster of scientific disciplines and technological outcomes: from computer science to life sciences, from economics, law, and education to engineering, logic, and mathematics, from medicine to physics, human sciences, and politics. It aims at covering all the challenging philosophical and ethical themes of contemporary society, making them appropriately applicable to contemporary theoretical, methodological, and practical problems, impasses, controversies, and conflicts. The series includes monographs, lecture notes, selected contributions from specialized conferences and workshops as well as selected Ph.D. theses.

Advisory Board

More information about this series at http://www.springer.com/series/10087

Donna E. West · Myrdene Anderson
Editors

Consensus on Peirce's Concept of Habit

Before and Beyond Consciousness

Editors
Donna E. West
Department of Modern Languages
State University of New York at Cortland
Cortland, NY
USA

Myrdene Anderson
Department of Anthropology
Purdue University
West Lafayette, IN
USA

ISSN 2192-6255 ISSN 2192-6263 (electronic)
Studies in Applied Philosophy, Epistemology and Rational Ethics
ISBN 978-3-319-45918-9 ISBN 978-3-319-45920-2 (eBook)
DOI 10.1007/978-3-319-45920-2

Library of Congress Control Number: 2016950368

Printed on acid-free paper

This Springer imprint is published by Springer Nature
The registered company is Springer International Publishing AG
The registered company address is: Gewerbestrasse 11, 6330 Cham, Switzerland

This volume is dedicated in loving tribute to John and Brooke Williams Deely, for their unceasing devotion and scholarly contributions nationally and internationally to the field of sign studies and Semiotics at large. Their legacy will endure for generations.

Preface

The 29 contributors featured in this volume make their home in no fewer than 12 countries in Europe and North and South America, and represent at least a dozen disciplines, including:

- philosophy,
- linguistics,
- psychology,
- cognitive science,
- biology,
- anthropology,
- sociology,
- communication studies,
- literature,
- pedagogy.

The co-editors themselves straddle expertise in semiotics with linguistics, psychology, and anthropology. We have long allowed ourselves the habit of being fascinated with Peircean habit, having also organized a number of symposia on the subject at national and international congresses.

One reason for habit commanding such a presence in all these fields of scholarship concerns its relevance to all conduct, beyond human and living realms. Three overarching themes are addressed: the etiology of habit, its characterization as continua, and the mental perpetuation of habit as a self-modifying system, incorporating unconscious and conscious patterns of behavior/reasoning.

Given the recent celebration of the centennial of Charles Sanders Peirce's death, it is especially fitting to present an integrated account of his concept of habit, since (above any other ontological, phenomenological, or semiotic issue) it demonstrates how Peirce ultimately consolidates individual physical and mental worlds into his continuum. In response to the recent outcry against psychologism among some Peircean philosophers, the present consensus provides a fresh perspective. Without violating Peirce's metaphysical or semiotic model, the present volume highlights the primary call beckoning Peirce (consonant with his commitment to forums of realism and pragmatism)—that all organisms (despite the particularities of their distinct physical and cognitive niches) realize individual features, while, at the same time, constitute expressions of Peirce's continua.

This anthology fills a substantial gap in the literature; it addresses, in a single volume, the breadth and influence of a topic heretofore left fallow. It offers a cross-disciplinary account of the continuity among fields of study which, on their face, appear disparate. The chapters in this volume illustrate how Peirce unifies the physical with the phenomenological, and the social with the psychological. It likewise showcases how (over the course of forty plus years) Peirce, via his construct of habit, integrates highly distinct sources of action and cognition—predetermined systems of knowledge and action (inclusive of a priori endowments) together with learning paradigms of implicit/explicit instrumentation. It is this latter application of habit which demonstrates that no perception or cognition originates from nothing. Earlier images, propositions, theorems, or models of belief constitute remnants for potential habit.

Cortland, NY, USA Donna E. West
West Lafayette, IN, USA Myrdene Anderson

Acknowledgments

We acknowledge all of the contributors to this volume, who have received no benefit but for significant renovation of an emerging field in Peirce Studies. In rallying to the cause of extending Peircean scholarship, they patiently revised multiple drafts of their respective chapters.

Special gratitude is accorded to Lorenzo Magnani, SAPERE Series Editor—for his support and ingenuity, and to Leontina DiCecco, Springer Editor, whose spirit of congeniality made the production process a pleasurable experience.

André de Tienne and Ken Ketner, together with their graduate assistants and staff at the Peirce Edition Project and the Institute for Studies in Pragmaticism, have constituted an invaluable source to access unpublished Peircean manuscripts. Their generosity permitted the inclusion of several manuscripts and chronology, which otherwise would have become a far more arduous process.

We likewise express our gratitude to Purdue University, for granting Myrdene Anderson a sabbatical leave during the editing of this volume.

Of particular mention are Donna West's and Myrdene Anderson's graduate research assistants and colleagues, for their tireless attention to technical and textual details.

Contents

Editors and Contributors

About the Editors

Donna E. West is Professor of Modern Languages and Linguistics at the State University of New York at Cortland (USA), teaching: Cognitive Linguistics, Language Acquisition, and Spanish Phonology/Grammar. For more than 25 years she has been presenting and publishing internationally in Semiotic studies using Peirce's sign system, and comparing it to semiotic properties in the works of Karl Bühler, Lev Vygotskii, and Jean Piaget. She is the first investigator to apply a developmental Psycholinguistic perspective (supplying fresh data) to Peirce's ten-fold division of signs; as such, her work offers empirical answers to phenomenological questions. Her 2013 book, *Deictic Imaginings: Semiosis at Work and at Play* (Springer-Verlag), investigates the role of Index in the acquisition of demonstratives and personal pronouns. The impetus for this anthology derives from her longstanding fascination with how Peirce's concept of habit relies chiefly upon index's influence on event processing.

Myrdene Anderson Associate Professor of Anthropology, West Lafayette, Indiana, USA, has served as president of the Semiotic Society of America and of the Central States Anthropological Society. Primarily an ethnographer, her settings and topics range from Saami reindeer-breeding in Lapland, to community gardening in the USA, to the artificial life movement in biology, and to cognitive and semiotic issues in mathematics education. Her publications include edited volumes on human-alloanimal ethnology, on ethnicity and identity, on the cultural construction of trash, on semiotic modeling, on mathematics education, on violence, and now on Peircean habit.

Contributors

Atocha Aliseda holds a Ph.D. in Philosophy and Symbolic Systems from Stanford University (1997) and did a postdoc at Groningen University (2000–2002). She is Full Professor at the Institute for Philosophical Research in the National Autonomous University of Mexico (UNAM). She has published and edited a number of books and articles on Logic and Philosophy of Science and specializes on abduction and the logics of scientific discovery, the topics of her first monograph *Abductive Reasoning: Logical Investigations into Discovery and Explanation* (Springer 2006).

Fernando Andacht is Full Professor at the Department of Theory and Methodology, Facultad de Información & Comunicación, Universidad de la República, Montevideo. He does research on Peircean semiotic and its application to visual media, culture, and society. His recent publication: "The lure of the powerful, freewheeling icon: On Ransdell's analysis of the distinctive function of iconicity" (2013).

Selene Arfini Department of Philosophy, Education and Economical-Quantitative Sciences, University of Chieti and Pescara, Chieti-Pescara, Italy.

Pedro Atã is a Ph.D. candidate and a member of the Iconicity Research Group (Juiz de Fora, Brazil). He has been working on the distributedness of meaning in material and social contexts, particularly intermedial phenomena and cognitive niches. His work also examines creativity and aesthetics. The title of his Master's thesis is "Niche Builders: Towards Art as Meta-semiotic Engineering". His recent publications include "Semiotic niche construction in musical meaning", "Creativity as niche construction and some examples in theatrical dance", and "Iconic semiosis and representational efficiency in the London Underground Diagram".

Francesco Bellucci currently a research fellow at the Tallinn University of Technology in Estonia, received a Ph.D. in 2012 from the University of Siena, Italy —dissertation on the history of diagrammatic thinking. His research focuses on Peirce's logic. His recent publications are in *Versus*, *Transactions of the Charles S. Peirce Society*, *Journal of the History of Ideas*, *History and Philosophy of Logic*, *British Journal for the History of Philosophy*, and *The Review of Symbolic Logic*.

Mats Bergman is a senior research fellow of the Academy of Finland, affiliated with the Department of Social Research of the University of Helsinki. He is the author of *Peirce's Philosophy of Communication* (Continuum, 2009) and the principal editor of *Commens: Digital Companion to C. S. Peirce*.

Tommaso Bertolotti Department of Philosophy and Computational Philosophy Laboratory, University of Pavia, Pavia, Italy.

Elize Bisanz is Professor of Culture and Communication, Faculty of Culture Studies, Leuphana University, in Luneburg, Germany. As a Charles S. Peirce scholar, she is an active research member of the Institute for Studies in Pragmaticism at Texas Tech University.

Sara Cannizzaro is Research Assistant at the Centre for Academic Practice Enhancement at Middlesex University. Her Ph.D. thesis at London Metropolitan University concerned "Biosemiotics as Systems Theory: an Investigation into Biosemiotics as the Grounding for a New Form of Cultural Analysis". Other works span biosemiotics, Peircean semiotics, cybernetics, systems theory and media theory, such as "Transdisciplinarity for the 21st century", or "biosemiotics as systems theory", "Where Did Information Go? Reflections on the Logical Status of Information in a Cybernetic and Semiotic Perspective", "The Line of Beauty": on Natural Forms and Abduction".

Vincent Colapietro is Liberal Arts Research Professor in the Departments of Philosophy and African American Studies at Penn State University. One of his main areas of historical research is classical American pragmatism, with special emphasis on C.S. Peirce's writings. One of the principal foci of his more strictly philosophical interests is the philosophy of mind. In his contribution to this volume, these two preoccupations are, yet again, brought together. While he has written on a wide range of topics (from music, in particular jazz, to cinema, from the philosophy of history to metaphysics, and from the philosophy of mind to metaethics), he is first and foremost a scholar of pragmatism devoted to showing in detail how the thought of Peirce especially can make an important contribution to contemporary thought.

John Coletta is Professor of English at the University of Wisconsin-Stevens Point, Stevens Point, Wisconsin 54481, USA (jcoletta@uwsp.edu). A former President (2010) and Vice President (2009) of the Semiotic Society of America, Professor Coletta's principal research interests include Peircean semiotics, eco- and biosemiotic criticism, the major British poet and minor naturalist John Clare, poet Samuel Taylor Coleridge and his theory of the symbol, and posthumanism and "critical animal studies". Professor Coletta sits on the Editorial Board of *The American Journal of Semiotics.*

Scott Cunningham approaches completion of Ph.D. in Educational Psychology at Texas Tech University in Lubbock, Texas. There, he is currently the Assistant Director at The Institute for Studies in Pragmaticism. His research area of interest concerns how cognitive models of higher order mental processes, specifically the making of valid inferences, relates to the underlying structures and neural circuitry of the developing as well as the mature human brain.

Dinda L. Gorlée is a semiotician and multilingual translation theorist with research interests in interarts studies. Her most recent academic function was Visiting Professor at the University of Helsinki. Gorlée is Research Associate of *Wittgenstein Archives* at the University of Bergen and Associate Editor of *American Book Review* at University of Houston, Victoria, TX. Gorlée is widely published internationally. Her books include: *Semiotics and the Problem of Translation: With*

Special Reference to the Semiotics of Charles S. Peirce (1994), *On Translating Signs: Exploring Text and Semio-Translation* (2004), *Song and Significance: Virtues and Vices of Vocal Translation* (2005), *Wittgenstein in Translation: Exploring Semiotic Signatures* (2012), and *From Translation to Transduction : The Glassy Essence of Intersemiosis* (2015).

Juuso-Ville Gustafsson is a research student at the Department of Behavioural Sciences and Philosophy, University of Turku.

Nathan Houser is Professor Emeritus of Philosophy at Indiana University in Indianapolis (IUPUI) and President of the Charles S. Peirce Foundation. He has served as Director of the Peirce Edition Project and the Institute for American Thought and as President of the Charles S. Peirce Society and the Semiotic Society of America. From 1993 to 2009 he was General Editor for the Indianapolis critical edition of Peirce's writings and he co-edited the two-volume *Essential Peirce* and *Studies in the Logic of Charles Sanders Peirce*.

Erkki Kilpinen is currently Adjunct Professor at his former alma mater, the University of Helsinki, where he previously has served as Professor and as Senior Lecturer. In 2009–2010 he held a Fellowship at the Swedish Collegium for Advanced Study (Uppsala). He has published internationally on semiotics, philosophy, and social theory and its history.

Lorenzo Magnani Department of Philosophy and Computational Philosophy Laboratory, University of Pavia, Pavia, Italy.

Winfried Nöth Professor of Linguistics and Semiotics and Director of the Interdisciplinary Center for Cultural Studies of the University of Kassel until 2009, is Professor of Cognitive Semiotics at the Catholic University of São Paulo and, from 2014 to 2015, Visiting Professor at Humboldt University zu Berlin. He is an honorary member of the International Association for Visual Semiotics and former president of the German Association for Semiotic Studies. Nöth's 300 articles and 31 authored or edited books (such as *Handbook of Semiotics, Self-Reference in the Media*) are on cognitive semiotics, semiotic linguistics, semiotics of literature, images, maps, the media, systems theory, culture, evolution, and Charles S. Peirce.

John Pickering has degrees from Edinburgh and Sussex universities in the UK. After postdoctoral fellowships in the US, at Rochester and Stanford, he has been at Warwick University, UK, where he lectures on psychology and environmental issues. His principal research interests are consciousness, process thought, ecological psychology, and biosemiotics.

Ahti-Veikko Pietarinen is Professor and Chair of Philosophy, Tallinn University of Technology in Estonia. Pietarinen works on pragmatistic philosophy of science, logic, mathematics, theories of diagrams, language and action. He has recently published in *Synthese*, *International Journal of the Philosophy of Science*, *Axiomathes*, *The Review of Symbolic Logic*, and *Journal of Logic, Language, and Information*. His books include *Signs of Logic* (2006) and *Game Theory and*

Linguistic Meaning (2007). Currently he is working on the edition of Peirce's Existential Graphs (*Logic of the Future*, forthcoming).

João Queiroz is Professor at the Institute of Arts and Design, Federal University of Juiz de Fora, Minas Gerais. He has taught courses on cognitive semiotics, Peirce's Philosophy, intermediality studies. He is the director of the Iconicity Research Group (Juiz de Fora, Brazil), a director member of the International Association for Cognitive Semiotics (IACS), member of the Linnaeus University Centre for Intermedial and Multimodal Studies (Vaxjo, Sweden), and member of the Group for Research in Artificial Cognition (Feira de Santana, Brazil). His research interests include cognitive semiotics, Peirce's semiotics and pragmatism, Brazilian and South American arts and literature. His most recent publications are "A Lógica de Diagramas de Charles S. Peirce", Commens—Digital Companion to C.S. Peirce (co-edited with Mats Bergman and Sami Paavola). He is currently preparing a book on creativity and distributed cognition with Pedro Atã.

Stanley N. Salthe has been Emeritus in Biology at the City University of New York, since 1992 as Visiting Scientist in Biological Sciences at Binghamton University. From biology, to systems science and semiotics, to natural philosophy, his developmental version is founded in the Big Bang concept of cosmology, but respects an understanding of qualia, rather than for a totalized materialist construct. His books range from *Evolutionary Biology* (Holt, Rinehart, Winston, 1972) through *Evolving Hierarchical Systems: Their Structure and Representation* (Columbia University Press, 1985) to *Development and Evolution: Complexity and Change in Biology* (MIT Press, 1993).

Lucia Santaella is Full Professor at São Paulo Catholic University. She has published 42 books, organized 14 books, and also published around 300 articles in journals and books in Brazil and abroad. She was awarded the Jabuti Prizes (2002, 2009, 2011, 2014) the Sergio Motta Prize in Art and Technology (2005) and the Luis Beltrão Prize (2010). She is a CNPq researcher 1-A.

Göran Sonesson is Professor of Cognitive Semiotics at Lund University. Apart from the book *Pictorial Concepts* (1989), he has published numerous articles about pictorial semiotics, semiotics of culture, urbanism, evolutionary theory and several experimental studies in these domains.

Frederik Stjernfelt is Professor of Semiotics, Intellectual History, and Philosophy of Science at the Department of Communication, University of Aalborg at Copenhagen.

List of Figures

List of Tables

List of Appendices

Chapter 1
Preamble—Peircean Habit Explored: Before, During, After; and Beneath, Behind, Beyond

Myrdene Anderson

Abstract Charles Sanders Peirce, far more than any scholar in recent centuries in the West, devoted much of his productive life to probing the idea of and behavior of "habit". In order to do so, he both narrowed and widened his focus on the notion. Otherwise an ordinary term in quotidian use in English, habit suggests regularity, usually pertaining to individual human behavior—but Peirce also focused on habit's utility for understanding behavior beyond the human, and even processes beyond the organic world. In pursuit of refining and operationalizing habit, Peirce drew on any number of disciplines, close to and far from his expertise—these spanning from logic/philosophy, biology and psychology to theology and cosmology. Peirce's foundational work on habit continues to be irresistible for contemporary humanities scholars, social scientists, scientists, and for practitioners beyond the academy diagnosing the ills of self and society, as Peirce's oeuvre in its infra-dialectical form (frequently in fragmentary paragraphs), cannot be satisfactorily appreciated through any rear-view mirror. Rather, one might say that, in recognizing the habits behind habits and habit-change—whether confirming them through belief or challenging them through doubt—Peirce still invites us to permute, expand, contest, and refine his explorations of a century ago.

Keywords Habit-taking · Habit of habit-taking · Habit-breaking · Culture · Addiction · Peirce's categories

Habit, a Rorschach

Over the most recent centuries, at least in the English language, any number of scholars have contributed to the liberation of the term "habit" from its quotidian usage for a regularity in behavior. Those philologically and philosophically adept trace the notion of habit back to or before the Greeks (cf. Aristotle, also Bernacer

M. Anderson (✉)
Purdue University, West Lafayette, IN, USA
e-mail: myanders@purdue.edu

D.E. West and M. Anderson (eds.), *Consensus on Peirce's Concept of Habit*,
Studies in Applied Philosophy, Epistemology and Rational Ethics 31,
DOI 10.1007/978-3-319-45920-2_1

and Murillo 2014). We cannot help but look forward to other linguïcultural discourses that will surely inflect this concept in manifold ways (cf. Anderson and Gorlée 2011; Durst-Andersen 2011), to consider a near-infinity of distinctly different—in degree and/or in kind—realms of habits, or perhaps to consolidate habit into a concept with more manageable elements. Meanwhile there is enough on our plate, just from Charles Sanders Peirce. Peirce, and indeed other 20th-century pragmastists including William James, George Herbert Mead, John Dewey, and Thorstein Veblen, made the pursuit of the study of habit a veritable habit in itself.

In contemporary parlance, consciously noticed and perhaps even labeled human habits may be good or bad, or better or worse, or—less probable—neutral. That is, habits are usually "marked": they are figures on a ground of unmarked behavior (cf. Jakobson 1970; Waugh 1982). That very unmarked, routine ground of behavior, though, will already be shaped by habit, as many phenomena potentially qualifying as habits may rest below the radar, being unmarked, unremarked, or all-but-imperceptible: whether we lead with right foot or left, how we braid hair, which side of our body we situate on which side of the bed.

Habit may involve either conscious or non-conscious states, when focusing on humans and other animals. Peirce asserted that, rather than persons having habits, it is habits that have us—that we dwell in and through habits. By extension, culture would be itself little beyond habit and likewise so would be individual personalities. This recursivity and layeredness makes demands on those scholars presumptuous enough to ponder the development of habit, as did Peirce throughout his life, 1839–1914. Reflecting on this observation, habit may be to humans as water is to fish.

Peirce, the Darwinian

Although Peirce typically used individual humans as units of analysis in his philosophizing about habit, he also adventured far from just *Homo sapiens*, even beyond animals in general. Peirce had no hesitance in assuming plants to have habits (CP 5.492, 1907; see CP 6.17, 1891; 6.300, 1893, as did Darwin (1809–1882)), his 30-year-older near look-alike, in his foundational *On the Origin of Species by Means of Natural Selection* (1859).

Doubtless Peirce as well as Darwin would have embraced fungi, monera, and protoctista along with animals and plants as habit-taking and habit-breaking, had they lived in the latter 20th century. That's when science and society were introduced to Robert Whittaker's and Lynn Margulis' revelation of the five-kingdom taxonomy in biology (cf. Margulis 1971) (Margulis herself was a semiotician, so convinced by Thomas A. Sebeok!).

Indeed, Peirce reveals himself to be a generalist when dealing with this generalizing phenomenon of habit, eventually extending the notion beyond the "living" part of the "natural" world, to consider fossils and crystals and landscapes as habit-taking. A river channel would then be the temporal response to the river's own habit of dancing to gravity. Or, the whole stone age of prehistoric human culture would be a response to flint's own habit of reliably breaking through conchoidal fractures into flakes affording their fashioning into tools and weapons. Such thinking also anticipated Uexküll's introduction of "*Umwelt*" (1957[1934]), from German, now well on the way of being naturalized, habituated, as "umwelt" in English, and also Gibson's ecological notion of "affordance" (1979: 127–143).

Habit in any event is predicated on a regularity in time and place, on endurance, on persistence, with or without palpable modification. For habit to be endowed with regularity, it must be at least implicitly trusted, or viable, for some period. Hence a key term used by Peirce to understand habit is belief, defined by him as "a state of mind of the nature of a habit" (EP 2: 12), "nature" meaning the primary or original nature of "mind". Following Peirce, belief has instinct as its ultimate cause (CP 5.358–387, W 3: 242–257, 1877); it is considered by Peirce as "…an intelligent habit upon which we shall act when occasion presents itself" (EP 2: 19, 1895).

In the inert world, Peirce accepted patterns, typically perceived visually in space, to align with explanation involving regularities through time—either in the material world observed or/and in the mental world of the observer. Our science continues to have the habit of visual metaphors and modeling (cf. Anderson and Pettinen 2014; Henry 2005). The processes behind these various units of analysis—persons, societies, landscapes, crystals—have habits, and are also constituted by habits. Peirce emphasized dynamical process over static units of analysis: that we have habits boils down to habits having us, that we dwell in and through habits, just as we are in thought rather than thoughts being in us (W 2: 227, 1868).

Although Peirce's attention did tilt toward the individual as a unit for philosophical and psychological analysis, he never sacrificed consideration of social collectivities or the myriad of other interpretable habits constituting if not steering the cosmos as a whole. His respect for the dynamics of biology, and of both the deeper structures of culture and of its surface manifestations in society, links most closely to the emerging discipline of psychology, without succumbing to the mechanistic determinisms occasionally endemic in that and many other fields.

Peirce, the Proto-evolutionist

Concerning the triadic nature of the Peircean categories of firstness, secondness, and thirdness—while habit will most prominently be referenced to the regularities, generalities, and laws of thirdness, Peirce allowed that habit manifests also in firstness, shared with sensation and instinct, and in secondness, with sensation and "the 'I will' of volition" (MS 930). Furthermore, regularities in empirically-accessed outward behavior have antecedents and analogues (though never causes) in inner behavior and experience, and vice versa. Relations and process must be respected. In contemporary semiotic discourse, whether Peircean or not, linear thinking is eschewed, along with any assumptions of closed loops of cause and effect. Peirce and his contemporaries refer to "determination" in a metaphorical, not literal, sense, not today to be taken literally. Likewise, Peirce's frequent mention of "evolution" will not strictly denote or connote the dynamics of the evolution of today—succinctly, as the accumulation of information underlying phylogeny—vis-à-vis the sibling notion of development, involving stagal unfolding of evolution in ontogeny (cf. Salthe 1993), both evolution and development subsumed in ecologies of significant surrounds, Umwelten (cf. Uexküll 1957[1934]).

Peirce did assume that good habits would grow, somewhat linked to their association with palpable pleasure in the individual and by extension and analogy, in the collectivity. Good and bad habits could only be so judged with reference to context, individual and social, or even imaginary ones. The standard of good habits was conceptually unlimited. In contrast, bad habits would be literally and figuratively self-limiting. Peirce relied on a Darwinian model of natural selection to explain the extinction of bad habits:

> For every kind of organism, system, form, or compound, there is an absolute limit to a weakening process. It ends in destruction; there is no limit to strength. The result is that chance in its action tends to destroy the weak & increase the average strength of the objects remaining. Systems… which have bad habits are quickly destroyed, those which have no habits follow the same course; only those which have good habits tend to survive. (EP 1: 223)

As an aside, it should be noted that by the late 20th century, the discourse within biology and semiotics—the Evo-Devo and now Evo-Devo-Eco discourses—have refined the classic model of natural selection, particularly in the discipline of biosemiotics. The International Society for Biosemiotic Studies, established in 2005, produces the flagship journal (*Biosemiotics*, since 2008) and holds Annual Gatherings in Biosemiotics (since 2001). These venues draw Peircean and other scholars and scientists from disciplines beyond biology and philosophy. This current volume, in part, feeds out of, and certainly back into, biosemiotics as well as many other fields, primarily philosophy.

Particularly since the formulations of 1965 Nobel Laureates François Jacob and Jacques Monod (cf. Monod 1971), the natural selection recognized by Peirce only pertains to the so-called terminally unfit who decease (but for any reason) before reproduction, along with the potentially fit but unlucky individuals who happen to be in the wrong place at the wrong time. There can be no literal "survival of the fittest", but only the admittedly teleological "survival of the not terminally unfit". Following these observations, note Monod's (1971) couplet, "necessity" and "accident" in the title of his celebratory book. "Necessity" pertains to the "why" of the removal of the terminally unfit, and "accident" describes the "how" of the stochastic events beyond prediction or control. In Peircean models, "necessity" resides in thirdness and/or secondness, and "accident" in firstness and/or secondness. These fuel the emergence, maintenance, and modification-unto-deletion of habit with respect to any medium, living or nonliving.

While scholars still tend to overemphasize and overgeneralize natural selection, even referring to "adaptation", few of them bother refining the parallel notion of artificial selection. Natural selection, as outlined above, largely relies on "selection-out" by necessity and accident; artificial selection largely operates within culture or any other medium through deliberate but perhaps unconscious "selection-in", hardly a viable long-term strategy. Think of faster horses or thornless roses: through positive artificial selection one may make some strides in those directions, albeit only motivated by cultural whims. At the same time, though, any number of other traits, whether beneficial or not, if linked with the speed of horses or the lack of thorns, will come along for the ride and risk becoming indelible; some such traits are bound to be deleterious, even terminally so.

Deleterious habits—so deemed by their incumbents or/and by their subsuming societies or/and by their underlying physiologies—persist, despite Peirce's evident hubris in this quote, unabated until and unless they are terminally unfit, that is, fatal, or until accident (as abductive inspiration) or necessity (intervening conscious self-control) might rise to the occasion.

Peirce, the Systems Scientist

There are obvious and even explicit links between habit, at any level and of any sort, with culture. Culture, the deeper-structure of a society, animated by the living things constituting it and afforded by both the living and the inert, is a dynamical system and far-from-equilibrium, hence, a dissipative structure (cf. Prigogine and Stengers 1984/1979). Semioticians recognize that, in the interpretant and notions of habit-taking, Peirce anticipated anticipatory theory itself, as eventually formulated by mathematician and semiotician Rosen (1985), and the same connections obtain regarding the attractor sink of far-from-equilibrium dynamical chaos theory, associated with Nobel Laureate Ilya Prigogine (cf. Prigogine and Stengers 1984/1979), and catastrophe theory by mathematician and semiotician Thom (1968, 1975). Cultural processes are both elastic (malleable, analogue, continuous) and plastic (fracturable, digital, episodic) in resonance with catastrophe theory. Culture is also an anticipatory system. Cultural processes dwell in and on habits, those that are social as well as those arising through individual behavior.

Peirce outlines many instances linking the regularity of habits at any scale with encompassing processes. These observations concerning habits in and of culture, together with the general notion of addiction, render culture ever more amenable to scientific analysis, in contrast to the earlier "habit" of relating culture to concrete instantiations and often material details and/or to ever more tenuous abstractions (cf., Kroeber and Kluckhohn 1952; Lovejoy 1936; Williams 1985[1981], 2014 [1985, 1976]). Culture saturates its humans (and their other creatures, their built environments, even their technologies [cf. Bryant 2014]) with habits. Yet habits also antecede culture and life itself, as Peirce recognized. Biological processes in league with culture endow living constituents with habit-like regularities, amongst them instincts and intuitions, however conceived.

While habit in its suspenseful developmental trajectory (cf. Salthe 1993) cannot be avoided in the logical analysis of semiosic systems; its dialogue with evolutionary emergence and surprise allows for nonlogical creativity to have the primary and definitive word—unpredictability, complexity, chaos, nothingness, paradox, ambiguity, inconsistency, uncertainty—short of any ultimate interpretant.

Belief and Doubt on the Playing Field of Habit

For Peirce, belief is opposite to doubt. Indeed "Doubt… [is] the privation of a habit… [that is] a condition of erratic activity that in some way must get superseded by a habit" (EP 2: 336–337). "Error" in this sense is a fortuitous event which brings about a possible increase in the variety of habits, and it proves fundamental in the evolution and development of habit. In obsession and addiction as well as in normative culture, an infinitely receding utopian completion of all things with their entrenched beliefs can swamp curiosity—and the "chance" (accident) as well as the "necessity" of creativity, inclusive of evolution.

For Peirce, habit-taking at any level, like life itself, grows, or not, dies with or without replacement. Habits that may be taken, affording general traits of a species or culture, for instance, render the receiver more general and analogous to a type, while at the same time individuating instances at the level of token. At the level of tone, habit might best be visualized as pre-actualized instinct. It is interesting that instinct, or inclination without any labeled affordance, seems to be reserved as an attribute to animals with central nervous systems, while other forms of life and the inert world do not have instincts, their agency-free habits emerging even in firstness.

In the 21st century, we are discovering that many of the distinctions assumed between life forms—for instance, between humans and alloanimals, and between animals generally and plants, and between any of the above and other forms of life —have been gradually lessened or even eliminated. This is the result of better semiotic understanding of life and living, and of increased detection of communication within and between units of analysis, whether that be individual organism, a group of conspecifics, or unrelated species in proximity or even at distances from each other.

Habit—Unto Addiction?

As mentioned, Peirce remained broad in probing habit, inferring habit in the inert physical world as well as in the natural living world, and in that paradigm case of human mental and behavioral practices. These pursuits lead naturally to a consideration of that marked manifestation of habit beyond just a "bad habit", to what is colloquially labeled "addiction". Peirce left himself open to the possibility of addiction, not only perhaps colored by his own habits-unto-addictions (Brent 1993), but because his models recognized that habit-taking was a recursive process—leading to habits of habit-taking, and eventually to the growth and spreading of viable habits, whether they be judged better or not.

Complementing Peirce's dynamical approach to habit-taking, he considered habit-breaking. In light of this dynamism, the popular idea that addiction might be either a "reversible" or "irreversible" habit seems closed to both context and reflexivity, and also appears oblivious to the fact that addictions—perhaps just extreme habits—will always be bound up with culture, language, and historical period. Given that Heraclitus himself would blush at either "reversible" or "irreversible" habits, it seems a deleterious semantic habit may be impeding both the understanding and the treatment of a recognized social problems, tamed under the single rubric of "addiction".

Both passive and active venues of habit may ratchet up to other levels of often involuntary compulsion, and display the quite distinct condition of self-patience, unto addiction. In fact, through the opaque looking-glass of habituation, the facile entrainment of habit loops back—sometimes autocatalytically if not grammatically —transforming actors not to objects but to literal and figurative patients (or abductive agents), as they are digested by the overarching linguïcultural system (Anderson and Gorlée 2011) to intermix good and bad experiences. Finally, in MS 930 (1913), Peirce introduces just once the term "habituescence," and succinctly ponders the correlates of habit-taking. For habituescence, Peirce pulled into the foreground the "consciousness of taking a habit" (MS 930: 31, 1913; cf. West 2014 for further discussion), that would be grounded in thirdness. This consciousness also obtains in maintaining a habit or addiction, and in the self-control consistently foregrounded by Peirce as an active ingredient in habit-change. In all these processes, habits do not reside in either agents or patients, but in all possible interactive relations uniting them. Habituality (or perhaps habituescence) then becomes ecological, as the threads connect through both time and space. Ecology, as the semiotic science of relations, finds habits-unto-addictions all the way from the proximal overdetermined pathologies of personal relationships all the way to the distal entitlements assumed to guarantee our nest in the cosmos.

Habit, Self, and Self-Control

In reflecting on habit, one returns time and time again to Peirce, and to other pragmatists, while at the same time one is thrust forward into the throes of contemporary science in cognition and neurology, not to mention up against the very pragmatic issues arising with respect to mental health, social dysfunction, and ecological stress. Peirce often captures our attention by forced, inverted, and inside-out analogies; perhaps the cartoonist Walt Kelly was inspired by Peirce when he had Pogo declaring, "…We have met the enemy and he is us."[1]

---

[1]Pogo's iconic remark appears in the "Pogo" daily strip upon finding nature sullied by trash, Earth Day, 1971.

The processes behind various units of analysis—persons, societies, landscapes, crystals—have habits, and are also constituted by habits. Peirce emphasized dynamical process over static units of analysis: that we have habits boils down to constellations of old and new habits having us, that we dwell in and through habits, just as we are in thought rather than thoughts being in us (W 2: 227, 1868). Peirce's eschewing of static models of necessarily dynamical realities, anticipated the contemporary mood in science, taking physics as the least likely type specimen to declare, as has Henry (2005): "The only reality is mind and observations, but observations are not of things. To see the Universe as it really is, we must abandon our tendency to conceptualize observations as things." This also argues against privileging the single sense of vision, and leaves open the distinctive shaping of sensation, perception, and conception (cognition) by our various human languages (cf. Durst-Andersen 2012).

Understanding "self" as a manifestation of a bundle of habits, implicates "self-control" as a player. It happens that contemporary research in the social and behavioral sciences seems to back into the study of self-control, though seldom via Peirce's insights. In the longitudinal Dunedin project continuing past 40 years in New Zealand, self-control became associated with "health" (absence of bad habits), "wealth" (presence of good habits, of living within one's means), self-ascribed satisfaction, and, coincidentally, living within the law (Moffitt et al. 2013).

In another project, an economist found himself forced to study linguistics and in particular languages to explain the disparity of economic behavior apparent even in aggregate statistics between nation-states (Chen 2013). Here, there were also associations between "health" and "wealth"—with languages free of any obligatory future tense. Speakers of languages with a handy future tense (interestingly, not associated with language families, but English and Greek would still be classified together), are not thinking, speaking, behaving as though there is a tomorrow, when in the present tense one can defer any consideration of the future.

Probably no single angle on habit will turn out to be so significant to the contemporary world as self-control.

References

Anderson, Myrdene, and Dinda L. Gorlée. 2011. Duologue in the familiar and the strange: Translatability, translating, translation. In *Semiotics 2010*, ed. Karen Haworth, Jason Hogue, and Leonard G. Sbrocchi, 221–232. Ottawa: Legas Press.

Anderson, Myrdene, and Katja Pettinen. 2014. Perception: Seeing is believing? In *Semiotics 2013*, ed. Jamin Pelkey, and Leonard G. Sbrocchi, 217–224. Ottawa: Legas.

Aristotle. 1926 [c.335 B.C.E.]. *The Art of Rhetoric* (Loeb Classical Library, 193) (translated from the Greek by John Henry Freese). London: Heinemann.

Bernacer, Javier, and Jose Ignacio Murillo. 2014. The Aristotelian conception of habit and its contribution to human neuroscience. *Frontiers in Human Neuroscience* 8: 883. doi:10.3389/fnhum.2014.00883.

Brent, Joseph. 1998 [1993]. *Charles Sanders Peirce: A Life*. Bloomington: Indiana University Press.

Bryant, Levi R. 2014. *Onto-Cartography: An Ontology of Machines and Technology*. Minneapolis: University of Minnesota Press.

Chen, M.Keith. 2013. The effect of language on economic behavior: Evidence from savings rates, health behaviors, and retirement assets. *American Economic Review* 103(2): 690–731.

Darwin, Charles. 1859. *On the Origin of Species by Means of Natural Selection, or the Preservation of Favoured Races in the Struggle for Life*. London: John Murray.

Dunedin Multidisciplinary Health and Development Research Unit (DMHDRU). (1972-present). Publications in numerous venues from New Zealand longitudinal study, cf. Moffit et al., 2013.

Durst-Andersen, Per. 2011. *Linguistic supertypes: A cognitive-semiotic theory of human communication*. Berlin: Mouton.

Durst-Andersen, Per. 2012. What languages tell us about the structure of the human mind. *Cognitive Computation* 4: 82–97.

Gibson, James J. 1979. *The ecological approach to visual perception*. Boston: Houghton Mifflin.

Henry, Richard Conn. 2005. Concept: The mental universe. *Nature* 436 [7047] (7 July 2005): 29.

Jakobson, Roman. 1970. Linguistics. *Main trends of research in the social and human sciences; Part 1: Social Sciences,* 419–463. Berlin: Mouton, for UNESCO.

Kelly, Walt. 1972[1971]. *Pogo: We have met the enemy and he is us.* New York: Simon and Schuster.

Kroeber, Alfred L., and Clyde Kluckhohn. 1952. Culture: A critical review of concepts and definitions. *Papers of the Peabody Museum of Archaeology and Ethnology*, Harvard University, 47.1. Cambridge: Peabody Museum.

Lovejoy, Arthur O. 1936. *The great chain of being: A study of the history of an idea*. Cambridge: Harvard University Press.

Margulis, Lynn. 1971. Whittaker's five kingdoms of organisms: Minor revisions suggested by considerations of the origin of mitosis. *Evolution* 25(1): 242–245.

Moffitt, Terrie E., Richie Poulton, and Avshalom Caspi. 2013. Lifelong impact of early self-control: Childhood self-discipline predicts adult quality of life. *American Scientist* 101 (September–October 2013): 352–359.

Monod, Jacques. 1971. *Chance and necessity: An essay on the natural philosophy of modern biology*. New York: Alfred A. Knopf.

Peirce, Charles Sanders. i. 1867–1913. *Collected Papers of Charles Sanders Peirce*. Vols. 1–6, eds. Charles Hartshorne and Paul Weiss. Cambridge: Harvard University Press, 1931–1935. Vols. 7–8, ed. Arthur W. Burks. Cambridge: Harvard University Press, 1958. [References to Peirce's papers will be designated by CP, followed by volume, period, paragraph number.].

Peirce, Charles Sanders. i. 1867–1893. *The Essential Peirce: Selected Philosophical Writing*. Vol. 1 (1867–1893), eds. Nathan Houser and Christian Kloesel. Bloomington: Indiana University Press, 1992. [References to this volume will be designated by EP 1, followed by colon, page number.].

Peirce, Charles Sanders. i. 1893–1913. *The Essential Peirce: Selected Philosophical Writing*. Vol. 2 (1893–1913), ed. the Peirce Edition Project. Bloomington: Indiana University Press, 1998. [References to this volume will be designated by EP 2, followed by colon, page number.].

Peirce, Charles Sanders. i.1867–1913. *Writings of Charles S. Peirce: A Chronological Edition*. Vols. 1–6 to date, ed. the Peirce Edition Project. Bloomington: Indiana University Press, 1982. [References to these volumes will be designated by W, followed by volume number, colon, page number.].

Prigogine, Ilya, and Isabelle Stengers. 1984[1979]. *Order Out of Chaos. Man's New Dialogue with Nature.* Translated from the French. Boulder, Colorado: New Science Library.

Rosen, Robert. 1985/2012. *Anticipatory systems: Philosophical, mathematical, and methodological foundations*. New York: Pergamon.

Salthe, Stanley N. 1993. *Development and evolution: Complexity and change in biology*. Cambridge: MIT Press.

Thom, René. 1968. Topologie et signification. *L'iige de la science* 4: 219–242.

Thom, René. 1975. *Structural stability and morphogenesis: An outline of a general theory of models.* London: Benjamin.

Uexküll, Jakob von. 1957[1934] A stroll through the worlds of animals and men: A picture book of invisible worlds. pp. 5–80 in *Instinctive behavior: The development of a modern concept*, edited and translated from the German by Claire H. Schiller. New York: International Universities Press, Inc.

Waugh, Linda R. 1982. Marked and unmarked—A choice between unequals in semiotic structure *Semiotica* 38.3/4: 299–318.

West, Donna E. 2014. From habit to habituescence: Peirce's continuum of ideas. *The SSA annual: Semiotics 2013*, eds. Jamin Pelkey and Leonard G. Sbrocchi, 117–126. Toronto: Legas Press.

Williams, Raymond. 1985[1981]. *The sociology of culture*. Chicago: University of Chicago Press.

Williams, Raymond. 2014[1985, 1976]. *Keywords: A vocabulary of culture and society*. New York: Oxford University Press.

Part I
Background: Eco-logical Systems

Chapter 2
On Habit: Peirce's Story and History

Dinda L. Gorlée

> We all know that real, genuine acting upon principle is exceptional; that at best what is so called is mostly only the working of a good habit, and at its worst of an odious inclination. Just so, what people dignify by the name of their reasonings are mostly mere passages from one judgment to another in a way in which natural bent, habit, experience, the example of the wise, half-consciously move them to think. (CTN: 2: 270, 1900).

Abstract Peirce's speculative story of the history of habit remains inferior to Thirdness, degenerating into individual belief (Firstness) with social rules (Secondness). The nervous sensation of habit is investigated by the physiological method of induction. The reactionary habit is caused by the emotional beliefs of the interpreter to remedy the discomfort of mind or body through energetic efforts. The variety of habits are emotional and energetic interpretants in search of logical interpretants. At a later time, Peirce developed habits into the psychological aspects, then a new discipline. Faced with the troublesome reasoning of habits, induction was re-considered as psychological abduction. Habits not merely reflect the moral truth of good habits but also provoke bad habits. Habit becomes habituality, ending in habituescence. Peirce's concept of habit interprets emotions and experiences to come to understanding (or misunderstanding). In Peirce's words, from desire (First) and pleasure (or displeasure) (Second), habits can grow into satisfaction (or dissatisfaction) (Third).

Keywords Habit-habituality-habituescence · Belief-doubt · Habit-interpretant · Induction-abduction

Introductory Words

The desire to solve mysteries is as old as humanity itself, and nothing is more intriguing than the story and history of a "simple" word, that has been "translated" by Peirce (1839–1914) into a "complex" word with a surprising meaning (Casti

D.L. Gorlée (✉)
Wittgenstein Archives, University of Bergen, Bergen, Norway
e-mail: gorlee@xs4all.nl

D.E. West and M. Anderson (eds.), *Consensus on Peirce's Concept of Habit*, Studies in Applied Philosophy, Epistemology and Rational Ethics 31,
DOI 10.1007/978-3-319-45920-2_2

1994: 1–42, 269–278). A familiar but perhaps surprising example is the abstract word "habit", etymologically derived from Latin *habitudo* (from *habere*), determining from the Middle Ages the actual state of having or holding something (a quality, interest, or property) to mark the formal fact of *tenere* (to hold) or *possidere* (to possess) an exclusive right. Peirce wrote that Aristotle's *διάθεσις* (disposition, tendency) was the original source to broaden it into the philosophical type of habit (MS 681: 23–25, 1913).

Aristotle's diathesis signified the ethical character of the good human life transpired in the Greek habits to live better (*Nicomachean Ethics* book 2, Aristotle [c. 335 B.C.E.]1996: 31–50). Aristotle's practical rhetoric in the set of rules *pathos, logos*, and *ethos*, from *The Art of Rhetoric* (Aristotle [c. 335 B.C.E.]1926: *passim*), has inspired the threeway hermeneutic of Peirce's triadic sign-action (semiosis) (CP: 1.215, 1902; 2.306, 1901; 2.553, 1901). Aristotle's early version of habituation has also influenced the Christian habit-changes in Peirce's good (and bad) effects of life, carrying from the virtue (or vice) to the habits of right (or wrong) behavior (CTN: 3: 278, 1906). Taking real habits is the "broader" and "complex" action of signs of Peirce's semiosis (CP: 5.484–485; 5. 488–489, 1907).

In its historical development, the story of habit is, in traditional philosophy, a neglected issue. Still, the general term of habit remains, sometimes overlapping or replaced by the word of custom, usage, use, or wont, which "may be tentatively defined as a norm of action (precept, rule of conduct) generally accepted and practiced by a group of people who regard it as sanctioned by the general tradition of the group" (Ladd 1969: 278; OED: 6: 993 [9b]).

In the *Oxford English Dictionary*, Peirce's notion of habit is not mentioned at all. Although Peirce's habit means the habitual rule, this rule does not refer to the collective reality of group habits. He firstly determined with habit the personal thought, embodying belief in a single action, which serves as guiding line "in terms of the habits they generate, sustain, and modify" (Colapietro 1993: 109, see 50–51). Personal habits are transposed to the pluralistic society. Peirce's notion of the "community" treasures the diversity of opinions, but he consciously decided to foreground the single entity in the social community. In Peirce's semiosis, the habit is a social sign. The frequent repetition of the same action and reaction is basic to the technical, religious, and other terms of the good habit. The plan of semiosis in the hands of human manners and customs is the interpretation of the semiosis of habit, mediating sign and object into the interpretant. Yet genuine semiosis can easily become limited to human practice, moving from the ideal habit as close to Thirdness to the pseudo-semiosic of Secondness and Firstness.

Habits are basically inferior to real Thirdness. The triad of Thirdness can easily become degenerated downward to the dyad of individual chance (Firstness) functioning with social rules (Secondness) for the community. In Peirce's terminology: "Chance is indeterminacy, is freedom. But the action of freedom issues in the strictest rule of law" (W: 4: 552, 1884). In the psychology of habit, Peirce described habit reversely as: "Chance is First, Law is Second, the tendency to take habits is Third" (CP: 6.32, 1891, see CP: 1.409, 1887–1888). Peirce's "tendency" displays the sensuous drive (Second) of the agent or interpreter (First) to imply the habit as the Third activity. Yet, in this process, "the trend of the driving force emphasizes the idea of the [...] variety of arbitrariness" of the interpreter's feelings (First). In early Peirce, the concept of law (Second) "receives exaggerated attention" (CP: 6.32, 1891) without achieving the epicenter of philosophical thought of culture (Third). Law is regarded as the mediating process between the habitual ideas of the accidental circumstances of the First and Second to make the cognitive habit (Third). At a later date, habits are threatened by feeling and law to become, in the psychological perspective of habits, degenerate signs. The Thirdness of law is the voluntary "repudiation of any Secondness" to have "for its principal component the conception of First" (CP: 6.32, 1891).

Peirce's notion of the degeneracy of signs (Gorlée 1990, 2007, 2014) reveals the story of the fragmentariness of the dyadic and monadic divisions and subdivisions of Peirce's categories, reduced from three kinds of signs into two or even only one. Moving from the simple habit to the number of various habits with different meanings, the degenerated habit-sign gives multiple habits in the continuity of the interpreter's action and passion, to be regenerated again in mental thought. Peirce followed Darwin's organic evolution of natural selection, in which primitive habits would be transformed to a diversity in the stream of habits. Applied to every purpose and every intention, the natural harmony of habituality attempted to govern the future action of society from sign and object through the work of Peirce's interpretants. If the personal habit has a set of habits through logical and illogical signs, habits are a set of social habits against a conservative force, producing a flow of new variants in habituality. The disordered variety of interpretants are also individual habits which embody the primordial tendency for people to repeat themselves in recurring habits. The unawareness of different and ambiguous forms of habituality is regenerated, in later Peirce, to the process of habituescence.

Peirce's theoretical and practical questions about the concept of habit seem split into three stories, mixing up Peirce's mathematical, logical, and scientific arguments into different moments of his life. First, habit expressed the fixed disposition of mind and soul to become the second nature of belief governing the primary habit. The concept of habit conducts the experience of human life calculated by geometrical behavior. Second, as suggested by Pascal's *Pensées* ([1669], published in folio [1977]1995), the realism of the geometrical procedures (*consuetudo*), educated in his youth, turned later into the uncertainty of mystical meditations (CP: 1.281, 1902). Creating a number of conscious and unconscious habits, there emerges the new perfection (and sometimes imperfection) of Peirce's habit. Habit originally meant incorporating new skills and style to make new habits. Third, this means that Peirce's habit did no longer symbolize the morality of right or wrong *cursus* of the straight road of reasoning, but seemed to drive the human soul into the via *rupta* or broken road (CP: 5.493, 1907; see also Colapietro 1989: 95–97) with regard to evil reasoning.

The current sense of the *Oxford English Dictionary* is "A settled opinion or tendency to act in a certain way, esp[ecially] one acquired by frequent repetition of the same act until it becomes almost or quite involuntary" (OED 1989: 6: 993 [9a]). Repeated in the synonyms of habituation as readaptation, the *Oxford Companion to the Mind* affirms that the reflex behavior of animals tend "to disregard irrelevant stimuli" since if "the snail came out every time its shell got tapped, it would get worn out!" (OCM 1987: 299). Reflexes require reinforcement in repeated experience (learning) for learning to settle in memory. Peirce would disagree with the acquired or learnt habits, he advised modes of intelligence in a much wider range, as follows:

> Learning is largely concerned with establishing the predictive value of stimuli that do not of themselves demand responses. We recognize other animals as intelligent when they predict effectively from these stimuli, and as highly intelligent when they begin to predict by analogy, generalizing to the point where they can take appropriate preemptive action without prior experience of the particular sequence of events to which they are reacting. (OCM 1987: 389)

The intelligent mind of the neurological "brain" acts as a "system of signs displayed, for example, as a physical network, or structure, of neurons [, which] is transcoded into our central nervous system" (Sebeok 1991: 133). This ecosystem sensitizes the emotional reflexes of emergences in alarm calls, necessary to handle potential danger for the survival of the animal or human individuals (Gorlée 2015: 38–40, 59–61). This chapter will argue that the flux of emotional interpretants generate quick and forced habits, and the energetic interpretants emerge with pulsant quickness. Yet the logical interpretant can take some critical time to act.

In the three-way scientific categories of experiment (First) and experience (Second) to make expertise (Third), Peirce dynamized the habit formation down into habit reformation to rebuild all types of different voluntary and involuntary habits, with positive and negative effects. The effect of the uncertainty of habits dealt with person-oriented habits with ambiguous forms and contradictory shapes attempting to form social (or asocial) behavior. For Peirce, good habits have "attractive" feelings (EP: 2: 432, 1907) for ethical and meaningful habits, which produce correct feelings and conduct to pleasure. Bad habits have "repulsive" feelings (MS 318: 182, EP: 2: 432, 1907) with Peirce's negative effect of pain as unethical and neurotic behavior to society. Peirce's religious habits assumed free will to live better lives. The young Peirce wrote that "bad habits are quickly destroyed, those which have no habits follow the same course; only those which have good habits tend to survive" (W: 4: 553, 1884). However, in later years, Peirce's rule of ethics experimented on the "living" habits to enjoy the pleasures and even suffer the displeasure of addiction.

Beyond some "philosophical" remarks about the inferential reasoning of drugs, interspersed among his works (CP: 8.12, 1871; 4.234, 1902; 4.463, c. 1903; 5.534, c. 1905), Peirce's manuscripts do not mention the real fact that he strongly suffered from acute neuralgia. To endure the migrainous attacks of pain in his face, Peirce was addicted to opium and morphine (Brent 1993: 39–40, 48, 351–352 fn.). As a patient, Peirce took drugs to cope with his neurological troubles. But, at the same time, as scientific laboratory-man, he brought habit and habituality into sharpened focus to be investigated by the community of scholars. Peirce, as religious man, had the mismatch of his personality and intellect to do justice to the positive and negative difficulties of his own beliefs and doubts. Peirce's struggle was to distinguish between punishment and blame, legal and illegal transactions. In semiotic terms, Charles S. Peirce was the rational-emotive interpreter (agent, inquirer, victim) of his own habit-interpretants.

Attention, Sensation, and Understanding (1865–1880)

After the summary account of Peirce's terminology of habit, the footsteps of the scientific habit start, when young logician Peirce lived with his wife Melusina (Zina) in Cambridge, Massachusetts. He worked at the U.S. Coast and Geodetic Survey, but also gave series of lectures at different institutions. Peirce allowed himself further hope for an academic appointment in logic at Harvard (Brent 1993: 82–135).

At age twenty-nine, Peirce in "Some Consequences of Four Capacities" (CP: 5.264–5.317, 1868) wrote that humans have logical thought as their intellectual instrument. The human mind is removed from the weakness of the individual's bodily experience. Peirce's community actively works through the working-tool of inferential reasoning of sign and object to achieve the truth. The three-step methods of explanatory reasoning are expected to yield truthful conclusions by Peirce's method: inferential reasoning of semiotic signs. Surprisingly, Peirce's method did not concentrate on the linearity of deductive logic to give the logical certainty and truth to the character of habit. Instead, Peirce choose the less certain or probable reasoning of inductive reasoning. Focusing on Aristotle's induction, he formed a causal chain of cause and consequence "to substitute for a series of many subjects, a single one which embraces them and an infinite number of others", adding that, "Thus induction is the species of the reduction of the manifold to unity" (CP: 5.295, 1868). The uniformity of the experimental investigation consists in taking a fair sample to achieve a statistical view of the total. Inductive reasoning assumes that the signs of habit points outside itself to the single object (Esposito 1980: 95).

Peirce's historical parallel is significant of expressing the "regularity of continua" (West 2014: 117) of habit. In his first story, the young Peirce focused on individual "habits as nervous associations" (CP: 5.297, 1868). The random "sensation" of habit is an experimental cognition of the interpreter as the habit-agent performing this "sensation". In the second story, Peirce's course of sensation in the single habit can repeat itself "in a larger context of thought and action" (Esposito 1980: 132), to create a multiplicity of the same habit, changed or exchanged into different ones. The "voluntary actions resulting from the sensations produced by habits, [are] instinctive actions result[ing] from our original nature" (CP: 5.297, 1868). The variety of habits is denominated by Peirce as habituality, consisting of the variety of habits. Habituality attempts to be Thirdness, but really interacts with the sensory stimuli of habits, not coming from the logical mind (Thirdness), but from the natural body (Secondness) or instinct (Firstness). Habits are inferior to Thirdness. But if habits are formed inductively (Secondness), they must for Peirce include the combination of all three cognitive elements to generate the true habit of habituality producing "true conclusions from true premisses" (CP: 5.367, 1877). Peirce's reasoning would mean that habit as type of "nervous association" can no longer be regarded as the unconscious sign, but must be regarded as the conscious sign, changing the single concept of habit into a stream of habits.

Peirce's basic formula for "Attention, Sensation, and Understanding" (CP: 5.298, 1868) claimed that habit is the conscious sign. If the role of the interpreter is to react immediately to the emotional or physical condition from the environment, he can directly sense, reason, and mediate the event in question as the "rational animal" (CP: 8.3, 1867; see also CP: 3.472, 1897; 5.585, 1898; 2.361, c. 1902; 1.591, 1903). Beside the close attention of human awareness, the interpreter needs for Peirce a state of "belief" (Firstness) that leads under the same circumstances to a habit of the same ideas and thought (CP: 7.332, 1873; 7.354–7.355, 1873). In Peirce's essay "The Fixation of Belief" (CP: 5.358–5.387, 1877), he emphasized that the interpreter's belief (Firstness) must end the doubt unsettling the inquirer and grow into new behavior (Secondness). The conscious inquirer can guess correctly the emotion (First) and information (Second) to have good habits (Third), but the unconscious agent can fall into the problematic state of bad habits.

Peirce's early concept of habit analyzed the mechanical process of logic, fixing the positive or negative meaning in the certainty of sign and object in the single situation. Peirce changed the concept to include the habits into the triadic relation of sign, object, and interpretant. The interpretant is the reactionary habit of interpreting immediately the previous sign, but the meaning of the interpretant is impossible to anticipate. While the interpretant is itself a reactionary sign, this interpretant-sign can be re-inventing itself in new habits, acquired or learned, based on new knowledge. Peirce stressed that a habit can hardly be considered as the same as the logical Third, but must include non-logical signs of an emotional and bodily nature (First, Second). Habits work in a multi-layered relationship between the three kinds of signs, indispensable for all reasoning in Peirce's semiosis. Habit remains a paradox.

In "How to Make Our Ideas Clear" (CP: 5.388–5.410, 1878), Peirce argued about the "false distinctions" of habits with "different beliefs" that require "different modes of action to which they give rise" (CP: 5.398, 1878). This "confusion" (CP: 5.398, 1878) of individual beliefs is Peirce's difference with the degenerated meaning of the three types of interpretants, which, as he argued now, could be right or wrong, suppressed or distorted interpretations (Gorlée 1990). The regularly connected series of Peirce's three interpretants concern the immediate, dynamical, and final interpretant-signs, also called emotional, energetic, and logical interpretants, connecting together the sign and object. The interpretants offer three kinds of reasoning, moving from illogical to logical. In the individual habit, the habit can give an immediate (emotional) interpretant-sign within the dynamical (energetic) sign of human conduct. The intellectual habit of the final or logical interpretant of the sign hardly seems to occur. The individual habit is expressed with vague or unresolved interpretants—but Peirce stated that hard habits could possibly happen in the semiosic process of habit-signs.

Although Peirce's classification of interpretants has provoked a controversy (for example, Savan 1987–1988: 40–72; Johansen 1993: 145–185), my argument after studying Peirce's work will suggest that the first trio—immediate, dynamical, and final interpretants—must be limited to the successive stages of Peirce's technical process of semiosis. But the second one—emotional, energetic, and logical interpretants—shows pseudo-semiosis from the limited perspective of the inquirer's quasi-mind. Although the human ability of pseudo-semiosis produces the habit involved in what can be considered good interpretants, Peirce has at this early stage not yet mentioned the similarity of habits classified in the variety of the interpretants.[1] One can conclude that Peirce's habit must be regarded as a "confused" type of interpretant, out of step with the explicit values of the emotional, energetic, dynamical, logical and final interpretants. The interpretants of habit secretly break new ground. Coming together in the details of other alternatives of interpretants, sporadically used in Peirce's work, the interpretants of habit suggest the vague, suggestive, ejaculative, imperative, indicative, usual, destinate, and normative interpretants (CP: 5.480, 1907; 8.369–8.374, 1908; and others). Peirce's idea of these emotional effects would mean that habit belongs and participates in such vaguely unresolved and undiscussed alternative approaches, claiming that Peirce's habit must now be regarded as the conscious and voluntary interpretant-sign.

At the age of forty-one, Peirce criticized his own reasoning in the crucial year of 1878, when he presented the pragmatic maxim to herald the evolutionary explanation of American pragmatism (Gorlée 1993, Gorlée in press). In this moment of transition of thought, Peirce questioned his reasoning in "Deduction, Induction, and Hypothesis" (CP: 2.619–2.644, 1878). The inductive decision hangs in the balance between the habitual use of rule, case, and result through the energetic inference of induction (Secondness). In his words:

> Induction infers a rule. Now, the belief of a rule is a habit. That a habit is a rule active in us, is evident. That every belief is of the nature of a habit, in so far as it is of a general character, has been shown … Induction, therefore, is the logical formula which expresses the physiological process of formation of a habit. (CP: 2.643, 1878)

Instead of the physiological method of induction, Peirce seemed to prefer the psychological perspective of the method of hypothesis involving abduction (Esposito 1980: 95). In other words, instead of "the habitual element", Peirce choose the "sensuous element" (CP: 2.643, 1878) bringing out the significant role of the habit-interpreter's feeling in the mode of reasoning. Hypothesis (abduction) sustained the vital role of the emotional interpretant (Firstness) to initiate the intuitive and instinctive form of reasoning, coming directly from the inquirer's good (or false) belief, without being indirectly treated.

---

[1]For example, CP: 1.553–1.555, 1867, 1.339, 1895, 2.228–2.229, c.1897, 1.541–1.542, 1903, 8.332–8.333, 1904, 8.337–340, 1904, 8.343–8.344, 1908, 8.334–6.347, 1909, 8.177–8.185, 1909, 8.314–8.315, 1909.

In "On the Algebra of Logic" (CP: 3.154–3.215, 1880), Peirce approached once again habit as scientific ideal of a highly calculated and mental habit. Habits fall under the mathematical or statistical actions producing "geometrical" procedures of human conduct. The mental state of habit replaced the progressive development of the "spontaneous" beliefs to possibly achieve the "cerebral habit of the highest kind" (CP: 3.160, 1880) of human thought. Thus the notion of habit has transformed the intuitive and instinctive sense of the "inborn" habit of the inquirer to the strong "belief-habits" (CP: 3.161, 1880) of Peirce's different habits, carrying new beliefs to give new information. The judicious mixture of uncertainty (suspense) and predictability (security), gives habits valid and invalid elements of the "purely cerebral activity" (CP: 3.158, 1880) of the mind.

Peirce's idea of habit started with the "spontaneous development" of individual belief (Firstness). Then, the experimental action of the inquirer (Secondness) ended with the leading principle (belief). Now, the inquiry (doubt) of scientific cases gives the "*judgment*" (truth) (Thirdness) (CP: 3.160, 1880) of critical habits.

Experiment, Experience, and Expertise (1880–1900)

Peirce's inquiry is no longer guided by the vital processes of the individual and practical sensations as described during his early thought. From the 1880s (Brent 1993: 136–202), Peirce was fully guided by non-personal but scientific beliefs to trust experimentally-verifiable judgments of the cultural rule of conduct to achieve ultimate logical interpretants (truth). Peirce's habitual conduct is realized through good habits by the truth of scientific experiments. In the pages of "A Guess at the Riddle" (CP: 1.354–1.416, 1887–1888), Peirce took the "strengthening habits" of "first, chance; second, law; and third, habit-taking" (CP: 1.409, 1887–1888). Peirce's goal was to reach the good habits (final or logical interpretants) by asking critical questions. The logical method was exemplified by controlled experiments in several habits and beliefs. The habit with the motive and the conditions was still seen "as a verbal definition [which] is inferior to the real definition" (Savan 1987–1988: 65).

Moving from habits in ethics and economics toward physics and biology, Peirce told the story of the mutilated frog used as the agent (or patient) of habits (CP: 1.390, 1887–1888). The speculative story is the laboratory experiment (Gorlée 2007: 263–267) experimented with

> … a frog whose cerebrum or brain has been removed, and whose hind leg has been irritated by putting a drop of acid upon it, after repeatedly rubbing the place with the other foot, as if to wipe off the acid, may at length be observed to give several hops, the first avenue of nervous discharge having become fatigued. (CP: 1.390, 1887–1888, later discussed in CP: 2.711, 1883; 6.286, 1893; 7.188, c. 1901) (Gorlée 2007: 263–267)

The inquirer, Peirce, notes that the poor frog makes some springy leaps. The animal has lost the excitability of the first nervous sensation of the habit. The nervous systems of the brain have now become inert, or practically dead, when the stimulus ceases; but when the source of irritation returned, the frog must find a new way of the pain. Peirce wrote that: "This is the central principle of the habit; and the striking contrast of its modality to that of any mechanical law is most significant" (CP: 1.390, 1887–1888).

The frog's alarm reaction is the habit of action. Does the same procedure happen to the law of conduct in the human organism? The habit-taking was the dynamical action (energetic interpretant) to remove the irritation produced on the frog's organism. Peirce concluded that a habit is not a coded piece of furniture (logical interpretant) available for mental use. Instead, the habit is considered as an experiential test (energetic interpretant) of the victim (emotional interpretant) to actively resist the uncomfortable situation. The reaction moves from the creative uncodedness to the recodification of the critical situation to create a manageable and coded form of the frog's well-being. In Peirce's letter to his pupil, the logician Christine Ladd-Franklin, he wrote that he had developed his cosmology. The past is the chaos of the world, while the pure chance grows to the future in the rules of taking "hyperbolic habits" (CP: 8.317, 1891). Peirce's habit is "purely physical" (CP: 8.318, 1891) law of habit, generating some psychical ease and comfort, meaning that the decision can be a good habit, with respect to the internal feelings. Habit exists to solve or cure the person of the bad habits of external pain or internal suffering to reach the relative idea of good "health".

To solve the tell-tale sign of the frog, Peirce's inductive reasoning transforms the uncertainty of interrogation and doubt by arriving at the human certainty of law. The logical experiment with the frog can be true or false. Peirce ended "Mind and Matter" (CP: 6.272–286, c. 1893) with the inductive statement that:

> A decapitated frog almost reasons. The habit that is in his cerebellum serves as a major premiss. The excitation of a drop of acid is his minor premiss. And his conclusion is the act of wiping it away. All that is of any value in the operation of ratiocination is there, except one thing. What he lacks is the power of *preparatory meditation*. (CP: 6.286, c. 1893, my emphasis; see Colapietro 1989: 109)

Of course, nonhuman animals cannot grasp the idea of the cognitive mind leading to rational uncertainty. In Peirce's experiment in "The Law of Mind" (CP: 6.102–6.163, 1892), the feeling mind of the decapitated frog takes the "power of exciting reactions" (CP: 6.145, 1892) as it comes. The frog's automatic habit of reaction to outside stimuli alleviates the immediate crisis; but Peirce emphasized that the unfortunate situation could be remedied by the human act of deciding and judging the misfortune with the frog's alarm cry (CP: 5.480, 1907). Peirce noticed that the logical reasoning of habit-taking had not (yet) become the general or universal habits of behavior in human words of alarm. The logical reasoning of habit-taking was "pretty vague on the mathematical side" (CP: 6.262, 1892), but Peirce attempted to develop the habits of conduct into the psychological aspects of belief and doubt.

In the following articles of "Man's Glassy Essence" (CP: 6.238–6.271, 1892) and "Evolutionary Love" (CP: 6.287–6.317, 1893), Peirce took an intuitive and poetical vision of life. This field of research was more in harmony with Peirce's own life, which had taken a chaotic turn. Since he turned away from service in the U.S. Coast and Geodetic Survey, he could in his final years, hardly survive within the meager income of the reviews for *The Nation*. His spirit seemed to change from academic reason to something like a mystical quest (Brent 1993: 203–268). The paradox of his own fundamental opinions enabled him to leave the mechanical reasoning of induction and taking the unknown impulse of abduction (emotional habits) as reasoning method. Continuing the letter to Christine Ladd-Franklin about the cosmology of the world (CP: 8.317–8.318, 1891), the elasticity of habits tend to "spring back again [as] an apparent violation of the law of energy" (CP: 6.261, 1892). Peirce suggested the growth of intuitive knowledge through hypothesis to first guesses (abduction) in habits. Habits do not clarify hard truth, but rather produce from the "primal chaos" of the experiments the soft "reaction from the environment simply by virtue of some chance novelty [...]" (Esposito 1980: 166).

Habits "depend upon aggregations of trillions of molecules in one and the same condition and neighborhood; and it is by no means clear how they could have all been brought and left in the same place and state by any conservative force" (CP: 6.262, 1892). Habits must be "operative" signs of physical and psychological energy, from inside to outside; but "wherever actions take place under the established uniformity, there, so much feeling as there may be, takes the mode of a sense of reaction" (CP: 6.266, 1892) coming from the outside to the inside. Peirce asserted, "[g]eneralization is nothing but the spreading of feelings" (CP: 6.268, 1892) to the habits. Peirce now emphasized that the general ideas of habits involve habit-taking, but also habit-leaving (CP: 6.300, 1893). Habits can also be replaced by other habits with new information and feelings to perfect themselves to reason. "Chance actions" (CP: 6.269–6.270, 1892) work as weaker phenomena of Peirce's habits, learning what he described as the "action of love" (CP: 6.300, 1893). In Peirce's evolutionary philosophy, "[l]ove, recognizing germs of loveliness on the hateful, gradually warms it to life, and makes it lovely" (CP: 6.289, 1893). For Peirce, love and hate must be understood as marginal but opposite "synonyms", working against different signs to "equivalent" habits. Growth takes changes and exchanges.

The double part of ordering habit-formation and disordering habit-leaving makes habits from automatic into non-automatic signs. Esposito (1980: 169) argued that habit-formation was later called negentropy, with complementary habit-breaking being the tendency of entropy. The biperspective view of nature and mind operates well in the growth of biological organisms. The human concept of habit has turned into "energetic projaculation" (CP: 6.300, 1893) of habits, afflicted by sane and insane types of equivalent habits. Habits are no longer regarded as "sundry thoughts into situations in which they are free to play" (CP: 6.301, 1893), but Peirce wrote that habits are lessons to learn from the new area, human psychology:

> This new thought, however, follows pretty closely the model of parent conception; and thus a homogeneous development takes place. The parallel between this and the course of molecular occurrences is apparent. Patient attention will be able to trace all these elements in the transaction called learning. (CP: 6.301, 1893)

Peirce's knowledge about human psychology was, as he wrote to his friend, psychologist and philosopher William James, insufficient. Peirce sent him in a letter, hoping for a reply to his vexing questions:

> The question is what passes in consciousness, especially what emotional and irritational states of feeling, in the course of forming a new belief. The man has some belief at the outset. This belief is [...] a habit of expectation. Some experience which this habit leads him to expect turns out differently; and the emotion of *surprise* suddenly appears. Under the influence of *fatigue* (is this right?) this emotion passes into an irritational feeling, which, for want of a better name, I may call *curiosity*. I should define it as a feeling causing a reaction which is directed to the invention of some *possible* account, or *possible* information, that might take away the astonishing and fragmentary character of the experience by rounding it out. [...] When such possible explanation is suggested, the idea of it instantly sets up a second peculiar emotion of "Gad! I shouldn't wonder!" Fatigue (?) again transforms this into a second irritational feeling which might perhaps be called *suspicion*. I should define it as a feeling causing a reaction directed toward unearthing the fault by which the original belief that encountered the surprise became erroneous in the respect in which it is now suspected to be erroneous. When this weak point in the process is discovered, it at once and suddenly causes an emotion of "Bah!" Fatigue (?) transforms this into the irritational feeling called *doubt*, i.e. [,] a feeling producing a reaction tending to the establishment of a new habit of expectation. This object attained, there is a new sudden emotion of "Eureka" passing on fatigue into a desire to find an occasion to try it. (CP: 8.270, 1902)

Peirce searched to rephrase the habit, belief, doubt, and other terms into the newer technical vocabulary, terminology, and phraseology of psychology. He attempted to place William James' emotional and energetic interpretants of religion into the cognitive frame of the logical interpretants. Yet Peirce's mathematical and scientific experiments did not persuade James' religious faith about the pragmatic habits of prayer, saintliness, and mysticism of *The Varieties of Religious Experience* ([1902]1997). Their common duties in pragmatism were not the universal theory of consciousness as a modern psychological thought for the new century. It seems that both scholars were afflicted with doubt and self-doubt to find expression in the truth and error of the new paradigm. To give a psychological name to the process of human habits, Peirce had come to a blind alley. Habits and certain topics of logic could at that time not be solved by new nomenclature of emerging psychology (CP: 3.571–3.577, c. 1903), which were by Peirce mathematically "translated" into the equally emerging, symbolic logic (CP: 3.578–3.608, c. 1903; see CP: 3.594, c. 1903). Unfortunately, Peirce's health situation around the turn-of-the-century also made further objective progress in the emerging psychology an impossible task.

Peirce lived in the same Darwin era as the Italian criminologist and physician, Cesare Lombroso. For Lombroso, "crazy" or "insane" criminals are atavistic individuals who have ceased to evolve progressively (CTN: 1: 143, 1892). Peirce hoped that the degenerate patient—designated by the pragmatic theologian William James as the "sick soul" ([1902]1997: 149–187)—could release prejudice and hope for God's grace for recovery. For Peirce, the degenerate patient could self-control the habitual conduct by managing the didactic course of rational processes, in the triadic course of semiotics (CP: 8.303–8.304, 1909). In concrete terms, Peirce meant that the passive brains of drunkards and criminals could through learning (that is through correcting and self-correcting themselves) be transformed into active brains (CTN: 1: 143, 1892; 2:167, 1898). Peirce's understanding (or misunderstanding) of the evolution of criminality was also influenced by his religious belief. He believed in self-control, in which he responded in religious faith to all vices in order to create virtues. While "pursuing pleasure and avoiding pain", there was "a connection more or less complete between pleasurability and wholesomeness, pain and harm" (CTN: 2: 111, 1895). By today, assumptions have changed several times over. Later in the 20th century, Western culture tended to orient patients to seek and treat bad habits in the self-analyzing "learning" of Freudian theory and practice of psychotherapy and psychoanalysis (discussed by Muller and Brent 2000), while now in the 21st century, assumptions and paradigms are again up in the air.

Habit, Habituality, and Habituescence (1900–1914)

In the final years after the end-of-the-century crisis, Peirce suffered both poverty and declining health. He continued to work on the integration of logic with semiotic theory, associating the existential graphs with pragmatics (Roberts 1973). Peirce stated that the logic of good (and bad) forms of reasoning can be extended into a general theory of semiotic signs. His interests were no longer the truth of general habits (Thirdness), but rather the uncertainty of the psychological belief (Firstness as major premiss) guiding the certainty of the physical action (Secondness as minor premiss) (Colapietro 1989: 59–60).

Habits are lessons of practical inference transpired in the reality (or imagined reality) of approaching danger. Habits of pseudo-semiosis—Peirce's quasi-reality (CP: 2.305, 1901)—are captured by the central but imaginary principle of the interpreter's quasi-inference of *his* or *her* forms of reality. In a review of *The Nation*, Peirce wrote that,

> To a man standing between the rails of a track on which a locomotive is approaching, it is successive sudden enlargements that he perceives. The sense of continuous change is an affair of quasi-inference. (CTN: 3: 189, 1904)

To protect himself, the man makes more or less spontaneous habits of reaction to the speedy action of the locomotive. The cry for help embodies Peirce's belief and conduct as to the victim's (interpreter's) truth of the urgent necessity of the immediate habit.

In this period, Peirce charged against fixed habits of Thirds to enter into the undertaking of quasi-certainty or pseudo-certainty to make new types of habits. In Peirce's words:

> Belief is not a momentary mode of consciousness; it is a habit of mind essentially enduring for some time, and mostly (at least) unconscious; and like other habits, it is (until it meets with some surprise that begins its dissolution) perfectly satisfied. Doubt is of an altogether

> contrary genus. It is not a habit, but the privation of a habit. Now a privation of a habit, in order to be anything at all, must be a condition of erratic activity that in some way must get superseded by a habit. (CP: 5.417, 1905)

Peirce argued against the major premiss of belief, which he had previously taken as a sign of honesty and credibility. Now he convinced that belief can also be the spontaneous and erroneous sign of habituality, changing from unconscious to fully conscious assent to the interpreter's identity. The volitional belief allows one to choose the good or bad conclusion of personal behavior. Doubt cannot be believed any more to lead to truth, because doubt creates suspense without valid conclusion of truth.

For Peirce, the knowledge of the opinions in the inquiry can never be genuinely believed, as it depends on the pragmatic inquirer's imagined situation, which can shift and change in pseudo-semiosis (sign-action) from dream to action. In the mechanism of the locomotive, the habit could be solved into performing the vague ideas of the sign-action (semiosis) in sign, object, and the interpretant. The "feeling of recognizing the sign as such" (emotional interpretant) "provokes efforts [...] against resistance, or of resisting a force [...] (energetic interpretants)" (MS 318: 36, 1907). Pseudo-semiosis gives the solution to the experimental action, which cannot at all times provide true habits, but can become good or bad imagined habits. In the process of critical "self-control" of the individual habits and beliefs, pseudo-semiosis centers on the "process of self-preparation" (CP: 5.418, 1905) to react to an action. Peirce meant to judge the validity of good and bad reactions in the "realism" of the agent's (or patient's) habits, seen as the good or bad basis for the personal conduct in habit-life (CP: 8.336, 1904).

Peirce's dilemma about the (un)certainty of habit interpreted the different roles of the three habit-interpretants in the famous MS 318 (1907; only part of the 368 pages with partial drafts has been published in CP: 5.11–5.13; 5.464–5.496; and EP: 2: 398–433, 1907). MS 318 was a highly complex document in alternative pages with different variants, with the primary aim to explain the method of pragmatism (or pragmaticism). MS 318 (1907) was the key to Peirce's logical semiotics, dealing with all kinds of semiotic visions. Peirce emphasized in MS 318 that the genuine habit from the inner feeling and effort (emotional and energetic interpretants) was transported to the outer activity of human "thought" (in Peirce's quotation marks). In Peirce's life history, he pointed out in MS 318 a path for the logical interpretants (MS 318: 80–81, 1907). He anticipated habit as law, moving law from Secondness to Thirdness. MS 318 has two stories of habit, as follows.

First, the habit was determined as logical interpretant to offer "mental fact" and "general reference" to the emotional and energetic interpretants. The logical interpretant was assigned to the conscious activity (sign-action) of, in Peirce's words, "some signs" (MS 318: 81, 1907). Peirce clarified habit, moving from a simple alarm cry (as argued before) up to the linguistic command of "Ground arms!" (MS 318: 150, 157; CP: 493, 475, 1907, repeated in CP: 8.178, 315, 1909). The habit of the military command "Ground arms!" obeys Peirce's triadic sign-action (semiosis) in that the officer of infantry gives the sign of command with reference to the object. The action of the object, the muscular movements of the military sign, always speaks louder than words. The soldiers' habits are not the individual habits of their emotional interpretants, but rather collective habits giving the energetic interpretants. The soldiers are merely affirmative agents (inquirers, receivers), so that the positive effect (or negative effect) of the officer's words of command depends totally on obeying (or not) the collateral observation of the military situation involved. The imperative utterance concerns the technical effort to ground arms (energetic interpretant), but this usual kind of collective habits does not (yet) approach the mental area of the final logical interpretant.

Logical interpretants include "conceptions, desires (including hopes, fears, ….), expectations, and habits" (MS 318: 89; CP: 5.486, 1907). Through the energetic and emotional interpretants, the logical interpretant announces the not-conditional expectation of the interpreter to check (First), control (Second), and even to restrain (Third) the repetitions of the single habits into meaning with truth conditions. The interpreter centers the quality of the habit. Peirce's new definition of habits was a realistic dream of modern psychology at that time. In his words, the concept of Peirce's habit included:

> […] *reiterations in the inner world—fancied reiterations—if well-intensified by direct effort, produce habits*, just as do reiterations in the outer world; *and these habits will have power to influence actual behaviour in the outer world*; especially, if each reiteration can be accompanied by a peculiar strong effect that is usually likened to issuing a command to one's future self. (MS 319: 94; CP: 5.487, 1907)

After the voluntary "confession" to the reader, Peirce went back to the energetic effect of the word of command, followed by the introduction of the logical habit, which in his words came from his vague speculations. He was aware that the introduction of the logical interpretant was, at this point,

> […] not a scientific result, but only a strong impression due to a life-long study of the nature of signs. My excuse for not answering the question scientifically is that I am, as far as I know, a pioneer, or rather a backwoodsman, in the work of clearing and opening up what I call *semiotic*, that is, the doctrine of the essential nature and fundamental varieties of possible semiosis; and I find the field too vast, the labor too great, for a first-comer. (MS 318: 96; CP: 5.488, 1907).

Peirce wrote in MS 318 that "while I hold that all logical, or intellectual habits are habits, I by no means say that all habits are such interpretants", only "*self-controlled* habits that are so, and not all of them, either" (MS 318: 180, EP: 2: 431, 1907). Habits must be regarded by Peirce as psychological (that is, conscious and voluntary) interpretant-signs, embedded with some

> […] critical feelings as to the result of inner and outer exercises stimulate to strong endeavors to repeat or to modify those effects […] in the one case to reproduce or continue them, or as we say, "attractive" feelings, and in the other case to annul and avoid them, or, as we say, "repulsive" feelings. (MS 318: 182, EP: 2: 431–432, 1907)

Peirce's doubts of habit as logical interpretant are self-controlled by himself (as the inquirer and agent of his own habits) to become social habits. Intellectually, Peirce thought of habit as logical interpretants giving general and universal signs to break bad habits to form a good society? But are habit and logical interpretants also individual signs with an unconscious memory? (CP: 5.492–5.493, 1907). Peirce remembered in MS 318 a little incident of his childhood, which narrates the memory of his brother's real habit:

> I well remember when I was a boy, and my brother Herbert, now our minister at Christiania, was scarce more than a child, 1 day, as the whole family were at table, some spirit from a "blazer", or "chafing-dish", dropped on the muslin dress of one of the ladies and was kindled; and how instantaneously he jumped up, and did the right thing, and how skillfully each motion was adapted to the purpose. I asked him afterward about it; and he told me that since Mrs. Longfellow's death, it was that he had often run over in imagination all the details of what ought to be done in such an emergency. It was a striking example of a real habit produced by exercises in the imagination. (MS 318: 95, CP: 5.488, 1907; see CP: 5.638, c. 1902)

Peirce was aware that in the living example of the brother's experiential habit, the "intellectual" Thirdness of his "higher" habit was mediated into the "lower" force of energetic and emotional interpretants (Secondness and Firstness). The judgment of the logical interpretant becomes easily degenerated into this realistic mixture of effort and belief through the force of energetic and emotional interpretants (Secondness and Firstness), as discussed before. But in MS 318 Peirce fully realized that the weakening process can be improved inversely by strengthening the "inward habit" (MS 318: 162; CP: 5.478, 1907) of feeling and the effort to grow into intellectual actions. In this habit-change, the logical interpretant remains the possible inferential element of the critical experiments of the habit.

Peirce questioned the logical interpretant, but adopted this as hypothesis characteristic for all "living" habits in the significative sign-action (semiosis). Within the pages of MS 318, Peirce wrote a rough note wrapping up the indefinable situation of the threeway idea of habit:

> Habit. Involuntary habits are not meant, but voluntary habits, i.e., such as are subject (in some measure to self-control). Now under what conditions is a habit subject to self-control? Only if what has been done in one instance with the character, its consequences, and other circumstances, can have a triadic influence in strengthening or weakening the disposition to do the like on a new occasion. This is as much to say that voluntary habits is conscious habit. For what is consciousness? In the first place feeling is conscious. But what is a feeling, such as blue, whistling, sour, rose-scented? It is nothing but a quality, character, or predicate which involves no reference to any other predicate or other things than the subject in which it inheres, but yet positively is. [...] Our own feelings, if there were no memory of them for any fraction of a second, however small, if there were no triadic time-sense to testify with such assurance to their existence and varieties, would be equally unknown to us. Therefore, such a quality may be utterly unlike any feeling we are acquainted with, but it would have all that distinguish all our feelings from everything else. In the second place, effort is conscious. It is at once a sense of effort on the part of the being who wills and is a sense of resistance on the part of the object upon which the effort is exerted. But these two are one and the same consciousness. Otherwise, all that has been said of the feeling consciousness is true of the effort consciousness; and to say that this is veracious means less if possible than to say that a thing is whatever it may be.
>
> There is, then, a triadic consciousness which does not supersede the lower order, but goes bail for them and enters bonds for their veracity.

> Experiment upon inner world must teach inner nature of concepts as experiment on outer world must teach nature of outer things.
>
> Meaning of a general physical predicate consists in the conception of the habit of its subject that it implies. And such must be the meaning of a psychical predicate.
>
> The habits must be known by experience which however exhibits singulars only.
>
> Our minds must generalize these. How is this to be done?
>
> The intellectual part of the lessons of experimentation consists in the consciousness or purpose to act in certain ways (including motive) on certain conditions. (MS 318: 183–184, 1907; EP: 2: 549–550, 1907 [with punctuation corrections], not published in CP)

In the correspondence to Victoria Lady Welby, Peirce argued further about the degenerate varieties of the Thirdness of habit. He broadened the previous perspectives (CP: 8.332, 1904, CP: 8.335, 1904; 8.361–8.376, 1908) to discuss the informational (now called communicative) functions of the sign: "It appears to me that the essential functions of a sign is to render inefficient relations efficient,—not to set into action, but to establish a habit, or general rule whereby they will act on occasion" (CP: 8.332, 1904). The occurrence of different habits has a symbolic significance, in Peircean sense, denoting a "convention, a habit, or a natural disposition of its interpretant or of the field of its interpretant" (CP: 8.335, 1904).

Second, Peirce's information or communication of the habit in the logical interpretants transforms dramatically the habit into the category of law. MS 318 emphasized that the habit must be logically determined for the future by the logical inference of the general law. Peirce insisted that the natural disposition of the logical interpretant must be understood as determining the "special Tendency" (CP: 8.381, 1908) of habit as a legal document, which "never expresses brute fact, but has some relation of an intellectual nature, being either constituted by the action of a mental kind or implying some general law" (MS 318: 19, 1907). Peirce demanded an appeal to Darwin's idea of natural selection in the metaphoric biodiversity of habits; at the same time Peirce, as a Christian believer, also recognized the faith in God as an ingredient in healing from bad habits to take on good ones (Brent 1993: 4). Understanding the "neglected argument" of habit integrated into the three categories and logical pragmaticism, the continuity of different habits in Peirce's habituality is no longer a moral law defining right and wrong conduct in the category of Secondness, as argued before. Now, habit is generalized into the fixed law of nature (*Black's Law Dictionary* 1999: 1025, 1049) as universalized Thirdness. The law of nature governs the habit-laws of gravitation, elasticity, electricity, and chemistry using technical reasons in Peirce's logical interpretants (Thirdness) (Turley 1977: 31–33). As examples of the laws of nature, Peirce used the general viscosity of gases, explicable by natural law, as opposed to the elastic properties of crystals, observed not by attractions or repulsions between particles, but by force and chance (MS 318: 22, 1907).

In "A Neglected Argument for the Reality of God" (CP: 6.452–6.491, 1908), Peirce's habit started as the "brute" compulsion, "whose immediate efficacy nowise consists in conformity with rule or reason" (CP: 6.454, 1908). Then, the irritation of doubt can become a meaningful dialogue with God to enlighten our insight, clarifying how the divine truth can change our habits of conduct (CP: 6.481, 1908). The religious wisdom gives "not merely scientific belief, which is always provisional, but also a living, practical belief [...]" (CP: 6.485, 1908). The true vision of belief ends the real doubt, giving the habit-interpreter "a state of *satisfaction,* [which] is all that Truth, or the aim of inquiry, consists in" (CP: 6.484, 1908). Peirce's pragmatic order of habits applied to religious belief generates this "Super-order" coming from the "pure mind, as creative of thought" to create a "super-order" or a "super-habit" (CP: 6.490, 1908) as absolute Third (Turley 1977: 29–31).

Peirce meant the effect of the logical interpretant as habit or habit-change. Although he continued the previous problem of classifying the personal habits, in which "there must have been a tohu bohu of which nothing whatever affirmative or negative was true universally", now he displayed the super-habits to show that the "tendency to take habits" has an "elasticity of volume", which is bound to grow (CP: 6.490, 1908) from single feeling with individual action to the social habituality of different streams of habits (Turley 1977: 76–78). Peirce's cosmology seems to violate the law of energy to be adopted into the habitual modes of legal activities governed by the "simulated, or *quasi*, chance, such as Darwin calls into produce his fortuitous variations from strict heredity" (CP: 6.613, 1893). For Peirce, the Darwin-like continuance of evolution worked by the rule of,

> [...] the law, or *quasi*-law, of growth [into] the law of habit, which is the principal, if not (as I hold it to be) the sole, law of mental action. Now, this law of habit seems to be quite radically different in its general form from [the earlier] mechanical law, inasmuch as it would at once cease to operate if it were rigidly obeyed: since in that case all habits would at once become so fixed as to give room for no further formation of habits. In this point of view, then, growth seems to indicate a positive violation of the law. (CP: 6.613, 1893)

Peirce argued here that the hypothesis of God and Darwin's evolution are the super-habit of law, growing into final habits. Substituting the argument about the abiding faith of religious wisdom and the reproductive success of Darwin's survival of the fittest, Peirce as experimentalist thinker solved the problems of evolution by using, as intelligent agent, the method of logical reasoning, the ability of Thirdness. The different ways of reasoning in habits clearly pointed out that belief and experience grounds the variety of actual habits. After Peirce's first definition of habit as in-betweenness of "*would-be*" (CP: 8.380, 1913) denoting real or unreal Third, signifying the conscious "*actually is*" (Second) as well as the unconscious "*may be*" (First) (CP: 5.171, 1903), habit has in 1909 turned into a kind of psychological and logical "inference by which we pass from belief in a surprising fact to giving some degree of credence to a pure hypothesis, whose truth would explain the fact and do away with its astonishing character" (MS 637: 2, 1909). This, in Peirce's words,

> [...] involves a complexity of conditions, one of which is how much time can be allowed for coming to one's conclusion. A general who during a battle must instantly risk the existence of a nation either upon the truth of a certain hypothesis or else upon its falsity, must perforce go upon his judgment at the moment; and his doing so is in so far logical that

> all reasoning is based upon a tacit assumption that Nature, in the sense of the aggregate of truth, is conformed, more or less, to something similar to the reasoner's Reason. [...] This will be subject to considerations of economy. (MS 637: 4–5, 1909)

The habit of the impulse gives active reasons to the general, but not necessarily the real truth. A habitual action without doubt, one can expect to be a good habit, but with doubt perhaps some bad course of habit. Peirce's somewhat bizarre example is that habitual associations are "often inconvertible, or irreversible, so that an ordinary Christian cannot say his *Pater noster* backwards, etc." (MS 637: 10, 1909). Habit has now been transformed by Peirce from the single habit into the reality of different beliefs in habituality dealing with different associations of practical truth. Habit has definitively moved from skilful hypothesis to give the pure logic of guessing right to explanatory or unreasonable conjectures based on impulse to guess wrong. Habit has become the "decision" of "thought to ourselves" phrased in different habits of soul and mind. The habit "needs the special science of psychology to discover" (MS 638: 5, 1909) the truth or untruth of the degrees of human performance for real or imaginary habits. Habit has not been determined larger as logical law, because all habits depend on the good or bad quality of the beliefs involved for the (re)action.

In Peirce's final struggle, he fortified his soul by announcing the law-like completeness of his logic and pragmati(ci)sm. Peirce introduced an abstract term, "habituescence" (MS 930: 18, 1913) without explaining the term in this fragmentary and unpublished manuscript (West 2014). After habit and habituality (in missing pages of MS 930), the habituescence as Thirdness refers to the inter-group process of habits collectively, forming the tendency or capacity toward Peirce's natural action. Peirce stressed that the definition of habituescence into the "Third mode of Consciousness" must move away from "the formation of every acquired habit [...] as constraints" of "good habits" to show the great "*skill*" of "natural habits" (MS 930: 18, 1913). Natural habits are used in the mechanical actions of practical sciences, as the "think-tank" of the physician, the chemist, the lawyer, or the artist. Their knowledge was not only having the scientific or artistic talent, but their activity was to "think" naturally the knowledge in any or every aspect of the theoretical and actual practices of their trade. Peirce stated that "every belief and inclination toward belief is a Habit" and,

> [...] not merely in acquiring a purely intellectual habit, but, as I have convinced myself by a long experimental inquiry, [there] is there the same peculiar kind of consciousness in acquiring any difficult dexterity, or the mastery of any art on language. It is the same interpretative consciousness in every case. It is the "twigging" of a new idea, and neither the "I will" of volition nor the mere feeling of pure sensation. (end of MS 930: 19, 1913)

Peirce moved the weakness and self-consciousness of habit as a basic factor to the analysis of the self-oriented monologue (First) and the dramatic dialogue of habituality (Second) to forcing the "new conception" (MS 681: 27, 1913) of habituescence (Third). Habituescence consists in the multilogue of natural reasoning showing itself in the routine of the mental procedure in professional habits (Third).

In his final years (from 1911 to his deathbed on 19 April 1914), Peirce was in ill health and hardly able to proceed further and render logical finesse of the unfinished concept of "our Feelings, our Energies, and our Thoughts" for the "science" of habit (MS 675, EP: 2: 459, c. 1911, in Peirce's quotation marks). For Peirce, the state of habit almost came full circle.

Final Words

Moving from ethics and economics to physics and biology, Peirce's works about the interdisciplinary complexity of habit reveal no consolidation of "equivalent" or "same" meanings. Habits show clear signs of improvement during each year of Peirce's early work. The fragments about habit seem to zigzag through the volumes of the *Collected Papers*. The original word habit comes from Aristotle's right and wrong behavior, but transfigured by Peirce into the term of acquired (learned) habits as inferior sign to Thirdness. In Peirce's pragmatic period, habits tend toward real Thirdness. Real semiosis becomes pseudo-semiosis. The dyadic sign-object relation becomes Peirce's triadic series of sign-object-interpretant. Habit is a special type of "vague", but "suggestive" and "ejaculative" variety of the special but alternative interpretant.

The single habit of the individual (personal) action is transformed into the habituality of general (social) conduct of habits. Now, habit-taking (of new habit) means habit-breaking (from old habit). Inductive reasoning becomes abductive reasoning. In psychology, moral law becomes the law of nature. Good habits become bad ones. Peirce's broad approach coincides with the interdisciplinary activities of Peirce's habituality. Habituality understands emotion and experience to understand (or misunderstand) the meaning of the habits. In Peirce's late terminology, habit, habituality, and habituescence show that habits take from belief (First) to form action (Second), but some habit could possibly grow into final logical habits (Third).

References

Aristotle. 1926[c. 335 B.C.E.]. *The art of rhetoric* (Loeb Classical Library, 193) (translated from the Greek by John Henry Freese). London: Heinemann.

Aristotle. 1996[c. 335 B.C.E.]. *The nichomachean ethics* (Wordsworth Classics of World Literature) (translator Harris Rackham). Ware, Hertfordshire: Worthworth Editions.

Black's Law Dictionary. 1999. Garner, Bryan A. (ed.). 7th ed. St. Paul, Minnesota: West Group.

Brent, Joseph. 1993. *Charles sanders peirce: A life*. Bloomington and Indianapolis: Indiana University Press.

Casti, John L. 1994. *Complexification: Explaining a paradoxical world through the science of surprise*. New York: Harper Collins.

Colapietro, Vincent M. 1989. *Peirce's approach to the self: A semiotic perspective on human subjectivity*. Albany, New York: State University of New York Press.
Colapietro, Vincent M. 1993. *Glossary of semiotics*. New York: Paragon House.
Esposito, Joseph. 1980. *Evolutionary metaphysics: The development of peirce's theory of categories*. Athens: Ohio University Press.
Gorlée, Dinda L. 1990. Degeneracy: A reading of Peirce's writing. *Semiotica* 81(1/2): 71–92.
Gorlée, Dinda L. 1993. Evolving through time: Peirce's pragmatic maxims. *Semiosis: Internationale Zeitschrift für Semiotik und Ästhetik* 71/72 (3/4): 3–13.
Gorlée, Dinda L. 2007. Broken signs: The architectonic translation of Peirce's fragments. (Special issue *Vital Signs of Semio-Translation*, ed. Dinda L. Gorlée). *Semiotica* 163.1/4: 209–287.
Gorlée, Dinda L. 2014. Peirce's Logotheca. *Charles Sanders Peirce in his own words: 100 years of semiotics, communication and cognition*, (Semiotics, Communication and Cognition, 14), eds. Torkild Thellefson and Bent Sørensen, 405–409. Berlin and Boston, Massachusetts: De Gruyter Mouton.
Gorlée, Dinda L. 2015. From words and sentences to interjections: The anatomy of exclamations in Peirce and Wittgenstein. *Semiotica* 205 (June 2015): 37–86.
Gorlée, Dinda L. (in press). From Peirce's pragmatic maxim to Wittgenstein's language-game. *Proceedings of the series Master in Semiotics of the 12th World Congress of the International Association for Semiotic Studies (IASS) (Sofia 2014)*, eds. Paul Cobley and Kristian Bankov. Berlin and Boston, Massachusetts: De Gruyter Mouton.
James, William.1997 [1902]. *The Varieties of religious experience* in *selected writings*. New York: Book-of-the-Month Club, 23–549.
Johansen, Jørgen Dines. 1993. *Dialogic semiosis: An Essay on signs and meaning (Advances in semiotics)*. Bloomington, Indiana: Indiana University Press.
Ladd, John. 1972[1967]. Custom. *The encyclopedia of philosophy*, 8 vols., ed. Paul Edwards. Volume 2: 278–280. New York: Macmillan Publishing and The Free Press, London: Collier Macmillan Publishers.
Muller, John, and Joseph Brent (eds.). 2000. *Peirce, semiotics, and psychoanalysis (Psychiatry and the Humanities, 15)*. Baltimore, Maryland and London: The Johns Hopkins University Press.
Pascal, Blaise. 1995[1977][1669]. *Pensées*, ed. Michel Le Guern. Paris: Gallimard (folio).
Peirce, Charles Sanders. i. 1867–1913. *Collected papers of Charles Sanders Peirce*. Vols. 1–6, eds. Charles Hartshorne and Paul Weiss. Cambridge: Harvard University Press, 1931–1935. Vols. 7–8, ed. Arthur W. Burks. Cambridge: Harvard University Press, 1958. [References to Peirce's papers will be designated by CP, followed by volume, period, paragraph number.].
Peirce, Charles Sanders. i. 1867–1913. *Writings of Charles S. Peirce: A Chronological Edition*. Vols. 1–6 to date, ed. the Peirce Edition Project. Bloomington: Indiana University Press, 1982. [References to these volumes will be designated by W, followed by volume number, colon, page number.].
Peirce, Charles Sanders. Charles Sanders Peirce: Contributions to "the Nation": Part IV: Index, edited by Kenneth Laine Ketner and James Edward Cook (Lubbock: Texas Tech University Press, 1987).
Peirce, Charles Sanders. i. 1893–1913. *The essential Peirce: Selected philosophical writing*. Vol. 2 (1893-1913), ed. the Peirce Edition Project. Bloomington: Indiana University Press, 1998. [References to this volume will be designated by EP 2, followed by colon, page number.].
Peirce manuscripts in Texas Tech University Library at Texas Tech University, Institute of Studies of Pragmaticism, beginning with MS—or L for letter—and followed by a number, refer to the system of identification established by Richard R. Robin in Annotated Catalogue of the Papers of Charles S. Peirce (Amherst: University of Massachusetts Press, 1967), or in Richard R. Robin, "The Peirce Papers: A Supplementary Catalogue," Transactions of the Charles S. Peirce Society.
Richard L. Gregory, and O. L. Zangwill, eds. 1987. *The Oxford Companion to the Mind*. Oxford: Clarendon Press (OCM 1987).
Roberts, Don D. 1973. *The existential graphs of Charles S. Peirce*. The Hague and Paris: Mouton.

Savan, David. 1987–1988. *An introduction to C.S. Peirce's full system of semeiotic* (Monograph Series of the Toronto Semiotic Circle). Toronto, Ontario: Toronto Semiotic Circle at University of Toronto.
Sebeok, Thomas A. 1991. Vital signs. *Semiotics in the United States*, 119–138. Bloomington, Indiana: Indiana University Press.
Simpson, J.A., and E.S.C. Weiner, eds. 1989. *The Oxford English Dictionary.*. 20 vols. 2nd ed. Oxford: Clarendon Press (OED 1989: vol# page# [paragraph#]).
Turley, Peter T. 1977. *Peirce's cosmology*. New York: Philosophical Library.
West, Donna E. 2014. From habit to habituescence: Peirce's continuum of ideas. *Semiotics 2013*, (Proceedings of the Semiotic Society of America), eds. Jamin Pelkey and Leonard G. Sbrocchi, 117–126. Toronto: Legas Publishing.

Chapter 3
Habits, Habit Change, and the Habit of Habit Change According to Peirce

Winfried Nöth

Abstract Peirce's "law of habit" extends the ordinary and scholarly concept of habit from human to nonhuman habits and to habits in the animate and the inanimate nature. It predicts that habits change by the habit of habit change and distinguished between habits, laws, rules, and norms. Human habits as habits of thought, action, and feeling and perception are phenomena of Firstness, Secondness, and Thirdness. With human habits, the habits of nature share the feature of plasticity. Peirce attributes the plasticity of habits also to the laws of cosmic and biological evolution and distinguishes laws as habits of nature from rigid laws that do not change. The paper also examines good habits in their contrast to bad ones and concludes with remarks on habit and habit change in semiosis.

Keywords Habits of thought · Habits of action · Habits of feeling · Habits of nature · Law · Rule · Norm

Habits in the Ordinary Sense of the Word, Habits of Action, and Habits in Scholarly Terminology

A habit, in its most common sense, is a usual way of doing something, a customary way of acting. Peirce has this sense in mind in a context in which he declares that he uses the word to denote "some general principle working in a man's nature to determine how he will act" ("Minute Logic", CP 2.170; 1902). In this sense, a habit presupposes agency. The predicate, *have the habit of,* is commonly used with a

This paper is a much expanded and revised version of the author's keynote lecture given at the 2015 Michicagoan Linguistic Anthropology Conference at the University of Michigan, Ann Arbor, on May 8, 2015. Thanks are due to the organising committee of the meeting, especially James Meador and Andrew Forster.

W. Nöth (✉)
Pontifícia Universidade Católica (PUC) de São Paulo, São Paulo, Brazil
e-mail: noeth@uni-kassel.de

D.E. West and M. Anderson (eds.), *Consensus on Peirce's Concept of Habit*,
Studies in Applied Philosophy, Epistemology and Rational Ethics 31,
DOI 10.1007/978-3-319-45920-2_3

grammatical subject denoting a human or a nonhuman animal. Only animals, human and nonhuman, *have habits*, but inanimate objects have none. Habits are more or less active ways of doing things, hardly passive modes of experience, such as dreaming, feeling disgust, or having bad luck.

The ordinary kind of action performed habitually is an action that Peirce circumscribes as "really modifying something" ("First Introduction", MS 671: 5; 1911). It is a "purely physical action open to outward inspection" ("Essay Towards Reasoning in Security and Uberty", EP 2: 464; 1913). The "habit of putting my left leg into my trouser before the right" ("Minute Logic II.2", CP 2.148; 1902) is an example. Habitual acts in this sense are not only acts performed consciously, as in the case of the habit of drinking coffee every morning. They may also be acts performed with little or no self-control, as in the case of the bad habit of nail biting.

Habit is a key term of philosophy, psychology, sociology, and cultural anthropology (Camic 1986; cf. Pietarinen 2005: 364). The term is often restricted to the sense of an acquired habit. The definitions listed in the *Oxford English Dictionary* (OED) show that many authors use it only in this sense. John Locke, for example, defines habit as a "power or ability in Man, of doing any thing, when it has been acquired by frequent doing the same thing" (*Essay* II.xxii: 134). The dictionary also quotes Peirce's contemporary William Hamilton as one who distinguishes between habits in this sense as opposed to dispositions. Both are "tendencies to action", but whereas "*disposition* properly denotes a natural tendency, *habit* [denotes] an acquired tendency" (*Lect. Metaphysics* I.x: 178, 1859). This restricted usage of habit contrasts with a now less common sense of the word habit that the OED defines as "temperament, or constitution of the body, whether obtained by birth, or manner of living".

Mere ideas and feelings are not ordinarily among the objects of habits. Expressions denoting mental habits or habits of feeling do not feature in the examples of usage quoted by the OED. It would sound strange to call someone's fixed idea, recurring dream, or fear of spiders as a habit, although the word *habit*, used without a specific object of the habitual doing, may refer to the general "way in which a person is mentally or morally constituted", or to a "mental constitution, disposition, [or] character" (ibid.). However, the expression *habit of thought* can be found in academic discourse. Shuger, for example, in *Habits of Thoughts in the English Renaissance* defines habits as "interpretive categories [...] which underlie specific beliefs, ideas, and values" (1990: 9), and the historiographer Grendler (2004) discusses "medieval habits of thought" criticized by Renaissance humanists.

The attribution of habits to nonhuman "agents" in physical and biological nature is also documented in the OED. Charles Darwin attributed habits to plants in the title of his book of 1875 *On the Movements and Habits of Climbing Plants*, and a handbook of *Crystallography* of 1895 defines certain molecular properties of crystals as habits. Nevertheless, the OED marks such categorial transgressions from the animate to the nonanimate as modes of "transferred usage". The compilers of the dictionary note that habits are "properly" only attributed to living beings and only occasionally to inanimate things.

Charles S. Peirce, the Bachelor of Science in Chemistry, by contrast, attributed habits both to plants and to crystals without marking his usage as metaphorical. An author quoted by the OED who refers to habits in the same broad way as Peirce did, is William James. In his *Principles of Psychology* (I.4: 104), of 1890, he writes, "The moment one tries to define what habit is, one is led to the fundamental properties of matter. The laws of Nature are nothing but the immutable habits which the different sorts of elementary matter follow in their actions and reactions upon each other." Peirce referred to James's extended usage of the word in his paper "Man's Glassy Essence" in support of his own usage of the term: "Professor James says, 'the phenomena of habit … are due to the plasticity of the … materials'" (CP 6.261, EP 1: 346; 1892).

The OED does not list machines as possible nonhuman agents to whom habits may be attributed, although expressions, such as, "My computer has the habit of rebooting" or "My car has the habit of braking suddenly" can easily be found in colloquial language or, as in these examples, on the internet. The reason for the lexicographic exclusion of machines from the domain of actors of habits is apparently that the dictionary compilers consider attributing agency to a machine a metaphor (cf. Glebkin 2013) usually excluded from their scope. Popular humanist writers criticize such ways of speaking and argue that only persons can act (e.g., Lanier 2010), but speculative realists and object-oriented philosophers are meanwhile defending theories to the contrary. In a line of argument adopted from the philosopher of technology Lewis Mumford, the speculative realists Levi Bryant argues that machines, due to their own technical exigencies "issue certain imperatives on their designer that run away from the[ir] intentions [… so that] the machine itself ends up contributing to the design not intended by the designer" (2014: 19). In fact, "matter imposes imperatives on designers at all levels" (ibid.), but what is true of matter is equally true of tools, machines, and our environment in general: tools and machines "impose their techniques upon us [… and] issue certain problems as imperatives to be solved" (20). They are "nonhuman actants [… with] a strange autonomy and vitality" of their own, says Steven Shaviro in his *Universe of Things* (2011: 3).

A distinguishing feature of a habit in any sense of the word is continuity in time. This makes continuity a key feature of pedagogical definitions of the term. "Habit introduces continuity into activity", writes Dewey in *Some Thoughts Concerning Education* (1932: 185), and Locke advises his readers, "Do not begin to make any thing Customary, the Practice whereof you would not have continue, and increase" (1705: 111). Despite Locke's advice to the contrary, habits evidently also need to change, and the ability to change habits needs to be learned and taught in what is now known as lifelong learning.

From 19th century psychology, the OED quotes a line from Alexander Bain's *Emotions and Will* of 1859 from which Peirce might have derived their thesis that "intellect consists in a plasticity of habit" ("Scientific Metaphysics", CP 6.86; 1898). Bain writes, "Some natures are distinguished by plasticity or the power of acquisition, and therefore realize more closely the saying that man is a bundle of habits." The behaviorists reduced the concept of habit to "an automatic, 'mechanical' reaction to a specific situation which usually has been acquired by learning and/or repetition" (OED). An influential 20th century sociological theory of habit is connected to Bourdieu's concept of *habitus*. The author restricts himself to the habits of human agents when he defines habitus as "society written into the body, into the biological individual" (Bourdieu 1990: 63).

From the history of philosophy, Aristotle's concept of habit deserves special mention since habit is the eighth of his ten categories (or predicaments)—substance, quantity, relation, quality, place, time, position, habit, action, and passion. The Greek word that corresponds to the Latin *habitus* is ἕξις (hexis). Its meaning is not only 'habit', but also 'having' and 'state'. In his *Metaphysics*, Aristotle defines ἕξις as follows:

> "Having" means (a) in one sense an activity, as it were, of the haver and the thing had, or as in the case of an action or motion; for when one thing makes and another is made, there is between them an act of making. In this way between the man who has a garment and the garment which is had, there is a "having". Clearly, then, it is impossible to have a "having" in this sense; for there will be an infinite series if we can have the having of what we have. But (b) there is another sense of "having" which means a disposition, in virtue of which the thing which is disposed is disposed well or badly, and either independently or in relation to something else. E.g., health is a state, since it is a disposition of the kind described. Further, any part of such a disposition is called a state; and hence the excellence of the parts is a kind of state. (*Metaphysics* 5.xx, 1022b)

Peirce does not follow Aristotle in adopting habit as a metaphysical category of thought.

However, as will be seen below, there is a parallel between the Aristotelian and Peircean categorial systems as Peirce considers habit a prototypical phenomenon of his universal category of Thirdness and occasionally even defined the category as such as the category of "the tendency to take habits" ("The Architecture of Theories", CP 1.32; 1891).

Habits According to Peirce

Habit is a key term of Charles S. Peirce's pragmatism, philosophy of mind, and even his cosmology, which has so far only been examined from specific points of view (Miller 1996; Santaella 2004; Erny 2005: 30–52; Colapietro 2009; Nöth 2010). Peirce's most comprehensive treatises of the topic are MS 318, "Pragmatism" (only in part: CP 1.536, 5.11–13, 5.464–496 and EP 2: 398–433; 1907), and MS 951, edited by Kenneth L. Ketner with comments by Hilary Putnam under the title *Habit* as Lecture 7 of *The Cambridge Conferences of 1898* (RLT) and in the *Collected Papers* (CP 7.468–517).

Peirce's usage of the word habit extends its ordinary sense in several directions. Firstly, not only practical or physical ways of doing things may be habits; there are also mental habits. Furthermore, there are also intermediate forms of habit, which are both physical and mental ones. Speaking is an example of such a habit since to speak means both to think about what you are saying and to articulate speech sound. Peirce defines these three kinds of habits as follows:

> While action may, in the first place, be purely physical and open to outward inspection, it may also, in the second place, be purely mental and knowable (by others, at any rate, than the actor) only through outward symptoms or indirect effects, and thirdly it may be partly inward and partly outward, as when a person talks, involving some expenditure of potential energy. ("Essay Towards Reasoning in Security and Uberty", EP 2: 464; 1913)

Secondly, Peirce also postulates habits of feeling. Thirdly, he extends the concept of habit from living beings to plants and the inanimate nature. Fourthly, even to say that humans have habits is in a certain respect an inadequate way of speaking, according to Peirce. Instead of saying that we *have* a habit, Peirce says that habits are in us: "That a habit is a rule active in us, is evident" ("Deduction, Induction, and Hypothesis", CP 2.643, EP 1: 199; 1878).

The perspective on agency developed in this formulation has a parallel in Peirce's semiotic theory of cognition, which postulates, rather similarly, that thoughts are in us, instead of stating that we have thoughts. In a footnote to his paper on "Some Consequences of Four Incapacities" of 1868, Peirce explains this theory of being in thoughts. We should not believe that we *have* thoughts but acknowledge that we *are in* thoughts, argues Peirce: "Just as we say that a body is in motion, and not that motion is in a body we ought to say that we are in thought, and not that thoughts are in us" ("Some Consequences of Four Incapacities", CP 5.289; 1868). Liszka's paraphrase of this idea, "thought thinks in us rather than we in it" (1996: ix), can thus be adapted to the present topic as meaning, "we should say that habit is in us rather than that we have habits".

One of Peirce's earliest definitions of the concept of habit is based on the Scholastic distinction between actual and habitual cognitive processes. With reference to Duns Scotus, Peirce writes, "There are two ways in which a thing may be in the mind,—*habitualiter* and *actualiter*. A notion is in the mind *actualiter* when it is actually conceived; it is in the mind *habitualiter* when it can directly produce a conception" ("Fraser's Berkeley", CP 8.7–38, W 2: 472; 1871). Only the habitual, not the actual, has continuity. Only habits have the power of evoking ideas.

The most general characteristic of a habit in any sense is that it "acts" as a "would-be" ("Pragmatism" CP 5.482, EP 2: 402, 410; 1907). Peirce also speaks of "the 'would-acts' [or] 'would-dos' of habitual behavior" (ibid., CP 5.467, EP 2: 401–402; 1907). "To define a man's habit [is] to describe how it would lead him to behave and upon what sort of occasion—albeit this statement would by no means imply that the habit consists in that action" ("Elements of Logic III.6", CP 2.664; 1910). Habits are general laws active in our minds: "Every habit has, or is, a general law. […] It is a potentiality; and its mode of being is *esse in futuro*" ("Minute Logic II.2", CP 2.148; 1902).

Habit and habit taking are prototypical phenomena of the category of Thirdness and, more specifically, the third mode of consciousness besides the ones of sensation (Firstness) and volition (Secondness). However, habits are not restricted to the phenomenological domain of Thirdness since Peirce also considers habits of feelings, which are phenomena of Firstness, and habits of bodily actions, which are phenomena of Secondness. When he thus extends his studies of habits to phenomena of Firstness and Secondness, Peirce does not contradict himself insofar as he is still considering habits as phenomena of Thirdness. More precisely, habits of feeling are phenomena of Thirdness of Firstness, and habits of action are phenomena of Thirdness of Secondness.

Peirce was well aware that this categorial definition has some affinity to Aristotle's definition of habit as a category itself (see above). In 1913, he explains his own notion of habit with several references to the Aristotelian category of habit:

> If you ask me what this [third] mode of consciousness is, I shall reply, in brief that it is that of being aware of acquiring a habit. [...] The Latin *habitus*, 'having, or possession' [...is] a term of philosophy, owing to the corresponding Greek words, ἔχειν and ἕξις, being favorite expressions with Aristotle that I shall be quite within the bounds of propriety as long as Aristotle sanctions my use of the word. [...In] the 20th chapter of Aristotle's 4th book of Metaphysics [...] we find it applied to any predicate of an object, to which its *remains* attached as much as wine remains in a pitcher, which is one of the instances given at the end of his Predicaments. It would therefore be quite contrary to good philosophical usage to make the distinction between Habit and Disposition to be that the former is acquired, as an effect of repetition; for both are Aristotelian terms, "disposition" being equivalent to διάθεσις, which is said by the Stagirite to be a Habit that is good or bad. ("A Study of How to Reason Safely and Efficiently", MS 681, 20–22; 1913)

Peirce's theory of habit is a dynamic theory in the sense that it does not restrict itself to describing why and how we have habits but also examines how habits increase and how they diminish. The law of habit, as Peirce calls it, includes the habit of habit change. Peirce speaks of the "habit of taking and laying aside habits" ("Uniformity, in Baldwin's *Dictionary*", CP 6.101; 1901).

The notion habit of habit change sounds paradoxical. It seems to be in conflict with the idea of habit itself since habit, as a tendency, presupposes continuity of the same, and not its change. This apparent self-contradiction is due to the self-referentiality of the concept of habit in combination with the law of the growth of habits. Habit change is itself a habit, and since it is a law that habits grow just as "the tendency to obey laws has always been and always will be growing" (CP 1.409, 1890), the tendency of all things to take habits is "the only tendency which can grow by its own virtue" ("Uniformity, in Baldwin's *Dictionary*", CP 6.101; 1901).

Peirce was aware of the apparent self-contradiction inherent in the notion of the habit of habit change. While habit implies continuity, habit change implies discontinuity. Both at the same time, seem to imply a paradox:

> This law of habit seems to be quite radically different in its general form from mechanical law, inasmuch as it would at once cease to operate if it were rigidly obeyed: since in that case all habits would at once become so fixed as to give room for no further formation of habits. In this point of view, then, growth seems to indicate a positive violation of law. ("Reply to the Necessitarians", CP 6.613; 1891)

One of the solutions to this paradox is in Peirce's evolutionary theory of laws and habits to be discussed below. Another is the uncertainty of mental action, which forbids habits to become an absolute necessity and accounts for the plasticity of habits, which in turn is the root of intellectual life:

> No mental action seems to be necessary or invariable in its character. In whatever manner the mind has reacted under a given sensation, in that manner it is the more likely to react again; were this, however, an absolute necessity, habits would become wooden and ineradicable and, no room being left for the formation of new habits, intellectual life would come to a speedy close. Thus, the uncertainty of the mental law is no mere defect of it, but is on the contrary of its essence. The truth is, the mind is not subject to "law" in the same rigid sense that matter is. It only experiences gentle forces which merely render it more likely to act in a given way than it otherwise would be. There always remains a certain amount of arbitrary spontaneity in its action, without which it would be dead. (CP 6.148, 1891).

Habits of Thought

Examples of habits of thought are beliefs, judgments, or leading principles of reasoning ("Algebra of Logic", CP 3.160; 1903; cf. Pietarinen 2005). Beliefs are habits, but at the same time feelings. They "guide our desires and shape our actions" so that "the feeling of believing is a more or less sure indication of there being established in our nature some habit which will determine our actions" ("Fixation of Belief", CP 5.371, EP 1: 114; 1877). A belief-habit, says Peirce, is a "cerebral habit" that "will determine what we do in fancy as well as what we do in action" ("Algebra of Logic", CP 3.160; 1903). While habits in general consist in a readiness "to act in a certain way under given circumstances", a belief is "a deliberate, or self-controlled habit" ("Pragmatism", CP 5.480; 1907). Knowledge is a habit because it guides us in our judgments ("Prolegomena to an Apology for Pragmaticism"; CP 4.531; 1907). Concepts are "mental habits, habits formed by exercise of the imagination" ("Pragmatism", MS 318, alt. version, p. 44 [=0746]; 1907).

The notion of continuity is essential to Peirce's definition of habits. The study of habit taking is a chapter of Peirce's doctrine of "synechism, or the doctrine that all that exists is continuous" ("Notes on Scientific Philosophy", CP 1.172; c. 1897). In the study of mind, the doctrine of synechism manifests itself in habit taking itself, in the tendency towards generalization, and in the tendency of ideas to grow and to spread (cf. Nöth 2014a). Beliefs are only true beliefs when they have continuity in time. Therefore, "a man does not necessarily believe what he thinks he believes. He only believes what he deliberately adopts and is ready to make a habit of conducts" ("Sketch of Dichotomic Mathematics", MS 4: 43; c. 1903). Habits of thought thus also determine what we do in imagination. "What particularly distinguishes a general belief, or opinion, such as is an inferential conclusion, from other habits, is that it is active in the imagination" ("Minute Logic II.2", CP 2.148; 1902).

Such definitions extend the notion of habit from practical to mental modes of behavior, from actions to thoughts. Peirce uses the binomial "habit of thought and conduct" to express this dichotomy ("Sketch of Dichotomic Mathematics", MS 4: 43; c. 1903). Without creating any fundamental opposition between the two, he emphasizes the close connection between thoughts and actions. Doing things in the inner world of thoughts prepares or anticipates practical ways of doing things in the outer world. Habits of thought also serve to control habits of action and to evaluate actual experience, and vice versa, perception and experience affect habits of thought:

> An expectation is a habit of imagining. A habit is not an affection of consciousness; it is a general law of action, such that on a certain general kind of occasion a man will be more or less apt to act in a certain general way. An imagination is an affection of consciousness which can be directly compared with a percept in some special feature, and be pronounced to accord or disaccord with it. Suppose for example that I slip a cent into a slot, and expect on pulling a knob to see a little cake of chocolate appear. My expectation consists in, or at least involves, such a habit that when I think of pulling the knob, I imagine I see a chocolate coming into view. When the perceptual chocolate comes into view, my imagination of it is a feeling of such a nature that the percept can be compared with it as to size, shape, the nature of the wrapper, the color, taste, flavor, hardness and grain of what is within. ("Minute Logic II.2", CP 2.148; 1902)

In this scenario, the habit of thought is the result of some previous practical experience that resulted in a habit of imagining that the same would happen again under the same circumstances. However, the opposite scenario is also possible. Habits of imagining may precede and eventually guide habits of doing things. Such habits of imagining are not only the source of practical creativity; they are also at the root of reasoning and intellectual conduct:

> A fancied conjuncture leads us to fancy an appropriate line of behavior. Day-dreams are often spoken of as mere idleness; and so they would be, but for the remarkable fact that they go to form habits, by virtue of which when a similar real conjuncture arises we really behave in the manner we had dreamed of doing. [...] People who build castles in the air do not, for the most part, accomplish much, it is true; but every man who does accomplish great things is given to building elaborate castles in the air and then painfully copying them on solid ground. Indeed, the whole business of ratiocination, and all that makes us intellectual beings, is performed in imagination. Vigorous men are wont to hold mere imagination in contempt; and in that they would be quite right if there were such a thing. How we feel is no matter; the question is what we shall do. But that feeling which is subservient to action and to the intelligence of action is correspondingly important; and all inward life is more or less so subservient. Mere imagination would indeed be mere trifling; only no imagination is *mere*. ("Grand Logic" 3.2, CP 6.286; 1893)

Peirce does not restrict his study of habits to frozen patterns of thinking or acting. He also defines habits in their development and growth. The description of how a belief comes into being and continues to develop is the following: "A belief-habit in its development begins by being vague, special, and meagre; it becomes more precise, general, and full, without limit. The process of this development, so far as it takes place in the imagination, is called thought" (ibid.). The logical method of habit formation is inductive reasoning:

> By induction, a habit becomes established. Certain sensations, all involving one general idea, are followed each by the same reaction; and an association becomes established, whereby that general idea gets to be followed uniformly by that reaction. ("The Law of Mind", CP 6.145, EP 1:327–28; 1892)

In Peirce's semiotics, the symbol is the prototype of a sign determined by a mental habit. The symbol's "special significance or fitness to represent just what it does represent lies in nothing but the very fact of there being a habit, disposition, or other effective general rule that it will be so interpreted" ("Logical Tracts 2", CP 4.447; 1903). Verbal knowledge and linguistic competence are the results of habits. Symbols are those signs that "represent their objects [...] because dispositions or factitious habits of their interpreters insure their being so understood" ("A Sketch of Logical Critics", EP 2: 460–462; 1911). This definition of the symbol as a sign acquired by a habit is reminiscent of the behaviorist doctrine according to which learning a new word means establishing the habit of associating it to its denotatum and significatum. Was Peirce a behaviorist *avant la lettre*? The reasons why he was not are the following.

First, behaviorism postulates that we learn when *outer* stimuli are repeated and reinforced. Learning is thus controlled by external agents. Peirce, by contrast, postulates the principle of self-control in the acquisition of habits. According to his principle of self-controlled habit formation,

> Every man exercises more or less control over himself by means of modifying his own habits; and the way in which he goes to work to bring this effect about in those cases in which circumstances will not permit him to practice reiterations of the desired kind of conduct in the outer world shows that he is virtually well-acquainted with the important principle that *reiterations in the inner world,—fancied reiterations,—if well-intensified by direct effort, produce habits*, just as do reiterations in the outer world; *and these habits will have power to influence actual behavior in the outer world*; especially, if each reiteration be accompanied by a peculiar strong effort that is usually likened to issuing a command to one's future self. ("Pragmatism" CP 5.487, EP 2: 413; 1907)

Second, the habits that determine symbols are not mechanical or automatized habits resulting from the repeated exposure to a stimulus. Habits are carried out according to a general "rule of conduct, including thought under conduct" ("Syllabus", CP 2.315; 1902) and in "conformity to norm" ("An Attempted Classification of Ends", CP 1.586, c. 1903). They are not "precepts" or actions performed "in obedience to a law" either, but mental patterns "copied" again and again (ibid.).

Third, the habit that controls an action need not only be acquired by learning; there are also inborn habits, habits by instinct. Even symbols may be signs by instincts. Peirce specifies that even the habit determining a symbol may be an "acquired or inborn" disposition (CP 2.297, EP 2: 9; 1894). Nothing is less compatible with the behaviorist theory of habit formation than the notion of an "inborn disposition".

The fourth reason why acquiring a habit, according to Peirce, differs from the behaviorist notion of habit acquisition concerns the role of the agent described as a learner by the behaviorists. Habit taking is more than a disposition of a learner; it is a law, the "law of habit", which testifies to "a primordial habit-taking tendency" ("Man's Glassy Essence", CP 6.262, W 8: 179; 1892). The habit that determines an action is thus not only the habit of an agent who learns; habit taking embodies the law of habit. The symbol, for example, is the result of the law of habit taking. Furthermore it embodies a purpose of its own, the purpose to be understood and to be interpreted as meaningful. This is what Peirce has in mind when he defines meaning as "the intended interpretant of a symbol" ("Harvard Lecture VI", CP 5.175, EP 2: 218; 1903) instead of defining it as the symbol user's purpose (cf. Nöth 2014a, b). The symbol has the purpose to be interpreted in an interpretant, and this purpose embodies a habit of its own.

Not only actions may be determined by habits, but also unintentional experience may be, if it is made repeatedly. For Peirce, the frequent experience of an event is a habit determined by the probability of its occurrence. For example, Cubans have "more habit of meeting and dealing with negroes than New Yorkers have, merely because they *do* meet with a greater proportion of them". Peirce clarifies that it would be inappropriate to say that the Cubans *have* this habit since this way of speaking would suggest "that the cause lies in them, in particular at least, and not in the greater proportion of negroes in the Cuban population" ("First Introduction", MS 671: 8–9; 1911).

Habits of Feeling

In 1911, Peirce applies his three phenomenological categories of Firstness, Secondness, and Thirdness to distinguish three kinds of consciously experienced habits. He declares, "[I] use the word 'Habit' to denote any state of mind by virtue of which a person would, under definite circumstances, —mostly, if not invariably, consisting in his experiencing conscious experience of some kind, —either think, or act, or feel in a definite way" ("A Logical Critique of Essential Articles of Religious Faith", MS 852: 8–9; 1911). The habits of thinking and acting are the ones discussed so far. They pertain to the phenomenological categories of Thirdness and Secondness, respectively. Feeling, by contrast, is a phenomenon of Firstness, the category of phenomena unrelated to anything else.

What is a habit of feeling? By feeling, Peirce does not only mean emotions, but also sensations and modes of perception. In this sense, feeling is "passive consciousness of quality, without recognition or analysis" ("One, Two, Three: Fundamental Categories of Thought and of Nature", CP 1.377, W5: 246; 1885). Habits of feeling are thus also topics of the study of perception (cf. Santaella 2012) and of aesthetics.

In his theory of perception, Peirce describes habits of feeling as *perceptual judgments*, the psychological counterpart to logical judgments. A perceptual judgment "is the first judgment of a person as to what is before his senses" ("Harvard Lecture IV", CP 5.115, EP 2: 191; 1903). From a logical perspective, a perceptual judgment is "a judgment asserting in propositional form what a character of a percept directly present to the mind is" ("Harvard Lecture II", CP 5.54, EP2 155; 1903). In contrast to logical judgments, perceptual judgments are beyond the control of the perceiving mind, in the sense that perceptual qualities are phenomena that cannot be doubted; they are whatever they are in their suchness. "Our perceptual judgments […] cannot be called in question. All our other judgments […] will be borne out by perceptual judgments. But the perceptual judgments declare one thing to be blue, another yellow—one sound to be that of A, another that of U, another that of I. These are the Qualities of Feeling" ("Harvard Lecture IV", CP 5.116, EP 2: 191; 1903).

Elsewhere, Peirce discusses the habits of hearing as habits of feeling to show how emotions arise from complex sound qualities of musical instruments:

> Now, when our nervous system is excited in a complicated way, there being a relation between the elements of the excitation, the result is a single harmonious disturbance which I call an emotion. Thus, the various sounds made by the instruments of an orchestra strike upon the ear, and the result is a peculiar musical emotion, quite distinct from the sounds themselves. This emotion is essentially the same thing as an hypothetic inference, and every hypothetic inference involves the formation of such an emotion. ("Deduction, Induction, and Hypothesis", CP 2.643, EP 1: 199; 1878)

In addition to habits of feeling in the form of perceptual judgments, Peirce considers habits of feeling determining aesthetic judgments. Aesthetics, for Peirce, is the study of "those things whose ends are to embody qualities of feeling" ("Harvard Lecture V", CP 5.129, EP 2: 200; 1903). It is "not the silly science […] that tries to bring our enjoyment of sensuous beauty" but a science of "meditation, ponderings, day-dreams (under control) concerning ideals" ("A Sketch of Logical Critics", EP 2: 460; 1911). Aesthetics deals with deliberate judgments guided by an ideal that "must be a habit of feeling which has grown up under the influence of a course of self-criticisms and of hetero-criticisms; and the theory of the deliberate formation of such habits of feeling is what ought to be meant by esthetics" ("Basis of Pragmaticism" CP 1.574, EP 2: 377–378; 1906).

The difference between habits of feeling and habits of action and thought is thus one of the degree to which the human mind has control over the respective habits (Colapietro 1989: 88; 2009: 350–351). Habits of feeling are "unreflectively established" ones, and they allow for no habit change, whereas habits of thought and action are modifiable and include the habit of habit change.

Nonhuman Habits—Habits of Nature

To illustrate Peirce's extension of the concept of habit from human to physiological and physical nature, two quotes from 1902 and 1907 will serve best:

> In its wider and perhaps still more usual sense, […the word habit] denotes such a specialization, original or acquired, of the nature of a man, or an animal, or a vine, or a crystallisable chemical substance, or anything else, that he or it will behave, or always tend to behave, in a way describable in general terms upon every occasion (or on a considerable

> proportion of the occasions) that may present itself of a generally describable character. ("Reason's Rules", CP 5.538; c. 1902)

> Empirically, we find that some plants take habits. The stream of water that wears a bed for itself is forming a habit. Every ditcher so thinks of it. Turning to the rational side of the question, the excellent current definition of habit, due, I suppose, to some physiologist, [...] says not one word about the mind. Why should it, when habits in themselves are entirely unconscious? ("Pragmatism", CP 5.492, EP 2: 418; c. 1907)

Peirce illustrates the way plants take habits more in detail in his "Syllabus" of 1902, where he describes the habit of a sunflower turning towards the sun as an instance of a semiotic agency determined by the biological purpose of self-reproduction:

> A Sign is a Representamen with a mental Interpretant. Possibly there may be Representamens that are not Signs. Thus, if a sunflower, in turning towards the sun, becomes by that very act fully capable, without further condition, of reproducing a sunflower which turns in precisely corresponding ways toward the sun, and of doing so with the same reproductive power, the sunflower would become a Representamen of the sun. ("Syllabus", CP 2.274; 1902)

Among the biosemiotic agents to which Peirce attributes semiotic habits or, as he sometimes specifies, habits determined by "quasi-signs" ("Pragmatism", CP 5.473; 1907), there are also neurons. From semiotic habits in botany, we thus proceed to semiotic or quasi-semiotic habits in neurology, where "nervous associations" exemplify "the habits of the nerves", for example, "the habit of the nerves in consequence of which the smell of a peach will make the mouth water ("Fixation of Belief", CP 5.373, EP 1: 114; 1877).

Habits in nature are above all habits of growth (cf. Nöth 2014a): "There are many circumstances which lead us to believe that habit taking is intimately connected with nutrition. Protoplasm grows [...by] attracting matter like itself, [...and] by chemically transforming other substances into its own chemical kind" ("Grand Logic", CP 6.283; 1893). From organic nature, Peirce proceeds to discuss habits of chemical growth in inorganic substances, without ignoring the difference between the organic and the inorganic: "Crystals also grow; their growth, however, consists merely in attracting matter like their own from the circumambient fluid" ("Man's Glassy Essence", CP 6.250; 1893).

In 1890, Peirce extends the concept of habit to physical laws of nature, such as the one of the "permanence of mass, momentum, and energy", and concludes, "The substances carrying their habits with them in their motions through space will tend to render the different parts of space alike. Thus, the dimensionality of space will tend gradually to uniformity; and multiple connections, except at infinity, where substances never go, will be obliterated ("Guess at the Riddle", CP. 1.415–416; 1887–1888). In 1891, Peirce sketches the outlines of a "molecular explanation of habit" to account for the habit determining "systems of atoms having polar forces" ("Man's Glassy Essence", CP 6.262, EP 1: 335–336; 1892). His thesis is now that the tendency to take habits begins in atomic physics:

> All things have a tendency to take habits. For atoms and their parts, molecules and groups of molecules, and in short every conceivable real object, there is a greater probability of acting as on a former like occasion than otherwise. This tendency itself constitutes a regularity, and is continually on the increase. ("A Guess at the Riddle", CP 1.409 and EP1: 277; 1887–1888).

In 1911, Peirce speaks "as readily of the 'habits' of oxygen or hydrogen, or of the 'habits' of electricity, as of the habits of bees or of classes of men" ("First Introduction", MS 671: 7; 1911).

This is the point where at least one of the most obvious objections against Peirce's extension of habit taking into the realm of the physical universe needs to be taken into consideration, one that Peirce attributed to the "necessitarians", the advocates of the "Doctrine of Necessity" (CP 6.35–65, 1892). The necessitarians postulate that there are only necessary laws in the physical universe, laws that do not allow for exceptions. Critics committed to the necessitarian doctrine still argue today that Peirce failed to recognize that the concept of habit is inapplicable to the evolution of the physical universe (e.g., Kull 2014).

Peirce knew the arguments of the necessitarians well. His characterization of their doctrine is the following: "The laws of physics know nothing of tendencies or probabilities; whatever they require at all, they require absolutely and without fail, and they are never disobeyed" (CP 1.390, c. 1890). Peirce, by contrast, defends the position

> … that all laws are results of evolution; that underlying all other laws is […] the tendency of all things to take habits. […]. If law is a result of evolution, which is a process lasting through all time, it follows that no law is absolute. That is, we must suppose that the phenomena themselves involve departures from law analogous to errors of observation. […] In so far as evolution follows a law, the law of habit, instead of being a movement from homogeneity to heterogeneity, is growth from difformity to uniformity. But the chance divergences from law are perpetually acting to increase the variety of the world, and are checked by a sort of natural selection and otherwise (for the writer does not think the selective principle sufficient), so that the general result may be described as "organized heterogeneity," or, better, rationalized variety. ("Uniformity, in Baldwin's *Dictionary*", CP 6.101; 1901)

Today's rigid laws of physics have thus emerged from less rigid ones and ultimately from a state of lawlessness. The laws of physics have only become fixed regularities in the course of their evolution, in a process during which chance divergences from uniformity, analogous to errors of observation, occurred, "perpetually acting to increase the variety of the world". These divergences, extremely rare and slow in their effects in physical nature, are increasingly frequent in living nature, culminating with the flexibility and instability of the human mind with its "remarkable degree a habit of taking and laying aside habits" ("Uniformity, in Baldwin's *Dictionary*", CP 6.101; 1901). Peirce considers the assumption that laws have evolved a necessary premise for any theory of cosmic evolution, for without it, it would be impossible to explain how the universe has evolved. His argument has a metaphysical and a cosmological foundation.

Peirce's cosmological argument is that the universe has evolved from a state of undetermined chaos, in which all possibilities of evolution were still open and will develop towards a future end state of immutability in entirely frozen order (for details, see Houser 2014). "We look back toward a point in the infinitely distant past when there was no law but mere indeterminacy; we look forward to a point in the infinitely distant future when there will be no indeterminacy or chance but a complete reign of law" ("A Guess at the Riddle", CP 1.409; 1887–1888). However, this law, too, the law of habit, is subject to the evolutionary interference of chance. In this process, chance events are the source of creativity that slow down the general tendency towards frozenness. In 1886, Peirce writes, "If the universe is thus progressing from a state of all but pure chance to a state of all but complete determination by law, we must suppose that there is an original, elemental, tendency of things to acquire determinate properties, to take habits" ("Nomenclature and Divisions of Triadic Relations", W2: 293; 1886).

Laws that do not change cannot explain why and how the universe has evolved. The laws of physics must have changed themselves for the universe to evolve, and the principles that explain why they have changed in the course of time are the ones of absolute chance (accounted for by the doctrine of tychism), habit taking, and continuity (accounted for by the doctrine of synechism). Not immutable laws, but only processes of habit taking, conceived of as "infinitesimal chance tendencies [...] towards generalization" can thus have been the "primordial principle of the universe" ("Man's Glassy Essence", CP 6.262; 1892). Infinitesimal deviations from the course of a physical law do occur, repeat themselves, and eventually give rise to a new law. The element of chance accounts for the difference between law and habit. Laws work without exception, but habits, like the ones that determine human behavior, know exceptions and deviations: "According to this, three elements are active in the world: first, chance; second, law; and third, habit-taking" ("A Guess at the Riddle", CP 1.409, EP 1: 277; 1887–1888). The sense in which the laws of physics are hence habits, which may involve deviations from the regularities they predict, is the following: "Try to verify any law of nature, and you will find that the more precise your observation, the more certain they will be to show irregular departures from law" ("The Doctrine of Necessity Examined", CP 4.46; 1892).

In light of the results of research in physics half a century after this attacks against the necessitarians, Potter comments on the modernity of Peirce's concept of the laws of nature as follows:

> Heisenberg did not enunciate his indeterminacy principle until more than a quarter of a century later (1927), but Peirce would not have been surprised if he had lived to see science so conclude. He certainly would have sided with those who have interpreted the principle to mean that atoms are endowed with a certain spontaneity in their movement against others who attribute the apparent indeterminacy to the intrinsic limitations of measurement techniques. (Potter 1967: 160)

Peirce's metaphysical argument to support his theory that nature takes habit is in the famous Schelling-inspired doctrine of "objective idealism" (see Ibri 2015). Its postulate, which cannot be examined in detail here, is that "that matter is effete mind, inveterate habits becoming physical laws" ("The Architecture of Theories", CP 6.25; 1891; cf. Santaella 2001). In other words, we must "regard matter as mind whose habits have become fixed so as to lose the powers of forming them and losing them" ("Uniformity, in Baldwin's *Dictionary*", CP 6.101; 1901). Hence, mind precedes matter, and matter has evolved from mind that has lost its flexibility, having eventually taken an immobile and frozen form. This argument explains perhaps best why Peirce uses the concept of habit despite all of its psychological connotations to explain the laws of nature. At its root, we find Peirce's conviction that there is continuity between the phenomena of physical and living nature. Peirce's metaphysical account of this continuity is, "I believe the law of habit to be purely psychical. But then I suppose matter is merely mind deadened by the development of habit. While every physical process can be reversed without violation of the law of mechanics, the law of habit forbids such reversal" ("Letter to Christine Ladd-Franklin", CP 8.318; 1891).

A more recent perspective on Peirce's anti-necessitarian views of the laws of nature in light of modern physics is the following:

> What the directly measured facts of scientific practice seem to tell us, then, is that, although the universe displays varying degrees of habit (that is to say, of partial, varying, approximate, and statistical regularity), the universe does not display deterministic law. It does not directly show anything like total, exact, non-statistical regularity. Moreover, the habits that nature does display always appear in varying degrees of entrenchment or "congealing". At one end of the spectrum, we have the nearly law-like behavior of larger physical objects like boulders and planets; but at the other end of the spectrum, we see in human processes of imagination and thought an almost pure freedom and spontaneity; and in the quantum world of the very small we see the results of almost pure chance. (Burch 2014, § 5)

Law, Habit, Rule, and Norm

The concept of habit requires a closer examination in its relation to the ones of rule, norm, and law (for the latter, see especially Potter 1967). Peirce often refers to laws as well as to norms and rules as phenomena of Thirdness. Among the characteristics of Thirdness that the four concepts have in common are generality, continuity, and regularity (cf. Nöth 2011). "Real regularity is active law […], Thirdness as Thirdness" ("Harvard Lecture V"; CP 5.121, EP 2: 197; 1903). The difference between law and habit is then that a habit is a kind of law. Peirce speaks of the law of habit ("The Architecture of Theories", CP 6.16, W 8: 105; 1890) and postulates, "Every habit has, or is, a general law" ("Minute Logic" III.5; CP 2.148; 1902). In 1903, however, Peirce does not refer to habit as a law but as the embodiment of a law: "A law can never be embodied in its character as a law except by determining a habit. […] A law is how an endless future must continue to be" ("Harvard Lecture III", 3rd draught, CP 1.536; 1903). When Peirce calls habits embodiments of laws instead of calling them laws, Peirce changes his focus from the idea of regularity inherent in habits and laws to the idea "that a habit is a rule active in us" discussed above in the second section of this paper.

By "embodiment", Peirce often means the instantiation of some general law or rule in an actual phenomenon of this principle (cf. Nöth and Santaella 2011). A symbolic legisign (or type), for example, is embodied in a symbolic sinsign (or token). The meaning of a word is embodied in an actual thought or in some dictionary definition. Peirce describes such forms of embodiment of the general in an instance of it as a case of "Secondness involved in Thirdness" ("Harvard Lecture III", 3rd draught, CP 1.530; 1903). However, when Peirce calls habit the embodiment of law, he cannot mean that habits are phenomena of genuine Secondness in the sense of a singularity determined by a general law. Habits are always and without exception phenomena of genuine Thirdness in Peirce's writings.

The expression "embodiment of a law in a habit" can thus only mean that a more general mode of Thirdness, a law, is embodied in a less general one, a habit. There is apparently a hierarchy of phenomena of Thirdness, according to which habits are less general than laws. Peirce develops this idea more explicitly in the context of his theory of "The Reality of Thirdness" (CP 1.343–49; 1903), where he argues that laws are real because they have real effects on the phenomena determined by them. In this context, Peirce writes, "whatever is real is the law of something less real" ("The Logic of Mathematics", CP 1.487; c. 1896). Applied to laws, which are real by this definition, this means that a more general law may be the law of a less general one—in our case, a habit. Hence, both the laws and the habits in which they are embodied may be phenomena of Thirdness.

That laws are more general than the habits embodying them implies that they may also be embodied in phenomena that are not habits. The most general manifestation of a law is in "the idea of law", which Peirce also circumscribes as "reasonableness" ("Telepathy", CP 7.687; 1903). The evolutionary principle of "becoming governed by laws" is but a phase "of the process of the growth of reasonableness" ("Pragmatic and Pragmatism", CP 5.4; 1902). High up in the hierarchy of the phenomena of Thirdness, the idea of law in the sense of "objective reasonableness" is embodied "in the laws of human nature" ("Logical Machines", W 6: 69–70, 1887). Further down, it is embodied in psychological and mental habits.

That a habit embodies a law thus also means that it shares a number of characteristics with laws besides the ones that characterize their mode of embodiment. One is that habits inherit from laws the characteristics of continuity and regularity. A habit as well as any "real regularity is active law" ("Harvard Lecture V"; CP 5.121, EP 2: 197; 1903). Another characteristic that habits share with laws is that both have the being of an *esse in futuro*. A law of nature, for example, has "a sort of *esse in futuro*" ("On Phenomenology", CP 5.48, EP 2: 153; 1903). Both habits and laws are realities in the Peircean sense in which "reality consists in regularity" and "real regularity is active law" ("Harvard Lecture V"; CP 5.121, EP 2: 197; 1903).

In contexts in which Peirce considers laws and habits as two phenomena of different degrees of generality, he often simply refers to habits as laws or rules without distinguishing between the three terms. A symbol, for example, which is a habit of thought, "is a law, or regularity of the indefinite future" ("A Syllabus of Certain Topics of Logic", CP 2.293, EP 2: 274; 1903), and a "Legisign is a law that is a Sign" ("Nomenclature and Divisions of Triadic Relations", CP 2.246, EP 2: 291; 1903). In the same context in which he calls the symbol a law, Peirce also refers to it as "a rule that will determine its interpretant" ("A Syllabus of Certain Topics of Logic", CP 2.292, EP 2: 274; 1903). Law, habit, and rule are thus quasi-synonyms in many contexts, and the expression "habit or rule" is common in Peirce's writings. A symbol, as quoted above, is a sign characterized by "there being a habit, disposition, or other effective general rule that it will be so interpreted" (see above). Savan (1976: 45) gives the following account of Peirce's usage of the term rule in this context:

> The rule is […] a habit of action, which may be expressed by words. The expression of the habit states that if certain actions are performed upon objects answering to a certain description, results of a general rule will be observable. The acts which the rule prescribes may be muscular and physical, or they may be imaginative acts experimenting as it were upon images and fancied diagrams. But in either case experimental acts re-act upon the rule, leading us to modify it in such ways as to make it more reliable in producing the predicted results. The rule or habit is a pattern of actions which would, under certain appropriate conditions, be repeated indefinitely in the future.

According to this description, the characteristics of rules seem hardly any different from the ones of laws and habits, as defined above. Indeed, laws are occasionally also defined in terms of rules. For example, Peirce explains why "future events are really governed by law" by saying that if predictions about the future have "a tendency to be fulfilled, it must be that future events have a tendency to conform to a general rule" (CP 1.26, "Harvard Lecture IIIa"; 1903).

On the other hand, Peirce also draws subtle distinctions between laws, rules, and habits. The term *rule*, and not law, is mostly used in the sense of a linguistic rule and in the sense of rules of social behavior. The nature of the verbal symbol *man*, for example, "consists in the really working general rule that three such patches seen by a person who knows English will effect his conduct and thoughts according to a rule" ("Logical Tracts 2", CP 4.447; c. 1903). Rules, and not laws of conduct, determine the course of life of human beings. They are "convenient and serve to minimize the effects of future inadvertence and, what are well-named, the wiles of the devil within" them ("Harvard Lecture I", 3rd draught, CP 1.592; 1903).

As to the differences between habit and law, one of them was already discussed above. Habits are embodiments of laws. Since interpreting symbols according to linguistic rules and acting according to rules of social behavior are both human habits, it can be presumed that rules differ from binding laws in the same way as habits do. A more specific difference is that habits, in contrast to laws, have varying degrees of strength:

> These grades are mixtures of promptitude of action […]. The habit-change often consists in raising or lowering the strength of a habit. Habits also differ in their endurance (which is likewise a composite quality). But generally speaking, it may be said that the effects of habit-change last until time or some more definite cause produces new habit-changes. ("Survey of Pragmaticism", CP 5.477, c. 1907)

That habits have varying degrees of strength is one of the characteristics of their plasticity. This plasticity of habits can evolve into two directions, increase or decrease. As predicted by the law of habit, the decrease of the plasticity has a consequence that habits become "inveterate" and finally freeze to fixed laws in a process of "strengthening habits into absolute laws regulating the action of all things in every respect" ("A Guess at the Riddle", CP 1.409 and EP1: 277; 1887–1888). The opposite tendency is the weakening of a habit, which can result in habit change. As far as human habits are concerned, one of the symptoms of the weakening of habits is the tendency to error, in which Peirce sees a source of creativity:

> This tendency to error, when you put it under the microscope of reflection, is seen to consist of fortuitous variations of our actions in time. But it is apt to escape our attention that on such fortuitous variation our intellect is nourished and grows. For without such fortuitous variation, habit-taking would be impossible; and intellect consists in a plasticity of habit. ("Scientific Metaphysics", CP 6.86; 1898)

Plasticity of Habits and Laws as Thirdnesses Versus Rigidity of Laws as Secondnesses

In most of what we have read so far, habit and law seemed to be equally typical phenomena of Thirdness. However, whereas habit is always the prototype of Thirdness in Peirce's writings, law is not. Law is sometimes even contrasted with habit. Above, in the chapter on habits of thought, we already encountered an instance of such a contrast in the context of Peirce's comments on symbols as habits of 1903. There, habits were contrasted with "precepts" and actions performed "in obedience to a law". In the domain of psychology, Peirce describes how we follow our habits by "mere custom", according to a general rule of conduct, and in "conformity to norm" in contrast to unreflected acting "in obedience" to a law ("An Attempted Classification of Ends", CP 1.586; c. 1903). The difference between the two is clearly the criterion of "plasticity" of habit, which is missing in unreflected mechanical modes of acting. Only the plasticity of habit makes habit a phenomenon of Thirdness, whereas unreflected modes of behavior (as they were later in the focus of the behaviorists) are phenomena of Secondness. Whereas the sphere of Secondness is determined by "blind laws", the domain of Thirdness is determined by laws of a different kind, such as habits, rules, and norms. One of the consequences of the plasticity of habits is their "peculiar characteristic of not acting with exactitude" ("Man's Glassy Essence", CP 6.260, EP 1: 346; 1892).

In several other contexts, Peirce elaborates this distinction. In the domain of psychology, Peirce emphasizes that the plasticity of habit is a characteristic of mental associations and criticizes psychologists of his time who interpreted them in a necessitarian way:

> There is a law in this succession of ideas. We may roughly say it is the law of habit. It is the great "Law of the Association of Ideas,"—the one law of all psychical action. Many psychologists hold that this law as strictly necessitates what idea shall rise on a given occasion as the law of mechanics necessitates how a body in a given relative position to other bodies endowed with given forces shall have its motion altered. This is a theory hard to disprove; but it is a mere forejudgment, or prejudice: no observed facts afford the slightest warrant for it. ("Grand Logic", CP 7.388–389; c. 1893)

There are hence two kinds of laws, "blind" laws acting by efficient causation in the domains of Secondness and "intelligent laws", as Peirce calls them in his "Reply to the Necessitarians" (CP 6.614; 1891), in the domain of Thirdness, acting by final causation (cf. "Uniformity, in Baldwin's *Dictionary*", CP 6.101; 1901). Since Peirce does not always say what kind of law he is referring to, the meaning of law that he has in mind must often be inferred from the context.

The ambiguity of the concept of law is particularly great when Peirce discusses the laws of nature. Physical laws are contrasted with habits when Peirce writes, "The law of habit exhibits a striking contrast to all physical laws […]. A physical law is absolute, [… but] no exact conformity is required by the mental law" ("The Architecture of Theories", CP 6.23; 1890). In such contexts, Peirce means "blind laws" and their brute or "active" force as opposed of laws that act by "legislation". Blind law "as an active force is second, but order and legislation are third" ("Fragment 'Third'", CP 1.337; 1875). More precisely, however, the blind law is not a phenomenon of genuine Secondness, but one of Secondness in Thirdness. After all, the law that obliges us to act without any alternative and knows no exception is still a law, and its characteristics are continuity and regularity.

At this point, it must be asked whether Peirce is not self-contradictory when he defines the laws of nature sometimes as habit-like and the laws of physics sometimes as blind. On the one hand, we read that "the universe is *not* a mere mechanical result of the operation of blind law" ("Fallibilism, Continuity, and Evolution, CP 1.162; c. 1897), that it is wrong to believe "that science has *proved* that the universe is regulated by law down to every detail" ("Cambridge Lecture VIII", CP 6.201; 1898), and "those dualistic philosophers" deserve to be blamed who commit themselves "to the proposition that law bears absolute sway in nature" since their "thought is marked by Secondness" (unidentified, CP 1.325; n.d.). On the other hand, we find Peirce writing elsewhere that the "laws of physics know nothing of tendencies or probabilities; whatever they require at all they require absolutely and without fail, and they are never disobeyed" ("A Guess at the Riddle", CP 1.366, EP 1:254–255; 1887–1888) or that "a physical law is absolute" and requires "an exact relation" so that "Law is Second" (i.e., a phenomenon of Secondness) ("Architecture of Theories", CP 6.23, 6.32, EP 1: 292, 297; 1891). According to the first line of argumentation, the laws seem to be phenomena of genuine Thirdness, whereas the second kind of analyses seems to suggest or even states that they are phenomena of Secondness in Thirdness.

An answer to the question whether Peirce was inconsistent in his phenomenology of laws may be that he changed his mind after his debate with the Necessitarians of 1891. It seems that the above quotes in which physical laws are characterized as blind date before the year 1891, whereas the quotes in which plasticity is attributed to them have a later date. However, it is not necessary to attribute a change of mind to Peirce. An equally plausible account of the changes in his argumentation is that he changed his focus, shifting from a synchronic perspective on the laws of nature to an evolutionary one. From this perspective, only the plastic laws that have developed in evolutionary history are phenomena of genuine Thirdness, whereas the blind laws of physics, as they act *hic et nunc* are phenomena of Secondness in Thirdness. "Generally speaking genuine secondness consists in one thing acting upon another,—brute action. I say brute, because so far as the idea of any law or reason comes in, Thirdness comes in" ("Letter to Lady Welby", CP 8.330; Oct. 12, 1904).

Habits, by contrast, have all of the essential characteristics of phenomena of genuine Thirdness, continuity, generality, and law (see above). A habit is general and lawlike because it determines the "general way in which it will act" and because "every habit has, or is, a general law" ("Minute Logic II.2", CP 2.148; 1902), but it is not rigid and blind. Thoughts, symbols and legisigns are habits (cf. Nöth 2010). However, habits are not thoughts in the sense of actual instances of thinking. In accordance with the *habitualiter-actualiter* dichotomy, actual thoughts are neither habits nor phenomena of Thirdness. Acts or instances of thinking are only manifestation of habits, in other words, determined by habits. Like time, another category of Thirdness, a habit mediates between the past and "potentialities *in futuro*" ("Minute Logic II.2", CP 2.148; 1902), and Peirce considers habits of nature as mediators in the sense that they mediate between the general tendencies of changes by chance and determination by unchanging mechanical laws:

> If the universe is thus progressing from a state of all but pure chance to a state of all but complete determination by law, we must suppose that there is an original, elemental, tendency of things to acquire determinate properties, to take habits. This is the Third or mediating element between chance, which brings forth First and original events, and law which produces sequences or Seconds. ("One, Two, Three: Kantian Categories", EP 1: 243; 1886)

How do feelings, which are phenomena of Firstness, combine with habits as phenomena of Thirdness? Peirce advises his readers not to confound the habit with the feeling in a habit of feeling, such as a perceptual judgment: "A quality is something capable of being completely embodied. A law never can be embodied in its character as a law except by determining a habit. A quality is how something may or might have been. A law is how an endless future must continue to be" ("Harvard Lecture III", 3rd draught, CP 1.536; 1903).

Habits of feeling, action, and thought thus exemplify how phenomena of Firstness (feeling), Secondness (acting), and Thirdness (thought) may be included in, subsumed under, or determined by a phenomenon of Thirdness (habit). Peirce distinguishes between Firstness in Secondness, Firstness (and Secondness) in Thirdness and Secondness in Thirdness (cf. Nöth 2011). These modes of inclusion illustrate the general principle of Peirce's phenomenology according to which categories of the lower orders are included in the higher order (but not vice versa) and cannot be separated (or “prescinded”, as Peirce calls it) from them: “The category of first can be prescinded from second and third, and second can be prescinded from third. But second cannot be prescinded from first, nor third from second” (“Notes on the Categories”, CP 1.353; c. 1880). Peirce formulates this principle as follows:

> Secondness is an essential part of Thirdness though not of Firstness, and Firstness is an essential element of both Secondness and Thirdness. Hence there is such a thing as the Firstness of Secondness and such a thing as the Firstness of Thirdness; and there is such a thing as the Secondness of Thirdness. But there is no Secondness of pure Firstness and no Thirdness of pure Firstness or Secondness. (“Harvard Lecture III”, 3rd draught, CP 1.530; 1903)

This means that habits of thought include habits of feeling and action, whereas mere habits of feeling are possible without habits of thought and habits of action. This inclusion of habits of feeling, which are uncontrolled habits, within habits of thought, which are self-controlled, also means that habits of thought contain elements that lack self-control. Hence, habits of reasoning cannot be entirely self-controlled given the habits of feeling and perception. As Colapietro interprets this Peircean insight, “our conscious deliberations are, ultimately, rooted in our unreflective, somatic competencies” (2009: 351).

Another example of how Firstness, the category of feeling and sensation, and Secondness, the category of reaction, is included within Thirdness, the category of generality and habit-taking, is abductive reasoning (“hypothetical inference”). Peirce describes it as follows: “Habit is that specialization of the law of mind whereby a general idea gains the power of exciting reactions. But in order that the general idea should attain all its functionality, it is necessary, also, that it should become suggestible by sensations. That is accomplished by a psychical process having the form of hypothetic inference” (“The Law of Mind”, EP 1:327–328; 1892). The Thirdness of abductive reasoning consists in the generality of inferential reasoning. Abductive reasoning involves imagination insofar as the inference must be “suggestible by sensations”, wherefore Firstness is involved, and the mental “reaction” thus excited in the process of abductive reasoning exemplifies the aspect Secondness involved in the process.

The Aesthetics, Ethics, and Logic of Good and Bad Habits

The notion of habit as such is value neutral. There are good and bad habits, but the law of habit is applicable to both. This reduces the explanatory power of Peirce’s law of habit. It cannot explain whether, why, or how good or bad habits grow. Peirce admits this restriction: “The law of habit is a simple formal law, a law of efficient causation” (“Uniformity, in Baldwin’s *Dictionary*”, CP 6.101; 1901). The question whether a habit is good or bad cannot be explained by efficient but only by final causation (ibid.), whose purpose needs to be evaluated by principles formulated by the three normative sciences: aesthetics, ethics, and logic (cf. Bergman 2012; Pape 2012). In other words, the final cause of the growth of a habit cannot be evaluated by the law of habit itself. It must be evaluated on grounds of the Normative Sciences of "esthetics [which] considers those things whose ends are to embody qualities of feeling, ethics [that studies] those things whose ends lie in action, and logic [that studies] those things whose end is to represent something" ("Harvard Lectures V", CP 5.129; 1903).

A habit that does not change "is mere inertia, a resting on one's oars" ("Evolutionary Love", CP 6.300, EP 1: 360; 1893). However, the norms and ideals studied by the Normative Sciences do exert a real influence in the habits of thought, action, and feeling. Bad habits of reasoning tend to be corrected by good ones under the influence of the self-control exerted by the norms and the critique of logic, which provide evidence of whether the habit produces a true or a false conclusion:

> The habit is good or otherwise, according as it produces true conclusions from true premisses or not; and an inference is regarded as valid or not, without reference to the truth or falsity of its conclusion specially, but according as the habit which determines it is such as to produce true conclusions in general or not. The particular habit of mind which governs this or that inference may be formulated in a proposition whose truth depends on the validity of the inferences which the habit determines; and such a formula is called a *guiding principle* of inference. ("Fixation of Belief", CP 5.367, W 3: 245; 1877)

The aesthetic guideline of the *summum bonum* exercises control over aesthetic habits in the way outlined above ("Habits of Feeling"). The norms of ethics tend to improve bad conduct, although not in the fashion of a science "not yet mature enough fully to comprehend its own purpose" since "it still sticks to the obsolete pretense of teaching men what they are 'bound' to do" ("A Sketch of Logical Critics", EP 2: 459, CP 5.129; 1911). Instead, they operate through the real influence of self-government and an "intense disgust with one kind of life and warm admiration for another" which have the result that "those who desire to further the practice of self-government ought to shape their teachings accordingly" (ibid.: 460). Self-control is the common denominator of habits of conduct and habits of thought. "The machinery of logical self-control works on the same plan as does moral self-control, in multiform detail. The greatest difference, perhaps, is that the latter serves to inhibit mad puttings forth of energy, while the former most characteristically insures us against the quandary of Buridan's ass. The formation of habits under imaginary action [...] is one of the most essential ingredients of both" ("Issues of Pragmaticism", EP 2: 347, 1905).

Nevertheless, the notion of habit tends to have more negative connotations than positive ones. Habits are more typically old than new ones, and old habits need to be broken. Habits connote immobility, immutability, conservativism, and stagnation rather than mobility, change, development, and evolution. Sticking to an old habit against better evidence only for the sake of the habit itself characterizes the unfruitful method of tenacity, which Peirce caricatures in his paper "The Fixation of Belief" of 1877. One of his examples is a man who resolutely continues to believe "that he would be eternally damned if he received his *ingesta* otherwise than through a stomach-pump" (CP 5.377, 1877). From sticking to old beliefs against all evidence, Peirce writes, such tenacious men and women may even derive great satisfaction, for it cannot "be denied that a steady and immovable faith yields great peace of mind" (ibid.). The insight echoes Michel de Montaigne's aphorism that ignorance and the lack of curiosity are a soft pillow ("L'ignorance et l'incuriosité font un doux oreiller", *Essais*, III, 13, 1580).

However, despite the tenacity of some individuals, old habits no longer felt to be adequate tend to be abandoned and become replaced by new ones. Just as the tendency to stick to good habits is a habit, the tendency to change bad ones is, too. Both habits exemplify the habit of habit change, but go in opposite directions. The logic behind the argument that "taking habits" includes "getting rid of them" (MS 671, 6) is the following: "The fact that any real subject, on any particular sort of occasion, A, would *not* behave in the particular manner, B, is just as truly a 'habit', in my sense of the word, as if he were sure to behave in the manner B. For a 'habit' is nothing but the reality of a *general fact* concerning the conduct of any subject" ("First Introduction", MS 671: 7; 1911).

Knowledge is the prototype of a habit associated with habit change. The self-organizational growth of knowledge in the sciences exemplifies the habit of habit change (cf. Nöth 2014a). Old knowledge is abandoned when new evidence leads to the insight that it is incompatible with experience and reality. In the formation of judgments,

> Logic is wanted, to pull to pieces our inferences, to show whether they are good or bad, how they can be strengthened, and by what methods they ought to proceed. Intermediate between the lesser and the greater inferences, lies an intermediate class, which are best governed by habits; yet by habits formed or corrected under conscious criticism. (CP 7.457)

The habit of habit change is thus a prerequisite for the advance of knowledge and scientific research. It is a habit guided by the principle of self-control, which states that, "Every man exercises more or less control over himself by means of modifying his own habits" ("Pragmatism", CP 5.487, EP 2:413; 1907). Its final causes of self-control are studied by the normative sciences. There we find the answers to the question whether a habit is good or bad. From the point of view of logic, for example, a habit is good "provided it would never (or in the case of a probable inference, seldom) lead from a true premiss to a false conclusion; otherwise it is logically bad" (CP 3.163, 1880).

Habits determined by principles of ethics and morality, by contrast, seem to withstand changes more tenaciously, but this tendency to conservatisms inherent in an ethics too immature to "fully to comprehend its own purpose" (as quoted in the beginning of this chapter) is a recurrent topic of Peirce's critique. Even in the work of scientists, who need undoubtedly to be guided by ethical principles, too, following moral principles tenaciously may not always be advisable, is Peirce's provocative argument of c. 1896 (CP 1.50). Although following moral principles is necessary in general, the problem with moral principles is their tendency to impede creativity and innovation:

> Good morals and good manners are identical, except that tradition attaches less importance to the latter. The gentleman is imbued with conservatism. This conservatism is a habit, and it is the law of habit that it tends to spread and extend itself over more and more of the life. In this way, conservatism about morals leads to conservatism about manners and finally conservatism about opinions of a speculative kind. [...] Hence it is that morality leads to a conservatism which any new view, or even any free inquiry, no matter how purely speculative, shocks. (CP 1.50, c. 1896)

However, habits are not conservative in principle, as we have seen since habit includes the habit for habit change. The habit of habit change leaves always the space for creativity that guarantees the growth of science and culture. It is therefore wrong to believe that habits are conservative forces wherever they may be found. This is the general advice that Peirce has to give to impatient youngsters of his days, averse to old habits in general, "Some undisciplined young persons may have come to think of acquired human habits chiefly as constraints; and undoubtedly they all are so in a measure. But good habits are in much higher measure powers than they are limitations; and the greater number even of acquired habits are good, like almost all those that can properly be called natural" ("On the Meaning of 'Real'", MS 930, 31; n.d.).

Remarks on the Role of Habit and Habit Change in the Interpretants of Symbols

The perspectives on Peirce's concept of habit presented in this paper are necessarily incomplete, for a full account of the topic requires its extension to the study of sign processes. Since this issue is being examined in detail elsewhere in this volume (Santaella), a few remarks on the topic must suffice.

Habit is the defining characteristic of the symbol (Nöth 2010), which Peirce defines as a sign "used and understood as such" because of a habit (CP 2.307, 1901) or as a sign that, "determined by its dynamic object" to be so interpreted, "depends" upon a habit of its interpretant (CP 8.335, 1904). Since only symbols depend on habits we can restrict ourselves, in the following, to these signs. Furthermore, only a special but prototypical kind of symbol can be examined here, namely "intellectual concepts", that is, terms "upon which reasonings may turn" ("Pragmatism", CP 5.8, 1907) or upon which "arguments concerning objective facts may hinge" ("Pragmatism", EP 2: 401; 1907). Interpreters of symbols are guided by habits in the process of interpreting a symbol, and these habits make up the conventions that determine what the symbol means. Let us remember, however, that Peirce does not speak of the interpreters' habits but of the habits that are in the interpreters when they interpret symbols.

One of Peirce's definitions states that an interpretant is "the entire mental effect which a sign of itself is calculated, in its proper significative function, to produce" ("Pragmatism", EP 2: 429, 1907). The interpretants of intellectual concepts are, first of all, their meanings. Now, since the meaning of a concept can be expressed in words that translate, paraphrase, or define it, the interpretant of a concept should be a symbol and hence a verbal habit, too. The view that the meaning of a symbol is another symbol that conveys the same meaning is widespread with linguists, who have attributed it, since Roman Jakobson, to Peirce (cf. Nöth 2012). In principle, this view is not against the Peircean theory of meaning. After all, a definition, according to Peirce, "even if it be imperfect owing to vagueness, is an intellectual interpretant of the concept of the term defined" ("Pragmatism", EP 2: 430; 1907).

However, Peirce is also convinced that "a verbal definition is inferior to the real definition" (ibid.: 418) and for this and other reasons, he rejects the view that the ultimate logical (or intellectual) interpretant of a symbol should again be a symbol. The interpretant that Peirce refers to here is one that he identifies elsewhere as 'meaning' *tout court* (cf. CP 5.494, 1907). Read in conjunction with Peirce's pragmatic maxim, the ultimate logical interpretant is "the kind of habit that would be the result of critical deliberation on its potential consequences", as Bergman (2012: 137) puts it. What Peirce rejects is that a generally or ultimately accepted verbal definition should be the ultimate interpretant of a symbol. The reason is that "signs which should be merely parts of an endless viaduct for the transmission of idea-potentiality, without any conveyance of it into anything but symbols, namely into action or habit of action, would not be signs at all ("Basis of Pragmaticism [in the Normative Sciences]", EP 2: 388; 1906).

In his pragmaticist analysis of symbols that are "intellectual concepts" of MS 318, Peirce reaffirms that a symbol cannot be an ultimate intellectual interpretant. The definition of a concept is not an ultimate logical interpretant because "it is itself a sign, and a sign of the kind that has itself an intellectual interpretant, which is thereby an intellectual interpretant of the term defined" ("Pragmatism", EP 2: 430; 1907). In conjunction with the thesis that the meaning of a concept is of the nature of a "mental effect" and a "would-be" that consists of its "significative capacity" (ibid.), Peirce arrives at the conclusion "all logical, or intellectual, interpretants are habits" (ibid.: 431; cf. Houser 1998: xxxv–vi). More in detail, Peirce describes the "most perfect account of a concept that words can convey" as one that consists "in a description of the habit which that concept is calculated to produce" (ibid.: 418). He concludes with the question, "But how otherwise can a habit be described than by a description of the kind of action to which it gives rise, with the specification of the conditions and of the motive?" (ibid.: 418). The "veritable and final logical interpretant" ("final" is here used as a synonym of "ultimate") of an intellectual concept is hence the "living definition that grows up in the habit" (CP 5.491). From this formulation, commentators since Savan (1976: 48) and Rosenthal (1994: 36–37) have concluded that the habit that characterizes the ultimate logical interpretant is a "living habit", a term not used literally by Peirce, but faithful to his ideas.

With this shift from a semantic to a truly pragmatic theory of meaning, Peirce succeeds, as Short put it, to "break out of the hermeneutic circle of words interpreting words and thoughts interpreting thoughts" (2007: 59). By offering a theory of real definition of a new pragmaticistic kind, Peirce conveys the insight that "it is only through the medium of purposeful action, even if only a potential action for a possible purpose, that words and thoughts relate to a world beyond themselves" (ibid.).

Peirce's MS 318 of 1907, from which the above quotes are taken, consists of several drafts of a paper under the common heading of "Pragmatism". The version from which the passages quoted above stem has been published under the title "Pragmatism" in EP 2. Another version of the same MS, published in CP 5.467–96 under the title "A Survey of Pragmaticism" seems to display a significant difference in the definition of the habit involved in the ultimate logical interpretant of an intellectual concept. While "Pragmatism" offers a version of MS 318 according to which the ultimate logical interpretant of a concept is the cause of a habit living in the "kind of action to which the concept gives rise" (EP 2: 418), the "Survey" characterizes the same habit as a habit of habit change. The "Survey" definition of the ultimate proper significate effect of a concept, alias ultimate logical interpretant, which is the living habit that is not another symbol, has the following wording:

> It can be proved that the only mental effect that can be so produced and that is not a sign but is of a general application is a habit-change; meaning by a habit-change a modification of a person's tendencies toward action, resulting from previous experiences or from previous exertions of his will or acts, or from a complexus of both kinds of cause. It excludes natural dispositions, as the term "habit" does, when it is accurately used; but it includes beside associations, what may be called "transsociations," or alterations of association, and even includes dissociation, which has usually been looked upon by psychologists (I believe mistakenly), as of deeply contrary nature to association. ("A Survey of Pragmatism", CP 5.476, 1907)

Are the two versions of the same MS self-contradictory, or did Peirce change his mind? Why should not a mere living habit instead of a habit change suffice as the ultimate logical interpretant of an intellectual concept? Commentators on Peirce's "Survey" argument of CP 5.476, which amounts to saying that the proper and ultimate conception of the meaning of an intellectual concept involves a habit change, have given different, but converging answers. Santaella (this volume), based on Kent (1987), argues that Peirce's formulation that habit change, and not a habit, is required as the ultimate logical interpretant of a symbol has its roots in the evolutionary dimension of Peirce's pragmatism. In and almost, but not entirely, Derridaean vein, Peirce's evolutionary pragmatism postulates that the ultimate meaning of a concept lies in a distant future, and that its discovery is always postponed. Kent's version of this argument is that the aesthetic ideal of "the *summum bonum* for pragmaticism is of an evolutionary nature, the denouement of which may be approached asymptotically, but might never be reached" (1987: 158). Evidently, grasping the meaning that changes in this way requires a habit of habit change.

Two other commentators on the same question are Savan and Bergman. Savan argues in reply to the question of what would happen if the habit created by a symbol would change never more. His answer is that the result would be "a simple succession of identical acts [that are merely] mechanical, dead, and an example of Secondness", which could never "interpret the significance of a sign" (1976: 48). Bergman's argument is similar although more abstract: "An intellectual interpretant, if it is allowed to run its due course, will result in a modification of the habitual character of the interpreter, his or her dispositions to feel, act, or think [… even though] the modification may be nothing but a strengthening or weakening of a previous habit" (2004: 375).

Bergman (2012: 136–140) develops this argument further by association the habit change involved in the ultimate logical interpretant with Peirce's notions of "concrete reasonableness" and "pragmatistic adequacy", defined in 1910 as "what ought to be the substance, or Meaning, of the concept or other Symbol in question, in order that its true usefulness may be fulfilled" (MS 649: 2). On these premises, the author's suggestion is that the ultimate logical interpretant in the sense of the "pragmatistic adequacy of symbols—what they should signify rather than what they do mean—is something that is discoverable or developable only by active experimentation in both the internal and the external world, that is, through intelligent habit-change" (2012: 140).

In sum, when Peirce characterizes the habit embodied in the ultimate logical interpretant as a habit of habit change, he is more precise that when he calls it only of a living habit, but he does not contradict himself since a living habit is by definition one that incorporates the habit of habit change. The reason why the embodiment of the ultimate logical interpretant cannot just have the form of another symbol but the form of a living and ever changing habit is the growth of symbols (Nöth 2014a), which is a "growth in idea-potentiality ("Basis of Pragmaticism [in the Normative Sciences]", EP 2: 388; 1906)". The ultimate logical interpretant of an intellectual concept must change in order to reflect the growth of knowledge about the nature of the objects represented by the symbols. Since symbols grow and never remain the same, the habits by which they are interpreted need to change while the symbols grow.

References

Bergman, Mats. 2004. *Fields of signification: Explorations in Charles S. Peirce's Theory of Signs*. Vantaa: Dark Oy.

Bergman, Mats. 2012. Improving our habits: Peirce and meliorism. In *The normative thought of Charles S. Peirce*, eds. Cornelis de Waal and Krzystof Piotr Skowrónski, 125–148. New York, NY: Fordham University Press.

Bourdieu, Pierre. 1990. *In other words: Essays towards a reflexive sociology*. Stanford, California: Stanford University Press.

Bryant, Levi R. 2014. *Onto-cartography: An ontology of machines and media*. Minneapolis: University of Minnesota Press.

Burch, Robert. 2014. Charles Sanders Peirce. *The stanford encyclopedia of philosophy*, Edward N. Zalta, ed. http://plato.stanford.edu/archives/win2014/entries/peirce/.

Camic, Charles. 1986. The matter of habit. *American Journal of Sociology* 91: 1039–1087.

Colapietro, Vincent. 1989. *Peirce's approach to the self: A semiotic perspective on human subjectivity*. Albany: SUNY Press.

Colapietro, Vincent. 2009. Habit, competence, and purpose: How to make the grades of clarity clearer. *Transactions of the Charles S. Peirce Society* 45(3): 348–377.

Dewey, John. 1932. *Later works VII: Ethics*, ed. Jo Ann Boydston. Carbondale: Southern Illinois Press.

Erny, Nicola. 2005. *Konkrete Vernünftigkeit: Zur Konzeption einer pragmatischen Ethik bei Charles S. Peirce*. Tübingen: Mohr Siebeck.

Glebkin, Vladimir. 2013. A socio-cultural history of the machine metaphor. *Review of Cognitive Linguistics* 11(1): 145–162.

Grendler, Paul F. 2004. Renaissance. In *Europe, 1450 to 1789: Encyclopedia of the early modern world*, vol. 5, ed. Jonathan Dewald, 177–185. New York: Scribner's and *Encyclopedia.com*, May 2015. http://www.encyclopedia.com/doc/1G2-3404900963.html.

Houser, Nathan. 1998. Introduction. In Peirce, Charles Sanders. 1998. *The essential Peirce*, Vol. 2, ed. Peirce Edition Project, xvii–xxxyiii. Bloomington: Indiana University Press.

Houser, Nathan. 2014. The intelligible universe. In *Peirce and biosemiotics: A guess at the riddle of life*, eds. V. Romanini and E. Fernández, 9–32. Heidelberg: Springer.

Ibri, Ivo Assad. 2015. *Kósmos noetós: A arquitetura metafísica de Charles S. Peirce*, 2nd ed. São Paulo: Paulus.

Kent, Beverly. 1987. *Charles S. Peirce: Logic and the classification of the sciences*. Kingston: McGill-Queen's University Press.

Kull, Kalevi. 2014. Physical laws are not habits, while rules of life are. In *Charles S. Peirce in his Own Words: 100 Years of Semiotics, communication and cognition*, eds. T. Thellefsen and B. Sørensen, 87–94. Berlin: Mouton de Gruyter.

Lanier, Jaron. 2010. *You are Not a Gadget*. New York: Knopf.

Locke, John. 1705. *Some thoughts concerning education*. London: A. and J. Churchill.

Miller, Marjorie C. 1996. Peirce's conception of habit. In *Peirce's Doctrine of Signs*, eds. Vincent M. Colapietro, and Thomas M. Olshewsky, 71–77. Berlin: Mouton de Gruyter.

Liszka, J. J. 1996. *A General introduction to the semeiotic of Charles Sanders Peirce*. Bloomington: Indiana University Press.

Nöth, Winfried. 2010. The criterion of habit in Peirce's definitions of the symbol. *Transactions of the Charles S. Peirce Society* 46(1): 82–93.

Nöth, Winfried. 2011. From representation to Thirdness and representamen to medium: Evolution of Peircean key terms and topics. *Transactions of the Charles S. Peirce Society* 47(4): 445–481.

Nöth, Winfried. 2012. Translation as semiotic mediation. In *Σημειωτική: Sign Systems Studies* 40 (3): 279–98.

Nöth, Winfried. 2014a. The growth of signs. In *Σημειωτική: Sign Systems Studies* 42(2/3): 172–192.

Nöth, Winfried. 2014b. The life of symbols and other legisigns: More than a mere metaphor? In *Peirce and biosemiotics: A guess at the riddle of life*, eds. V. Romanini and E. Fernández, 171–182. Heidelberg: Springer.

Nöth, Winfried, and Lucia Santaella. 2011. Meanings and the vagueness of their embodiments. In *From first to third via cybersemiotics*, eds. T. Thellefsen, B. Sørensen, and P. Cobley, 247–282. Copenhagen: SL forlagene.

Pape, Helmut. 2012. Self-control, values, and moral development: Peirce on the value-driven dynamics of human morality. In *The normative thought of Charles S. Peirce*, eds. Cornelis de Waal and Krzystof Piotr Skowrónski, 149–171. New York: Fordham University Press.

Peirce, Charles Sanders. i. 1867–1913. *Collected Papers of Charles Sanders Peirce*. Vols. 1–6, eds. Charles Hartshorne and Paul Weiss. Cambridge: Harvard University Press, 1931–1935. Vols. 7–8, ed. Arthur W. Burks. Cambridge: Harvard University Press, 1958. [References to Peirce's papers will be designated by CP, followed by volume, period, paragraph number.].

A Comprehensive Bibliography of the Published Works of Charles Sanders Peirce with a Bibliography of Secondary Studies (second edition, revised), edited by Kenneth Laine Ketner with the assistance of Arthur Franklin Stewart and Claude V. Bridges (Bowling Green, Ohio: Philosophy Documentation Center, Bowling Green State University, 1986). A microfiche edition of Peirce's extensive lifetime publications is available from the same source. References to Peirce's publications begin with P, followed by a number from this bibliography.

Peirce, Charles Sanders. 1982-. *Writings of Charles S. Peirce: A chronological edition*. Vols. 1–6 to date, ed. the Peirce Edition Project. Bloomington: Indiana University Press. [References to these volumes will be designated by W, followed by volume number, colon, page number.].

Peirce, Charles Sanders. i. 1867–1893. *The Essential Peirce: Selected Philosophical Writing*. Vol. 1 (1867–1893), eds. Nathan Houser and Christian Kloesel. Bloomington: Indiana University Press, 1992. [References to this volume will be designated by EP 1, followed by colon, page number.].

Peirce, Charles Sanders. i. 1898. *Reasoning and the logic of things: The cambridge conferences lectures of 1898*, ed. Kenneth Laine Ketner. Cambridge: Harvard University Press, 1992. [References to this volume will be designated by RLT, followed by lecture number, colon, page number.] Introduction, and comments, by Kenneth Laine Ketner and Hilary Putnam: 1992: 1–102.S.

Peirce, Charles Sanders. i. 1893–1913. *The essential Peirce: Selected philosophical writing*. Vol. 2 (1893–1913), ed. the Peirce Edition Project. Bloomington: Indiana University Press, 1998. [References to this volume will be designated by EP 2, followed by colon, page number.].

Pietarinen, Ahti-Veikko. 2005. Cultivating habits of reason: Peirce and the logica utens versus logica docens distinction. *History of Philosophy Quarterly* 22(4): 357–372.

Potter, Vincent G. (1967) 1997. *Charles S. Peirce: On Norms and Ideals*, 2nd ed. New York: Fordham University Press.

Rosenthal, Sandra B. 1994. *Charles Peirce' pragmatic pluralism*. Albany: State University of New York Press.

Santaella, Lucia. 2001. Matter as effete mind. *Sign System Studies* 29(1): 49–61.

Santaella, Lucia. 2004. O papel da mudança de hábito no pragmatismo evolucionista de Peirce. *Cognitio* 5(1): 75–83.

Santaella, Lucia. 2012. *Percepção: Fenomenologia, ecologia, semiótica*. São Paulo: Cengage.

Santaella, Lucia. 2016. The originality and relevance of Peirce's concept of habit. In *Consensus on Peirce's concept of habit: Before and beyond consciousness*, ed. Donna E. West and Myrdene Anderson. (Studies in Applied Philosophy, Epistemology and Rational Ethics [SAPERE]) New York: Springer.

Savan, David. 1976. *An introduction to C.S. Peirce's complete system of semiotics: A manuscript for the Toronto semiotic circle*. Toronto.

Shaviro, Steven. 2011. The universe of things. *Theory and Event* 14(3) (doi:10.1353/tae.2011.0027.

Shuger, Debora Kuller. 1990. *Habits of thought in the english renaissance: Religion, politics and the dominant culture*. Berkeley: California University Press.

Chapter 4
The "Irrealevance" of Habit Formation: Stjernfelt, Hofstadter, and Rocky Paradoxes of Peircean Physiosemiosis

John Coletta

Abstract Using examples of geophysical processes, I discuss the implications for semiotics (physiosemiotics) of Peirce's statement that "habit is by no means exclusively a mental act ... The stream of water that wears a bed for itself is forming a habit" (CP 5.492). Geophysical phenomena are habit-producing processes that open up, not close down, the future and are always already recording phenomena that take advantage of their past and that reveal themselves as "natural propositions", in Stjernfelt's terms (2014). Further, the relationship of habit structure to habit structure, as concerns especially the emergence of "semiotic freedom" (Hoffmeyer 1996, 2010), can best be understood in terms of how habits simultaneously make themselves necessary AND necessarily irrelevant: here I follow Hofstadter (2007) and his concept of "responsible irrelevance" (2007), what I call "irrealevance". The question is this: how has nature been able to seal itself off from that for which it is responsible in order to open up its own future?

Keywords Physiosemiosis · "Natural propositions" · Syntax · Emergence · Stjernfelt · Hofstadter · Hoffmeyer · "Irrealevance"

Overview

Using examples of geophysical processes, I discuss the implications for semiotics of Peirce's statement that "habit is by no means exclusively a mental act... The stream of water that wears a bed for itself is forming a habit" (CP 5.492). Now I should say, with respect to my purpose here, that the perspective that drives my research is a desire to undermine the prejudice that geophysical processes are somehow "mechanical" in nature—the "received" cultural perspective, one that represents a perverse reading backward into nature of the biases of the industrial age. Geophysical phenomena, rather, are always already recording phenomena that

J. Coletta (✉)
University of Wisconsin, Madison, USA
e-mail: jcoletta@uwsp.edu

D.E. West and M. Anderson (eds.), *Consensus on Peirce's Concept of Habit*,
Studies in Applied Philosophy, Epistemology and Rational Ethics 31,
DOI 10.1007/978-3-319-45920-2_4

take advantage of their past, that open up the future, and, as Deely writes, that shape their "past on the basis of future events, a shaping that can be discerned even in the rocks and among the stars—a veritable physiosemiosis" (2001: 27). Accordingly, I entertain the idea in this chapter that geophysical systems are, to appropriate Terrance Deacon's term, "teleodynamic" (2012). This is to say, essentially, that mountains heal themselves, a requirement, according to Professor Deacon, of teleodynamic systems, though he would not, if I understand correctly, ascribe such dynamism to "mountains".

As concerns the specific argument that I will make here, let me say that I will show how geophysical processes give evidence of what Peirce calls "Thirdness-based interpretants" (West 2014: 117). Of course, I necessarily must address also Marc Champagne's (2013) cogent call for evidence of the interpretant function within the realm of the physiochemical, that is, if physiosemiosis is ever to be granted full status within a Peircean consideration of the (necessarily triadic) structure and function of the sign.

In order, then, to demonstrate "physiosemiosis" as an open-ended, habit-producing process involving "Thirdness-based interpretants," I have chosen to show how habit formation in the geophysical realm reveals structures that qualify as "natural propositions", in Stjernfelt's terms (2014), AND how the relationship of habit to habit, of habit structure to habit structure, can best be understood by recognizing that the power of habit formation is in how habits simultaneously make themselves relevant AND necessarily irrelevant (Hofstadter 2007).

Habits free us from the concerns of one level of existence and allow us to attend to another level. The universe has evolved habit structures not (merely?) to accommodate our thinking about nature (though how much easier to describe the physics of an avalanche in Newtonian terms or in the terms of a geologist than in terms of the underlying atomic phenomena) but to free itself from the tyranny of its own constitution: thus, for example, because atoms and molecules have sealed themselves off from interacting with the route a snow ball takes down a hill though they constitute the snow ball, snow balls are free to "snowball" into avalanches or to be picked up by people before they "snowball" and thrown at others by habit or made into snowmen by convention (habit). Here I follow Hofstadter and his concept of "responsible irrelevance" (2007), what I call "irrealevance". "Responsible irrelevance" is, as Hofstadter writes, "[t]he idea—that the bottom level, though 100 % responsible for what is happening, is nonetheless irrelevant to what happens" (2007 emphasis added). Indeed, this could be a pragmatic definition of the effect of habit-producing processes: "habit" as the production of "responsible irrelevance", of "irrealevance". In fact, "irrealevance" is a key dimension of the operating system of the semiosphere, a dimension that helps support greater "semiotic freedom", a key concept for Hoffmeyer (1996, 2010). The question is this: how and when has nature been able to seal itself off from that for which it is responsible? (Sounds a lot like the habit formation of one necessary skill for good parenting?!)

First, let me discuss how habit formation in the geological realm reveals structures that qualify as "natural propositions" in Frederik Stjernfelt's terms. That is, let me discuss how rocks have inclinations and gravitas! Perhaps Donna West

(2014: 117) best lays down a bridge to understanding the Peircean natural history (Secondness, fact or brute actuality) of "natural propositions" and of "habituescence" (Thirdness, law):

> Nonetheless, without the raw material of expression in Secondness, most especially in participatory engagement not merely with the entity but with the utility of the rudimentary effect of the entity, Thirdness-based Interpretants would be hard pressed to materialize (West 2014: 93–94). This is so because conceptions across phenomena would be but unembodied and hence irrelevant phantasms, absent any possibility of actualization.
>
> …
>
> Were regularities so regimented that they need to be obeyed at all cost, "intelligence would be cut off at the outset; the virtue of Thirdness would be absent" (CP 1.390). In other words, without change in the course of seeking after Final Interpretants, habit is vitiated; and regular, mechanistic happenings are nothing short of blind application of unmanageable laws. Essentially, without the possibility of semiosis or intelligent seeking, the regularity would be devoid of Thirdness—as a defining attribute of habit. Any regularity which "cuts off" the pursuit of logic construction, precludes any possibility of qualifying as habit.

Thus, as Donna West argues, "Thirdness-based Interpretants"—such as, in this chapter, geophysical phenomena such as the "[s]triking circular, labyrinthine, polygonal, and striped patterns of stones and soil [that] self-organize in many polar and high alpine environments" (Kessler and Werner [2003])—only "materialize" and "matter", only have *gravity* or *gravitas*, only give evidence of an "inclination" (of the physical equivalent of "will", "to be so inclined", so to speak), and can evolve *forms* (self-enclosures, in this chapter) in advance of *function* (if rocks can enclose themselves in walls of their own making then why not the molecules that self-organize into biological patterns of organization such as cell walls?), because *Secondness* is a domain of what West calls "corporeal enactment"—not mere corporeal reaction, as generally assumed (2014: 122); because *matter* is, as Bataille states, "the non-logical difference" (1985), difference being able to be recorded and used to make a difference being a necessary dimension of even geophysical processes; and, because matter is, as Peirce, of course, writes, "effete mind" (1891: CP 6.24). As I hope to show in this chapter, all of us have mystified these teleodynamic geophysical processes through, aptly, our mechanistic bias. But matter was never mechanical until we arrived. This is to say, then, that habit formation in the realm of brute actuality involving matter is a process whereby material bodies enact themselves rather than merely reacting to other bodies in a mechanical fashion.

Even at the most elemental level, literally, in what I call below "material memory", "material irony", or "hard habit", a stone (a rolling rock) carries its history encoded in its surface (viruses have only further formalized the process) in the nicks and dings that the stone acquires (in the chipping away that tends toward the telos of sphericality), which history, like a bar-code, can be used by geologists to identify and trace the history of the stone but also literally serves to "shape" and direct the stone's own rolling and interaction with other objects: stones with different shapes roll differently and interact with other objects as a function in part of their surface features—all the while widening and deepening the trough they define in their rolling, entrenching, so to speak, a habit and making it more likely that other stones will follow in their path down the widening trench (see Peirce's habit-taking stream above that wears a bank for itself).

The stone's nicks and dings, as I hint at above, especially those chippings away of the surface of a stone that tend toward the creation of a stony sphere and thus toward the more efficient rolling of that stone and thus toward an increasing likelihood of its creating a trough down which other stones may roll, emerge as *representamens* to the extent that this surface chipping "calls out", as Peirce was wont to say (cf. 1902: CP 1.220), the greater likelihood of other stones rolling down the trough, of other stone's having taken on the inclination of both the theoretical first stone, of the literal slope or inclination down which it rolls, which inclination represents always already a material interpretation of gravity's effect, the stones and the slope altogether of course having created the very gravitational field within which they represent accumulating outcomes (see my Einstein reference below).

Thus, a chip in the surface of a stone that makes it rounder and more likely to roll and create a trough is a *representamen* in the interpretive context of the inclination of the slope and of the gravitational field that it, in part, helps create; that is, the chip (the chipping) is a representamen of a particular type, specifically, an iconic index, an iconic index of its *objects*, immediate and dynamic: in this case, the *immediate object* is the shape of the chip, which stands for the iconic inverse of the shape of the surface feature that caused the chipping, and the *dynamic object*, the evolving inclination and nature of the surface feature itself, the physical slope with which the chipped stone interacted so as to take on a particular shape and trajectory, the physical slope that shaped its shape and set its speed and direction and that is changed by the chipped stone. *Interpretants* are proper significate outcomes that stand *to* the *dynamic object* (in this case, the actual inclined geophysical surfaces that chipped in the first place the surfaces of the rolling stone, the *representamen*) in an equivalent manner to how, or in a more developed manner than does, the *representamen* (the indents caused by the exdents of the dynamic objects, to which the former stand as iconic indices of the latter).

These *interpretants*, then, are both the features of the trough and the trajectories of future rolling stones that are now literally more inclined to follow the first stones down the trough that they in fact carved out, thus helping impart inclinations (habits) to future rolling stones. In other words, talking in generalities, the tendency toward sphericality of a given stone—which sphericality is a sign (*representamen*) of the roughness of the inclined surfaces, a sign of the Peircean *object* that imparted to the *sign* both their natures—is precisely that which enables the rolling stone to get up the momentum necessary to carve out a trough down which future stones will be more inclined to roll and so on. The *interpretant*, then, is the resultant tendency or habit now of subsequent stones to behave in a similar manner, only more fully so, flowing over time with greater ease, standing to the inclined surface (the Peircean *object*) as a more developed sign of the initial trajectory of the first rolling stone—widening the first stone's trough and thereby broadening the habit.

Indeed, as Peirce suggests, speaking of "habit," "it causes actions in the future to follow some generalization of past actions; and this tendency is itself something capable of similar generalizations; and thus, it is self-generative" (1887–1888: CP 1.409; from "A Guess at the Riddle"). And habits, at all levels, in consolidating futures, do so by means of "that kind of causation whereby the whole calls out its parts" (1902: CP 1.220). In this sense, then, we may say that the future influences the present: gravity creates the very conditions of its own future role (roll) of interpretation in other ways ("possibility is modally real," writes Robert Corrington, 1994). The surface features of stones, nicks and dings—which features represent a record of the stones' past—are also interpreted by the slope as potential futures at each moment that the stone reacts with a surface that, in registering the shape of the surface of the rolling stone, offers that stone a new inclination, both certain AND only broadly determinate.

The Nucleus of a Stone

We are of course used to thinking in biosemiotic terms when thinking about teleodynamic systems and thus of the power of the nucleus (and DNA) as a steering mechanism. However, even the function of a nucleus, absent of course in stones, may be understood as operating in a stone "in effect" as the stone's "center of gravity," which "center" serves an interpretive function in the context of the potential route a stone takes down an inclination, the inclination itself an interpretation of gravity, the ur-interpretant of stones; that is, gravity is the original field phenomenon within which objects create the very fields that then influence and determine (more or less) the trajectories of those objects. Haley (1999) defines the Peircean interpretant as "the proper significate outcome of any Sign-action, which thereby mediates (ratifies or validates) the Sign-Object relation, revealing it for what it is (i.e., a significant relation) and contributing to the actual emergence or unfolding of its potential meaning" (personal communication). This Peircean definition of the interpretant function sounds structurally like Einstein's characterization of the interpretive nature of nature. Einstein (1919) writes, "The geometrical behavior of bodies and the motion of clocks … depend on gravitational fields, which in their turn are produced by matter" ("What Is the Theory of Relativity"). Here we see that matter creates the conditions for its own mattering, and does so in regular, lawful, and even quantifiable ways. However, as we shall see, the regularizations and stabilizations brought about by the habits of even a rolling stone can actually *increase* not merely decrease differentiation, as when the forces that added all together tend to elide all difference in the production of a predictably (theoretically) perfect spherical rock produce as an outcome greater *un*predictability—as rounder (than flatter) rocks have a much greater (and less predictable) potential range of movement, a clear Peircean example of the openness of regularity and habit; thus, physical processes are habitual (prone to error and differences that makes a difference) not mechanical, the latter being a human invention that we have read back into nature.

Habit can increase differentiation because it fully participates in triadic Peircean semiosis. Even at the geophysical level, habit can increase differentiation because it involves the interpretant function. This is to say, can rocks be signs of each other? If a sign is anything (x) that stands for something (y, its object) to someone or something else with respect to some quality (z) and in such a way as to bring a third thing (x^1), an equivalent or more developed sign, to stand to (y) as did x, then what is the "something else" to which the sign stands if the sign is a stone? My answer is gravity, the Ur-interpretant, and the various inclinations that interpret gravity while themselves being interpreted.

Let us continue to see how stones become more developed signs. For Peirce "… every comparison requires, besides the related thing, the ground, and the correlate, also a mediating representation which represents the relate to be a representation of the same correlate which this mediating representation itself represents. Such a mediating representation may be termed an interpretant, because it fulfills the office of an interpreter, who says that a foreigner says the same thing which he himself says" (1867a, b: "On a New List of Categories," EP 1:5). In the geophysical system under consideration, the role of the "somebody", of the "idea", or of the "mediating representation" to which the sign stands is the "*inclination* of the interface", which is to say gravity, the existence of which makes possible and thus gives meaning to the effects; indeed, the inclination has meaning in a geological process only as that inclination represents an interpretation of gravity's effect, which is to say that gravity calls out, or is the enabling context for, the effects generated by a rolling rock, and just as the inclination may be understood as an interpretation of gravity, the rolling rock may be seen as an interpretation of the inclination. In synoptic terms, the main point is exemplified as follows: since the effects of gravity on a stone are determined or mediated by a surface's inclination, a surface may be said to "interpret" gravity; furthermore, a slope's inclination is transferred to a stone's inclination to roll, a kind of will "in effect". Significantly, (1) gravity has already played a major role in determining the nature of the slope that interprets it, and continues to do so; also, (2) gravity, like an *interpretant*, keeps mass and weight (like sign and object) open to productive difference, which difference is the material instantiation of the distinction between a *representamen* and a sign, a thing and an experienced thing (a Peircean object).

The Openness of Habit

First, let us engage in a little review. As I outlined in "The Semiosis of Stone" (2009), and which I draw from generously here, Kessler and Werner (2003) describe two feedback mechanisms (habit structures like Peirce's "stream of water that wears a bed for itself") that are crucial to the process of geophysical self-organization and that help us understand Peircean habit. In their article, Kessler and Werner explain how "[s]triking circular, labyrinthine, polygonal, and striped patterns of stones and soil self-organize in many polar and high alpine environments. These forms emerge because freeze-thaw cycles drive an interplay between two feedback mechanisms." Of course, the key term used here by Kessler and Werner is "emerge", the 1 + 1 = 3 idea (see the doctrine of "emergence" in biology) that is part of the fabric of Secondness (as West (2014) writes above, "without [the possibility of] change in the course of seeking after Final Interpretants [emergence], habit is vitiated") and that is expressed as Thirdness. According to Terrence Deacon, "teleodynamics" (the process by which mere goal direction becomes goal intention) "is constituted by the co-creation, complementary constraint, and reciprocal synergy of two or more strongly coupled morphodynamic process" (2012: 552). This is indeed the process described by Kessler and Werner, who show how, in Deacon's words, two "morphodynamic" processes ("feedback mechanisms") interact to provide for the emergence of complex "teleonomic" (goal directed) and "teleogenic" (goal intended) forms and patterns—*forms*, for example, as in the case described by Kessler and Werner, that anticipate functions (such as cell walls) that only later evolve.

Kessler and Werner describe, then, two feedback mechanisms, that is, two "morphodynamic systems" of mutual "complementary constraint" and "reciprocal synergy", feedback mechanism that are involved in the production, in alpine regions, of those "striking circular, labyrinthine, polygonal, and striped patterns of stones and soil" patterns that have every appearance of having been the result of intentional design—and so the stories of fairy-circles and ancient astronauts.

Describing the "first feedback" mechanism (again, Deacon's "morphodynamic" process), Kessler and Werner write, "A freezing front (0 degrees isotherm) descending from the ground surface mimics the morphology of the stone-soil interface … Consequently, where the interface is inclined, frost heave … pushes soil down and toward soil-rich regions and pushes stones up and toward stone-rich regions, eventually giving rise to distinct stone and soil domains" (2003: 380). Note first that "frost heave" only has the effect that it does in areas "where the interface is inclined" (or "so inclined", as we would say in the domain of human intentionality), which is to say where gravity may be understood to function as an interpretive medium, but more on that later. Describing the "second feedback" mechanism, Kessler and Werner write, "squeezing and confinement stabilize the vertical thickness of stone domains, because uplift increases with thickness, causing stones to avalanche from regions of high to low thickness. Similarly, squeezing and confinement stabilize the width of stone domains, because wider sections which are deeper and more easily deformed, experience greater uplift than do narrower sections." Here, "stabilization", one kind of habit structure, increases uplift: stabilization *increases* not decreases differentiation, a clear Peircean example of the openness of regularity and of what Deacon calls "morphodynamism".

In semiotic terms, then, in the "first feedback" mechanism or morphodynamic system, the "freezing front" is the Peircean *sign* or *representamen*; the "morphology" of the stone-soil interface is the Peircean object that the "freezing front" "mimics" (Kessler and Werner's word, which literally is a figurative word, that is,

which describes Peircean iconicity). In other words, the “freezing front” is an iconic sign of (or stands for)—that is, comes to resemble in its shape or form—the actual morphology (the shape or form) of the “stone-soil interface” with respect to some quality, in this case the quality being receptivity to penetration, that is, the freezing front’s differential rate of descent through the different media (stone regions vs. soil regions) that constitute the morphology of the stone-soil interface. The “freezing front”, we are told, “descends faster in overlying stone regions … than in fine-grained soils.” The shape of the “freezing front”, then, which is a function of the different rates of descent of freezing, stands for that morphological distinction in the “stone-soil interface” between “stone regions” and “fine-grained soils”, and so that morphological distinction is the Peircean *object* that, while an “other” to the “freezing front”, nonetheless “imposes certain parameters that [the freezing front] must fall within” (Zalta et al. 1995)—the Peircean object in one of its key roles, that of offering constraints. Habits, however, are constraints on constraint, the effects of which produce freedom from constraint at another level. This is the process Deacon describes in his discussion of the emergence of teleodynamic systems.

Significantly, it is only “where the interface [or object] is inclined” (that is, angled away from a right angle to the perpendicular of gravity, gravity’s rectilinear rectitude) that the “frost heave” mechanism can have its “proper significate effect” (a Peircean definition of the Interpretant function), that is, can give rise to “distinct stone and soil domains.” It is only by means of the “inclination”, then, that the “frost heave” becomes a difference that generates differences or distinct identities: “distinct stone and soil domains.” Note again that even geophysical processes may serve an interpretive function (be so “inclined”), that is, may serve as “difference engines”, inclinations making “distinction” possible. Indeed, we see that regularity in geophysical domains can produce distinction and difference, not more of the same, a key dimension of Peircean habit.

Material Irony

Playing off and extending the example from the previous paragraphs, since the effects of gravity on, say, a stone are determined or mediated by a surface’s inclination (thus a slope, like a person, as I have said, may be “so inclined”), a surface may be said to “interpret” gravity; furthermore, a slope’s inclination is transferred to a stone’s inclination to roll, a kind of will “in effect”. Of course, significantly, gravity has already played a major role in determining the nature of the slope that interprets it, and continues to do so. As with matter and gravitational fields, a sign also draws other signs to it and makes things matter. Something matters when it makes a difference. (Recall above how through Kessler and Werner’s work we observe that “frost heave” becomes a difference that generates differences or distinct identities: “distinct stone and soil domains”.) And regularities and habits in nature both allow nature to black-box itself (i.e., to free itself from the tyranny of its own constitution) as well as to allow for the regularization of irregularity, as when, again, the weathering of a blockish-looking stone leads to the relatively more open future of the *spherical* stone, the regularly weathered sphere having less "bias" (see "bias" in bocce ball) than the stone had before being subjected to the regularity of physical process. Something spherical is always more ready to move in any direction (has greater response-ability) than something of irregular shape (a biased bocce ball). Indeed, Peirce, as West (2014: 123) tells us, "addresses the critical place of Thirdness in defining habit when he maintains that readiness (physical and cognitive) is necessary for developing sensations, volitions [,] foundational to producing and recognizing regularities. Here matter and mind… become integrally bound."

"Readiness", as I have mentioned, is not just a physiological or cognitive state; it is produced and even magnified over time as a rolling rock, becomes more spherical with rolling or with the action of weathering, thus increasing its readiness to roll in any direction. Thus, in what I call **material irony**, the geophysical forces that, added all together, tend to elide all difference in the production of a predictably (theoretically) perfectly spherical rock, produce as an outcome greater unpredictability—as rounder (than flatter) rocks have a much greater (and less predictable) potential range of movement—and thus its semiotic freedom increases. So, even regularity both creates habits, a rock slide carving out a space that makes it more likely other rocks will follow AND through generating roundness creates radical instabilities. Grave habits emerge through the interstices of these layers of interpretation.

The Paradox of Close Capture

Further, in what I call the "Paradox of Close Capture", we again see that geophysical nature is a paradoxical, interpretive, freedom-generating phenomenon. For example, here on Earth, rolling rocks roll as a function of the gravitational fields of the larger mass, the Earth, of which they are a part but within which field their own gravitational effects upon each other are negligible. However, the very act of being closely caught at the surface of an immense gravitational field (of which you yourself are a part with respect to something extraterrestrial) is what paradoxically gives rocks their "freedom", their potential energy (consider the endless procession of a ring of rocky debris in outer space, endlessly circling the sun, comprised of rocks that are differentially irresponsive to each other). This ring of rocky debris is not a habit structure in Peirce's terms because semiotic freedom does not obtain; only through close capture is freedom made possible. Greater structure generates greater freedom.

Quantum Sweeps and "irrealevance": Actual Phenomena Are Real yet Irrelevant to the Signing Action of the Epiphenomena

Douglas Hofstadter has coined the phrase "responsible Irrelevance." For Hofstadter, "responsible irrelevance" is "[t]he idea—that the bottom level, though 100 % responsible for what is happening, is nonetheless irrelevant to what happens." For stones, and their interaction in the self-organization of sorted patterned ground (the epiphenomenon), "the bottom level" (the phenomenon) represents the physics of the atomic interactions within the rock, interactions that make the stone possible but that are irrelevant to the sorting process. One could attempt to describe the epiphenomenon of sorted patterned ground in terms of the underlying atomic phenomenon, but this would introduce untold complexity into the description and create for all intents and purposes a descriptive barrier. Just as the phenomenon and the epiphenomenon are insulated from each other representationally, there is a parallel interactive barrier too in that the epiphenomenon, while it is literally made possible by the phenomenon, is irrelevant to it in what amounts to a perfect disappearing efficiency, a "quantum sweep", whereby a whole level of interaction is in effect swept away with respect to relevance, but swept away in such a way that it makes possible a clean slate for a whole new level of interactivity. For stones, and their interaction in the self-organization of sorted patterned ground (the epiphenomenon), "the bottom level" (the phenomenon) represents the physics of the atomic interactions within the rock, interactions that make the stone possible but that are irrelevant to the sorting process. When one level of interactivity can make itself so efficient and necessary as to disappear in relevance, thus freeing the next level from the reality below it, I call the process, after Hofstadter, "irrealevance". In Peircean terms, an epiphenomenon is a proper significate outcome of, an Interpretant of, a phenomenon, if the phenomenon creates a reality, a domain of greater semiotic freedom, that is, one for which it is necessary but irrelevant.

Writing about the "self-organization of sorted patterned ground" as described by Kessler and Werner, indeed, writing about their very article, Mann, in "On Patterned Ground", writes (2003),

> Hence, smaller, faster processes are *slaved* by larger, slower ones within the patterned ground system. Such interactions can have unexpected results, and self-organizing systems typically have emergent properties that are not predictable from the physics of their fundamental particles. There is nothing in the physics of a shovelful of stony mud that can predict the emergence of an intricate pattern of interlaced, stone-bordered polygons covering many square meters. According to the self-organization paradigm, many geomorphic phenomena on Earth's surface are responsible for their own development and maintenance. A landform is not just a by-product of processes operating at the scale of its fundamental particles; it can only be understood at its own greater-than-sand scales. (Emphasis added)

Stones are the epiphenomena of atoms; various patterns of sorted ground are the epiphenomena of stones. Each domesticates ("slaves") the other, epiphenomena having always already reduced—by having made irrelevant to their activity—the internal complexities of the underlying phenomena (the atoms for which gravity is internally irrelevant even as their numbers make the outward response of stones to gravity possible) and reducing those complexities to an energy source for itself. If stones are the epiphenomena of atoms, and various patterns of sorted ground are the epiphenomena of stones, a "toad" that "folds into a stone" in the mind of its predator (the toad making the stone's patterns its own) is an epiphenomena to the pattern-making of stones. Habit structures unfold, even becoming digitalized in the interpretants generated in the minds of toad predators: toad or stone; food or no food; strike or refrain from striking.

The Habits of Streams

Peirce wrote, "habit is by no means exclusively a mental act… The stream of water that wears a bed for itself is forming a habit" (CP 5.492). Peirce also writes,

> [A]ll things have a tendency to take habits. For atoms and their parts, molecules and groups of molecules, and in short every conceivable real object, there is a greater probability of acting as on a former like occasion than otherwise. This tendency itself constitutes a regularity, and is continually on the increase. In looking back into the past we are looking toward periods when it was a less and less decided tendency. But its own essential nature is to grow. It is a generalizing tendency; it causes actions in the future to follow some generalization of past actions; and this tendency is itself something capable of similar generalizations; and thus, it is self-generative." (1887–1888: CP 1.409; from "A Guess at the Riddle")

Let us look in the preceding terms at the seemingly simple hydrological phenomenon of stream-bank formation more closely: Peirce is referring to the fact that water, during the process of its carving out a bank, has made it more likely that more water will flow in the direction of and in the manner befitting, literally fitting, the bank carving; this increasing likelihood of a process getting itself to repeat itself in a more and more predictable manner is what Peirce means by "habit," and while this process initially seems almost beneath notice, that is only because we have become inured to the self-habituating nature of physical nature.

Similarly, as I mention above, when a rock rolls down a hill (let's just envision a sandy hill where the effects are more distinct), it clearly makes it more likely that another rock, rolling in the same general direction and area, will follow the path at least in part of the previous rock, widen or deepen it, and thereby make it more likely that the next rock will have even a greater chance of following it, widening, and deepening the route. Since for Peirce "habit" need not be a psychological phenomenon, any general that generates particular effects that increase the generality of the general and thus the likelihood of a predictable effect is a habit. The key here is that in the development of habit, a general, even a bend in a stream, is a general by virtue of the fact that it engenders reactions of a particular type (water flowing through the bend or a stone creating a trough in the sand) and in so doing makes it more likely through the widening of the bend or trough (the general) that a greater number of water molecules or stones will follow suit. In the rolling stones example, the route carved is for the second stone an indexical sign in effect in as much as it conditions (and literally stands for by removing what stands in the way) that second stone's future pattern of movement, and then the second conditions the third in even a statistically more likely fashion, while the third is an icon of the second which is an icon of the first, which stone is an icon of the process recorded in the indexical trough.

The Syntax of Streams; or, Predation as Predication

As Ibri (2008) writes, "natural abductions are evidenced by the insertion of diversity and growing complexity of the universe, while the laws of nature act deductively on its pertinent events, that is, enabling them to be conducted out, by necessity, from the rules that constitute them" (2008: 45). Stjernfelt writes as well about this process of the generation of "natural abductions" and of what Peirce calls the creation of the "pictorial predicate" (Stjernfelt 2014: 64), a concept that links spatial relationships to syntactic ones (part of his diagrammatic logic) and that illustrates precisely how these "insertions" are accomplished and how deduction serves to "conduct out" these insertions, that is, how the "whole calls out its parts". In what follows I show that even physiosemiotic phenomena may be understood as proto-dicent signs, as "natural propositions", structurally. As Stjernfelt also says, "Collocation as a primitive semiotic phenomenon may be viewed as part of Peirce's overall spacialization-of-logic hypothesis. In a certain sense, it is connected to the so-called localism hypothesis in linguistics—that basic semantic and grammatical relations are fundamentally motivated in spatial relations" (2014: 108). West cogently makes a similar argument:

> Peirce's sense of habit encompasses both temporal and spatial continuity, given that regularity is expressed as physical and functional resemblance across instantiations. Habit as regularity likewise requires coherence of co-occurring entities, states of being, or events. As such, habit (as a system of continuous existents) is the essence of Thirdness—it governs how the instantiation of one phenomen[on] implies the presence or relevance of another (reagents). In fact, habit embodies the very core of logic-based meaning relations intrinsic to Peirce's Interpretant, housed in the Logical Interpretant. (2014: 117)

West continues,

> Nonetheless, without the raw material of expression in Secondness, most especially in participatory engagement not merely with the entity but with the utility of the rudimentary effect of the entity, Thirdness-based Interpretants would be hard pressed to materialize (West 2013: 93–94). This is so because conceptions across phenomena would be but unembodied and hence irrelevant phantasms, absent any possibility of actualization. Some foundation in Secondness must pre-exist more elevated forms of habit: recognition of regularities of natural phenomena, notice of spatial contiguity between entities/events, or

> awareness of the significance of patterns of operation. Regularity actualized in Secondness then is paramount to establishing a course of conduct or event structure sufficiently systematic to qualify as habit, without strict adherence to absolute regularity and without precluding reliable speculative assertions from which inferences flow. (2014: 117)

Both Stjernfelt and West make it easy for readers to connect Peirce's notion of how "matter is effete mind" to the concept put forth in Clark and Chalmers' (1998) well known "The Extended Mind". Indeed, I would maintain that there is a proto-syntax in the widening trough's ability to bind rolling stones to itself and to give them a headedness (see "phrasal heads" in linguistic structures), that is, to condition the direction of rocks rolling in a certain way: this is the "temporarily and spatial continuity" of which West speaks above. Also, her phrases "course of conduct" and "from which inferences flow" supports the Peircean contention that "without the raw material of expression in Secondness, most especially in participatory engagement not merely with the entity but with the utility of the rudimentary effect of the entity, Thirdness-based Interpretants would be hard pressed to materialize" (West 2014: 117).

Further, as in paradigmatic relations in language, where what can be substituted for what in a given syntactic position is a function of the semantic qualities of the word, what can roll down the trough is limited by the semantics (the qualities) of the object. The syntagmatic dimension then has to do with the continuity of the trough and the passing on of the continuity of the trough to the continuity of the stone. "Physiosemiosis", a term proposed by Deely in his *Basics of Semiotics* (1990), refers to the process whereby the geophysical universe, the universe of matter—of atoms and molecules, of substances—organizes itself, the phrase "organizes itself" referring to that process whereby matter makes itself more readily available for subsequent organization, that is, makes "itself capable, through the action of signs, of recording and then taking advantage of its own history" (Coletta 2015). Physiosemiosis may also be defined as that process whereby matter organizes itself (as "irrealevant" habit structures) so as to make itself more readily available for inclusion in biosemiosis and anthroposemiosis (processes for which physiosemiosis is 100 % responsible but to which it is essentially irrelevant). Indeed, as concerns the syntactic and semantic dimensions of rolling stones and the physiosemiotic propagation of proto-dicent signs, of "Peirce's overall spatialization-of-logic hypothesis" (Stjernfelt 2014: 108), I argue that,

- the widening trough in the sand left by a rolling rock in an avalanche of rolling rocks makes it more likely (the syntagmatic aspect) that subsequent rocks (of a particular size and shape—the paradigmatic aspect) will follow in its "wake", and then that
- the fact that the route carved by the first stone for the second stone is an indexical sign "in effect" in as much as it conditions (and literally stands for by removing what stands in the way) that second stone's future pattern of movement (further syntagmatic functions, "Peirce's spatialization of logic hypothesis"), and then that

- the second stone conditions the third in even a statistically more likely fashion, while the third is an icon of the second which is an icon of the first, in that they roll more and more like each other and tend toward sphericality, and then that
- the smoothing out of the sides of the trough by the rolling stones is the same event that smooths and rounds the stones' surfaces themselves (a mutual shaping), the movement of which stones (their "inclinations") representing, re-presenting, in the context of their Ur-interpretant gravity, a transference of the literal inclination of the slope into an "inclination" to roll, that is, to roll in a way that translates the shape and inclination of the trough into motion, motion being an interpretation of the slope of the trough, and then that
- this widening trough becomes a *habit* structure, like a Peircean stream bank, one that water might use as a future stream, the sides of the troughs becoming stream banks, spawning a new ecology, … or that
- the proto-syntax that emerges in the widening trough's ability, as I mention above, to bind rolling stones to itself and to give them a headedness (see "phrasal heads" in linguistic structures), that is, to condition the direction of rocks rolling in a certain way, and that
- these actions observed by humans over time have become "Let's get rolling", or "I'm on a roll", or "I'm in a rut", or, prior to those specific locutions, that these observed actions carved out in our minds iconic indices of inclinations to see the world as in itself and representable as a contiguity (a "collocation" [see below] in the theory of syntax and predication) of cause, effect, and resemblance (see Peirce's "pictorial predicate" (Stjernfelt 2014: 64), all linked by self-generated habit structures with a goal, "It's all downhill from here",

are, I believe, examples of a physical process "recording [in the sand] and then taking advantage of its own history." Indeed, the emergence of habit structures is the emergence of a recording system that takes advantage of the record, the "record" being the syntax of the physics of thought itself.

However, let me now argue backwards to the habit or syntax of a rock rolling down a sandy hill by looking at the syntactic structure of a fish called the "rock beauty" (*Holocanthus tricolor*), a fish that helps us understand "collocation" as the "basis of syntax" (Stjernfelt 2014: 108). The "rock beauty" is a fish that always appears to be hiding behind a rock (or coral head); or, one might say that it appears to be hiding behind itself. How is this so? The rock beauty has a small yellow head and a small yellow tail as well as a large, dark (rock-like or coral-head-like) midsection—therefore, it presents itself as a smaller, yellow fish hiding behind a rock or coral head. (Note that rocks have succeeded in getting themselves represented in biological codes, a "quantum leap" of representation that I have termed, after Hofstadter's "responsible irrelevance", "irrealevance": the atomic and Newtonian physics of rocks, while 100 % responsible for the rocks that get represented, are irrelevant to the ecology of perception between rock beauties and their predators.)

The rock beauty, then, is itself an ongoing proposition, that is, a dicent sign, the fish *embodying* a predication that is the visual equivalent (as in "pictorial predicate") of the following statement: "this yellow fish [is] behind the rock" (Coletta 1996). The copulant "is" is a function of the visual juxtaposition (or collocation) of a dark rounded "rock" and yellow head and tail, the effect (interpretant) on a predator being, as I have said, a juxtaposition that is read as a superimposition: the yellow fish is *behind* the rock. If the rock beauty hopes to avoid an attack, predators must read the juxtaposition as a superimposition, must believe in the continuity of the yellow headed and tailed fish and that it is actually behind a rock—and thus hard to attack. The rock beauty, then, hopes to turn itself into a dicent sign in the mind of a barracuda and create a "thirdness-based interpretant" or habit, that being the barracuda's habitual turning away from the rock beauty as a source of food, or, rather, from the elusive "yellow fish" always (*habitually*) inaccessible. Thus, the rock beauty, in its collocation (or juxtaposition) of an *index* (the yellow look-at-me head and tail) and an *icon* (the rock or coral head behind which the "yellow fish" "hides") is what Peirce would call a "pictorial predicate" (Stjernfelt 2014: 64).

"Collocation" is the basis of syntax, predation = predication in this case, and has even an earlier instantiation in the headedness of heading down hill, the trough created being a kind of spine along which run stones that make the route easier for subsequent rocks and thus instantiate habit, habits running along a spine. In Stjernfelt's terms, the dicisign or proposition, say, "The yellow fish [is] behind the rock," is constituted of an *index* (or subject, the literal head of the fish and the "head" of the phrase) and an *icon* (or predicate). As Peirce writes in his "Basis of Pragmatism", "Thus every proposition is a compound of two signs, of which one functions significantly, the other denotatively" (1905: MS 284: 4, qtd. in Stjernfelt 2014: 58).

In the fishy phrase, the yellow indexical head functions *denotatively* (This yellow fish…) and the rock-like iconic sign of the body functions significantly (… [is] behind a rock), that is, as a predicate, with the copulant "is" supplied by collocation. This denotative → significant, index → icon, subject → predicate, head → body, and perception → action pattern or habit structure, one that defines the dicent sign that the fish wishes habitually to create in the mind (or *head*) of its predators (dicent signs as "thirdness-based interpretants"), is present even in the rock *headed* (rolling) down a sand hill (or a stream carving out a bank), as stones headed downhill compel other stones to follow until, in rounding each other out, the telos of sphericality makes it more likely that stones will roll out of the trough they helped create: habit producing freedom, even here at a literally elemental level (or slope). Or, we might say, the propositional structure comprised of two signs, an index and an icon, we find even in a trough, wherein the trough is an index and the stone headed in the direction indicated by the trough is an icon of the path, but the iconicity increases as the path is worn more deeply, and the trough has a headedness like a phrase while the inclination of the slope is itself an interpretation of the interpretive framework, gravity, and the stones are interpretations of the inclination.

Again, in Hofstadter's terms—in those of "responsible irrelevance" (what I call, after Hofstadter, "irrealevance")—the atomic forces that make stones possible are 100 % necessary to but irrelevant for the Newtonian physics of rolling stones headed down an inclination and the headedness of the argument structure of the trough, and those Newtonian forces are 100 % necessary to but irrelevant for the semiotic freedom achieved in a rock getting itself represented in the biological code represented in the dicent sign of the rock beauty. This is to say that the semiosphere has the habit of creating "irrealevant" habits as a means of creating both an underlying hierarchical scaffolding as a means of liberating itself from itself: To reiterate, the relationship of habit to habit, of habit structure to habit structure, can best be understood by recognizing that the power of habit formation is in how habits simultaneously make themselves relevant AND necessarily irrelevant. Habits free the process of semiosis from the concerns of one level of existence and allow it to attend to another level.

The rocky productive paradox here, though, is that even as semiosis achieves freedom from itself through the production of "irrealevant" habit structures, is not the isomorphism between levels (it is an argument structure all the way down) a sign that something is passed on, that one level is not "irrealevant" to the next? Stjernfelt provides a solution to this seeming paradox, I think, for, while the universe can achieve in its "irrealevant" scaffolding freedom from itself, as Stjernfelt also writes, "The process from simple to complex should not be conceived of as a process of composition: the overall arc of the semiotic argument process structure is there from the metabolic beginning, only in a undifferentiated, general shape—and semiotic evolution rather takes the shape of the ongoing subdivision, articulation, and sophistication of primitive signs, an ongoing refinement of parts and aspects acquiring still more autonomy" (2014: 159). To this I would add only that "the overall arc of the semiotic argument process structure is there from the [physiosemiotic] beginning." Thus isomorphism and irrealevance are not mutually exclusive phenomena.

In conclusion, as Stjernfelt also writes,

> biologically simple signs could not be isolated icons or indices, only later to be composed into more complex signs—nor simple unstructured "signals", associations or stimulus-response mechanic reactions. Biologically simple signs, rather, are full-fledged perception-action arguments—only lacking explicit internal articulation—but bearing with them, due to their double function, the possibility of later segmentation, articulation, autonomization, adaptation to further purposes, making them flexible, potentially giving semiotic structures still more plasticity and eventually making the compositional combination of different Dicisigns, of predicates and subsets, possible. (2014: 157–158)

This is true also of geologically simple signs, which are full-fledged perception-action argument—only lacking explicit internal articulation.

References

Bataille, Georges. 1985. *Visions of excess*. Minneapolis: University of Minnesota Press.
Clark, Andy, and David J. Chalmers. 1998. The extended mind. *Analysis* 58: 10–23.
Champagne, Marc. 2013. A necessary condition for proof of abiotic semiosis. *Semiotica* 197: 283–287.
Coletta, W.John. 1996. Predation as predication. *Semiotica* 109(3/4): 221–235.
Coletta, W. John. 2015. Semiotic modeling: A pragmaticist's guide. In *International handbook of semiotics*, ed. Peter Pericles Trifonas, 951–980. Heidelberg New York: Springer.
Coletta, W. John, Dometa Wiegand, and Michael C. Haley. 2009. The semiosis of stone: A "rocky" rereading of Samuel Taylor Coleridge through Charles Sanders Peirce. *Semiotica* 174: 1/4: 69–143.
Corrington, Robert. 1994. *Ecstatic naturalism: Signs of the world.* Advances in semiotics. Bloomington: Indiana University Press.
Deely, John. 1990. *Basics of semiotics*. Bloomington: Indiana University Press.
Deely, John. 2001. Physiosemiosis in the semiotic spiral: A play of musement. *Sign Systems Studies* 29(1): 27–48.
Deacon, Terrence. 2012. *Incomplete nature*. New York: W. W. Norton and Company.
Einstein, Albert. 2008 [1919]. What is the theory of relativity? In *The Oxford Book of Modern Science Writing*, ed. Richard Dawkins, 314–317. Oxford: Oxford University Press.
Haley, Michael Cabot. 1999. Personal Communication. 28 June 1999. (Also published [page 113] of W. John Coletta, Dometa Wiegand, and Michael C. Haley, The semiosis of stone: A "rocky" rereading of Samuel Taylor Coleridge through Charles Sanders Peirce. *Semiotica* 174(1/4): 69–143, 2009.).
Hoffmeyer, Jesper. 1996. *Signs of meaning in the universe*. Bloomington: Indiana University Press.
Hoffmeyer, Jesper. 2010. Semiotic freedom: An emerging force. In *Pierce and biosemiotics: A guess at the riddle*, eds. Davis, Paul and Nils Henrik Gregersen, 185–204. Cambridge: Cambridge University Press.
Hofstadter, Douglas R. 2007. *I am a strange loop*. New York: Basic Books.
Ibri, Ivo A. 2008. The continuity of life: On Peirce's objective idealism. In *Pierce and biosemiotics: A guess at the riddle*, eds. Vinicius Romanini and Eliseo Fernandez., 33–49. Heidelberg, New York: Springer.
Kessler, M. A. and B. T. Werner. 2003. Self-organization of sorted patterned ground. *Science* (17 January 2003) 299[5605]: 380–383.
Mann, Daniel. 2003. On patterned ground. *Science* (17 January 2003) 299[5605]: 354–355.
Peirce, Charles Sanders. i. 1867–1913. *Collected Papers of Charles Sanders Peirce*. Vols. 1–6, eds. Charles Hartshorne and Paul Weiss. Cambridge: Harvard University Press, 1931–1935. Vols. 7–8, ed. Arthur W. Burks. Cambridge: Harvard University Press, 1958. [References to Peirce's papers will be designated by CP, followed by volume, period, paragraph number.].
Peirce, Charles Sanders. i. 1867–1893. *The Essential Peirce: Selected Philosophical Writing*. Vol. 1 (1867–1893), eds. Nathan Houser and Christian Kloesel. Bloomington: Indiana University Press, 1992. [References to this volume will be designated by EP 1, followed by colon, page number.].
Stjernfelt, Frederik. 2014. *Natural propositions: The actuality of Peirce's doctrine of dicisigns*. Boston: Docent Press.
West, Donna E. 2013. *Deictic Imaginings: Semiosis at Work and at Play*. Heidelberg: Springer-Verlag.
West, Donna E. 2014. From habit to habituescence: Peirce's continuum of ideas. *Semiotics* 2013: 117–126.
Zalta, Edward N., Uri Nodelman, and Colin Allen, eds. 1995. *Stanford encyclopedia of philosophy*. http://plato.stanford.edu/contents.html (21 October 2008).

Chapter 5
Habit-Taking, Final Causation, and the Big Bang Theory

Stanley N. Salthe

Abstract Peirce's habit-taking relates primarily to formal causes. Leaving aside human purposes, finality in nature has been identified as both natural tendency and function. I propose that true finality should be viewed as a pull from the future rather than as a tendency to maintain a continuing present. This disqualifies function as a finality. I claim that true finality in nature is exemplified by the Second Law of thermodynamics. Assuming that the universe is a thermodynamically isolated system, the Second Law qualifies in the context of the Big Bang scenario because it references the pull of a continually receding future of ever sparser energy density. The attractor here is universal thermodynamic equilibrium, which could only be attained in an ever-receding long run.

Keywords Aristotle · Formal cause · Material cause · Efficient cause · Final cause · Second law

Introduction

In several passages Charles Peirce proposed that "nature takes habits", and in 1892 (CP 6.36) he identified this concept as being related to Aristotle's final causation. Bacon (cf. Urbach and Gibson 1994) and others argued successfully that finality ought to be attributed to human projects only, thereby eliminating consideration of possible tendencies in Nature that might run contrary to human projects, some of which tendencies might be raised to stand in the way of some project's realization. Not being entrained by any social projects, I have been suggesting that we could more fully understand the natural world if we follow Peirce in taking finality to be an aspect of Nature in general.

My basic trope in this context is that: Whatever we attribute to humanity must have been present in the natural world before the origin of human interests. That is,

S.N. Salthe (✉)
Emeritus, City University of New York, Binghamton University, Binghamton, USA
e-mail: ssalthe@binghamton.edu

D.E. West and M. Anderson (eds.), *Consensus on Peirce's Concept of Habit*,
Studies in Applied Philosophy, Epistemology and Rational Ethics 31,
DOI 10.1007/978-3-319-45920-2_5

in biological systems generally, as well as in wider Nature (Salthe 2012). I maintain, both on materialist and evolutionary grounds, that nothing—however unique it may seem to be or feel—can come from nothing, and that every materially-based thing must therefore have had a precursor in an earlier nature, if not as an adumbration without function (Salthe 2010a, b), then as a palimpsest. (It seems that the origin of the genetic system of biology, being the introduction of mechanicism into Nature, will be difficult to explain in this way!) Thus, the structure of human interests had to have had some sketch in the natural world as it was before primates had evolved. This includes human purpose, or teleology (Salthe and Fuhrman 2005). The framework for this view is my developmental systems theory outlined in 1993. The development of purposefulness would then have followed the basic developmental trajectory from vaguer to more definite embodiment (Salthe 2014), as (in this case):

$$\{\text{tendency} \rightarrow \{\text{function} \rightarrow \{\text{purpose}\}\}\} \text{ or } \{\text{teleomaty } \{\text{teleonomy } \{\text{teleology}\}\}\}$$

(see Salthe and Fuhrman 2005, for elaboration.)

How Do Each of the Four Causes (Fig. 5.1) Relate to Habit-Taking?

How might efficient push relate to habits? If there is a timing schedule, in fact that would tie the pushes to a habit. If the pushes, no matter in what pattern they occur, are always the same kind, then that indicates a kind of habit as well. But any structures like these would more cogently be assigned to formal cause. Any given trigger or push need not be entrained by habit; it could be a stand-alone event (an aspect of Firstness), with no relation to habit or kinds at all. Both Peirce, and Brenner (2008), link efficient and final causes as discussed below. But efficient cause could stand alone as embodied in unique episodes resulting from unclassifiable contingencies.

Does material cause (what is required for anything to happen) have any relation to habits? If there is a reliable source of reactants—that is, a continuous supply—or a required and reliable timing pattern (yearly monsoons, snow melts, tides ...), these would be habitual. But such patternings might, again, be considered instances of formal cause. An accidental co-occurrence of required materials would alone be sufficient to provide material causation for a one-time event.

Formal cause (the setup) clearly refers to a contextual, patterned, system of existents and habits, tending to function as suites of reliable components of entrainments or affordances in any "situation". This would include the fixed Laws of Nature (physical laws), which, if really unchanging, might be viewed in the Big Bang scenario as having "precipitated" at various early moments of the Universe, and are now reliable contexts (at least as perceived at our scale) for our activities. These would be parts of any setup whatever, while the rest of a setup would be

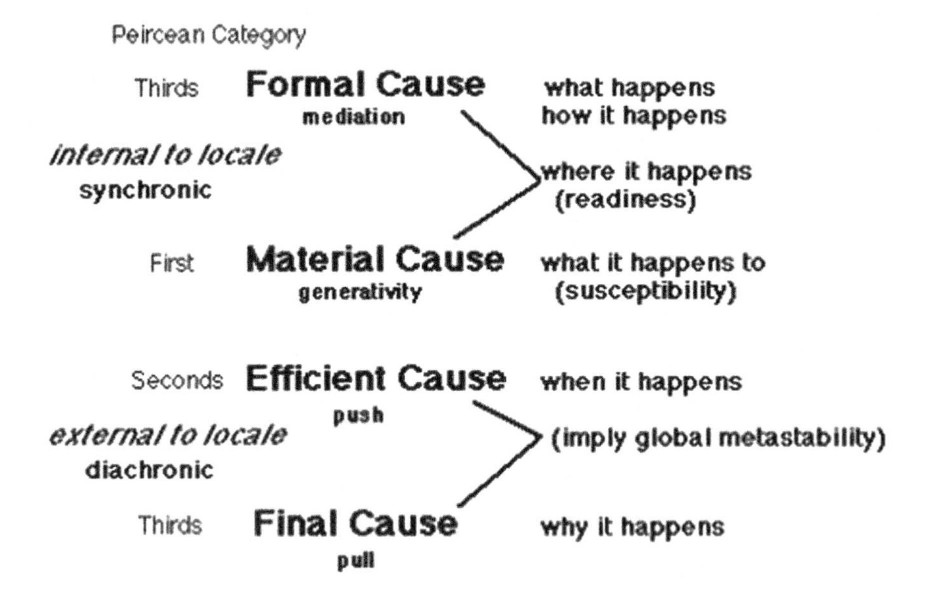

Fig. 5.1 The Aristotelian causal categories, with a hint about their Peircean aspects (From Salthe 2006)

historically assembled configurations and conformations. To consider final causes, we pass on to ...

The Nature of Finality

I believe that all four causes are required in order to understand the animation of any complex system—in particular, of dissipative structures. Three of the causes are generally acknowledged, at least implicitly, by most scientists. Efficient cause is usually foregrounded in experimental studies influenced by the physical approach to Nature, and material cause by investigations influenced by the chemical outlook. All scientists acknowledge formal cause implicitly, in the context imposed by the activities of measurement-taking—typically reported in a "materials and methods" section in published papers. Final cause, however, is more difficult to understand in a general sense, including its relationship to natural science. Is it (a) the future calling—a pull? This has made its appearance theoretically in dynamical systems theory, as the entrainment of dynamical variables by an attractor. That makes it conceptually available as a possibility in Nature. Or is it (b) a propensity resulting from some current lack which demands fulfillment, or a potential that tends to be realized? Or is it (c) a tendency resulting from the predictable continuity of previously established relations or forces? Perhaps it partakes of all of these in some degree in different examples.

Peirce advocated the (c) perspective. He viewed the laws of nature and of matter to have resulted from habit-taking by the universe. Once the habits were set, they continued unabated, and we can be sure that we can count on them—the force of gravity, the charge on the electron, the genetic code—to be in place at any future time in our universe. They would be integral aspects of any future. The earlier they were established, the longer into the future one might expect their continuity (if not their actual values in dynamic cases like the force of gravity). Perhaps there might be habits of this kind still in the process of being established? This once-and-for-all perspective is not an especially challenging view of finality. And it might better be assimilated to a "continuing present" concept.

Next we consider (b): examples of potentials that tend, or even demand, to be realized—necessarily at a time after "now". These are many. The requirements of organisms for energy and material supplies are an obvious example: hunting carnivores are end-directed; they currently lack something and seek it. Their continuity in the moving present requires that they project into the near future. This is not, however, the "pull" of the future, but just the desired/required maintenance of the *status quo ante*. It can be momentarily satisfied, but would be expected to arise again. All dissipative structures can be assimilated to this perspective during their active embodiment in the "now". They move forward, if successful, in an enduring present—until some kinds of them lose contact with sufficient energy gradient, and until others senesce and begin to fail. Then, might senescence itself be a pull into the future? It is not usually so considered, nor is it in my own theory (Salthe 1993). As I see it, it is just the failure of a dissipative structure to "keep up", resulting basically from information overload. Development ends in senescence (unless terminated by lack of sustenance) as a result of continuing developing past the stage of maturity (which *might* perhaps be worth considering as its "goal" in a still developing immature system). This kind of finality emerges from an urgency within "now" to propel itself forward, to continue as it is; eventually it fails.

Then, considering finality (a), where in Nature can we find an actual pull into the future? I have suggested (Salthe 2008, 2010a, b; Salthe and Fuhrman 2005) that the Second Law of thermodynamics imposes such a pull. Recently, I interrogated Joseph Brenner's 2008 *Logic In Reality* (*LIR* below) wherein finality is dealt with extensively, to see whether the Second Law satisfies its properties as described therein, and so might qualify as a pull. Thus:

LIR: Finality relates intimately with efficient cause, both actual and potential.

Local energy dispersions result directly from efficient actions, and each instance has the potential to trigger more (at present potential) dispersion elsewhere if the locale has any adjacent less energy-dense regions. This latter condition is generally satisfied near any energy gradient.

LIR: Once actualized, the result would appear locally as efficient cause.

As a sudden trigger. Once destabilized, an energy gradient releases energy flows. Efficient cause determines only when something will happen, meaning that the potential for its happening must be present—I would say—in part as a finality.

LIR: It will precipitate efficient causes in the future.

This is guaranteed by the Universal expansion of space in the Big Bang scenario, creating ever more regions of space with lower energy density.

LIR: Results of final causes are never fully completed.

Since the expanding space (ether) in the universe will always provide an adjacent less energy-dense region into which energy will flow, the process of energy dispersion cannot be completed. Even at thermodynamic equilibrium, we know from experiments that evanescent local energy gradients will form as fluctuations, generating local potentials that will eventually—usually quickly—get damped out.

Thus, I suggest that the Second Law represents a pull because it references a continually receding future state of universal thermodynamic equilibrium. Thermodynamic equilibrium is a universal apophatic principle (Ulanowicz 2011), being necessarily always missing in Nature. As well, there cannot even be local thermodynamic equilibria (except in the adiabatic chambers of physicists), because there will always be less energy-dense regions adjacent to any natural energy gradient (Annila 2010). And so the Big Bang model of the universe has elevated the Second Law of thermodynamics from a principle of experimental chambers to a universal principle in Nature—given the reasonable assumption that our universe is a thermodynamically isolated system. I think that denying this assumption given present knowledge would have to involve a much less parsimonious theoretical scenario than is entailed by accepting it.

Thus, the finality of the Second Law refers to a "pull" of ever less energy dense regions. Indeed, it is this geometry which generates the instability of local energy gradients, making them susceptible to various efficient causes. All four causes are required to understand the situation at any locale. Local energy supplies tend towards depletion by insatiable global energy demand—demand-side insatiability!

It is worth at this point to inquire whether the Second Law of thermodynamics actually functions in the "natural" world. After all, it has only been rigorously demonstrated (and always corroborated) in adiabatic chambers. Why should we think that it rules in the natural world? We know that the opposite kind of change is definitely *not* natural. That is, things do not spontaneously assemble into (what we would accept as) organized configurations; rather, organized things generally tend to crumble and scatter unless energy is exerted to keep them intact, as, for example, dissipative structures. Spontaneous (non-driven) occurrences tend to be dispersing/scattering, and when there is a natural process going the other way—like the separation of cream within raw milk—we can discover explanations consistent with a dispersing tendency. In this case it is the fact that fatty material does not form stable aqueous relationships when dispersed in water. In the case of the growth and maintenance of living things, we understand that internal processes are constrained by genetic information in such a way that growth may be derived from dissipative energy flows.

Discussion

The world system may be said to tend toward stability (Pascal and Pross 2014). In abiotic nature this is taken to be a continuing tendency, by way of energy dispersion, toward a stable thermodynamic equilibrium everywhere, while in biological systems it is a temporary, local dynamic stability based upon energetically-afforded and genetically-guided growth. Here stability is gained by propagation of the living system. Then, in the end we want to know if the universe really is a thermodynamically isolated system. We cannot know this; however, it is a reasonable assumption given the general scattering tendencies everywhere on Earth. If it actually is an isolated system, then we can confidently assert that the Second Law of thermodynamics does indeed rule therein. It may be noted that this natural finality is not in conflict with the more local finalities of human projects. No matter what the project (save perhaps meditation), the Second Law abets it by urging hard work—the harder the better for universal equilibration, given the poor energy efficiency of any significant work (Odum 1983: 102, 116).

Acknowledgments I thank Arto Annila and Edwina Taborsky for comments.

References

Annila, Arto. 2010. The 2nd law of thermodynamics delineates dispersal of energy. *International Review of Physics* 4: 29–34.

Bacon, Sir Francis. 1994[1620]. *Novum organum scientiarum.* (Translation from the Latin by Peter Urbach and John Gibson.) Chicago: Open Court.

Brenner, Joseph E. 2008. *Logic in reality*. Berlin: Springer.

Odum, Howard T. 1983. *Systems ecology: An introduction*. New York: Wiley Interscience.

Pascal, Robert, and Addy Pross. 2014. The nature and mathematical basis for material stability in the chemical and biological worlds. *Journal of Systems Chemistry* 5, doi:10.1186/1759-2208-5-3.

Peirce, Charles Sanders. i. 1867–1913. *Collected papers of Charles Sanders Peirce*. Vols. 1–6, eds. Charles Hartshorne and Paul Weiss. Cambridge: Harvard University Press, 1931–1935. Vols. 7–8, ed. Arthur W. Burks. Cambridge: Harvard University Press, 1958. [References to Peirce's papers will be designated by CP, followed by volume, period, paragraph number.].

Salthe, Stanley N. 1993. *Development and evolution: Complexity and change in biology*. Cambridge, MA: MIT Press.

Salthe, Stanley N. 2006. On Aristotle's conception of causality. *General Systems Bulletin* 35: 11.

Salthe, Stanley N. 2008. Purpose in nature. *Ludus Vitalis* 16: 49–58.

Salthe, Stanley N. 2010a. Development (and evolution) of the universe. *Foundations of Science* 15: 357–367.

Salthe, Stanley N. 2010b. Maximum power and maximum entropy production: Finalities in nature. *Cosmos and History* 6: 114–121.

Salthe, Stanley N. 2012. On the origin of semiosis. *Cybernetics and Human Knowing* 19: 53–66.

Salthe, Stanley N. 2014 A mode of "epi-thinking" leads to exploring vagueness and finality. *Modes of explanation: Affordances for action and prediction*, eds. M. Lissack and A. Graber, 115–119. New York: Palgrave MacMillan.

Salthe, Stanley N., and Gary Fuhrman. 2005. The cosmic bellows: The Big Bang and the second law. *Cosmos and History* 1(2): 295–318.

Ulanowicz, Robert E. 2011. Towards quantifying a wider realty: Shannon Exonerata. *Information* 2011(2): 624–634.

Urbach, Peter, and John Gibson. 1994[1620]. *Novum Organum*. (Translation from the Latin of Francis Bacon, 1620). Chicago: Open Court.

Chapter 6
Is Nature Habit-Forming?

John Pickering

Abstract The term "habit" as used in ordinary speech means a wide range of things. However, as used by C. S. Peirce, "habit" is generalized to such an extent that it seems to require a radical change in our worldview. Such a change is sketched by reviewing some developments in philosophy, physics, and the life sciences that seem to question the axioms of their disciplines in significantly similar ways. Panpsychism is once more being given serious consideration. Physicists are groping towards a phenomenological treatment of time. Biologists are turning towards a systems view, and psychologists are developing theories of cognition that do not separate mind from the body. These developments are brought together with Peirce's radical notion of habit, Whitehead's organic metaphysics, Gibson's theory of affordance, and biosemiotics, which blends Peirce's treatment of signs with the rational biology of Uexküll. The result is an organic worldview with intrinsic ethical entailments.

Keywords Panpsychism · Phenomenology · Rational biology · Embodiment · Whitehead · Process · Organic metaphysics · Affordances · Biosemiotics · Aldo Leopold

The Oddity of Habit Talk

What kinds of things can acquire habits? We know from direct experience that people acquire habits and it doesn't sound at all odd to say so. Nor does it sound particularly odd to talk about animals acquiring habits. In classic studies of learning carried out at the end of the nineteenth century, Edward Thorndike, a psychologist concerned with animal learning, provided a demonstration of how it happens. An animal, usually a cat, was placed in a cage locked by a simple mechanism but could see and smell food outside, such as a piece of fish in the cat's case. Initially, the cat would explore the cage without a specific aim. It might, for instance, play with a bit

J. Pickering (✉)
Warwick University, Coventry, UK
e-mail: j.a.pickering@warwick.ac.uk

D.E. West and M. Anderson (eds.), *Consensus on Peirce's Concept of Habit*,
Studies in Applied Philosophy, Epistemology and Rational Ethics 31,
DOI 10.1007/978-3-319-45920-2_6

of wood dangling on a string, as cats do. But this bit of wood was actually part of the unlocking mechanism. So eventually the unlocking mechanism would operate by chance allowing the cat to get out of the cage and eat the fish. Over about twenty repetitions of this process, the cats needed less and less time to escape, and their actions appeared to become more goal-directed. This preservation and improvement of chance events with positive outcomes can stand as a model for how habits in the usual sense of the word are acquired.

But habit can be used in wider senses too. We say plants have habits, and although there what's meant are characteristic patterns of growth not acquired by an individual organism in one lifetime but by a species over the course of evolution. Nonetheless, habit used in this way bears a fairly straightforward relationship to what habit means when applied to people and other animals, namely, a typical way that an organism acts.

But around that point on the continuum between living and non-living things, which is broadly a scale of complexity, it becomes increasingly odd to talk of habits. Organs can be described as behaving in ways that seem responsive to conditions. For example, when a digestive system adapts to a particular diet, or the liver adapts to prolonged alcohol use, this could just about be described as the system or organ acquiring a habit. But the adaptations are to conditions created by what a person habitually eats or drinks. It's in this latter sense that "habit" feels most natural to use. It would be odd to use it to describe adaptations of the body and its organs since they're better seen as adjustments to habits at a higher behavioral level.

When it comes to matter organized at less complex levels than bodily systems or organs, the term habit doesn't seem appropriate at all. It would sound particularly odd to describe sodium and chlorine as being "in the habit" of forming ionic bonds. Scare quotes would be mandatory from here on down the living-to-non-living, or complex-to-simple, continuum. It would sound even odder to say that bodies are in the habit of attracting each other according to Newton's law of gravity. That light travels at the speed it does is a fundamental and constant property of the physical world and to describe it as having the habit of doing so would sound absurd.

And yet, as used by Charles Sanders Peirce, "habit" appears to mean something very much like this. Peirce proposed that what we call the "laws of nature" are not primordial, necessary features of the cosmos. Instead he suggested they had evolved and, hence, were contingent, had a history and so could be called habits in some extended sense. Moreover, since habits can be acquired, modified, and discarded, he held that nature is open to change. Rather than having to obey ineluctable laws, the cosmos sports, producing spontaneous variations whose causes are untraceable. He said this and related things in various places and for reasons that changed as his ideas developed. But one persistent underlying reason was his rejection of mechanistic determinism. The cosmos in his view is not enclosed within a prison of mechanism, but is open and able to produce true novelty.

In advancing this at the end of the nineteenth century, when confidence in a mechanistic world view was at a high point, he was swimming against a strong tide. Herbert Spencer and Thomas Huxley had made the theory of evolution into a materialist metaphysic. Ernst Haeckel had announced that mechanistic reduction would not only solve the riddle of the universe but also, almost in passing as it were, show how mental life belonged in it: "The great abstract law of mechanical causality … now rules the entire universe, as it does the mind of man…" (Haeckel [1900]2013: 336).

Peirce's "Guess at the riddle" was very different: "… all things have a tendency to take habits. For atoms and their parts, molecules and groups of molecules, and in short every conceivable real object, there is a greater probability of acting as on a former like occasion than otherwise. This tendency itself constitutes a regularity, and is continually on the increase. In looking back into the past we are looking toward periods when it was a less and less decided tendency. But its own essential nature is to grow. It is a generalizing tendency; it causes actions in the future to follow some generalization of past actions; and this tendency is itself something capable of similar generalizations; and thus, it is self-generative" (EP 1: 245, 1887).

A crucial problem facing readers of this passage, both at the end of the nineteenth century and now, might be how to understand what Peirce had in mind when he wrote about matter, even at the smallest levels, "acting". Matter is acted upon, it does not act in and of itself. But if it is accepted that matter can act, then we can also accept that it can acquire habits. As Thorndike's work above shows, habits are acquired when patterns of action are preserved because they yield good outcomes. But to believe that matter can act in and of itself is very difficult, given the metaphysics of our time which, broadly speaking, are still those of the late nineteenth century.

How then can we approach this aspect of Peirce's work at the present time? Although complete reduction has been shown to be impossible, confidence in mechanism remains high to this day. Many popular books present contemporary scientific discoveries, especially in physics and biology, as showing how the human phenomenon fits into the wider order of the cosmos. A reviewer of one such book says: "The laws of physics have not changed in 13.8 billion years. In some unimaginable cosmic future, the speed of light in a vacuum will be the same, and the mechanics of waves–water, seismic, and light–will be as they were in the beginning" (Radford 2014).

The echo of biblical language in the phrase, "as they were in the beginning", serves as a reminder that the cultural dynamics of the last few centuries have resulted in science having thrust upon it the role once played by religion. This probably would not have been a problem for Peirce. His early encounter with Swedenborg appears to have remained with him and informed much of his thought, especially the blending of tychism and agapism in his later writings. For Peirce the source of form and permanence in the cosmos was not the fiat of a benign, transcendent creator. Rather, it is the immanent creativity of the cosmos itself which, agapism suggests, is also benign.

Why cosmic creativity might be benign will be considered again at the end of this chapter. First we need to consider more fully what we are to make of "cosmic creativity", if we allow ourselves to call Peirce's ideas that, in the context of contemporary metaphysics. Such an idea clearly doesn't fit within the conceptual framework around science and philosophy that has emerged over the last few centuries. Within that framework, physical laws are necessary features of the cosmos that will always be what they are now. Moreover, given those laws it is possible to explain how, under the right conditions, living systems can arise and evolve. Broadly put, the assumption is that physical laws produce evolution.

Peirce's view appears to be the opposite: evolution produces physical laws. This is a radical challenge to what we might call the Materialist Neo-Darwinian Conception of Nature, to borrow part of the sub-title of Thomas Nagel's book *Mind and Cosmos* (Nagel 2012). The full subtitle is actually: "Why the Materialist Neo-Darwinian Conception of Nature is almost certainly false." But this conception of nature has been so productive that to question it is often taken as a relapse into pre-scientific supernaturalism. Indeed, Nagel's book was criticized for giving comfort to nonsense such as Intelligent Design. In fact, the book is not anti-scientific but merely points to some shortcomings in scientific accounts of how living things, especially human beings, come to have intentions and experiences.

In doing so, the book can stand as one of a number of signs that there may be something distinctive about the way science is developing at the present time. What seems to be happening, in different disciplines and apparently independently, is that technology has so amplified our powers to observe, investigate, and experiment that we now know a great deal more than we can explain. Consequently there is a kind of metaphysical groping towards a new worldview. Science proceeds something like this all the time, but there seems to be something more radical going on at present. The new metaphysics takes experience, intentionality, and all other aspects of mental life to be fundamental features of the cosmos (Chalmers 2013). This is strongly reminiscent of Alfred North Whitehead's surmise that at its most fundamental level, the cosmos is an organic process; the ultimate and irreducible parts of nature are subjects, not objects. This is panpsychism, and although not deriving from Whitehead, a number of philosophers have begun to rehabilitate panpsychism in various forms and for various reasons (Strawson 2006; Skrbina 2009). This holds out the prospect of replacing mind in the cosmos as part of a more inclusive scientific world view in which sentience, intentionality, and qualia are seen as natural kinds rather than as anomalies that require special explanations or over-stretched conceptual tools, like emergence.

Although this is a radical break with the past, examples of changes in science that reflect this tectonic shift in metaphysics are not difficult to find. In biology, the idea of the organism is returning after having been rendered almost invisible by reduction to genes in one direction and by being lost in populations in the other (Nicholson 2014). Advances in genetics have had the unintended consequence of forcing biologists to abandon entrenched ideas about genes, which are now seen as necessary but not sufficient contributors to the creation of phenotypes. They are just one component in a systems approach where epigenetic factors, especially development, learning, and niche-creation must all be taken into account (Oyama et al. 2003; Odling-Smee et al. 2013).

Elsewhere in biology there are signs of a renaissance of the rational biology of Goethe and D'Arcy Thompson. An early example was the work of Brian Goodwin (2001), who studied with Conrad Waddington, who in turn was influenced by

Alfred North Whitehead. Goodwin's work, which helped with the return of rational biology, was marked by a consistent and reasoned opposition to strict, that is, reductive, Neo-Darwinism. Under that view, the structure of organisms arose from little more than genetic roulette. Instead, he developed a radical systems view in which evolutionary forces act on phenotypes, not genotypes. Active organisms, as they develop and learn and through meaningful actions, seek to harmonize themselves with their living environment and hence help generate their own selective pressures. This is effectively Lamarckism by other means.

Although rational biology is certainly more Lamarckian than conventional Neo-Darwinism, it does not reject the biochemical understandings of genetic activity on which neo-Darwinism rests. It does, however, take organisms to be active intentional centers of agency who investigate their surroundings for opportunities to carry out those intentions. Instead of the reductive views of neo-Darwinists, and, for that matter, behaviorists and computational cognitivists, this approach has at its core the notion of meaning. As the usage of "meaning" shows, it is a concept with a Janus-like quality. It points both to and from the organism. An intentional organism *means* to act; the environment of that organism provides objects and situations that *mean* to it that particular actions can be carried out. For biology to deal in meaning requires combining two important approaches which, taken together, promote a shift towards a panpsychist metaphysics.

One is James Gibson's radical approach to perceiving and acting. Gibson was a consistent critic of the computational metaphor for cognition. He also rejected the idea that what the senses provided was impoverished and ambiguous and thus some form of inference was needed in order to perceive accurately. It was assumed that this inference took the form of internal computation, that worked out what objects and situations were available as the basis for action. Gibson's approach, known both as ecological psychology and as the theory of direct perception, offers a different view. It is based on the idea that animals had evolved sensory systems that matched their capacities to act. The central concept is that of "affordance". An affordance is a directly-perceivable opportunity for actions of which the perceiver is capable. Thus chairs afford sitting for people and cats but not for bats or horses; flowers afford feeding to bees and humming birds, but not to snakes or sloths, and so on. The idea here is that meaningful action drives the mutual evolution of active perceivers and the environments, including other organisms, towards which they act, which is very much in the spirit of rational biology.

The second important approach is biosemiotics, which combines Pierce's semiotics with Jacob von Uexküll's meaning-based biology. Peirce's triadic notion of the sign as "something that stands for something to someone in some capacity" goes beyond Saussure's dyadic notion, which is synchronic and hence static. Peirce allows the "standing for", which he calls the interpretant, to be itself a sign that can lead to further interpretants, thus making possible a diachronic chain of meaning and signification more applicable to living processes.

Uexküll's approach to living processes was likewise meaning-based. He was dissatisfied with Darwinism since, having been strongly influenced by the

Naturphilosophie of Hegel and Schelling, he preferred to see progressive change as the unfolding of a plan rather than the accumulation of useful accidents. He saw environments as integrated systems of living things harmoniously interacting with one another on the basis of meaning. For Uexküll life itself was based on meaning: "… life can only be understood when one has acknowledged the importance of meaning" (Uexküll [1940]1982: 26). Biosemioticians bring these two approaches together to offer a picture of the organic world as perfused, and hence structured, by the exchange of signs (Hoffmeyer 2009; Romanini and Fernandez 2014). The resemblances here to rational biology are clear. Moreover, we can view the patterns of meaningful signs that harmonize the animals, plants and possibly the material components of ecosystem, as habits in the Peircean sense that have developed through mutual co-evolution.

Bringing together biosemiotics and Gibson's theory of direct perception, we can view affordances as signs that pass between mutually evolved organisms in the course of interacting with each other. Affordances also have an affective dimension. For example, the bared fangs and deep growls of dogs, both look and sound dangerous, while the smell of ripe peaches is delicious in itself. This recalls Spencer's view that such affective charges of perceptual experiences are psychological adaptations evolved to keep organisms away from danger and attract them to things which are beneficial.

Affordances are the behavioral and perceptual currency exchanged within ecological systems. They are meaning-based and integrate the three fundamental areas of psychology: cognition, conation, and affect. Moreover, they are not confined to any particular level of the living world. The affordances involved in the lives of long-lived social animals such as chimpanzees will be complex and liable to change when compared with those that matter to an organism like a parasitic tick, which will be simple and more stable. In particular, when organisms are able to learn, the repertoire of affordances will be open to progressive change. But complex or simple, conceptually, all affordances are the same. They are mutually evolved signs that guide the interaction between organisms and their surroundings. Affordances in Peirce's terms are habitual patterns of exchange of meaningful action. Seen in this light, the evolution of the living world is not so much the preservation of accidents as the perseveration and elaboration of habits that have proved beneficial (or at least not terminally deleterious) to active forms of life as they seek new ways to engage with their surroundings on the basis of meaning.

Conceptually tracing this process back in time raises the metaphysical issue of how life began, or, recasting this question in terms of the rational biology above, when did meaning-based interactions appear? Karl Popper commended the work of Wächtershäuser on the origins of life (Popper 1987; Wächtershäuser 1987). Wächtershäuser suggested that early self-sustaining chemical systems, so early that they predated the appearance of true organisms that replicated via genes, might have been capable of simple forms of trophic behavior based on sensitivity to light.

While Popper dismisses panpsychism, if it could be shown that there could have been light-driven action in early forms of organized matter, this would seem to be what is required for something like what Chalmers (2013) calls panprotopsychism to be true. It can stand as another example of the shift towards a less mechanistic and more life-friendly metaphysics noted above. A further example of a move away from mechanism can be seen in cognitive science where the computer metaphor for the mind, a lingering echo of nineteenth-century reductive mechanism, has been discarded. It was productive in the 1950s, but its limits are now clear and it has been replaced by an approach that recognizes that cognition is not detachable from the bodies of active organisms nor the situations in which they act (Rowlands 2010).

These changes, and others like them, not only hint at a new metaphysical picture but also reflect the postmodern shift in scientific epistemology (Griffin 1988). The shift challenges boundaries, methodologies become less conservative, single-factor explanations are weakened, and over-arching meta-theories, such as the computer metaphor or reductive neo-Darwinism, are replaced by a pluralistic synthesis of views and approaches, some of which may come from outside what is conventionally regarded as science.

Among the sciences though, physics is in some sense bedrock. When reductionism is discussed—either as something to be aimed at or as something to be avoided—the reductive chain and hence the explanatory buck generally stops at physics. So, while a postmodern reappraisal of methods and theories may influence sciences like psychology and biology, perhaps we would expect physics to be immune.

But in physics too there are signs that limits may have been reached and that radical challenges are being proposed. Despite the success of theories of matter at very small dimensions, it remains impossible to bring quantum theory together with relativity, and a grand unified theory appears no closer than ever it was. One notable response to this has been the work of Lee Smolin and colleagues (Unger and Smolin 2015). Although their work does not have Peirce's metaphysical breadth, there are some striking resemblances. Smolin takes as his central hypothesis that: "… the laws of nature evolve …" (2015: 355); and that how we think about time determines what we think a physical law actually is: "The notion of a law of nature is much changed if one thinks that the present moment and its passage are real or are illusions hiding a timeless reality. If one holds the latter view, then laws are part of the timeless substance of nature; whereas on the former view this is impossible, as nothing can exist outside of time" (2015: 361). As time is the essence of experience, such a view is much more mind-compatible than the timelessness of both classical and modern physics.

It was Peirce's deeper conviction that the cosmos was indeed mind-like in a fundamental and creative sense that prompted his proposal that the laws of nature are contingent and have evolved. Although creativity implies the appearance of something new and habit implies the re-enacting of something old, paradoxically, in Peirce's view the core of creativity is habit. More precisely, it is the capacity to acquire and to elaborate habits, as illustrated by these excerpts from the quotation above: "… all things have a tendency to take habits … every conceivable real object … actions in the future to follow some generalization of past actions; and this tendency is itself something capable of similar generalizations; and thus, it is self-generative" (EP 1: 245, 1887).

Peirce's view of the origins of the creativity of the cosmos appears to lie in the crucial last section of this quote. Habit in and of itself in not creative. But habits in and of themselves will be subject to variation along with the physical and mental events that are their vehicles, this being an important aspect of Peirce's tychism. If the tendency to take habits both varies and generalizes, it opens the way for habitual patterns of activity to evolve and spread from one ontological level to another.

When active organisms explore the world and find affordances of value, in a sense they play with the world. Thus a pattern that was pre-figured in playful abductive variation at the mental level could be enacted at the physical level. If it proves useful (or at least not selected out), it may then be re-enacted and eventually preserved in habitual patterns of action and the objects that afford that action. Thus, to put this idea in the form of another "just-so" story, protohumans may have discovered that flint-like stones offering certain affordances could be useful in breaking or striking. When the properties of the stones to flake in a certain way provided further affordances such as scraping or cutting, protohumans may have noticed and assimilated them into their habitual patterns of action, thus changing them. Here, habitual patterns can arise at any level and generalize up or down the ontological continuum stretching from the predominantly mental to the predominantly physical. The use of "predominantly" here is meant to indicate that no purely mental or physical levels of nature actually exists. The continuum is just that: the single dimension of the plenum of nature. What is being suggested here is akin to what David Bohm may have had in mind when he suggested that events and objects at all level of reality had both a mental pole, in a sense that seems close to Peirce, that he identified with signification, and a somatic or physical pole (Bohm 1985).

But it still sounds odd to say that all things, from human beings to atoms can acquire habits. To make it sound less so, the next section will approach the notion of habit from more familiar ground. Animals, and especially human animals, clearly do acquire habits and develop them in unique ways. Perhaps reflecting on how this has come about may help to understand what habits are and how they exist at different levels of nature.

The Odd Habits of Human Beings

Peirce's sense of habit appears to be that of a tendency for events of any sort and at any level to repeat themselves or to persist in some way that carries the history of previous events with them. The repetition is not perfect, else we would be trapped in what Whitehead called the "repetitious mechanism of the universe". This opens up a path into the future where variation and selection occur at all levels, from the physical building blocks of the cosmos to the supposed higher reaches of human consciousness, thus eliminating the boundary between biological and pre-biological evolution.

But in the human case, habit can become paradoxically detached from its physical and biological vehicles. Reflexive consciousness, of which human beings appear to have a monopoly, means that habits of any kind, mental or physical, can be noticed and modulated. We are able to recognize habitual patterns of thought and action, evaluate and manage them. Stopping smoking, trying to use gender neutral language or not to think about something that worries us are all familiar examples. The basis of change in habits here is some sense of their worth. Other animals acquire habits and change them too, but the process is not reflexive. It has more to do with the pragmatics of survival, which in the case of social animals might involve imitating those habits observed in conspecifics that lead to good outcomes and eliminating those that don't.

Both humans and other animals develop habitual ways of interacting with the affordances of the objects and events in their surroundings. In many cases, these are relatively fixed and we talk of animals depending on instincts to survive. In other cases animals develop relative novel ways of interacting with their surroundings, usually through social learning and development. This is part of the distinction made originally by Ernst Mayr between, respectively, "closed" and "open" evolutionary strategies (1974).

The human case is special by virtue of being uniquely open. The affordances of the human environment are mostly human made. Many species alter the affordances of their niche but the human species has taken this to such a degree as to make a qualitative break with the rest of the living world. Activity theorists such as Leontiev saw that what made the human mind unique was guided development within an environment "… transformed by the activity of generations" (Leontiev 1981). This transformation and the process of guided re-invention of habits originating from previous generations, is a human monopoly. It produces a culture of material affordances and the social practices that go with them. This idea was taken up by psychologists seeking to extend Gibson's ecological approach to the social domain by socializing affordance (Costal 1995; Heft 2001). The social practices developed by humans and presumably proto-humans, allow habitual ways of acting, and by implication, of perceiving, to become objects of attention and hence improved, perhaps initially by the preservation of useful accidents.

It is this that has allowed the human mind to become the unique creative force that it is. Michael Tomasello, in exploring this process, points to the enormous differences between the cognitive resources of humans and apes despite their remarkable genetic similarity (Tomasello 2008). As there has not been enough time for these differences to evolve genetically, the explanation must be epigenetic. Here he singles out the ability to co-operate and the cultural accumulation of human-made artefacts and the practices that go with them, which are then subject to progressive change through the efforts of successive generations. This is something that is observed in other animals, but it is vestigial when compared with the technologized environment created by human beings over the last few 1000 years.

How artefacts and practices are accumulated and improved is clearly important, but perhaps just as important is how they appeared in the first place. Many animals, especially social apes like chimpanzees, are tool-makers, but they are not tool-improvers to anything like the extent that human beings are. Here the notion of affordance might be useful in proposing how human beings may have come to be so adept at the making and progressive re-making of tools and other artefacts. Affordances, in Gibson's original formulation, concern what is perceived to be doable in the world as it presents itself to the senses. They are about perceiving the world "as is", so to speak. They are evolutionary habits of perception and action. But, as Peirce proposes, habits can vary and develop. What may have occurred in human evolution is the appearance of a new and distinctive habit, that of seeing things "as if". That is, when dealing with the affordances of an object or situation, humans may have developed the ability for metaphorical or counterfactual perception and, presumably, counterfactual thought as well.

To illustrate: when picking up a rock, a chimpanzee might notice, seeing it "as is", that it afforded the opening of hard-shelled nuts through pounding. A human or protohuman might also notice that if parts of the rock were to be removed to create a sharp edge, it would afford cutting and scraping. This requires the ability to perceive the rock counterfactually, that is, "as if" it were other than it actually is. Perception though is not enough, and to produce the edge would require removing bits of the rock. This could happen accidentally in the course of using the rock as a pounder. To an animal only able to see things "as is", the accident might pass unremarked. But an animal, or protohuman, able to see or imagine things as other than they are might also be able to notice the new affordances of this accidental outcome and perhaps intentionally reproduce it. Once intentional reproduction is possible, social learning (either by imitation or guided participation) will preserve this habit and most likely modify it so as to make it more effective. This process of externalization and improvement, something that Tomasello calls the "Ratchet Effect", when allied with reflexive consciousness and the ability to perceive counterfactually may well be something like the evolutionary process that led to human beings being able to acquire and improve habits in a manner not found anywhere else in the living world.

Such "just-so" stories about human evolution are easy to invent and virtually impossible to test, but the scenario above doesn't seem too improbable. More specifically, it points to the significance of being able to break habits or to explore, perhaps metaphorically or playfully, variations on habitual patterns of action. Children are very much inclined to exploratory play in order to find out what objects can be made to afford. Imaginary or mimetic play is a particularly rich case. Mimesis is rare in the animal world and when it is observed it seems to be a relatively fixed pattern of behavior, such as when birds mimic frequent sounds in their surroundings of no significance for their own survival, like the ringing of telephones. True mimesis, that is, the conscious reproduction of sounds or actions intended to communicate to a conspecific that something is being referred to, is only observed in humans. Indeed, Merlin Donald proposes that the capacity for true mimesis in this sense was a crucial developmental stage in the evolution of the human mind (Donald 1991). Prior to that stage, perhaps the noticing of new affordances sketched above could be seen as pre-figuring the capacity for mimesis. If, for example, the sharp edges produced by flaking a flints are seen to have the same affordances as human fingernails or teeth, then something akin mimesis has occurred. It is an early form of imaginary play expressed in perception and action but depending on mental processes somewhat like abduction. In this case, to make it more concrete, scraping needs to be done. Teeth and nails will do to some extent, but more effective means are sought. Now, the affordances of stone flakes become salient and, after an imaginary leap, are explored. Having been found to be effective, they are more likely to be remembered, used, reproduced, and improved. Thus new habitual patterns of perception and action appear and, in social species, will be imitated and as they spread will be developed by further exploration, again driven by more imaginative play.

While we can't know about the imaginary lives of animals, if there is any, we do know that they are able to find the affordances of things by exploratory and playful-seeming actions. In captivity chimpanzees can learn to operate quite complex devices and to pass the skills on socially. In the wild, orangutans are known to use a variety of tools but, since they are mostly solitary, will be less likely to develop tool using skills by imitation.

To devise or discover how to use tools is to learn or create affordances. Species with stable patterns of ecological habitation, might not appear to be discovering anything about the environment to which they are adapted, but in fact they are adapted because of affordances discovered by previous generations. These discoveries will have been preserved and passed down either by virtue of natural selection or, in more complex cases, by genetic assimilation. Organisms sharing the environment in question will also be adapted, not only to the fundamental physical features of it, such as climate, but also to each other. Animals that habitually graze will co-exist with grass species whose habits of growth are adapted to being grazed. Thus the interacting habits of species create a system of mutual affordances which, if useful, will over time become integrated, or to use a term much favored by Uexküll, harmonized.

This recalls some of Peirce's discussions of habit, especially mental habits, and the pragmatist approaches to value and truth. Habits of thought either survive and develop or disappear by virtue of how well or ill they fit, or harmonize, with other habits and with experience. Habits of action will *a fortiori* be the same. The habits of species and systems of species must likewise be compatible and over evolutionary timescales will have become so. However, as conditions are never static there must also be the possibility of generating or discovering new habits and new affordances. When conditions change so as to make old habits ineffective or harmful, new ones must be found. Here, Peirce's view is that this constitutes a variety of contradiction or problem that stimulates an abductive effort after a resolution. Again, this could be said both of patterns of thought and, perhaps in a more concrete sense, of patterns of action. For example, if we are trying to thread a needle and the thread is not rigid enough to get it through the eye, we will most likely cast around for some means to make it more rigid and so change what we can do with it, that is, to alter its affordances in the service of a particular end. Even though there is a single target affordance, rigidity, the various means could be quite different, such as doubling it, twirling it, moistening it, and so on.

There are echoes of Heidegger here. When a tool is being used it is, in Heideggerian terms, ready-to-hand. Its affordances are expressed in carrying out actions for which it's designed, and neither the tool nor its affordances are actually present in consciousness. What will be conscious is the task itself or perhaps the object of the task. If, however the tool is being examined, perhaps with a view to repairing it, improving it, or using it in a novel way, then it becomes present-to-hand. Now it is an object of conscious scrutiny and new affordances may be discovered or created by modification.

Apart from the human case, conscious and purposeful modification of affordances is likely to be rare. Playful modification with the preservation of affordance discovered incidentally might be more common though. On a much broader scale, the co-evolution of species and their environments can be seen as a kind of reciprocal exploration of a mutual affordance space. While it might sound fanciful to call it playful, the actions of animals who employ, in Mayr's terms, an open evolutionary strategy, are often exploratory and investigate their surroundings in ways that are focused on specific outcomes such as becoming familiar with new locations or discovering whether something can be eaten or not. Simpler species with closed strategies will presumably be less capable of exploratory behavior. However, here we might note that Popper's enthusiasm for Wächteshäuser's theory of the origins of life was principally because it attributed something like exploratory behavior even to very early, simple one-celled organisms.

The testing and breaking of habitual ways of perceiving and acting is the means by which forms of life can extend their ecological niche. How quickly this happens will depend where the form of life lies on Mayr's open-to-closed continuum. In closed forms it will be slow. A stable pattern of interaction with stable surroundings is not going to be improved by experimenting with new habits. An example here might be sharks, some species of which the fossil record shows to have been anatomically, and hence behaviorally, stable over very long periods. In open forms, which will typically have extended social interactions and relatively long developmental periods in the life span, exploratory behavior and the testing of new habits is likely to pay off. The classic work on the spread of food-washing in groups of monkeys can stand as an example here (Itani 1958).

Harmonious Habits and Benign Panpsychism

The examples above are a very small and highly speculative survey of how habitual patterns seen in the living world might develop and change. But, as suggested at the start of this chapter, the point in making it is to help to understand Peirce's vastly more ambitious surmise that "… all things have a tendency to take habits. For atoms and their parts, molecules and groups of molecules, and in short every conceivable real object, there is a greater probability of acting as on a former like occasion than otherwise." There is a large gap in both scale and credibility between treating habits at the scale of animals and plants and the idea that every constituent of the cosmos has a tendency to "take habits". The scare quotes here seem justified since the proposal presents such a challenge to the implicit metaphysics of the present time. The challenge becomes more radical still if, reflecting Peirce's agapism, we add the idea that habits may have value.

But rather than abandon or dilute the challenge, we might instead strengthen it by bringing it together with another radical challenge, that made by Alfred North Whitehead. Whitehead's organic metaphysics also sought to bring together matters of fact and matters of value in a fundamental way. How familiar Whitehead was with Peirce's work isn't clear. He played some role in editing Peirce's papers at Harvard, but he makes little or no reference to Peirce in his writings. However, both Whitehead and Peirce rejected a mechanistic worldview, which Whitehead refers to as the materialistic worldview in some places. Both offer a variety of panpsychism and both take the structure of the cosmos to be the product of evolution. They see little value in making a distinction between areas of science that deal with what is conventionally seen as the physical, or non-living, world and those that deal with living processes.

Whitehead expresses this clearly in a number of places. In his *Science and the Modern World*, he says "Science is taking on a new aspect which is neither purely physical nor purely biological. It is becoming the study of organisms. Biology is the study of the larger organisms; whereas physics is the study of the smaller organisms" (Whitehead 1926: 125). Like Peirce, Whitehead rejected a purely mechanistic, or materialistic, view of nature as patently inadequate to account for living processes including subjective mental life. True novelty and the progressive change seen in the evolutionary emergence of the living orders could not, in his view, be properly accounted for by a mechanistic metaphysics based upon insensate matter and timeless, unchanging laws. Aim, purpose, and intentionality, all qualitative aspects of mental life, cannot be understood if the only way they could have come to exist, according to the mechanistic view, is to somehow appear, *ex nihilo*, from a cosmos that is in reality totally dead.

Instead what Whitehead proposes is a living cosmos. There are no dead parts or inactive levels of nature. No part of nature lacks an organic connection to every other part: "… we should reject the notion of idle wheels in the process of nature" (Whitehead 1938: 214). Organic connection, both within an organism and between the organism and what it encounters in its surroundings, is of the essence of all organisms, which, as the quote above shows, are the ultimately real constituents of the cosmos in Whitehead's view. The structure of the cosmos, what enduring objects we take to be there, is in this view a matter of evolution. Moreover, even what we commonly assume to be the very nature of objects, is also in need of radical revision. In place of the massy indestructible but dead particles of Newton's universe, or the more lively particles in the standard model of contemporary physics, Whitehead offers processes and structured activity that has aim. Whitehead is quite explicit in his radical application of this organic view: "… the emergence of organisms depends on a selective activity which is akin to purpose. … the enduring organisms are the outcome of evolution; … beyond these organisms there is nothing else that endures. On the materialistic theory, there is material—such as matter or electricity—which endures. On the organic theory, the only endurances are structures of activity, and the structures are evolved" (Whitehead 1926: 130). Here, enduring evolved structures of activity seems to be very like Peirce's extended notion of habit.

Now the patterns of complementary affordances that underlie the harmony of an ecosystem are exactly that: structures of activity. They endure, but are not constant. In a continual process of historically constrained change, patterns of mutually evolved affordance will arise and persist so long as they have a sufficient degree of compatibility with other patterns around them to do so. When they do not, they will fade and be replaced by others that do. Here there is more than just resemblance between this view of the evolution of biological order and Peirce's view of how thought develops. In fact, if we are to take Peirce's surmise that "… all things have a tendency to take habits" as universally as he seems to have intended, it is identity. But if, to follow both Peirce and Whitehead, no distinction is to be made between living and, supposedly, non-living processes, then what mental things can do and what physical things can do is identical in some very deep sense. Because "things" can be both physical and mental, or any mixture of both in any proportion, the enduring patterns in which they participate and by which they endure are the same. Within such a panpsychist framework, it is possible to see how nature could form habits at any level. To accept this is to cross the credibility gap referred to above.

But to accept a panpsychist worldview like this requires crossing wider credibility gap since it runs so strongly counter to ideas about the physical world that have been regnant for the past four centuries or so. Even for the open-minded, the proposition that all levels of nature have something mind-like about them is virtually impossible to take seriously at first encounter. In discussions with skeptics, who are the norm, anyone defending panpsychism is likely to be told that "giving atoms minds" or something like that is absurd and un-parsimonious. This is understandable. The only minds human being know are their own (albeit partially) and so the idea that every part of nature is mind-like in some way, panpsychism is easily taken to mean just that—that things as simple as atoms are able to make decisions, have thoughts, feelings and so on.

This difficulty is not only encountered when trying to get on terms with Peirce but also with Whitehead. Both use common terms in radically broadened ways, "habit" in Peirce's case and "experience" in Whitehead's. The everyday meanings of words are hard to leave behind, so on hearing "habit" we tend to think of the human habits we know by direct acquaintance. But Peirce applies the term habit to any and all levels of reality. Likewise when hearing "experience" we think first of the percepts and thoughts of which we're conscious. But, again, Whitehead's radical proposal is that not all experience is conscious and that in this wider sense it is to be found at all levels of reality. Objections to panpsychism derive, at least in part, to the difficulty of relinquishing habitual meanings of terms.

But when more fully thought through and rigorously presented, panpsychism, in its contemporary manifestations avoids these problems. The re-appearance of more informed versions of panpsychism is another manifestation of the contemporary shift in metaphysics. That panpsychism fell from fashion reflected the constriction on the scientific imagination that followed the swing towards positivism in the early twentieth century. But that is not typical of Western thought considered in the longer term (see Skrbina 2005; Sprigge 1984), nor is it typical of worldviews found outside the Western cultures, Daoism and Hinduism being clear examples. Contemporary panpsychists are not attributing minds like those of animals or human beings to elementary particles and the like. What they are doing is proposing that every level of nature, has both physical and mental characteristics. This idea was developed in some detail by Bohm, who, like Peirce, sees signification as a crucial aspect of how activity at different levels of nature is actually the same activity mediated by the flow of meaning (Bohm 1985: Chap. 3). Whitehead does not discuss signification per se so much as symbolism, which is an elaborated type of signification found only in human beings. Even so, as a preparation to discussing symbolism proper he notes that its origins lie in primitive elements of experience that are probably shared with what he calls "low-grade organisms" (1927: Sect. 3). This recalls Popper's enthusiasm for Wächtershäuser's suggestion that proto-organisms may have had some form of perceptual system.

Panpsychism is not a scientific proposition, but a metaphysical one. Presently, metaphysics is rarely found in mainstream philosophy but it is all too easy to find appeals to panpsychism in quasi-mystical efforts to repair modernism's disenchantment of the cosmos. But these, by simply attributing too higher a grade of mental life to the material world, are little more than the description of a problem masquerading as a solution and actually explain nothing.

The grounds for taking some form of panpsychism seriously are in fact quite simple. Rather than being unparsimonious, it is in fact the reverse, since it is a solution to an enduring and important problem—the mind-body problem. The material world clearly exists, albeit that we may have to accept some form of Kantian limit to what we can know about it. Experience, in the form of qualia, also exists, even more clearly since the fact of conscious experience is what human beings are most certain about. As Nagel points out, qualia are what make the mind-body problem intractable. If our worldview, what Nagel terms the Materialist Neo-Darwinian Conception of Nature, offers only insensate ultimate elements, then metal experiences become inexplicable. Without adopting a panpsychist position of some sort, the emergence of mental life is rendered mysterious. It requires the assumption that things which are essentially dead, and which can only be known quantitatively, can give rise to living qualitative experience. If anything is unparsimonious, it is this.

Moreover, a panpsychist worldview is fundamentally relational. That is, the interactions between different levels and parts of the cosmos are based on meaning and on the inner natures of the interacting parts. This idea is clearly expressed by Peirce, Whitehead, Bohm, Uexküll, and other advocates of panpsychism old and new. Whitehead, like William James, criticizes the destructive analysis advanced by

Hume and his followers that would deny the relations between things any ontological significance. To Whitehead, and to rational biologists like von Uexküll, the interrelatedness of the organic world, was patent. It could not be properly understood as the mere accumulation of accidents or the chance encounters of atoms in the void. Rather it was the result of mutually evolved patterns of actions which survived, that is, became habits, because they were beneficial. Here beneficial could be defined as promoting Uexküll's harmonized patterns of reciprocal signification. This perhaps helps to fit Peirce's notion of habit more fully into the move towards panpsychism which, it is being suggested here, seems to be in progress at the present time.

If there is such a move, why is it happening now? The character sketch of science offered at the start of the chapter is one of rapidly developing techniques for investigating the world allied with a sense of having reached various conceptual limits. Knowledge of the physical world has reached a methodological peak, perhaps symbolized by the large hadron collider and its massive instruments buried in underground chambers the size of cathedrals, again reflecting the religious status that science has had thrust upon it. This status often leads to popular treatments giving scientific findings far more significance than the scientists who make them would. Issues in quantum physics, especially those concerning observation and non-locality are too quickly taken to have demonstrated that consciousness directly influences physical events or to support paranormal phenomena. In fact, it doesn't appear that scientific findings, for all that they are penetrating ever more deeply into the nature of the physical world, are bringing us any nearer to an understanding of the mental world.

A panpsychist view of the cosmos, which takes it to be creative, benign, and in some sense sacred, appears so commonly in all the world's cultures that it may be considered a human universal (McLuhan 1994; Gottlieb 2003). Something like this view is now to be found, not only in popular accounts (e.g., de Quincey 2002), but also in the work of scientists themselves who are exploring science's ethical and even spiritual dimensions in order, as Stuart Kauffman puts it, to "re-invent the sacred" (Thompson 2010; Kauffman 2010).

Within the panpsychist worldview that seems to be re-appearing, Peirce's surmise that nature forms habits becomes more acceptable or at least sounds less odd. Naturally enough, since Peirce advanced the notion of habit in the way he did as part of his particular version of panpsychism. Also part, and possibly a fundamental part, of that version was that evolution was not shaped by physical and biological forces alone, but was also an expression of selfless love (Peirce 1893). This proposal sounds as odd, or perhaps more odd, to contemporary ears as does his radical extension of the notion of habit. The idea that he called agapism, he is happy to acknowledge, derives from his appreciation of Swedenborg for which he thanks William James.

Agapism is a more developed expression of the pragmatist notion of truth. Truth is what works, what survives through being compatible with what surrounds it. This, allowing there is continuity between all levels of being, which is Peirce's notion of synechism, there is causal continuity between the real of mental life, that is between ideas and logical interactions and the more embodied realms of organic and physical being. His view is that "... matter is not completely dead, but is merely mind hidebound with habits. It still retains the element of diversification; and in that diversification there is life" (EP 1: 312, 1892). Diversification and intrinsic variation, tychism, brings the opportunity to develop new habits. Thus an idea, an organism, or a particular configuration of matter will persist so long as it fits with what surrounds it. With agapism, Peirce adds a spiritual dimension to the notion of "fit". Habits of mind or matter will survive if they mesh with what is around them, but survive here is not a Darwinian competition for existence, but more like the search for harmony found in Uexküll and in Goodwin. Harmony in and of itself is positive. It opens up the way to novel and more developed patterns of harmonious existence. Here we find a view of evolution, perhaps akin to that of Teilhard de Chardin's or to Bergson's, which takes evolution to be purposive and to progressively increase what is of value. This re-insertion of value into nature is not so much to "re-invent" the sacred as to place it at the heart of the cosmos.

This blending of scientific and religious or spiritual matters cannot, of course, fit with Hume's division of the factual from the normative. Yet it may not seem as inappropriate as it might have done in the past. Given the dark geopolitics of our time, a metaphysical shift of the sort that has been sketched here, along with its ethical implications, is sorely needed. It is vital that we move on from the mechanistic metaphysics of the nineteenth century that has helped human beings to damage the biosphere. Some form of panpsychism that combines Peirce and Whitehead would be intrinsically evolutionary and would be the basis of a reasoned environmental ethic. While it would be scientific, it would also permit what we might call the re-sacralizing of the cosmos. To do so would be to recover the intuitive surmise that the cosmos is perfused with value and that value has to do with inter-relatedness, what Uexküll called harmony.

Environmentalists such as Arne Naess and Aldo Leopold likewise recognized what is needed to avoid damaging the living systems on which human life depends. It is to have a value ethic of harmony at the heart of our implicit metaphysics. Leopold was particularly clear on this: "A thing is right when it tends to preserve the integrity, stability, and beauty of the biotic community. It is wrong when it tends otherwise" (Leopold 1949: 262). This is no new insight, it can be found in the religions of the world. For example, the Sanskrit phrase *Vasudhaiva kutumbakam* is found in the earliest Vedic hymns. It is translated as, "The earth is one family", with the implications of co-operation, cherishing, and harmlessness. In contemporary Judaism too, we find Abraham Joshua Heschel saying something very like this: "The good does not begin in the consciousness of man. It is being realized in the natural cooperation of all beings, in what they are for each other. Neither stars nor stones, neither atoms nor waves, but their belonging together, their interaction, the relation of all things to one another, this constitutes the universe. No cell could exist alone, all bodies are interdependent, affect, and serve one another" (Heschel and Rothschild 1997: 106).

Panpsychism is a stimulus to thought rather than a completed testable system. It leaves a lot to do. For example, critics often note what is called the "combination problem". This problem is that while it is all very well to propose that every element of the cosmos is mind-like in some way, as both Peirce and Whitehead do, how are these myriad minds to get together to form the larger minds, like that of human beings? The insights of the likes of Heschel and Leopold suggest that there is an evolved harmony of all levels of existence. Accepting this hints at a solution to the combination problem.

But a metaphysics of evolved harmony is not merely the means to solve philosophical problems. It offers a chance of relinquishing an engrained habitual way of conceiving, and perceiving, the cosmos in order to develop a new one. Strongly engrained habits are hard to break and new ones feel odd at first, especially if you are able to consciously reflect on them, as human beings are. If you are not able, as Thorndike's cats were not, then new habits may just have felt slightly ineffective until practiced. If there is no capacity to reflect at all, as will have been the case with Wächtershäuser's proto-organisms, if they existed at all, any feeling involved in acquiring new habits would have been vestigial. But not absent, given Whitehead's maxim that not all feeling is conscious or following Bohm's view that all events have both a mental and a physical pole. Taking Peirce's metaphysics in something like this spirit makes habit talk when applied outside its usual realm sound less odd.

What has been proposed here, and elsewhere (Pickering 2016) is that such a metaphysics will help renew our experience of the world as coherent and that coherence is in some sense benign. Such a change will help repair the ecological damage presently being done to the living systems of the world. Choosing to see the cosmos under this aspect will require practicing a new habit of mind. Although it would feel odd at first, with time it could become natural. And if it were done, it could just be a difference that will make a difference.

References

Bohm, David. 1985. *Unfolding meaning*. London: Routledge.

Chalmers, David. 2013. Panpsychism and panprotopsychism. *The Amherst Lecture in Philosophy* 8: 1–35.

Costal, Alan. 1995. Socialising affordances. *Theory and Psychology* 5(4): 467–481.

De Quincey, Christian. 2002. *Radical nature: Rediscovering the soul of matter*. Chicago: Invisible Cities Press.

Donald, Merlin. 1991. *Origins of the modern mind: Three stages in the evolution of culture and cognition*. Cambridge: Harvard University Press.

Goodwin, Brian. 2001. *How the leopard changed its spots: The evolution of complexity*. Princeton: Princeton University Press.

Gottlieb, Roger. 2003. *This sacred earth: Religion, nature, environment*. London: Routledge.

Griffin, David (ed.). 1988. *The reenchantment of science: Postmodern proposals*. Albany: Suny Press.

Haeckel, Ernst. 2013[1900]. *The riddle of the universe at the close of the nineteenth century*. London: Forgotten Books. (Translated from the German original published 1900.).

Heft, Harry. 2001. *Ecological psychology in context: James Gibson, Roger Barker, and the legacy of William James's radical empiricism*. Hillsdale, NJ: Lawrence Erlbaum.
Heschel, Joshua, and Fritz Rothschild. 1997. *Between god and man*. London: Simon and Schuster.
Hoffmeyer, Jesper. 2009. *Biosemiotics: An examination into the signs of life and the life of signs*. Chicago: University of Chicago Press.
Itani, J. 1958. On the acquisition and propagation of new food habits in the troop of Japanese monkeys at Takasakiyama. *Primates* 1: 84–98.
Kauffman, Stuart. 2010. *Reinventing the Sacred: A new view of science, reason, and religion*. New York: Basic Books.
Leontiev, Aleksei. 1981 [1947]. *Problems of the development of mind*. Moscow: Progress Press. (Translated from the original Russian by M. Kopylova.).
Leopold, Aldo. 1968[1949]. *A sand county almanac and sketches here and there*. Oxford: Oxford University Press.
Mayr, Ernst. 1974. Behavior programs and evolutionary strategies. *American Scientist* 62: 650–659.
McLuhan, Teri. 1994. *The way of the earth*. London: Simon and Schuster.
Nagel, Thomas. 2012. *Mind and cosmos: Why the materialist neo-darwinian conception of nature is almost certainly false*. Oxford: Oxford University Press.
Nicholson, Daniel. 2014. The return of the organism as a fundamental explanatory concept in biology. *Philosophy Compass* 9(5): 347–359.
Odling-Smee, John, Douglas Erwin, Eric Palkovacs, Marcus Feldman, and Kevin Neville Laland. 2013. Niche construction theory: A practical guide for ecologists. *The Quarterly Review of Biology* 88(1): 3–28.
Oyama, Susan, Russell Gray, and Paul Griffiths (eds.). 2003. *Cycles of contingency: Developmental systems and evolution*. Cambridge: MIT Press.
Peirce, Charles Sanders. i. 1867–1893. *The essential Peirce: Selected philosophical writing*. Vol 1 (1867–1893), eds. Nathan Houser, and Christian Kloesel. Bloomington: Indiana University Press, 1992. [References to this volume will be designated by EP 1, followed by colon, page number.].
Pickering, John. 2016. Signs in the flesh: Whitehead and evolutionary metaphysics. In *Through a prism: A.N. whitehead's thought*, eds. H. Maassen, and A. Berve. Berlin: Springer.
Popper, Karl. 1987. Natural selection and the emergence of mind. In *Evolutionary epistemology*, ed. G. Radnitzky, and W. Bartley. Chicago: Open Court Publishing.
Radford, Tim. 2014. Review of the copernicus complex: The quest for our cosmic (In)significance, by Caleb Scharf. Published by Scientific American Books. The review appeared in the Guardian, 10 Sept 2014.
Romanini, Vinicius, and Eliseo Fernández (eds.). 2014. *Peirce and biosemiotics: A guess at the riddle of life*. London: Springer.
Rowlands, Mark. 2010. *The new science of the mind: From extended mind to embodied phenomenology*. Cambridge: MIT Press.
Skrbina, David (ed.). 2009. *Mind that abides: Panpsychism in the new millennium*. Amsterdam: John Benjamins.
Skrbina, David. 2005. *Panpsychism in the west*. Cambridge: MIT Press.
Sprigge, Timothy. 1984. *A vindication of absolute idealism*. Oxford: Oxford University Press.
Strawson, Galen. 2006. Realistic monism: Why physicalism entails panpsychism. *Journal of Consciousness Studies* 13(10-11): 3–31.
Thompson, Evan. 2010. *Mind in life: Biology, phenomenology, and the sciences of mind*. Cambridge: Harvard University Press.
Tomasello, Michael. 2008. *Origins of human communication*. Cambridge: MIT Press.
Uexküll, Jakob von. 1982 [1940]. The theory of meaning. (Translated from the German.) *Semiotica* 41(1): 25–82.
Unger, Roberto, and Lee Smolin. 2015. *The singular universe and the reality of time*. Cambridge: Cambridge University Press.

Wächtershäuser, Gunter. 1987. Light and life: On the nutritional origins of sensory perception. In *Evolutionary epistemology*, eds. G. Radnitzky, and W. Bartley. Chicago: Open Court Publishing.
Whitehead, Alfred North. 1926. *Science and the modern world*. London: Pelican Books.
Whitehead, Alfred North. 1938. *Modes of thought*. London: MacMillan.
Whitehead, Alfred North. 1985[1927]. Symbolism: Its meaning and effect. Revised. New York: Fordham University Press.

Chapter 7
Habit in Semiosis: Two Different Perspectives Based on Hierarchical Multi-level System Modeling and Niche Construction Theory

Pedro Atã and João Queiroz

Abstract Habit in semiosis can be modeled both as a macro-level in a hierarchical multi-level system where it functions as boundary conditions for emergence of semiosis, and as a cognitive niche produced by an ecologically-inherited environment of cognitive artifacts. According to the first perspective, semiosis is modeled in terms of a multilayered system, with micro functional entities at the lower-level and with higher-level processes being mereologically composed of these lower-level entities. According to the second perspective, habits are embedded in ecologically-inherited environments of signs that co-evolve with cognition. Both descriptions offer a novel approximation of Peirce's semiotics and theoretical findings in other areas (hierarchy theory, evolutionary biology), suggesting new frameworks to approach the concept of habit integrated with its role in semiosis.

Keywords Semiosis · Hierarchical multi-level system · Niche construction · Peirce

Introduction

We present here two different approaches of habit in semiosis: as a macro-level in a hierarchical multi-level system, where it functions as boundary conditions for emergence of semiosis; and as a cognitive niche produced by ecological and environmental inheritance of cognitive artifacts. According to the first approach, Peirce's semiosis can be modelled in terms of a hierarchical multi-level system of constraints. In our description, semiosis is modeled in terms of a multilayered system, with micro functional entities at the lower level, and higher-level processes mereologically composed of these lower-level entities (Queiroz and El-Hani 2006a, 2012). According to the second approach, habits are embedded in

P. Atã · J. Queiroz (✉)
Federal University, Juiz de Fora, Brazil
e-mail: queirozj@pq.cnpq.br

D.E. West and M. Anderson (eds.), *Consensus on Peirce's Concept of Habit*,
Studies in Applied Philosophy, Epistemology and Rational Ethics 31,
DOI 10.1007/978-3-319-45920-2_7

ecologically-inherited environments of signs that co-evolve with cognitive advances.

Peirce's semiotics is grounded on a list of logical-phenomenological categories —Firstness, Secondness, Thirdness—which corresponds to an exhaustive system of hierarchically organized classes of relations (Houser 1997). This system makes up the formal foundation of his model of semiosis as a process and of his classifications of signs (Murphey 1993: 303–306). Firstness as a mode of being is related to the modality of possibility. It is the category of vagueness and novelty—"the mode of being which consists in its subject's being positively such as it is regardless of anything else. That can only be a possibility." (CP 1.25) Secondness is the mode of being "which is as it is relatively to a Second but regardless of any Third." (CP 6.200) It is a kind of reaction. Like Firstness, Secondness can be related to a modality, namely, the modality of actuality (CP 6.455; Parker 1998). The actuality of a thing is simply its occurrence. Rephrased, actuality is the realization of a possibility, without thereby making reference to something larger, be that a general law or an interpretation. Peirce considered "the idea of any dyadic relation not involving any third as an idea of secondness". (CP 8.330) Thirdness is the category of mediation, habit, generality, and conceptualization (CP 1.340). The example par excellence is Peirce's semiotic process (semiosis) in which a sign is related to an object by mediation through an interpretant.

According to Peirce, any description of semiosis should necessarily treat it as a relation constituted by three irreducibly connected terms (sign-object-interpretant, S-O-I), which are its minimal constitutive parts (CP 5.484; EP 2: 171; Atkin 2016: 131). Peirce also defines a sign as a medium for the communication of a form or habit embodied in the object to the interpretant, so as to constrain (in general) the interpretant as a sign or (in cognitive systems) the interpreter's behavior (De Tienne 2003; Hulswit 2001; Bergman 2000; Queiroz and El-Hani 2006b). The notion of semiosis as form communicated from the object to the interpreter through the mediation of the sign allows us to conceive meaning in a telic, processual, non-substantive way, as a constraining factor of possible patterns of interpretative behavior through habit and change of habit.

Stanley Salthe's Model and Semiosis

Queiroz and El-Hani (2012, 2006a, b) have modelled semiosis through a hierarchical multi-level system model (Stanley Salthe's hierarchical structuralist model). Salthe's model separates complex processes in a hierarchical structure. He emphasizes that, in order to describe the fundamental interactions of a given process, we need to: (i) consider it at the level where we actually observe it (focal level), (ii) investigate it in terms of its relations to its parts, at a lower level (usually, but not necessarily always, the next lower level (micro-level), and (iii) take in due account entities or processes at a higher level, in which the focal entity or process is embedded (macro-level). The processes described at the focal level are constrained

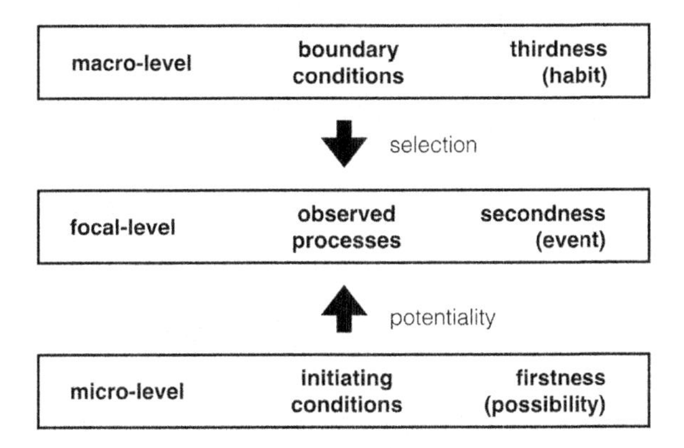

Fig. 7.1 A scheme of the determinative relationships in Salthe's basic triadic system in relation to Peirce's categories. A habit (Thirdness) is associated with the macro-level of the observed phenomenon, performing a selection, through boundary conditions, of potentialities which exist in the micro-level (associated with Firstness). The interaction between micro- and macro-level lead to the emergence of processes as observed in a focal-level (associated with Secondness)

by the influences of processes described at the higher and at the lower levels. These constraints allow us to explain the emergence of processes (e.g., semiosis) at the focal level.

At the micro-level, the constraining conditions amount to the *possibilities* or initiating conditions for the emergent process, while constraints at the higher level are related to the role of a (selective) environment played by the entities at this level, establishing the boundary conditions that regulate the dynamics at the focal level. In this model, an emergent process at the focal level is explained as the product of an interaction between processes taking place at lower and higher levels. The phenomena observed at the focal level should be "… among the possibilities engendered by permutations of possible initiating conditions established at the next lower level" (Salthe 1985: 101). Nevertheless, processes at the focal level are embedded in a higher-level environment that plays a role as important as the role of the lower-level and its initiating conditions. Through the evolution of systems at the focal level, this environment or context selects, among the states potentially engendered by the components, those that will be effectively actualized. Figure 7.1 shows a scheme of the determinative relationships in Salthe's basic triadic system.

Habit and Semiosis

The notion of "habit" can be characterized in several different ways: as a "pattern of constraints", a "conditional proposition" stating that certain things would happen under specific circumstances (EP 2: 388), a "rule of action" (CP 5.397, CP 2.643), a disposition to act in certain ways under certain circumstances, especially when the agent of the habit is stimulated, animated, or guided by certain motives (CP 5.480), or, simply, a "permanence of some relation" (CP 1.415). Its scope is broad: "all things have a tendency to take habits", that is, "every conceivable real object" (CP 1.409). Although habits are to be found in all things, it is not reducible to any number of instantiations: "no agglomeration of actual happenings can ever completely fill up the meaning of a 'would-be'" (EP 2: 402; CP 5.467). Habits participate in a self-generative development: "[A Habit] is a generalizing tendency; it causes actions in the future to follow some generalization of past actions; and this tendency is itself something capable of similar generalizations; and thus, it is self-generative." (CP 1.409)

Semiosis can be defined as the mediation of the self-generated regularity of habits, and a Sign can be defined as the vehicle of such mediation. That is, a Sign is something that communicates a "form", or habit which is embedded in another thing (an Object), generating a constraining factor in interpretative behavior (called an Interpretant) (see Queiroz and El-Hani 2006a). Note that in Peirce's work, the notion of habit is very similar to "form". Peirce separates a similar notion into two (Stjernfelt 2007: 37–38), with form being a "mere possibility", "anterior to anything actual", and habit referring to an already generalized possibility that governs actual occurrences. That is, "form as mere possibility in Firstness, anterior to anything actual, and form as realized possibility in Thirdness, where it governs Secondness in the shape of habits." (Stjernfelt 2007: 37–38)

> [...] a Sign may be defined as a Medium for the communication of a Form. [...]. As a medium, the Sign is essentially in a triadic relation, to its Object which determines it, and to its Interpretant which it determines. [...]. That which is communicated from the Object through the Sign to the Interpretant is a Form; that is to say, it is nothing like an existent, but is a power, is the fact that something would happen under certain conditions (MS [R] 793:1–3, "On Signs". See EP 2: 544, n. 22, for a slightly different version).

We refer to this irreducibly triadic relation as S-O-I (see Fig. 7.2). The irreducibility indicates a logical property of this complex: the sign process is not decomposable into any simpler relation (CP 5.484), and must be regarded as

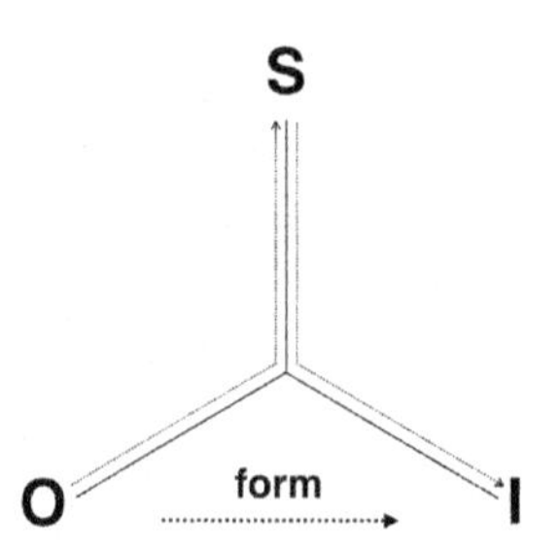

Fig. 7.2 Semiosis as a relation between three irreducibly connected terms (sign-object-interpretant, S-O-I). This triadic relationship communicates/conveys a form from the object to the interpretant through the sign (symbolized by the *horizontal arrow*). The other *two arrows* indicate that the form is conveyed from the object to the interpretant through a determination of the sign by the object, and a determination of the interpretant by the sign

associated with the interpretant, as an ongoing process of interpretation. For Peirce, "what a thing means is simply what habits it involves" (CP 5.400). It is a form embedded in the Object which allows a semiotic system to interpret the sign as indicative of a class of entities or phenomena (Queiroz and El-Hani 2006b: 183). Meaning is conceived, without any reference to psychological entities, as a constraining factor (S) in possible behavior (I) determined by a regularity of behavior previously embedded elsewhere (O). These are functional, interchangeable, roles: that which functions as a Sign in a given analytical description of the semiotic process could possibly be described as an Object, or an Interpretant, in another analysis. Note that the effect that characterizes the Interpretant does not necessarily act on an individual mind, but also, for example, on a social group or a culture (Bergman 2005: 218).

Processism and Emergence in Semiosis

This notion of semiosis as the mediation of a regularity of action allows us to conceive meaning in a processual, non-substantialist way. Processism and substantialism here refer to approaches that give more prominence to either substances or processes as basic explanatory units. Substances in substance metaphysics are ontologically basic entities, internally undifferentiated, bearers of properties and subjects of change, which are independent and durable (Seibt 2016: paragraph 4; see also Robinson 2014). In opposition to such notion, processes in process metaphysics are coordinated and systematically causally or functionally linked occurrences of changes in the complexion of reality (Rescher 1996: 38). While substance metaphysics take unchangingness as a default condition and emphasizes the need to explain changes, process metaphysics understands change as the default condition and emphasizes the need to explain stability (Bickhard 2011: 5). While substance metaphysics considers unchanging substances (e.g., atoms sensu Democritus) as the sole bearers of properties and causal powers (thus precluding emergence of new properties), process metaphysics is inherently relational and considers that some properties are presented by processes by virtue of organization, so that the emergence of new organizations may generate new properties, including causal powers (see Bickhard 2011: 5–7).

To explicate how semiosis allows us to conceive meaning in a processual way, consider the following example of process and interpretation of a process, given by Bickhard:

> If a cloud vortex produces a tornado, which then retracts, and then a funnel descends from the same cloud vortex, how many instances of a tornado process are involved? In terms of criteria of ground level damage, there are two (or more), but, in terms of criteria of locus of self-organization, there may be only one (the wind shear and consequent roll that produces the cloud vortex). (Bickhard 2011: 8)

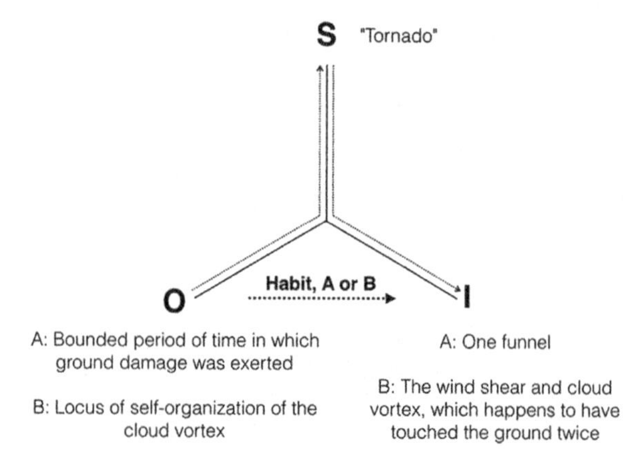

Fig. 7.3 For the same sign ("Tornado") and the same observed phenomenon, there are two distinct possibilities for what constitutes the Object and the Interpretant of the semiotic process. However, these possibilities cannot be freely combined: the fixation of an Object determines an Interpretant and vice versa, by virtue of a habit

Consider "tornado" as a sign (S). It can either refer to the funnels produced by a cloud vortex (let's term it possibility A, so that the O of S is A), or to the cloud vortex itself (let's term it possibility B, so that the O of S is B). Take a dyadic description (S-A, or S-B) of the tornado sign. Such description has no explanatory consideration whatsoever of why S is connected to either A or B. It is not sufficient to say that the sign tornado is dyadically connected to some entity in the world: the connection itself depends on criteria which are irreducible to the explanation of the relation of meaning. A dyadic account of the meaning of the sign "tornado" is not explanatorily powerful. Now consider Peirce's pragmaticist model as described above: in this case the meaning process is Interpretant-dependent, that is, it produces effects. In this case A and B are described as follows in Fig. 7.3.

This example illustrates the emergence of a semiotic process: the selection of either A or B is dependent on a habit (A or B) embedded in the Object, which constrains the semiotic process. The emergence of semiosis can be modelled according to Salthe's model. First, consider that semiosis necessarily involves chains of triads (see Merrell 1995). As Savan (1986: 134) argues, an Interpretant is both the third term of a given triadic relation and the first term (Sign) of a subsequent triadic relation (see Fig. 7.4).

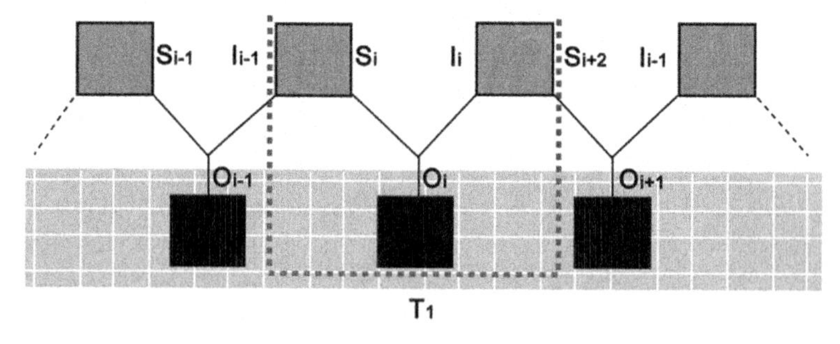

Fig. 7.4 The triadic relation S-O-I forming a chain of triads

Following Salthe's model, this dynamical semiotic process can be described at three levels (see Queiroz and El-Hani 2006b). The focal-semiotic level is the level in which a given semiotic process is observed. Semiotic processes at the focal level are described as chains of triads themselves. The micro-semiotic level concerns the relations of determination that may take place within each triad S-O-I. The relations of determination provide the way for the elements in a triad to be arranged in semiosis: the Interpretant is determined by the Object through the mediation of the Sign (I is determined by O through S) (MS 318: 81, 1907). Finally, the macro-semiotic level concerns the historically-constructed environment of networks of chains of triads, in which each individual chain is embedded. Focal-level semiosis will emerge as a process through the interaction between micro- and macro-semiotic processes, i.e., between the relations of determination within each triad and the embeddedness of each individual chain in a whole network of Sign processes, which take the form of habits in individual semiosis.

Habits exert a downward effect on the spatiotemporal distribution of lower-level semiotic items. In the "Tornado" example above, both possibilities of S-O-I triads regarding the sign "Tornado" are present at the micro-semiotic level. At the macro-semiotic level there is a historically established network of chains of triads which constitutes boundary conditions for the use of the "Tornado" sign according to certain situations (whether the situation calls for the use of 'Tornado' in the sense of a funnel, or 'Tornado' in the sense of a wind shear). This corresponds to a habit, a regularity of action (the tendency to regularly use 'Tornado' in one or another sense according to the situation), embedded in the Object.

Evolution of Habits as Cognitive Niche Construction

The notion of habit emphasizes that the emergence of meaning is dependent on context: this is a necessary condition for—and a constitutive part of—semiosis. Emergence entails that habits historically unfold and evolve, so that meaning is a temporally-situated evolutionary process. As habits can be embedded within the material properties of signs, such process is also materially-distributed.

The distributed cognition and extended mind approach (cf. Clark and Chalmers 1998; Clark 1998) have questioned the legitimacy of skin and skull to serve as criteria for the demarcation of the boundaries between mind and the outside world. According to this approach, various tools such as pen and paper, calculators, calendars, maps, notations, models, computers, shopping lists, traffic signals, measurement units, and so forth are considered non-biological extensions of the cognitive system that allow cognitive operations external to the skull (Hutchins 1999, 1995).

Accordingly, these are qualified as cognitive artifacts. On the one hand, cognitive artifacts impact cognitive performance: they may reduce the cognitive cost of an operation (such as when using a calculator to perform a division), increase its precision and efficiency (such as using a ruler to measure an object instead of just guessing its dimensions), or allow new capabilities that would be impossible to be performed by the brain alone (such as using a graphical diagram to represent the simultaneous relation between a large number of entities and infer specific visual patterns from it). On the other hand, cognitive artifacts also influence environmental opportunities and demands for certain types of cognitive performance—they participate in the creation of new problems and problem-spaces. Language, for example, is a powerful cognitive artifact (Clark 2006) that sets new demands and opportunities related to memory, perception, navigation, forms of generalization and categorization, modes of inference, and more. We treat here all kinds of cognitive artifacts as signs: they all necessarily constrain interpretative behavior (for example, performance in a materially-extended cognitive task) according to possessed regularities of actions.

Cognitive artifacts are ecologically inherited: they become part of the environment where cognition takes place, and changes in such an environment of artifacts are relegated to later generations. This cumulative process is evolutionary in an ecological (and not genetic) sense. Evolution in this case is a matter of (semiotic) niche construction. In biology, the niche of an organism indicates its ecological role and way of life. A niche is an imaginary n-dimensional hypervolume whose axes correspond to several ecological factors affecting the welfare of the organism (see Hoffmeyer 2008: 12). Clark (2006: 370) suggested that we are immersed in cognitive niches structured by language: "by materializing thought in words, we structure our environments, creating 'cognitive niches' that enhance and empower us in a variety of non-obvious ways".

A niche is dynamic, it develops and transforms over time. This transformation is often caused by ecosystem engineers (Jones et al. 1994) that alter their environment and ecosystem. Niche Construction Theory (Scott-Phillips et al. 2013; Odling-Smee et al. 2003) stresses that the transformation of niches by organisms performs a major role in evolution, establishing a non-genetic system of inheritance that shapes selective pressures creating a feedback loop between organisms and niches. For Laland and O'Brien (2011: 192–193) niche construction "should be thought of as the dynamical products of a two-way process involving organisms both responding to 'problems' posed by their environments and solving some of those problems, as well as setting themselves some new problems by changing their environments through niche construction". The notion of *cognitive* niche construction emphasizes the ecological evolutionary nature of cognition. For Clark (2008: 62–63), cognitive niche construction is a process of transformation of problem spaces by building physical structures that, combined with appropriate culturally transmitted practices, enhances problem solving or even makes possible new forms of thought.

In our approach, the cognitive niche is the locus in which habits become available for semiosis.

To exemplify the relevance of niche construction as locus of habit, consider the widely known London Underground Diagram (LUD) (Atã and Queiroz 2014) (see Fig. 7.5). Today, adapted versions of this diagram are present in virtually every major city in the world. The LUD has established an international paradigm instructing the public how to perform simple decision-making tasks regarding

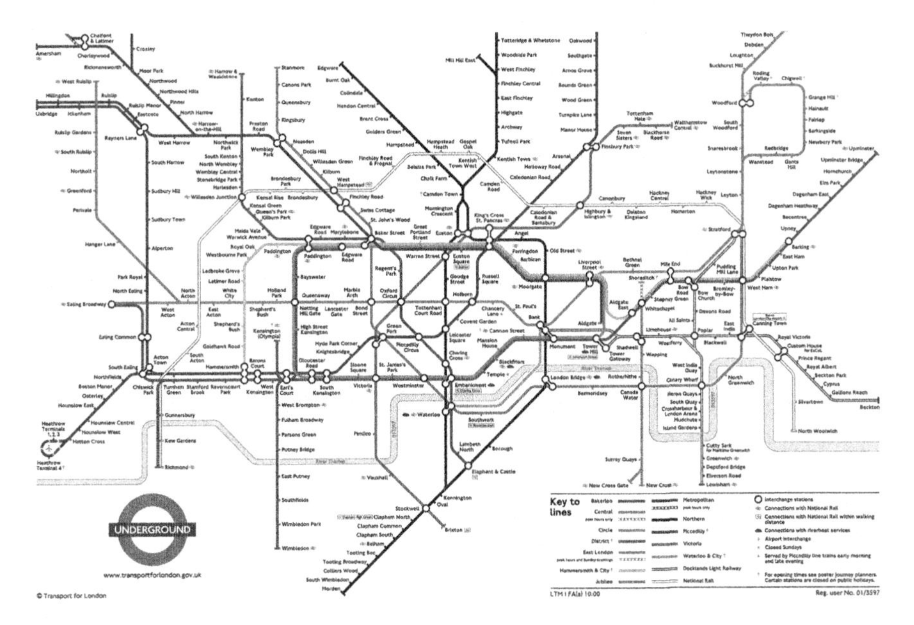

Fig. 7.5 The London Underground Diagram as we know it today, is an adaptation of Beck's original 1930's design which introduced the straight lines meeting at 45 and 90° angles and a representation of subway lines and regularly spaced blobs and ticks as a representation of stations. This diagram communicates a habit of action for thousands of underground users everyday. It is part of the cognitive niche of Londoners

networks of stations and lines. The original version of the LUD was created by Henry C. (Harry) Beck in 1933. Beck's design was based upon electrical circuit diagrams, which omit or falsify information about the real physical distribution of wires in order to convey information about connectivity. Beck noticed a similarity with the underground railway network in that it was possible to ignore the geographical information altogether and as such remove some of the sources of confusion in more literal maps. Some of the strategies, needs, and preferences of users may not be supported by the design choices of the LUD: trying to figure out which station is closer to a particular street, for example. The set of potentialities for action that a representation designed specifically for solving problems of navigation in the underground system embeds is only one between many other possible sets that might be derived from the system: that of a mechanic trying to locate a particular electrical fault in the system, for example. The set of potentialities that the LUD offers is a crucial part of any description or characterization of how thousands of commuters and tourists employ the London Underground System everyday. This habit of action is a constitutive part of the cognitive niche of Londoners. The LUD is a mediator of this habit of action and an important artifact in Londoners' niche construction process.

Conclusion

We have presented two different perspectives of habit in semiosis: as macro-level in a hierarchical structure where it functions as boundary conditions for emergence of semiosis; and as a cognitive niche produced by ecological and environmental inheritance of cognitive artifacts. The first perspective constitutes a model to study the emergence of semiotic processes, allowing a better understanding how habits participate in semiosis. The second perspective relates habit with niche construction. In this case, habit functions as an explanatory component for the co-evolution of environment and cognition. We have elsewhere stressed Peirce as an early proposer of situated and distributed cognition (Atã and Queiroz 2014). The ecological mechanism of inheritance conceptualized by the notion of cognitive niche is a necessary requisite for cognitive processes in a similar sense that the regularity of action conceptualized by the notion of habit as a requisite for semiosis. Both constitute processual strategies to approach meaning phenomena that emphasize their temporal situatedness and distributedness.

Acknowledgments P.A. acknowledges the financial support of CAPES Foundation, Ministry of Education of Brazil, Brasília—DF 70040-020, Brazil.

References

Atã, Pedro, and João Queiroz. 2014. Icon and abduction: Situatedness in Peircean cognitive semiotics. *Model-Based Reasoning in Science and Technology*, ed. L. Magnani, 301–313. Berlin, Heidelberg: Springer.

Atkin, Albert. 2016. *Peirce*. London: Routledge.

Bickhard, Mark H. 2011. Some consequences (and enablings) of process metaphysics. *Axiomathes* 21(1): 3–32, 301–313.

Bergman, Mats. 2000. Reflections on the role of the communicative sign in semeiotic. *Transactions of the Charles S. Peirce Society: A Quarterly Journal in American Philosophy* 36(2): 225–254.

Bergman, Mats. 2005. C.S. Peirce's dialogical conception of sign processes. *Studies in Philosophy and Education* 24(3–4): 213–233.

Clark, Andy. 1998[1997]. *Being there: Putting brain, body, and world together again*. Cambridge: MIT Press.

Clark, Andy. 2006. Language, embodiment, and the cognitive niche. *Trends in Cognitive Sciences* 10(8): 370–374.

Clark, Andy. 2008. *Supersizing the mind*. Oxford: Oxford University Press.

Clark, Andy, and David Chalmers. 1998. The extended mind. *Analysis* 58: 7–19.

De Tienne, André. 2003. Learning qua semiosis. *Semiotics, Evolution, Energy, and Development Journal* 3: 37–53.

Hoffmeyer, Jesper. 2008. The semiotic niche. *Journal of Mediterranean Ecology* 9: 5–30.

Houser, Nathan. 1997. Introduction: Peirce as a logician. *Studies in the logic of Charles Sanders Peirce*, eds. Nathan Houser, D. Roberts, and J. Van Evra, 1–22. Bloomington: Indiana University Press.

Hutchins, Edwin. 1995. *Cognition in the wild*. Cambridge: MIT Press.

Hutchins, Edwin. 1999. Cognitive artifacts. In *The MIT encyclopedia of the cognitive sciences*, eds. R.A. Wilson, and F.C. Keil, 126–127. Cambridge: MIT Press.

Hulswit, Menno. 2001. Semeiotic and the cement of the universe: A Peircean process approach to causation. *Transactions of the Charles S. Peirce Society: A Quarterly Journal in American Philosophy* 37(3): 339–363.

Jones, Clive, John Lawton, and Moshe Shachak. 1994. Organisms as ecosystem engineers. *Oikos* 69: 373–386.

Laland, Kevin N., and Michael J. O'Brien. 2011. Cultural niche construction: An introduction. *Biological Theory* 6: 191–202.

Merrell, Floyd. 1995. *Peirce's semiotics now*. Toronto: Canadian Scholar's Press.

Murphey, Murray G. 1993. *The development of Peirce's philosophy*. Indianapolis: Hackett.

Odling-Smee, John, Kevin N. Laland, and Marcus W. Feldman. 2003. *Niche construction: The neglected process in evolution*. Princeton: Princeton University Press.

Parker, Kelly. 1998. *The continuity of Peirce's thought*. Nashville: Vanderbilt University Press.

Peirce, Charles Sanders. i. 1867–1913. *Collected papers of Charles Sanders Peirce*. Vols. 1–6, eds. Charles Hartshorne, and Paul Weiss. Cambridge: Harvard University Press, 1931–1935. Vols. 7–8, ed. Arthur W. Burks. Cambridge: Harvard University Press, 1958. [References to Peirce's papers will be designated by CP, followed by volume, period, paragraph number.].

Peirce, Charles Sanders. i. 1867–1893. *The essential Peirce: Selected philosophical writing*. Vol 1, eds. Nathan Houser, and Christian Kloesel. Bloomington: Indiana University Press, 1992. [References to this volume will be designated by EP 1, followed by colon, page number.].

Peirce, Charles Sanders. i. 1893–1913. *The essential Peirce: Selected philosophical writing*. Vol 2, ed. the Peirce Edition Project. Bloomington: Indiana University Press, 1998. [References to this volume will be designated by EP 2, followed by colon, page number.].

Peirce manuscripts in Texas Tech University Library at Texas Tech University, Institute of Studies of Pragmaticism, beginning with MS—or L for letter—and followed by a number, refer to the system of identification established by Richard R. Robin in Annotated Catalogue of the Papers of Charles S. Peirce (Amherst: University of Massachusetts Press, 1967), or in Richard R. Robin, "The Peirce Papers: A Supplementary Catalogue," Transactions of the Charles S. Peirce Society.

Queiroz, João, and Charbel El-Hani. 2006a. Semiosis as an emergent process. *Transactions of the Charles S. Peirce Society: A Quarterly Journal in American Philosophy* 42(1): 78–116.

Queiroz, João, and Charbel El-Hani. 2006b. Towards a multi-level approach to the emergence of meaning in living systems. *Acta Biotheoretica* 54: 179–206.

Queiroz, João, and Charbel El-Hani. 2012. Downward determination in semiotic multi-level systems. *Cybernetics and Human Knowing—A Journal of Second Order Cybernetics, Autopoiesis, and Semiotics* 19: 123–136.

Rescher, Nicholas. 1996. *Process metaphysics: An introduction to process philosophy*. Albany: State University of New York Press.

Robinson, Howard. 2014. Substance. *The stanford encyclopedia of philosophy,* Edward N. Zalta, ed. http://plato.stanford.edu/archives/win2012/entries/davidson/.

Salthe, Stanley N. 1985. *Evolving hierarchical systems: Their structure and representation*. New York: Columbia University Press.

Savan, David. 1986. Response to T.L.Short. *Transactions of the Charles S. Peirce Society: A Quarterly Journal in American Philosophy* 22(2): 125–143.

Scott-Phillips, Thomas, Kevin N. Laland, David M. Shuker, Thomas E. Dickins, and Stuart A. West. 2013. The niche construction perspective: A critical appraisal. *Evolution* 68(5): 1231–1243.

Seibt, Johanna. 2016. Process philosophy. *The stanford encyclopedia of philosophy*, Edward N. Zalta, ed. http://plato.stanford.edu/archives/win2012/entries/davidson/.

Stjernfelt, Frederik. 2007. *Diagrammatology—An investigation on the borderlines of phenomenology, ontology, and semiotics*. Heidelberg: Springer.

Part II
Habit as Action Schema

Chapter 8
Is Ethical Normativity Similar to Logical Normativity?

Juuso-Ville Gustafsson and Ahti-Veikko Pietarinen

Abstract To expose a shortcoming in the study of logic and ethics, Hintikka (1999) draws an analogy that takes logic to have been taken over by a "defensive attitude": the avoidance of logical mistakes. A similar attitude, he claims, is prevalent in ethics, which approaches the subject from the perspective of the study of moral mistakes. He uses a distinction between definitory and strategic rules to examine this shortcoming and its consequences in logic. Hintikka does not examine the other side of his analogy, namely the distinction in theories of ethics, however. In this paper we examine those ethics-related aspects of his analogy that have previously gone unnoticed. They are: (1) the possibility of introducing and applying a novel distinction to ethics to distinguish two fundamentally different kinds of ethical rules, the definitory and the strategic rules; (2) the use of these rules to illustrate a fundamental shortcoming in the modern conception of normative ethics; (3) the possibility to separate two conceptions of ethics from each other based on the type of rules that they aim to formulate; (4) the radically different yet unexplored idea of treating ethical rules as strategic rules; and (5) taking Peirce's habits as strategic rules of interaction at work in both ethical and logical conduct.

Keywords Ethics and logic · Normativity · Strategic rules · Game theory · Hintikka · Toulmin · Pragmaticism

J.-V. Gustafsson (✉)
University of Turku, Turku, Finland
e-mail: juuso-ville.e.gustafsson@utu.fi

A.-V. Pietarinen
Tallinn University of Technology, Tallinn, Estonia
e-mail: ahti-veikko.pietarinen@ttu.ee

A.-V. Pietarinen
Xiamen University, Xiamen, China

D.E. West and M. Anderson (eds.), *Consensus on Peirce's Concept of Habit*, Studies in Applied Philosophy, Epistemology and Rational Ethics 31,
DOI 10.1007/978-3-319-45920-2_8

Introduction

Leibniz, in his commentary on Spinozian themes, once wrote that "perfection of the spirit is achieved with logic and ethics".[1] In this short statement Leibniz parallels logic and ethics, and hints at their close allegiance towards each other, even one that places logic before ethics. Such a rationalist thought might have meant that a person whose thinking is guided by the principles of logic will inevitably come to think correctly, and a person who thinks correctly will then also act correctly. Such an idea is understandable, since ethics and logic do have similar aims: as *normative sciences* they both aim at guiding our action by providing us with rules (or principles) of discernment—logic with regard to truth and falsehood, and ethics with regard to right and wrong conduct. But are the rules of discernment similar both in logic and in ethics? What, after all, is meant by such rules? Can the normativity of logic be taken as a model for ethical normativity? And, above all, is ethical normativity similar to logical normativity?

Those willing to give precedence to logic—and in order to keep matters simple let us confine logic here to deductive, non-ampliative logic—might think that if a rule-based model of instrumental rationality, provided by deductive logic, can guide us well with regard to truth and falsehood, then it will also guide us in matters regarding right and wrong action. That is to say, if such logic can tell us how to reason correctly, and if it also provides a procedure for doing so, then one might be tempted to think that it is very much a similar procedure that is also to be followed as regards matters of ethical concern. Nonetheless, giving into such a thought comes at a price—the price being the assumption that *ethical normativity is sufficiently similar to logical normativity*. That is, the assumption is that ethical normativity is grounded on similar types of rules as deductive logic. This assumption, however, contains a serious mistake. Our task in this paper is to spell out what the mistake is and why it should be considered as a reason to undermine the entire assumption. We then propose that the normativity of logic is based on rules that predominantly are of a strategic type, and that grounding logical principles on ethical ones would necessitate ethical rules to be of a strategic type, too. We find this conception of similarity between the two in Peirce's philosophy of pragmaticism, which is fundamentally reliant on habits.

[1]"Perfectio mentis obtinetur Logica, et Ethica.// Logica ostendit modum ita ratiocinadi, ut ad Felicitatem obtinendam convenit.// Ethica id agit ut mens a bene ratiocinado et a laetitia per affectus non impediatur" (Leibniz 1903: 527). "Perfection of the spirit is achieved with logic, and ethics. Logic indicates what mode of reason is pertinent in order to obtain joy. The aim of ethics is not to let affections impede mind from reasoning well" (our translation).

The Assumption Outlined

To notice the assumption at play one needs to look no further than the nearest elementary book in ethics dealing with such topics as "normative ethics" or "applied ethics". There one will come across a variety of definitions such as the following:

> Normative ethics [...] involves substantive proposals concerning how to act, how to live, or what kind of person to be. In particular, it attempts to state and defend the most basic principles governing these matters. [...] Is there, then, a single ultimate moral principle from which all other moral principles can be derived? The debate over whether there is, and if so, what might it be, is the concern on normative ethics. (Kagan 1998: 2)

Normative ethics studies principles about how we ought to live. It asks questions like: What are the basic principles of right and wrong? [...] Normative ethics has two levels. Normative ethical theory looks for very general moral principles, like "We ought always to do whatever maximizes the total pleasure for everyone." Applied ethics studies specific moral issues like abortion or lying, or moral questions in areas as business or medicine. Both levels formulate and defend moral principles. (Gensler 2011: 3)

> Normative ethics involves arriving at moral standards that regulate right and wrong conduct. In a sense, it is a search for an ideal litmus test of proper behavior. The Golden Rule is a classic example of a normative theory that establishes a single principle against which we judge all actions. Other normative theories focus on a set of foundational principles, or a set of good character traits. The key assumption in normative ethics is that there is only one ultimate criterion of moral conduct, whether it is a single rule or a set of principles. (Fieser 2015, Sect. 2)

> Normative ethical theories are theories that propose a criterion or multiple criterion [sic] for the assessment of the moral permissibility of actions. (Düwell 2013: 66)

Similar textbook examples of normative ethics, defined through its rule-giving function, abound in the literature.[2] Each of them defines normative ethics through a procedure. Regardless of the author, the story in all cases is typically the same: answering normative questions requires that we first establish and defend some universal principle(s), which we then apply to some particular cases in order to derive some practical, action-guiding conclusions—or in the very least to see whether some action conforms to such principle(s), that is, whether or not the conduct in question is permitted by the principle(s). This approach bears a striking similarity to deductive logic in that it advises us, first, to figure out the rules that govern the domain in terms of defining the legitimate inferential steps, and then, to perform reasoning according to those inferential rules. Assuming this similarity in procedures between ethics and logic, moral reasoning becomes the matter of finding the right rules that one is then to correctly apply to particular cases at hand.

[2]Cf. Stewart (2009: 11), Pietarinen and Poutanen (1998: 14), Tännsjö (2002: 6), Fischer (2011: 1–2), Beauchamp and Childress (2002: 81), Bunning and Yu (2004: 228), Miller (2003: 2), Gensler (2004: 1).

Moreover, what is right and what is wrong becomes a matter of what is or is not permitted by some rule or set of rules.

An illustrative case in point comes from Lafollette (2000), who has noted that most moral theories have been in the past and in fact still are "criterial" in the sense that they assume that answering normative moral questions requires us to look for generally applicable criteria. He describes the nature of such criteria (rules) as follows:

> When I say that most moral theories are *criterial,* I mean that the theories hold, at least in some attenuated form, that *the relevant criteria are (a) logically prior, (b) fixed, (c) complete, and (d) directly applicable.* Although many philosophers might deny that their views are criterial on this account, the character of most discussions in ethics suggests this view is still influential if not dominant. Thus, although the principle of utility might be revealed through experience, its truth is thought *(a) to be logically prior to experience and (b) to provide a measure for determining what is moral for all people, at all times. Moreover, this principle (c) does not need to be supplemented, and (d) can be directly applied to specific cases.* Likewise for deontological theories. Using the model of law, they envision a set of external rules or principles that tell us how we ought to act. To this extent, most deontologists share certain presuppositions with divine command theorists, namely, that *if morality* is to be binding, its source must be independent of those whom it "binds." (Lafollette 2000: 400, added emphasis)

Regardless of the kind of normative ethics one chooses to champion, it seems that rule-dependence and rule-dependent normativity in the above-mentioned sense cannot be altogether avoided. In both consequentialist and deontologist theories of ethics this is apparent, since both are approaches are clearly criterial.[3] The same might also be said of virtue ethics—for even a person of virtue is still a person of principle.

Thus, if the aforementioned descriptions of normative ethics are representative and correct, then a procedural similarity between ethics and logic follows. To accept this similarity we also need to acknowledge a similarity on the level of rules formulated, that is, between ethical normativity and logical normativity. This similarity in the rules of discernment follows from the assumed procedural similarity between normative ethics and logic, since both are conceived in a like manner and both are aimed at similar kinds of results. We take this view, however, to contain a mistake. In order to highlight what the main shortcoming in this assumption is, we look at Hintikka's suggested analogy between logic and ethics, and in particular his distinction between definitory and strategic rules.

Definitory and Strategic Rules—From Logic to Ethics

In his article "Is logic the key to all good reasoning?"[4] Hintikka briefly explores an analogy between logic and ethics, and claims that a misguided shift of emphasis has occurred both in the study of logic and ethics. He notes that ethics and logic both

---

[3]Cf. Fieser (2015), Sect. 2.

[4]Hintikka's paper was originally published in his book *Inquiry as Inquiry: A Logic of Scientific Discovery* (Hintikka 1999). It was also published twice in 2001, first with minor modifications in

began as forms of art and virtuosity: ethicians were interested in how to act well and logicians in how to reason well. In other words, they were first and foremost interested in how one ought to do something in order to do it well. He claims that both ethicians and logicians have, however, forgotten this important normative aspect of their subject and have instead shifted their emphasis to the *avoidance of mistakes*: logicians to the avoidance of logical mistakes in deductive inferences, and ethicians to the avoidance of moral mistakes in conduct and behavior.

Hintikka's claim is that this game change has been surprisingly similar both in ethics and in logic. It can be seen especially well in logic textbooks, appearing as what he terms the "defensive attitude". Why such a change happened has manifold reasons; he mentions in passing the Victorian morality in ethics and the developments of modern logic, especially that of the development of Frege's notion of logic. Hintikka relates the shift in ethics to the Victorian conception of morality as the preservation of one's moral virtue, that is, to the conception of ethics that has to do with sustaining the skill of blunder-avoidance, in particular those having to do with social and moral blunders. In logic, he relates the shift to the emergent conception of logic manifested in Frege's work, which came to emphasize the necessitarian and infallible character of logical reasoning, that is, to emphasize the avoidance of logical blunders and thus to the preservation of "one's logical virtue" (Hintikka 1999: 1–2).

But if your energies are focussed on such kinds of conduct and dodging fallacies, you will never become a truly virtuous man or woman or a truly innovative reasoner. One could even go further and claim that it is precisely the making of right kinds of mistakes in guesses that has paved the way to progress in the sciences (Pietarinen 2014a, b). After all, the core virtues in actual scientific discoveries can be likened to the rational procedures of making right kinds of guesses much more than they can to the preservation of the consistency of one's belief closures. At any event, in order to illustrate the consequences of this change of emphasis in logic, Hintikka employs the distinction between definitory and strategic rules. A well known proposal is that the distinction can "be generalized to all goal-directed activities which can be conceptualized as games in the sense of the mathematical theory of games" (Hintikka 1999: 2). Hintikka adds, however, that actual game-theoretical analysis of goal-directed activities is not as important as those key ideas and concepts, above all the notion of strategic rules, that can be gathered from that analysis and applied in novel contexts.

(Footnote 4 continued)

(*Argumentation* 15: 35–57, 2001) and later as a Finnish translation in his book *Filosofian Köyhyys ja Rikkaus* (Art House 2001). There are some minor but significant differences between these papers: e.g., in the 1999 version Hintikka uses the term "infallibility", but in the 2001 English version he replaced it with the word "cogency". However, in the 2001 Finnish translation the term appears as "erehtymättömyys" (meaning "infallibility"). It is also worth pointing out that the Finnish version of his paper is titled as "Is logic the key to all creative reasoning?". The Finnish version is less poignant than the two English versions, indicating choices and changes during the translation.

An illuminating definition of definitory and strategic rules is given in the logic textbook by Hintikka and Bachman (1991):

> **Definitory rules define the basic moves in the game.** They tell us what is and what is not admissible in a game. Following them, however, does not in the least guarantee that you play the game well, it only assures that you are playing the game correctly. For example, in a board game like chess or monopoly, the definitory rules tell you when and how you may move the different pieces. These rules do not tell you how to move the pieces in such a way as to make you successful against your opponent. [...]
>
> **Rules that tell you how to play a game well can be called strategic rules or strategic principles.** They are usually much more difficult to formulate and learn than are the definitory rules. Strategic rules have to take into account your overall goal, your opponent's goals and tendencies, your own strengths and weaknesses Strategic rules must also take into account the entire course of the game, not just particular moves one by one. In many games successful players follow strategic principles which they themselves are not clearly aware of. Strategic rules often have to be learned and developed by practicing the game in question and reflecting on how one plays the game. (Hintikka and Bachmann 1991: 32)

The important difference between definitory and strategic rules is that the former are by their nature descriptive whereas the latter are normative. If we think of these rules, as Hintikka thinks is the right way, in terms of different kinds of questions, we could take definitory rules as answers to descriptive questions regarding what is permissible and what is not permissible, whereas strategic rules are answers to normative questions regarding how we ought to do something in order to do it well. The important thing about strategic rules is that they are rules that can only be learned through practice, repetition, and reflection. They are not the matters of definition, choice or agreement, but instead arise from how the games are played, such as having been learned from playing actual games, repetitions of those games in a variety of contexts, or by contemplating on the various possible consequences of moves in a game. If expressed in natural language, they would have a conditional structure of counterfactuals and express subjunctive moods. The interesting thing to note here is that normative rules, understood as strategic rules, are bound both to the actual and the possible consequences of a move. What is more, it follows that also the zero-probability actions have to be counted among such possible moves that the good strategies have to accommodate.

Hintikka proposes that it is the game-theoretical concept of payoffs, and the punishment and reward structures of certain classes of games, that are employed to define the value of possible strategies, that is, to determine what counts as a good move within certain classes of games. The payoffs in a game itself and their possible consequences are what define the value of a strategic move. This means that the value of a strategic rule is not fixed but instead is determined by an interpretation. This, in turn, makes the value of a strategic rule open to differing interpretations in different situations.

With this distinction in place, Hintikka traces the consequences of the aforementioned shift of emphasis in logic and claims that logician have largely come to forget the normative aspect of logic, namely the idea of logic as the study of how one ought to reason, and have instead focused on inferential rules and their correct application. According Hintikka, the problem nevertheless is that definitory rules only tell us what the correct reasoning is and how that correctness can be maintained. These rules say nothing about how we ought to reason in order for us to reason well. Hintikka writes:

> The so-called rules of inference are definitory rules, not strategic ones. At each stage of a deductive argument, there are normally several propositions that can be used as premises of valid deductive inferences. The so-called rules of inference will tell you which of these alternative applications of the rules of inference are admissible. They do not say anything as to which of these rule applications one ought to make or which ones are better than others. For that purpose you need rules of an entirely different kind, viz. strategic rules. The so called rules of inference are merely permissive. They are rules for avoiding fallacies. They are not "laws of thought" either in the sense that they would tell us how people actually draw inferences or in the sense that they would tell us how we ought to draw inferences. (Hintikka 1999: 3)

Then he adds that:

> Without knowing the definitory rules of the game, one cannot understand its strategic rules, either. But if all or most attention is devoted to the definitory rules, the enterprise will remain predominantly defensive, an exercise in avoiding fallacies rather than discovering proofs or finding out new truth by means of deductive inference. [...] This is analogous to the fallacy of conceiving ethics as the art of avoiding moral mistakes. In concentrating their teaching on the so called rules of inference logic instructors are merely training students in how to maintain their logic virtue, not how to reason well. (Hintikka 1999: 3)

Hintikka's view is that in order for us to address the normative side of reasoning we need to focus on strategic rules: rules of an altogether different kind from those of definitory rules, formulated through altogether different processes. Recognizing this difference, Hintikka stresses, is especially important, since so many logicians have indeed attempted to express strategic ideas, but they have at the end done it in terms of definitory rules, a task he dubs "theoretically dubious" (Hintikka 1999: 4).

The reason why strategic rules cannot be formulated in terms of definitory rules is twofold. On the one hand, as noted before, strategic rules express normative ideas, whereas definitory rules express descriptive ideas regarding permissibility. On the other hand, as Hintikka notes, the former deal with "several alternative possible sequences of moves" (Hintikka 1999: 4–5), the value of which is determined through practice and reflection, whereas the latter only deals with what is allowed in some particular situation by the application of some rule or rules. Hintikka's further suggestion is that since we cannot formulate strategic rules in a step-by-step fashion, we might be better off calling strategic rules principles rather than rules.

Strategic Rules in Ethics

The other side of Hintikka's analogy deals with ethics. Unfortunately, he does not say much about ethics at all.[5] He only claims that the situation in logic is analogous to the situation in ethics. Here, however, lies a possibility that has gone unnoticed:

---

[5]Just before his death, we asked Hintikka (personal communication, July 2015) what he now thinks about this proposed analogy in ethics. He promised to think about it, as he mentioned not to

the introduction and application of the distinction between definitory and strategic rules to ethics. Hintikka does not suggest anything like this in his article. But if we are to understand his analogy well, then we must apply his distinction to ethics as well. It is this application that constitutes a novelty, since such a distinction between two different kinds of rules has not been made before in ethical theories. And even though Hintikka does not explicitly suggest how this line of research is to be pursued, the need to do so nevertheless is strongly implied by his analogy.

As noted above, the logician mistake was to confuse permissibility with normativity. Definitory rules told us what inferential moves we are permitted to make, but said nothing about the kind of inferential moves we ought to make. For this task we need strategic rules. This is why focusing solely on definitory rules was seen by Hintikka as a mistake that resulted in the skewed conception of logic as "the art of avoiding logical mistakes", which itself is a mistake "analogous to the fallacy of conceiving of ethics as the art of avoiding moral mistakes" (Hintikka 1999: 3). But what does all this actually mean?

Hintikka equates definitory rules with the rules of inference in logic. Definitory rules are rules that define the game, rules that define what counts as a permissible move in a game and what the range of moves are that are allowed in a given game. The mistake in logic was to think of normativity as the following of definitory rules; a mistake since it leads us to the understanding of 'oughts' in terms of permissibility. In ethics, the mistake is to emulate this conception of normativity. To formulate normative ethical principles in terms of definitory rules is to confuse normativity with the following of ethical principles in their definitory senses. Thus, just as the logicians' mistake has been to lay an excessive focus on the rules of inference and their correct application in reasoning, the ethicians' mistake is to likewise focus on ethical principles (understood as definitory rules) and their correct application in moral reasoning.

If we look at how normative ethics has been characterized (above), it does not take much effort to notice that it has been geared towards the formulation of ethical principles in terms of definitory rules. Maxims in systems of normative ethics can be compared to the definitory rules of logic, because the aim of normative ethics has been to establish rules of permissibility that are intended to be applied in rule-following practices to determine correct behavior. This is a corollary to the emulation of logical normativity in ethics.

When oughts are defined in terms of permissibility, rule-following becomes the measure of virtue. Normativity, however, cannot be reduced to rule-following practices. Hintikka (2007: 58) explains this such that evaluations of efficiency can only be attached to strategic rules, which concern how the entire histories and sequences of moves are constructed and how complete plans are formulated upon

(Footnote 5 continued)

have done much work on that analogy since the original proposal was written, adding though that there are many interesting issues in ethical theories in relation to logical theories that he has indeed been thinking about. He promised to revisit the proposal and get back to it.

them. We take the reason to be that in weighing any single action against a principle we can only give descriptions of permissibility. We have little means of inferring anything about "is it good for X?" If, however, the "fallacy of thinking correctly" is given primacy in moral deliberations, then acting correctly becomes nothing more but a description of correctness in rule-following. This in turn requires that we have already established some rules that we may apply. Normative evaluations, however, require taking into account both the actual and the possible consequences of a multiple sequence of moves in order for us to evaluate a course of action as being better or worse. This kind of evaluation is much more than a mere evaluation of correctness. It is selecting and recommending some course of action as better than others in some situation or case. In other words, we cannot say that acting in accordance with a principle of normative ethics, for example some formulation of consequentialism or deontologism, is better or worse, since such principles do not admit of these types of evaluations. They only admit of descriptive evaluations of correctness. Properties such as betterness or worseness, however, can only be attributed to entire strategies and indirectly the solution concepts and outcomes that follow from them.

It is this conception of normativity grounded on definitory or institutional rules that has affected the way we have come to think about ethical normativity. If we assume a similarity between ethical and logical normativity, and if we think of the oughts solely in terms of permissibility, then ethical reasoning becomes the venture of finding the right universal rules to apply correctly, and goodness and badness becomes the matter of finding out what is and is not permitted by such definitory ethical rules. It is this conception of ethics that we fear will lead to a "defensive attitude", as is illustrated in contemporary logic and which effectively transforms ethics, to use Hintikka's expression, into "the art of avoiding moral mistakes".

The problem with such a conception of ethics is that in it normativity (or the oughts) is based on definitory rules in which such oughts are understood merely in terms of what is permissible. In other words, the mistake of the traditional conception of normative ethics has been to understand normativity as a deontic concept, whereas it in the view of the distinction between the two kinds of rules instead should be seen as an evaluative concept.

Normative or strategic ideas (the oughts) cannot be expressed in terms of, or reduced to, definitory rules. And this is what has led Hintikka to suggest that there is the altogether different type or class of rules, the strategic conception of rules. In order to see, for instance, what is at issue in the normative nature of logic, one needs foreground that strategic conception (see also Pietarinen 2011a, b). What that means is that in order for us to avoid falling into the same trap as outlined above concerning logical reasoning, which would consist merely of the demonstrative aspect of deductive inferences, we should understand normativity in ethics principally in terms of strategic rules. Only then, we believe, can the 'excellence in conduct' thrive again. This is our second novel suggestion, since (i) it grounds ethical normativity on the similar kinds of strategic rules as they occur in logic, and (ii) it pushes the oughts towards the evaluative notion of norms and away from the deontic types of permissibilities.

On a related note, Jonsen and Toulmin (1989: 2) have criticised rule-based approaches to ethics and moral reasoning based on universal principles on the grounds that their "practical implications … can be free of exceptions or qualifications". They think that blind adherence to universal moral principles without making "equitable allowances" leads to the "tyranny of principles":[6]

> [O]nce we move far enough away from the simple paradigmatic cases to which the chosen generalizations were tailored, it becomes clear that no rule can be entirely self-interpreting. The considerations that weigh with us in resolving the ambiguities that arise in marginal cases, like those that weigh with us in balancing the claims of conflicting principles, are never written into the rules themselves. (Jonsen and Toulmin 1989: 8)

This passage reaffirms the casuist idea that following universal moral rules in ethics is insufficient, since such rules fail to capture the most important aspects of normativity. In order for a general moral rule to be applicable to a specific case the rule needs to be supplemented with a recipe of interpretation regarding its application to such particular cases. But interpretations can hardly be written into universal rules themselves.[7]

A similar, yet more general point can be found in connection to issues related to rationality and conceptual change:

> Universal authority may be claimed for an abstract, timeless system of "rational standards", only if it has first been shown on what foundation that universal and unqualified authority rests; but no formal schema can, by itself, prove its own applicability. (Toulmin 1972: 63)

Even though Toulmin and Hintikka have reached their conclusions through very different lines of thinking, they both are in agreement over a closely related point: that rule-following, whether in ethics or in logic, does not equal normativity. As we have argued above, following definitory rules is related to descriptions of permissibility, not normativity. If normativity is associated with rule-following, and the rules fail to distinguish between definitory and strategic rules, we end up equating normativity with permissibility, where permissible rules lack the evaluative dimension necessary for normativity.

It is worth noting that Toulmin's criticism of rule-based moral reasoning has its origins in his criticism of philosophy in which he has accused philosophers, especially epistemologists, for adopting formal or formalist ideals as the basis of human understanding. This, he claims, has lead to equating the two quite different notions of rationality and logicality.[8] Toulmin's proposal thus gives support to our point that attempts to explain why there is more to the similarity between ethical and logical normativity that typically has been assumed.

It is also worth noting that the tyranny of principles need not be limited to criticizing only definitory kinds of permissible rules. Also normative rules that

---

[6]See Toulmin (1981), Jonsen and Toulmin (1989: 5–11, 341).

[7]This well-known dilemma is in its clearest in trying to incorporate some lock-in procedures into constitutional rules, for example.

[8]See especially Toulmin (1972: 44, 83, 478, 484–488).

prioritize certain values over some others can fail to resolve ambiguities or balance conflicting perspectives. The precautionary principle serves as an instructive example: as it values environmental and human health factors over some other values, it may also prevent progress, development and investment that may—albeit perhaps only through unexpected coincidences—result in even higher environmental and health values than abiding to it would have. Think, for example, prospects for some large-scale global, trans-governmental and generational projects, say terraformation, that ultimately would aim at conditions that can establish permanent settlements on Mars for large-scale human habitation. As presently formulated, the precautionary principle might actually prevent governments to actually undertake such projects.

How Pragmaticism Emerges: Strategies as Habits of Action and Reasoning

It is tempting to adopt the assumption that there is a close methodological similarity between the two normative sciences, namely that of ethics and that of logic, from the thought that what rational ethical decisions are would somehow be best be modelled on those of what the rational steps in logical inferences are. This prevalent blunder derives from the negligence of the distinction between rules that define and the rules that are strategically advantageous. If we fail to draw that important distinction, rational ethical decisions appear to us to be of the same kind as what logical deduction would command them to be.

What are the main lessons to be learned from our result? Our considerations suggest that the failure to draw the distinction between these two kinds of rules has even deeper roots than the mere negligence. That is to say that there is a dependence or influence between ethics and logic which remains hidden without right kinds of conceptualizations and theoretical developments. This neglected relationship can now be brought to the view to a larger extent and from the wider perspective than has been the case in the literature.

If we agree it worthwhile to develop upon Leibniz's Spinozian views on the matter, logical principles may indeed turn out to repose on ethical principles. But it is important to understand the nature of those principles. Ethical principles or rules of conduct are not, according to this view, well modelled on how rational logical inferences are thought to be drawn in the typical cases of logical, namely demonstrative deductive inference. A full account of ethical principles cannot be limited to the permissive ones that define the conduct, just as good reasoning cannot be reduced to patterns of inference that follow the definitory rules of demonstration in deductive reasoning.

Nothing limits our argument to the class of deductive inference, either. The definitory/strategic distinction applies to all forms of reasoning, ampliative and non-ampliative alike. But whatever the details and the manifold interdependencies between logic in the wider sense and ethics may turn out to be, the upshot of our way of putting the relationship between the two in order should nonetheless be clear.

Moreover, this conclusion cannot be altogether new. Indeed, in developing upon the Leibniz-Spinozian view, Peirce's perennial classification of the sciences respects the same ordering from ethics to logic (Pietarinen 2005a). In his classification, logic, as one of the three normative sciences, depends on the ethical and also on those of esthetical ideals. The way in which it does so is a complex one and is explored in a greater length elsewhere (Pietarinen 2011a, b). In brief, aesthetic and ethical ideals have generality which their principles reflect, and therefore logical principles must follow up on those more general principles of ethics.

The strategic principles that Peirce talks about are the "habits of reasoning" (Pietarinen 2005b). This association of strategies and habits involves, crucially, the idea of habits that prefigure the game-theoretic concept of strategies, both in the historical and systematic senses of that anticipation (Pietarinen 2003, 2006). In habits of reasoning, we find "habits of conduct" that are in operation in ethics conceived quite analogously to moral principles: they exhibit general logical conduct in arguments concerning how we infer conclusions from premises according to some leading principles of reason. And even in esthetics we find Peirce applying the idea of the presence of such generalising principles, which he dubbed "habits of feeling" (Pietarinen 2005b, 2009). The nature of such general esthetic principles remains quite sketchy under Peirce's own account, but they are crucial pieces of the overall puzzle concerning the completeness of his semeiotic.

But what ultimately justifies our crucial linkage between Peirce's notion of habits and that of the later game-theoretic formulation of strategies and rules of strategic interaction is important. Now the textual evidence from Peirce has been explored in detail (Pietarinen 2003, 2006, 2013), so let us only focus on a couple of aspects in the systematic side of the connection. In game theory, different kinds of games are solved by identifying the connection between the payoff structures and the strategy profiles of the players with what the kinds of games they are that are being investigated. Those solutions are possible because there is some accepted purpose for these games to address certain phenomena: that purpose could concern finding stable points or equilibria where everyone would be worse off leaving those equilibria, or the purpose could be about defining the core of a coalitional game, or finding the values of the players (Shapley values), and so on. None of these solutions is the best or right in its own terms, as all of them indicate a specific purpose and a method to attain an improved understanding of the phenomena in question. Strategies yield a piecemeal examination of particular domains, but they do it in a generalized way; they have what Peirce terms a "tendency toward generalization,—a generalizing tendency" (RLT: 241; 1898). It is this *tendency* to behave, learn and imitate in certain ways in certain kinds of situations—in a word, to engage in habit taking, that is the crucial phenomenon. A rule that is thus generalized into a habit of action is capable of recommending courses of action whenever certain kinds of circumstances arise. It is that tendency to behave, and not the actions, as the contemporary extended evolutionary synthesis is now beginning to appreciate well, that over time may become a genetically inherited trait.[9]

Building solution concepts works bottom-up, from a multiplicity of examples and cases that the players could encounter over the courses of plays and with the passage of time. Such solutions are important, since they generalize the information that the players can gain from the games. Rather than grounding theories of games invariably on theories of rational decision, then, what the modern approach to modelling and solving complicated games is really doing is to echo some elements of casuistry; incorporating a range of issues that is of practical relevance to human cultural, cognitive and social action. Moreover, games can be repeated, be of indefinite length and size, and played in varying contexts, cultures and situations. In each of these cases it is important to investigate the nature and amount of information that is transmitted from one game situation into the next game situation. None of these features support the idea that players' strategies are rules fixed once and for all. What is more, varying the evolutionary dynamics, thus resulting in different and possibly altogether novel ideas of learning and solution concepts, is a commonplace thing to happen in evolutionary games.

For these reasons, rules that are conceived in these strategic and habitual senses are immune to the kind of tyranny Toulmin attributes to principles and rules that may be familiar from research in formal epistemology and formal logic. Unlike strategic rules, tyrannical principles are provoked to give authority not by individual cases and circumstances but by the very nature of how those rules are formulated. But such attempts to define the rules nevertheless fail to make them self-interpretable. And for that reason, namely lacking the self-interpretational mechanisms to enforce themselves, such principles can indeed result in social dilemmas, ethically questionable conduct and dire human and environmental consequences.

On balance, however, we also need to point out that, as soon as the formulation of rules takes place in strategic, interactive contexts as set out by modern game

[9]See e.g. Pigliucci and Müller (2010). Such is the case with the Baldwin Effect. This effect derives its name from James Mark Baldwin, Peirce's colleague and correspondent who after having read Peirce's 1893 Evolutionary Love article in the *Monist* published his article on the "new factor" in evolution (Baldwin 1896). The evolutionary love piece already contains the idea of the effect; that it is the *tendencies to learn* and behave in certain ways that become inheritable, not the learning and behavior per se. How Baldwin's theory of organic selection, ontogenetic adaptation, and social heritability resembles Peirce's agapastic, non-Lamarckian and anti-Spencerian evolution sketched in his "Evolutionary Love" has been sketched in Pietarinen (2011b). The general mechanism of learning specific behavioral traits, which variously has been argued to be manifested for instance in genetic assimilation or in niche construction, is in Peirce's agapastic terminology a "generalising tendency" that can be "energetically projaculated" though not congenitally inherited. Such tendencies are the "habits of acting", where the "self control" that organisms perform upon the habits can result in phenotypic rigidity favoured by biological evolution: learning is, after all, a costly phenomenon. Therefore Peirce's "Evolutionary Love" article is not, contrary to allegations often voiced on it, a white elephant or an isolated offshoot of his architectonic: properly understood, its generalising tendencies are what Peirce's logical theory of the meaning of intellectual signs, or pragmaticism, also relies on.

theory, Toulmin's charge of formalist or absolutist approach to rules loses its edge. For at the end, Toulmin's critique misses its intended target. He did not observe the possibilities opened up by having the strategic conception of rules at one's disposal in ethical decisions. For that reason his critical project was ultimately unsuccessful in carrying out what it was set out to do in any substantially positive manner, namely in the manner that could have reconciled his approach with that of contemporary logic, reasoning and game theory. The deep reason for Toulmin's project missing its target is that there is nothing inherently subjective in strategic rules—save perhaps how to define the prior probabilities in Bayesian games—on the contrary, the information concerning the effects of those rules are public and intersubjectively testable. Strategic rules are public responses to the actions of other players and do not exist in isolation from how others would think, act and develop judgments of value.

Toulmin is not alone here, since it is quite commonplace for the philosophers to lose sight of the actual developments that have packaged the modern game theory in those new and emerging procedural, exploratory, and cognitive outfits concerning boundedly rational and non-rational action. Such rapid advancements have made the theory of games a far cry from the early-day axiomatic, non-dynamic, and non-evolutionary approaches to strategic interaction.[10]

The Place of Habits in Pragmaticism

Or was it not early enough? The most interesting consequence for our purposes here is how strikingly some of the basic notions of modern game theory resemble what was going on in Peirce's philosophy of *pragmaticism*, especially as he came to develop it during the last years of his life. This resemblance continues to surprise (see Pietarinen 2006, 2013b for details), so much so that what game theory at bottom continues to reveal is virtually *that pragmatistic theory*—the theory according to which the meanings of actions, including linguistic actions, thoughts, reasoning and asserting boils down to the formulation of *conceivable conceptions* concerning our *general resolutions to act* on the basis of such assertions. This is also what the solution concepts achieve in their reliance on counterfactual formulations of general courses of action defined for all possible histories of the game does, including the zero-probability ones.[11] General resolutions to act are meant to

[10]Literature in game theory is rife with suggestions about the ways in which the obsolete "maximization of expected utilities" could be replaced with goals and ends that make less-than-ideal assumptions concerning, for instance, the rationality of the players (cf. Gintis 2009; Rubinstein 1998).

[11]Knockdown evidence that habits are built on the same fundamental notions as strategies are in Peirce's later, 1893 marginal addition to the revised version of his "How to Make Our Ideas Clear" (1878), which adds to the statement that the identity of habits depends not only on those circumstances that are likely to arise, but also on those that might possibly occur, no matter how

control a range of events that may happen; such events as those manifested in perturbed game situations including trembling hands when players make unintended mistakes concerning their actions. The trembles can concern entire processes of strategy discovery and selection, too, in which case the relevant solutions are known as quantal response strategies. Strategies, by their very nature of being formulated in interactive settings with incomplete information are fallible, and typically created and discovered as the game is being played, as a response to a multiplicity of cases: responses to payoffs of those strategies, or to others' actions and information derived from the histories, or to earlier plays of the games, or to the overall structure of the game-playing situations.

Moreover, viewing habits as strategic rules has further benefits, including making them applicable to analysing the meaning of many social and behavioral issues such as cooperation, altruism, trust, punish-and-reward, and so on. These issues thus turn out to be much less the kinds of subjective features of human psychology or mind that may commonly be thought. Peirce would have taken a great interest in these issues, remarking on their collective qualities, for example, and would have wanted, we believe, to investigate how, if at all, to make them part of his overall logical scheme. That is, he could have attempted to translate what is going on in the behavioral and experimental side of the social and cognitive psychology of games not as topics of "vital importance" but those pertaining to his logic, conceived in the wide sense of being co-extensive with semeiotic, the theory of signs. In fact he did some work along those lines, as can be perceived in his writings on habits in connection to logic, reasoning and leading principles (Pietarinen and Bellucci 2014, This volume), his game-theoretic definition of the common ground and common knowledge (Pietarinen 2006), and his abductive notion of scientific reasoning and discovery as a dynamic interrogative process, in which the strategic factors of economy of research are embedded in the very logical form of abduction (Ma and Pietarinen 2015). He also reviewed and criticised the early authors on sociology, including Ross and his *Foundations of Sociology*. In the unpublished pages of that *Nation* review, Peirce discussed the scientific role of sociology (MS 1496, "Notes on Ross", 1905). Can analogical reasoning be applied when comparing "social facts" with other kinds of facts? Peirce tells that "an argument from mere similarity, like those under consideration is a pure Abduction and as such concludes in the interrogative mood, only. … It is necessary to know what is meant by a social group. So far as I know it has not been intelligibly defined by any writer, and certainly is not defined in Ross's book. It is odd that three comrades should make a group, but two not. For the essential relation between members of a group is the intercommunication of ideas. Now only two persons are essential to dialogue". He then goes on to insist "that no better definition of that

(Footnote 11 continued)

improbable they are, the marginal note: "—no matter if contrary to all previous experience". When Pietarinen (2007) presented this definition of habits and Peirce's later addition to Hintikka, he remarked—what we certainly already knew well—that "this is exactly what the strategies in game theory are".

conception [a social group] is possible than that a social group is a collection of persons psychically influencing one another to collective action", hinting at the irrational character of such collective behavior and decision-making. He ponders on the question of whether there are properties of social groups not found among individuals, and states, "there is no reasoning or self-control" taking place in mobs, and nothing to be called a "purpose": rather the mob's behavior is automatic according to its first objective. The lack of self-control and thus the absence habits of reasoning confirms that a collective is prone to make irrational decisions.

Given our strategic perspective to Peirce's notion of habits, then, we can approach the questions of how to analyse normative meaning in a similar sense in which we approach the question of how to perform logical analysis: as the question of what the best plays of evaluation games are when the meaning and signification of logical assertions is the focal question (Pietarinen 2011a, b). Obviously, we cannot even begin to address what "the best play" could mean unless we take the strategic approach. How could we then address what the "best courses of action" would be based on rules and principles that aim at general definitions?

At the heart of the theory of such evaluation games (or semantic games if you will) nevertheless lies the notion of the players which, despite being theoretical constructs nonetheless must have, Peirce remarks, "all the characters of personal intellects possessed of moral natures" (MS 280: 32; cf. Pietarinen 2011a, b, 2013a, b). This highly revealing remark proposes that the logical theory, in the normative and pragmatistic senses in which Peirce was developing it, has an ethical heart. Peirce took this as a common sense that it must be so. There are deeply rooted moral sentiments even in such abstract theories as logic and semantics that give meaning to their own rules and principles. Just as there are such sentiments in ethical theories that are indispensable in making us real moral beings, logical sentiments in inference give meaning to what the good processes of reasoning are. Those sentiments contribute to the *summum bonum* of logic, the increase of the concrete reasonableness of intellectual thought.

But how to uncover the details of what the nature of habits is? According to Peirce, the true nature of habits is revealed as soon as we look away from what the observable actions are and focus instead on how logical analysis is to be performed.

Notational Analysis to the Rescue?

This last point concerning the nature of habits will be treated here only briefly. We remark on one important issue that contributes to such habits, especially in so far as the similarities between habits of reasoning in logic and habits of conduct in ethics are concerned. What we want to emphasize is that our success in developing, discovering and inventing those novel thoughts that contribute to the reasonableness of logical reason tend to hinge on having *good or the best possible notational means for analysis* at our disposal. In particular, such notational demand means involve success in representing and preserving essential aspects of meaning that one's logical assertions aim at capturing. After all, there is a prevalent pragmatic obstacle in, say scientific communication when having an idea of the meaning in mind but no good representation or formulation for it as yet. And it is in this junction in Peirce's semeiotic theory that the idea of *icons* emerges. In particular, it is by special kinds of icons, namely diagrams, and a special kind of diagrammatic imagination that one aims at representing and capturing the essential aspects of signification of logical notations. Pasteur was right: "Fortune favours the diagrammatic mind".

And now the analogy looks like this. Just as such diagrams as good notations give rise to the precepts concerning how to best go about in interpreting one's logical assertions, normative rules of conduct have to do, first and foremost, with such strategies that follow the general tendencies and resolutions the subjects have formed in view of managing the requisite social or moral brilliance in their conduct. Just as the diagrams show—and unlike what the discursive-symbolic repertoires of language can only try to tell—what those general forces are according to which we go about the task of selecting the right kinds of rules to be applied in our inferential proceedings, so do such *diagrams as general resolutions to act* aid in showing how to choose the right rules of conduct. Such diagrams, as general plans of action concerning conceivable events, may fail to apply in practical cases that aim at solving singular questions faced by individual decision makers. The diagrams accomplish such feats due to their self-same nature that necessitates experimentation on them which does not blanket the possibilities. How to transform diagrams into other diagrams is on this view of secondary concern, as transformations are matters of deductive inference and what is permissible in those transformations.

That the meaning of assertions is, first and foremost, inherently connected to such conceivable resolutions to act and as expressed in diagrammatic forms is merely to emphasize the important character of those general resolutions to act: rational decision makers must conceive or imagine also those cases and states of affairs that actually did not or even would not come to pass.[12] On this point Peirce would have agreed with cognitivism, that ethical statements have propositional content and that they can be evaluated by virtue of that content. However, a necessary qualification is that propositional content needs to be taken in a broader sense than what is customary in contemporary meta-ethics and philosophy of language. This broader sense is encapsulated in his semeiotic and naturalized notion of propositions as "dicisigns". We leave it merely as a reference to the future study that Peirce's extended notion of propositions as dicisigns may become a helpful novel instrument in meta-ethical analyses and in philosophy of language (Stjernfelt 2014; Pietarinen 2014b).

[12]Thinking through scenarios is a useful practice that invites us to think in terms of strategic modalities that in the usual probabilistic setting might be excluded from normal and non-fat-tailed distributions.

Conclusions

That such unactualised yet real possibilities can exert counterfactual influence on our reason and conduct is, we believe, the true perfection of the spirit that the meeting of the logical and ethical normativity can bring about. While Leibniz's Spinoza did not have and could not have anything like the modern concept of a strategy at his disposal, his commentary is well aligned with that pragmatist spirit which, as Dea (2006) discovered, Peirce well recognized in Spinoza's *Ethics*.

Breaking strategic rules does not entail having done something wrong. After all, in complex combinatorial situations not all the consequences of an action can be computed; in science and technology unexpected consequences ensue which no one can foresee. Are they for the good or for the bad? Strategic rules aim at improving one's moral ground in the generalizable, counterfactual and forward-looking senses, so that the tendencies to generalize are improved and the habit-taking tendencies that lead to the establishment of those habits could serve as a multidisciplinary method for a piecemeal exploration and analysis of complex moral and conceptual situations. This result is encouraging, as it suggests that the correct method that aims at a resolution of questions concerning ethical conduct is the scientific method.

References

Baldwin, Mark James. 1896. A new factor in evolution. *American Naturalist* 30(441–451): 536–553.

Beauchamp, Tom L., and James F. Childress. 2002. Morality and ethical theory. In *Applied ethics: Critical concepts in philosophy*, ed. Ruth Chadwick, and Doris Schroeder, 75–95. London: Routledge.

Bunnin, Nicholas, and Yu. Jiyan. 2004. *The Blackwell dictionary of western philosophy*. Oxford: Blackwell.

Dea, Shannon. 2006. Merely a veil over the living thought: Mathematics and logic in Peirce's forgotten Spinoza review. *Transactions of the Charles S. Peirce Society: A Quarterly Journal in American Philosophy* 42: 501–517.

Düvell, Marcus. 2013. *Bioethics: Methods, theories, domains*. London: Routledge.

Fieser, James. 2015. Ethics. *The internet encyclopedia of philosophy*, http://www.iep.utm.edu/. Accessed 12 Sept 2015.

Fischer, Andrew. 2011. *Metaethics: An introduction*. Durham: Acumen Publishing.

Gensler, H. 2004. Moral philosophy. In *Ethics: Contemporary readings*, ed. H. Gensler, E.W. Spurgin, and J. Swindal, 1–24. London: Routledge.

Gensler, Harry. 2011. *Ethics: A contemporary introduction*. London: Routledge.

Gintis, Herbert. 2009. *The bounds of reason: Game theory and the unification of the behavioral sciences*. Princeton: Princeton University Press.

Hintikka, Jaakko. 1999. Is logic the key to all good reasoning? In *Inquiry as inquiry: A logic of scientific discovery*. Jaakko Hintikka Selected Papers, Vol 5, 1–24. Dordrecht: Kluwer Academic Publishers.

Hintikka, Jaakko. 2007. *Socratic epistemology: Explorations of knowledge-seeking by questioning*. Cambridge: Cambridge University Press.

Hintikka, Jaakko, and J. Bachmann. 1991. *What if?...towards excellence in reasoning*. Mountain View, California: Mayfield Publishing Company.
Jonsen, A.R., and S. Toulmin. 1989. *The abuse of casuistry. A history of moral reasoning*. Berkeley: University of California Press.
Kagan, Shelly. 1998. *Normative ethics*. Boulder: Westview Press.
Lafollette, Hugh. 2000. Pragmatic ethics. In *The blackwell guide to ethical theory*, ed. Hugh Lafollette, 400–419. Oxford: Blackwell.
Leibniz, G. W. 1903. *Opuscules et fragments inédits de Leibniz. Extraits des manuscrits de la Bibliothèque royale de Hanovre*, ed. Louis Couturat. Paris: Félix Alcan. Available online at BnF (Bibliothèque nationale de France) Gallica: http://gallica.bnf.fr/ark:/12148/bpt6k68142b.
Ma, Minghui, and Ahti-Veikko Pietarinen. 2015. A dynamic approach to Peirce's interrogative construal of abductive logic, *IfCoLog Journal of Logics and their Applications*, in press.
Miller, Alexander. 2003. *An introduction to contemporary metaethics*. Cambridge: Polity Press.
Peirce, Charles Sanders. i. 1898. *Reasoning and the logic of things: The Cambridge conferences lectures of 1898*, ed. Kenneth Laine Ketner. Cambridge: Harvard University Press, 1992. [References to this volume will be designated by RLT, followed by lecture number, colon, page number.] Introduction, and comments, by Kenneth Laine Ketner and Hilary Putnam: 1992: 1–102.S.
Pigliucci, M., and G.B. Müller (eds.). 2010. *Evolution: The extended synthesis*. Cambridge: MIT Press.
Pietarinen, Ahti-Veikko. 2003. Peirce's game-theoretic ideas in logic. *Semiotica* 144: 33–47.
Pietarinen, Ahti-Veikko. 2005a. Interdisciplinarity and Peirce's classification of the sciences: A centennial reassessment. *Perspectives on Science* 14: 127–152.
Pietarinen, Ahti-Veikko. 2005b. Cultivating habits of reason: Peirce and the Logica Utens versus Logica Docens distinction. *History of Philosophy Quarterly* 22: 357–372.
Pietarinen, Ahti-Veikko. 2006. *Signs of logic: Peircean themes on the philosophy of language, games, and communication (Synthese Library 329)*. Dordrecht: Springer.
Pietarinen, Ahti-Veikko. 2007. *To Peirce Hintikka's thoughts. The epistemology and methodology of Jaakko Hintikka*. Copenhagen: Carlsberg Academy.
Pietarinen, Ahti-Veikko. 2009. Esthetic interpretants: Pragmaticism, semiotics, and the meaning of art. *Chinese Semiotic Studies* 2(1): 223–229.
Pietarinen, Ahti-Veikko. 2011a. Why Is the normativity of logic based on rules? In *The normative thought of Charles S. Peirce*, eds. C. De Waal, and K. P. Skowronski, 172–184. Fordham: Fordham University Press.
Pietarinen, Ahti-Veikko. 2011b. *The Peirce-Baldwin effect and its contemporary significance*. Lund: Nordic Association for Semiotic Studies.
Pietarinen, Ahti-Veikko. 2013a. Logical and linguistic games from Peirce to Grice to Hintikka (with comments by J. Hintikka). *Teorema* XXXIII/2: 121–136.
Pietarinen, Ahti-Veikko. 2013b. Pragmaticism revisited: Co-evolution and the methodology of social sciences. *Cognitio* 14(1): 123–136.
Pietarinen, Ahti-Veikko. 2014a. The science to save us from philosophy of science. *Axiomathes* 25: 149–166.
Pietarinen, Ahti-Veikko. 2014b. Natural propositions naturalized? *Cognitive Semiotics* 7(2): 297–303.
Pietarinen, Ahti-Veikko, and Francesco Bellucci. 2014. New light on Peirce's conceptions of retroduction, deduction, and scientific reasoning. *International Studies in the Philosophy of Science* 28(2): 1–21.
Pietarinen, Ahti-Veikko, and Francesco Bellucci. (this volume). Habits of reasoning: On the grammar and critics of logical habits. *Consensus on Peirce's concept of habit: Before and beyond consciousness*, eds. Donna E. West, and Myrdene Anderson. (Studies in applied philosophy, epistemology and rational ethics [SAPERE]) New York: Springer.
Pietarinen, Juhani, and Seppo Poutanen. 1998. *Etiikan teorioita*. Helsinki: Gaudeamus.
Rubinstein, Ariel. 1998. *Modeling bounded rationality*. Cambridge: MIT Press.
Stewart, Noel. 2009. *Ethics: An introduction to moral philosophy*. Cambridge: Polity Press.

Stjernfelt, Frederik. 2014. *Natural propositions: The actuality of peirce's doctrine of dicisigns*. Boston: Docent Press.
Tännsjö, Torbjörn. 2002. *Understanding ethics: An introduction to moral theory*. Edinburgh: Edinburgh University Press.
Toulmin, Stephen. 1972. *Human understanding. Volume 1: The collective use and development of concepts*. Oxford: Clarendon Press.
Toulmin, Stephen. 1981. The tyranny of principles. *The Hastings Center Report* 11(6): 31–39.

Chapter 9
Belief as Habit

Atocha Aliseda

> The essence of belief is the establishment of a habit (1878) [CP 5.398]

Abstract In this paper we analyze the thesis according to which *belief is a habit of conduct*, one purely of thought or leading to action, basing our analysis on the notion of abduction interpreted as an epistemic process for belief revision, all of this within the frame of Charles Peirce's Pragmatism. The notion of abduction in his work is entangled with many aspects of his philosophy. On the one hand, it is linked to his epistemology, a dynamic view of thought as logical inquiry, and corresponds to a deep philosophical concern, that of studying the nature of synthetic reasoning. On the other hand, abduction is proposed as the underlying logic of pragmatism: "If you carefully consider the question of pragmatism you will see that it is nothing else than the question of the logic of abduction." (1903) [CP 5.196]. Two natural consequences of this analysis are the following: the interpretation of Peirce's abductive formulation goes beyond that of a logical argument, especially when viewed as an epistemic process for belief revision and habit acquisition. Moreover, the requirement of experimental verification goes beyond hypotheses verification, for it also requires the calculation of their effects; those that produce new habits of conduct, being these theoretical or practical.

Keywords Pragmatism · Abduction · Habit · Pragmatic maxim

Introduction

Here we analyze Peirce's thesis according to which *belief is a habit of conduct*, one purely of thought or leading to action. We base our analysis on Peirce's notion of abduction, interpreted as an epistemic process for belief revision. Our discussion takes place within the framework of pragmatism, which according to Peirce, is a method of reflexion with the ultimate goal of clarifying ideas and for belief fixation. Indeed, in his 1877 essay "The fixation of belief" [CP 5.358–5.387], Peirce takes up the task of reviewing various methods for belief fixation: tenacity, authority, the a

A. Aliseda (✉)
National Autonomous University, Mexico, Mexico
e-mail: atocha@filosoficas.unam.mx

D.E. West and M. Anderson (eds.), *Consensus on Peirce's Concept of Habit*,
Studies in Applied Philosophy, Epistemology and Rational Ethics 31,
DOI 10.1007/978-3-319-45920-2_9

priori, and the scientific method, showing that all of them except for the last one, are deemed to fail as methods for belief fixation leading to the settlement of opinion, the object of inquiry. It is only the fourth method that delivers conclusions by which opinions and facts coincide.

Regarding abduction, three aspects determine whether an abductive hypothesis is promising: it must be *explanatory*, *testable*, and *economic*. It is the first of these aspects that accounts for the abductive logical formulation and what is related to the epistemic transition between the states of doubt and belief. The second of these aspects is what relates directly to pragmatism, for this doctrine provides a *pragmatic maxim*, in this case, serving as a guide to what counts as an explanatory hypothesis based on its being subject to experimental verification.

When abduction is interpreted as an epistemic process for belief revision, the process by which the view that beliefs are habits becomes salient, it runs as follows: when a belief and its associated habit are disrupted by a surprise, a state of doubt and irritation emerges and inquiry begins by triggering abductive reasoning. The aim of this type of reasoning is to soothe the doubt and produce an abductive hypothesis, one leading to a state of belief, which in turn establishes a habit of mind or one of action.

Our analysis suggests that there is more to the testing of an abductive hypothesis. The very end of a belief supporting a hypothesis is to establish a habit, achieved as a result of testing those consequences the belief in question produces. It is in the testing of these consequences where the pragmatic maxim is at work.

This paper is divided into four parts. After this introduction, in the second part, we present Peirce's epistemic view as well as the development of his notion of abduction, and we close it by a connection between the two, arguing for an interpretation of abduction as a process for belief revision. In the third part, we present Peirce's notion of pragmatism, the pragmatic maxim and its connection to abduction. We show that the abductive criterion of experimental verification goes beyond hypotheses verification, for it demands as well a calculation of their possible consequences or effects—those that produce new habits of conduct. In the fourth and final part of this chapter, we put forward our conclusions relating all these three aspects of Peirce's philosophy reviewed in this paper: the epistemological, the abductive, and the pragmatic, all contributing to our understanding of what Peirce meant when he put forward the thesis that beliefs are habits of conduct.

Abduction and Epistemology in Peirce

In Peirce's epistemology, thought is a dynamic process, essentially an interaction between two states of mind: *doubt* and *belief*. While the essence of the latter is the "establishment of a habit which determines our actions" (1878) [CP 5.388], with the quality of being a calm and satisfactory state in which all humans would like to stay, the former "stimulates us to inquiry until it is destroyed" (1877) [CP 5.373], and it is characterized by being a stormy and unpleasant state from which every human struggles to be freed:

> The irritation of doubt causes a struggle to attain a state of belief. (1877) [CP 5.374]

Thus, the pair "doubt-belief" is a cycle between two opposite states. While belief induces a habit, doubt puts it on hold. Doubt, however, Peirce claims, is not a state generated at will by raising a question, just as a sentence does not become interrogative by putting a special mark on it, there must be a *real* and *genuine* doubt:

> … genuine doubt always has an external origin, usually from surprise; and that it is as impossible for a man to create in himself a genuine doubt by such an act of the will as would suffice to imagine the condition of a mathematical theorem, as it would be for him to give himself a genuine surprise by a simple act of the will. (1905) [CP 5.443].

Just in the same way as doubt generation cannot be controlled at will, neither the breaking of a belief or the suspension of a habit is caused as a result of self-control. For a belief and its associate habit to be disrupted, an external stimuli is needed; it is surprise that breaks habits:

> For belief, while it lasts, is a strong habit, and as such, forces the man to believe until some *surprise* breaks up the habit. ((1905) [CP 5.524], my emphasis).

And Peirce distinguishes two ways to break a belief:

> The breaking of a belief can only be due to some *novel experience* (1905) [CP 5.524] or … until we find ourselves confronted with some experience *contrary to those expectations.* ((1881) [CP 7.36], my emphasis).

On the one hand, a belief establishes a habit, one which may be of mind or of action. On the other hand, what breaks up a habit is a surprise, either when a novel or an anomalous situation is encountered, both triggering abductive reasoning. Therefore, in order fully to understand the dynamics of habit formation and suspension, we shall first describe Peirce's theory of abduction.

Peirce's Abduction

When it comes to abduction, as conceived by Pierce, the epistemological, the cognitive, and the logical are intertwined. Peirce proposes abduction to be the logic for synthetic reasoning, that is, a method to acquire new ideas. However, he gave not a single characterization of abduction, and varied scholarly interpretations will be found in the literature. We will be reviewing two characterizations, the syllogistic and the inferential, favoring the latter one, when interpreted as an epistemic process for belief revision.

The development of a logic of inquiry occupied Peirce's thought since the beginning of his work. In the early years he thought of a logic composed of three modes of reasoning: deduction, induction, and hypothesis, each of which

corresponds to a syllogistic form, illustrated by the following, often quoted example (1878) [CP 2.623]:
DEDUCTION

Rule.—All the beans from this bag are white.
Case.—These beans are from this bag.
Result.—These beans are white.

INDUCTION

Case.—These beans are from this bag.
Result.—These beans are white.
Rule.—All the beans from this bag are white.

HYPOTHESIS

Rule.—All the beans from this bag are white.
Result.—These beans are white.
Case.—These beans are from this bag.

Of these, deduction is the only reasoning which is completely certain, inferring its "Result" as a necessary conclusion. Induction produces a "Rule" validated only in the "long run" (1903) [CP 5.170], and hypothesis merely suggests that something may be "the Case" (1903) [CP 5.171]. The evolution of Peirce's theory of abduction is also reflected in the varied terminology he used to refer to abduction; beginning with *presumption* and *hypothesis* (1901) [CP 2.776], (1878) [CP 2.623], then using *abduction* and *retroduction* interchangeably (1896) [CP 1.68], (1901) [CP 2.776], (1902) [CP 7.97].

Later on, Peirce proposed these types of reasoning as the stages composing a method for logical inquiry, of which abduction is the beginning:

> From its [abductive] suggestion deduction can draw a prediction which can be tested by induction. (1903) [CP 5.171]

In addition, his theory of abduction covers a broad spectrum of cognitive tasks. Abduction plays a role in direct perceptual judgments, in which:

> The abductive suggestion comes to us like as a flash. (1903) [CP 5.181]

As well as in the general process of invention:

> It [abduction] is the only logical operation which introduces any new ideas. (1903) [CP 5.171]

In all this, abduction seems to be both "an act of insight and an inference" as has been suggested (cf. Anderson 1986). These explications do not fix one unique notion. Peirce refined his views on abduction throughout his work. He first identified abduction with the syllogistic form above, to later enrich this idea by the more general conception of "the process of forming an explanatory hypothesis" (1903) [CP 5.171] and also referring to it as "the process of choosing a hypothesis" (1901) [CP 7.219]. In any case, this later view gives place to the logical formulation of abduction. Peirce was indeed the first philosopher to give to abduction a logical form, represented in the following argument-schema (1903) [CP 5.189]:

The surprising fact, C, is observed.
But if A were true, C would be a matter of course.
Hence, there is reason to suspect that A is true.

In addition to this formulation which makes up the first aspect of abductive inference, the *explanatory* one, there are two other aspects to consider for an explanatory hypothesis, namely its being *testable* and *economic*. While the second one sets up a requirement in order to give an empirical account of the facts, the third one is a response to the practical problem of having innumerable hypotheses to test and points to the need of having a criterion to select the best explanation amongst the testable ones. Therefore, a hypothesis is an explanation if it accounts for the facts and its status is that of a suggestion until it is verified.

Moreover, in his theory of inquiry, Peirce recognized not only different types of reasoning, but also several degrees within each one, and even merges between the types. In the context of perception he writes:

> The perceptual judgements are to be regarded as extreme cases of abductive inferences. (1903) [CP 5.181]

Abductory induction, on the other hand, is suggested when some kind of guess work is involved in the reasoning (1901) [CP 6.526]. He further distinguished three kinds of induction (1901) [CP 2.775], (1901) [CP 7.208], and even two kinds of deduction. (1901) [CP 7.224]. Anderson (1987) also recognizes several degrees in Peirce's notion of creativity. A clear and concise account of the development of abduction in Peirce, which distinguishes three stages in the evolution of his thought, is given by Fann (1970).

Therefore, it comes as no surprise there are several and varied interpretations of Peirce's abduction. Our own renders abduction as *an epistemic process for belief revision*, which we proceed to describe.

Abduction as Belief Revision

The connection between abduction and the epistemic transition between the mental states of doubt and belief is clearly seen in the fact that a surprise is both the trigger of abductive reasoning—as indicated by the first premise of the logical formulation of abduction—as well as that of the doubt state when a habit has been broken.

The overall cognitive process showing abductive inference as an epistemic process for belief revision can be depicted as follows: a novel or an anomalous experience gives way to a surprising phenomenon, generating a state of doubt which in turn disrupts a belief and its associated habit, all of which triggers abductive reasoning. The goal of this type of reasoning is to *soothe* the state of doubt. Note that it is "soothe" rather than "destroy", for an abductive hypothesis has to be put to test before converting itself into a fixed belief. It must also be economic, attending the further criterion Peirce proposed.

The interpretation of abduction as an epistemic process for belief revision is a familiar one in the area of Artificial Intelligence (AI). Under our own interpretation, Peirce's epistemic model proposes two varieties of surprise as the triggers for every inquiry, labelled as *novelty* and *anomaly*. Although this interpretation is based on the inferential perspective, its representation in an argumentative schema falls short of our interpretation as an epistemic model for belief revision. The changes these two varieties of surprise and its associated habits effect cannot be accounted for in an inferential model of abduction. In order to have a complete picture of the process, the testability criterion should be incorporated and analyzed within the connection between abduction and pragmatism, which we describe in what follows.

Abduction and Pragmatism, Pragmaticism and the Pragmatic Maxim

William James reports it was C.S. Peirce who engendered the philosophical doctrine known as *Pragmatism*, which Peirce preferred to call *Pragmaticism*. Pragmaticism is a philosophical method of reflexion with the aim of clarifying ideas and guided at all moments by the ends of the ideas it analyzes, being these practical or purely of thought. It is conceived as a method in logic rather than as a metaphysical principle:

> I make pragmatism to be a mere maxim of logic instead of a sublime principle of speculative philosophy. (1901) [CP 7.220]

As we shall see, abduction is the underlying principle of the pragmatic maxim. This maxim, in its original formulation, reads as follows:

> Consider what effects that might conceivably have practical bearing you conceive the object of your conception to have. Then your conception of those effects is the whole of your conception of the object. (*Revue philosophique* VII (1903) [CP 5.18]

For our purposes, the core of this maxim is that the conception of an object relies on its conceivable practical effects manifested in habits of action. Let us analyze each item in turn. The "conception of the concept" in these pragmatic terms (cf. (1908) [CP 6.481]) means that it is acquired through the following conditions by which a mastery of its use is attained. In the first place, it is required to learn to recognize a concept in whatever of its manifestations, and this is achieved by an extensive familiarization with its instances. In the second place, it is required to carry out an abstract logical analysis of the concept, getting to the bottom of its elemental constitutive parts. These two requirements however, are not yet sufficient to grasp the nature of a concept in its totality. For this, it is in addition necessary to discover and recognize those habits that the belief in the truth of the concept in question naturally generates, that is, those habits which result in a sufficient condition for the truth of a concept in any theme or imaginable circumstance.

Let us now move to illustrate the connection between abduction and pragmatism.

Abduction and Pragmatism

In Peirce's writings we find notes for a conference "Pragmatism—Lecture VII" (of which there is evidence that it was never delivered). This conference is composed by four sections, of which the third one, "Pragmatism—The Logic of Abduction" (cf. (1903) [CP 5.195–5.206]) is the relevant one for our discussion. In this section of the conference, Peirce states that the question of pragmatism is nothing else than the logic of abduction, as suggested by the following quote:

> Admitting, then, that the question of Pragmatism is the question of Abduction, let us consider it under that form. What is good abduction? What should an explanatory hypothesis be to be worthy to rank as a hypothesis? Of course, it must explain the facts. But what other conditions ought it to fulfill to be good? The question of the goodness of anything is whether that thing fulfills its end. What, then, is the end of an explanatory hypothesis? Its end is, through subjection to the test of experiment, to lead to the avoidance of all surprise and to the establishment of a habit of positive expectation that shall not be disappointed. Any hypothesis, therefore, may be admissible, in the absence of any special reasons to the contrary, provided it be capable of experimental verification, and only insofar as it is capable of such verification. This is approximately the doctrine of pragmatism. (1903) [CP 5.198]

What we see here is that Peirce puts forward the pragmatic method for analyzing the admissibility of an abductive hypothesis beyond its being explanatory. What becomes relevant here is the second criterion: every hypothesis should be subject to experimental verification. In view of the fact that the epistemic status of the conclusion of the abductive formulation is only tentative (...there is reason to *suspect* that A is true), this conclusion should be subject of experimental verification. But this quote shows that there is more to the testing of the abductive hypothesis. The very end of a belief supporting an abductive hypothesis is to establish a habit, and it is in the testing of consequences that habits are manifested. This seems to complete the picture of the second criterion for the admissibility of an abductive hypothesis. Moreover, when a surprise is encountered and its corresponding habit is disrupted for failing to meet positive expectations, then abductive reasoning is triggered and the process starts running all over again.

Belief as a Habit of Conduct

We are now in a position to analyze in full Peirce's thesis, namely, that belief is a habit, one which may be manifested in the realm of ideas or in the empirical world, and accordingly, induces a habit of mind or one of action. This is what Peirce states on the former kind:

> That which determines us, from given premises, to draw one inference rather than another is some *habit of mind*, whether it be constitutional or acquired. The habit is good or otherwise, according as it produces true conclusions from true premises or not; and an inference is regarded as valid or not, without reference to the truth or falsity of its conclusion specially, but according as the habit which determines it is such as to produce true conclusions in general or not. ((1877) [CP 5.367], my emphasis)

Therefore, in the case of habits of mind, these are associated with actions of mind as well, as when drawing inferences. Indeed "different beliefs are distinguished by the different modes of action to which they give rise" (1878) [CP 5.397]. In any case, all habits are habits of *conduct*. Regarding this notion, Peirce states:

> It is necessary to understand the word *conduct*, here, in the broadest sense. If, for example, the predication of a given concept were to lead to our admitting that a given form of reasoning concerning the subject of which it was affirmed was valid, when it would not otherwise be valid, the recognition of that effect in our reasoning would decidedly be a habit of conduct. ((1908) [CP 6.481], my emphasis)

And the recognition of effects links us directly to the pragmatic maxim:

> For the maxim of pragmatism is that a conception can have no logical effect or import differing from that of a second conception except so far as, taken in connection with other conceptions and intentions, it might conceivably modify our practical conduct differently from that second conception. (1903) [CP 5.196]

It becomes now clear that beliefs are habits of conduct, under an interpretation of abduction as an epistemic process within the pragmatic framework. But once a belief with its associated habit has been established, what makes them change? And how does a new belief-habit emerge? It is surprise that generates a state of doubt, one which may be of two kinds: either a surprise arises because there is a new phenomenon to account for and new habits are acquired or else, an anomalous experience causes a belief and its associated habits be disrupted, in which case these may be retracted before new beliefs and habits are conformed. The establishment of a habit is achieved as a result of testing those consequences the belief in question produces. It is then in the testing of these consequences where the pragmatic maxim is at work.

Discussion and Conclusions

In this paper we analyzed the thesis according to which *belief is a habit*. We based our analysis on Peirce's epistemology in its connection to abduction within the frame of pragmatism. According to Peirce, pragmatism is a method for reflexion with the ultimate goal of clarifying ideas and for belief fixation. Pragmatism is guided by the pragmatic maxim and abduction is its underlying logic.

We have illustrated our thesis within the abductive formulation when interpreted as an epistemic process for belief revision. The overall process may be described as follows: when a belief and its associated habits are disrupted by a surprise, a state of doubt and irritation emerges and inquiry begins by triggering abductive reasoning. The aim is then to soothe the doubt and produce an abductive hypothesis, one leading to a state of belief, which in turn establishes a habit of mind or one of action. The belief state with its associated habits will remain as such until another surprising fact is encountered, thus continuing the epistemic doubt-belief cycle.

Two natural consequences derive from our analysis: the interpretation of Peirce's abductive formulation goes beyond that of a logical argument, especially when viewed as an epistemic process for belief revision and habit acquisition. Moreover, the requirement of experimental verification goes beyond hypotheses verification, for it also requires the calculation of their effects; those that produce new habits of conduct, being these theoretical or practical.

When Peirce states the pragmatic maxim as a *logical gospel*—his colleagues then and his critics nowadays—did not understand that Peirce's notion of logic is much broader than what we conceive even nowadays. For Peirce, every thought is realized in signs and all reasoning consists of a logical inference with the ultimate goal of acquiring beliefs, which in turn endures the creation of habits of mind or of action.

References

Aliseda, Atocha. 1997. *Seeking explanations: Abduction in logic, philosophy of science and artificial intelligence*. Ph.D. Dissertation, Philosophy Department, Stanford University. Published by the Institute for Logic, Language and Computation (ILLC), University of Amsterdam (ILLC Dissertation Series 1997–4).

Aliseda, Atocha. 2000. Abduction as epistemic change: A peircean model in artificial intelligence. *Abduction and induction: Essays on their relation and integration*. Applied Logic Series, Vol 18, eds. Peter A. Flach and Antonis C. Kakas, 45–58. Dordrecht, The Netherlands: Kluwer Academic Publishers.

Aliseda, Atocha. 2005. The logic of abduction in the light of Peirce's pragmatism. *Semiotica* 153 (1/4): 363–374.

Aliseda, Atocha. 2006. *Abductive reasoning. Logical investigations into discovery and explanation*. Springer, Synthese Library, Vol 330. Heidelberg: Springer.

Aliseda, Atocha. 2009. Logic and knowledge: Expectations via induction and abduction. In *The many sides of logic*. Studies in Logic 21, eds. W. Carnielli, M.E. Coniglio, and I.M. D'Ottaviano, 497–510. College Publications: United Kingdom.

Anderson, Douglas R. 1986. The evolution of Peirce's concept of abduction. *Transactions of the Charles S. Peirce Society* 22(2): 145–164.
Anderson, Douglas R. 1987. *Creativity and the philosophy of C.S. Peirce*. Philosophy library, Vol 27. Martinus Nijhoff Publishers.
Campos, Daniel G. 2011. On the distinction between Peirce's abduction and Lipton's inference to the best explanation. *Synthese* 180: 419–442.
Fann, K.T. 1970. *Peirce's theory of abduction*. The Hague: Martinus Nijhoff.
Flach, Peter A., and Antonis C. Kakas, eds. 2000. *Abduction and induction. Essays on their relation and integration*. Applied Logic Series, Vol 18. Dordrecht, The Netherlands: Kluwer Academic Publishers.
Peirce, Charles Sanders. i. 1867–1913. *Collected papers of Charles Sanders Peirce*. Vols. 1–6, eds. Charles Hartshorne and Paul Weiss. Cambridge: Harvard University Press, 1931–1935. Vols. 7–8, ed. Arthur W. Burks. Cambridge: Harvard University Press, 1958. [References to Peirce's papers will be designated by CP, followed by volume, period, paragraph number.].
Peirce, Charles Sanders. i. 1867–1893. *The essential Peirce: Selected philosophical writing*. Volume 1 (1867–1893), eds. Nathan Houser and Christian Kloesel. Bloomington: Indiana University Press, 1992. [References to this volume will be designated by EP 1, followed by colon, page number.].
Peirce, Charles Sanders. i. 1893–1913. *The essential Peirce: Selected philosophical writing*. Vol 2 (1893–1913), ed. the Peirce Edition Project. Bloomington: Indiana University Press, 1998. [References to this volume will be designated by EP 2, followed by colon, page number.].

Chapter 10
The Originality and Relevance of Peirce's Concept of Habit

Lucia Santaella

Abstract After 1900, Peirce engaged himself in the development of his theory of signs, particularly in the theory of interpretants and even more specifically in the theory of the logical interpretants, since the latter represented the touchstone for linking pragmatism to the theory of signs. In 1907, he declared that the problem of what the "meaning" of an intellectual concept is could only be solved by the study of the interpretants, or the proper significate effects of signs. It was within this concept that Peirce developed his famous subdivision of interpretants into emotional, energetic and logical. Peirce stated in 1868 that the interpretant of a thought is another thought, and that this process, theoretically, is infinite. Many authors impressed with this assertion, and without bothering to follow the progress of this concept throughout Peirce's works, were favorably inclined toward infinite semiosis, as it is so often labeled. Umberto Eco, for instance, was one author who made extensive use of this notion of infinitude. The aim of this paper is to discuss the transformation that this concept of interpretant has undergone in Peircean works, particularly after 1907, when Peirce introduced his notion of the logical interpretant. This notion would come to change the idea—which unfortunately continues widespread—that semiosis is an abstract infinite process, unconnected with human action. Were it so, semiosis would bear no relation with pragmatism. When Peirce discovered the role of the logical interpretant in habit, and of the ultimate interpretant in the change of habit, he combined the processual nature of semiosis with pragmatism. From this synthesis derived the evolutionist character of his pragmatism.

Keywords Logical interpretant · Evolution · Synechism · Semiosis · Habit · Pragmaticism · Law of mind

L. Santaella (✉)
Pontifícia Universidade Católica (PUC), São Paulo, Brazil
e-mail: lbraga@pucsp.br

D.E. West and M. Anderson (eds.), *Consensus on Peirce's Concept of Habit*,
Studies in Applied Philosophy, Epistemology and Rational Ethics 31,
DOI 10.1007/978-3-319-45920-2_10

Preamble

When most of Peirce's commentators were still convinced that his thought was made of gaps and inconsistencies, Potter (1997 [1967]) coherently defended that Peirce's work in fact presents a considerable unity and a great number of the so called inconsistencies are only apparent since great portions of his work are remarkably interconnected and interdependent. For Potter, Peirce's revision of pragmatism, around 1903, brought with it the conviction that pragmatism was inextricably linked with logic, ethics, and aesthetics—to the extent that these normative sciences get us "upon the trail of the secret of pragmatism" (CP 5.129). In fact, "Peirce's realization of the place of these sciences put in his hands the capstone which unified all that he had been trying to do more or less successfully for some 40 years" (Potter 1997: 3). As the normative sciences were founded on the three universal categories, and as pragmatism, in its turn, depended on synechism, as much as this latter was based on Peirce's radical realism, Potter stated that the categories, the normative sciences, pragmatism, synechism, and realism are of a piece (Potter 1997: 3; Santaella 2001: 177).

I fully agree that the sequence shown above is the most consistent if we want to take Peirce on his own terms, that is, if we intend to understand his position from within. Furthermore, I believe that the originality of the concept of habit, formulated by Peirce along the trajectory of his work, is largely responsible for the coherent link between its parts, or rather between the scales of his complex work. These scales can be perceived in the passages from the phenomenological categories to the normative sciences, in which the signs and inquiry theories where developed, then to evolutionary pragmaticism, all flowing into metaphysics, radical realism, and especially into synechism. This article aims to discuss the passages from one scale to another, passages which are largely due to the ubiquitous notion of habit, habit-taking, and habit-change that run through all of them.

The notion of habit evolved concurrently with the evolution of Peirce's thought. Thanks to these developments, habit was gradually articulated and integrated more and more harmoniously into Peirce's conceptual network, to the point of being one of the main topics that were responsible for the complex and dynamic unity between the parts of such network.

The Tendency of the Universe to Acquire Habits

It was in the first pragmatism period in 1878, particularly in his essays on "How to make our ideas clear" and "The fixation of belief", that the notion of habit arose in the context of the explanations that Peirce was developing for the belief and doubt binomial. From the beginning, that is, since 1868, Peirce conceived of belief and doubt as modes of action. Now, actions that tend to be repeated in accordance with uniform patterns, under specific conditions, he called habits, of which belief is the most legitimate example. Therefore doubt is the deprivation of a habit of action. In fact, belief is a habit whose results can be expressed in a proposition.

> Belief does not make us act at once, but puts us into such a condition that we shall behave in some certain way, when the occasion arises (CP 5.373).

> A belief, in short, is a rule of action—a habit. The essence of thought is to establish a habit. That means that different beliefs are distinguished by the different modes of action they produce. (CP 5.397–398)

At this time, the notion was still rough, lacking the density that it would acquire later. In the Harvard lectures, in 1903, Peirce himself criticized that definition, describing it as obscure and psychological (CP 5.28). In fact, the idea of habit was then restricted to anthropological realms, a limitation that would be outdated when Peirce expanded his conception of thought, law, and habit in light of his mature version of the third category as it is explicit in the following quote:

> In a still fuller sense, Thirdness consists in the formation of a habit. In any succession of events that have occurred there must be some kind of regularity. Nay, there must be regularities strictly exceeding all multitude. But as soon as time adds another event to the series, a great part of those regularities will be broken, and soon indefinitely. If, however, there be a regularity that never will be and never would be broken, that has a mode of being consisting in this destiny or determination of the nature of things that the endless future shall conform to it, that is what we call a law. (MS 478: 32–33; EP 2: 269)

It was the broad notion of Thirdness as law that, since 1886, in his manuscript entitled "One, two, three: Kantian categories" (W 5: 293) had allowed Peirce to extend the notion of habit to the realm of nature. "We must therefore suppose an element of absolute chance, spontaneity, originality, freedom, in nature". And he added:

> We must further suppose that this element in the ages of the past was indefinitely more prominent than now, and that the present almost exact conformity of nature to law is something that has been gradually brought about. We have to suppose that in looking back into the indefinite past we are looking back towards times when the element of law played an indefinitely small part in the universe.

> If the universe is thus progressing from a state of all but pure chance to a state of all but complete determination by law, we must suppose that there is an original, elemental, tendency of things to acquire determinate properties, to take habits. This is the Third of mediating element between chance, which brings forth First and original events, and law which produces sequences of Seconds. Now the tendency to take habits is something essentially finite in amount, an infinitely strong tendency of this sort [unlike an absolute conformity to law] is inconceivable and self-contradictory. Consequently this tendency must itself have been gradually evolved; and it would evidently tend to strengthen itself. (W 5: 293)

In 1887, 3 years later, in his "A guess at the riddle" (W 6: 166–210), the tendency of "habit taking" did not "introduce something which is categorially distinct from law. This tendency is itself a law which explains the evolution of laws, including itself" (Hookway 1997: 20). At this point, Peirce could find his explanation for the evolutionary character of all laws, a character that comes from their being subject to growth and change.

> The tendency to obey laws has always been and will always be growing. [...] Moreover, all things have a tendency to take habits. [...] This tendency itself constitutes a regularity and is continually on the increase. In looking back into the past we are looking towards periods when it was a less and less decided tendency. But its own essential nature is to grow. It is a generalizing tendency; it causes actions in the future to follow some generalization of past actions; and this tendency is something capable of similar generalization; and thus it is self-generative. We have therefore only to suppose the smallest spur of it in the past, and that germ would have been bound to develop into a mighty and over-ruling principle, until it supersedes itself by strengthening habits into absolute laws regulating the action of all things in every respect in the indefinite future. According to this, three elements are active in the world, first, chance; second, law; and third, habit-taking.
>
> Such is our guess at the secret of the sphynx. (W 6: 208)

This guess suggests that habit-taking or continuity, thirdness, is the bridge, that is, mediation between possibility or chance, firstness, and actuality or operative law, secondness. Such a juncture becomes clear when one explores the notion of synechism giving expression to one of the basic concepts in Peirce's thought, in the heart of habit, that is, the concept of continuity.

Synechism, the Key to Continuity

Peirce's notion of synechism appears in his "The law of mind", a paper included in the 1890–1893 Monist series (CP 6.102–106, 163). Synechism, a Greek word that means continuity is the complementary opposite of Tychism, also a Greek word that means chance. Esposito (1973: 63) says that in later life, Peirce came to believe he had outlined a philosophical system that could serve as a matrix for his entire thought. The name he gave to that metaphysical system was synechism (CP 6.202). In a letter to William James, on 25 November 1902, when Peirce spoke of his "completely developed system, which all hangs together and cannot receive any proper presentation in fragments, he went on to describe synechism as the keystone of the arch" (CP 8.255–257; Potter and Shields 1977: 20).

Metaphysics is the first science in Peirce's architectonic classification of the sciences. It inquiries into the nature of the objective world rather than into the structure of thought as his semiotics does. This means that there is a difference between thought and the world (Parker 1994: 52). Peirce's synechism rejects this difference as being one of kind, but considers it instead as a difference only of degree.

Besides the development of his synechistic ideas, Peirce also gave ample thought to tychism or absolute chance. This latter was proposed because Peirce considered mechanistic and deterministic explanation insufficient in the light of his doctrine of categories. Despite its importance, tychism could not be taken as central to his metaphysics, since this centrality was due to synechism. That is why Peirce objected at having his metaphysical system as a whole called tychism. He explained that,

> Although tychism does enter into it, it only enters as subsidiary to that which is really, as I regard it, the characteristic of my doctrine, namely, that I chiefly insist upon continuity, or Thirdness, and, in order to secure to thirdness its really commanding function, I find it indispensable fully [to] recognize that it is a third, and that Firstness, or chance, and Secondness, or Brute reaction, are other elements, without the independence of which Thirdness would not have anything upon which to operate. Accordingly I like to call my theory Synechism, because it rests on the study of continuity. I would not object to Tritism. And if anybody can prove that it is *trite*, that would delight me [in] the chiefest degree (CP 6.202).

Synechism is defined as "that tendency of philosophical thought which insists upon the idea of continuity as of prime importance in philosophy". The continuum, in its turn, is defined as "something whose possibilities of determination no multitude of individuals can exhaust" (CP 6.169–170; see Noble 1989; Myrvold 1995). A rudimentary form of continuity is generality, since continuity is nothing but perfect generality of a law of relationship (CP 6.172).

Peirce frequently remarked that his pragmatism was intimately related to synechism, that is, his version of pragmatism leads to synechism in the sense that synechism includes pragmatism as a step. That is why Peirce emphasized the methodological aspect of synechism when stating that synechism is not "an ultimate and absolute metaphysical doctrine, but like the pragmatic maxim itself "is a regulative principle of logic" (CP 6.173). While this maxim deals with the meaning of concepts, the synechistic principle prescribes what sort of hypothesis is fit to be entertained and examined (CP 6.173; Potter 1997: 71–72). Despite the relevance of the methodological aspect of synechism and despite Peirce's statement that synechism is not an ultimate metaphysical doctrine, the principle of continuity involves other aspects which are no less relevant. These are the ontological and the metaphysical aspects of synechism.

In 1891, in his paper on "The Doctrine of Necessity Examined" (CP 6.35–6.65; W 8: 111–125), Peirce rejected the universality of the uniformity of nature and its consequent mechanism. According to Cosculluela (1992: 743), against the suggestion that the observation of nature proves that determinism is true, Peirce claimed that observation merely shows that there is an element of uniformity in nature; it does not show that such regularity is "exact and universal" (CP 1.55). "No observation or set of observations which human beings are physically capable of making can prove that every fact is precisely determined by law" (Cosculluela 1992: 743). In sum—facts do not conform precisely and uniformly to law. Peirce did not deny that there are laws in nature. On the contrary, he asserted that laws of nature are real generals. This means that there is an element of regularity in nature. The regularity of the laws, however, is constantly being violated to some degree.

> In the *Monist* for January, 1891, and in the number for April, 1892, I attacked the doctrine that every event is precisely determined by law. Like everybody else, I admit that there is regularity. I go further. A maintain the existence of law as something *real* and *general*. But I hold there is no reason to think that there are general formulae to which the phenomena of nature *always conform*, or to which they precisely conform (CP 6.59, 6.588).

Peirce's tychism resulted from the imperfect regularity of nature provoked by the "infinitesimal departures from law" with which nature is literally infected. The more precise our observations become, the more likely it is that we shall encounter facts which seem to depart from laws (CP 6.46). This is proof that chance is an objective feature of nature.

Hookway (1997: 18–21) remarks that, since 1884, in his "Design and Chance" (W 4: 544–554), Peirce was aware of the sporadic violation of the laws of nature in some infinitesimal degree. Noticing that chance is governed by the laws of the probability calculus, he argued that chance "has the property of being able to produce uniformities far more strict than those from which it works" (W 4: 551). From the indication that certain laws of nature are "statistical facts", Peirce concluded that all known laws are statistical facts, although some laws are so well established that the deviations they do undergo are so rare and minute as to be unnoticed. Peirce's further step, which was taken in a supplement to "Design and Chance" (W 4: 553), was to propose that the laws of physics may be "habits gradually acquired by systems". This anthropomorphic suggestion of habits of nature as an analogue of the processes whereby human beings acquire habits of conduct was not new, since it had already been endorsed in Peirce's manuscript "Methods of Reasoning" of 1881 (see Hookway 1997: 20).

From 1884 on, habits of nature became the central concept in Peirce's synechism at the same time that he became a defender of the relevance of anthropomorphic concepts in philosophy. "In fact, habits, from the mode of their formation necessarily consist in the permanence of some relation, and, therefore [...] each law of nature would consist in some permanence, such as the permanence of mass, momentum, and energy. In this respect, the theory suits the facts admirably" (W 6: 210).

Hence, Peirce's insistence on the importance of absolute chance was appropriately counterbalanced by the role that habits perform in nature. Therefore as stated above, habit-taking or continuity, thirdness, mediates between possibility or chance, firstness, and actuality or operative law, secondness. Peirce's categories should be understood here as categories of relation and modality rather than of substance and quality. They are neither limited within the mode of being of possibility alone nor within the mode of an individual thing or actual fact alone. According to synechism, there is nothing about actuality that just *is*. On the one hand, actuality always retains an element of arbitrary chance, an element of sporting which disposes it to be something other than what it is (Wells 1996: 233).

On the other hand, the law of habit prescribes that actual events cannot escape the governance of laws. However, the regularity of the laws are constantly being violated to some infinitesimal degree by the element of arbitrary chance. Hence, "in a dialectic of becoming, actual fact or existence, secondness, is only partially real; its destiny lies within the wider context of Thirdness" (Esposito 1973: 67). A thorough-going synechistic evolutionism implies that nothing escapes the guiding hand of habit-taking or thirdness.

In the light of synechism, thirdness means continuity, that is, relational thirdness (CP 6.190), which implies the interrelation of the three categories and their coexistence inside thirdness. Thus, continuity should not be understood as generalization fully spread out or taken to the limit of generalization. Continuity is rather a dispositional state that infinitely tends toward such spreading out (Wells 1996: 234). This is possible because continuity possesses within it the principle of discontinuity, since the originality of chance may violate the conformity of an event to the strict guidance of the law. That is why laws are approximations which retain a dispositional propensity for habit taking or continuity.

Habit in the Context of Pragmaticism

The presence of habit in human mind follows the same synechistic logic, which is basically an anthropomorphic logic, but with the characteristics proper to humans, as will be examined at greater length in the following.

According to Peirce, there are two kinds of knowledge, perceptual and conceptual. Besides being involuntary, the former are also strictly memories of what happened in a recent past, while all the conclusions of reasoning share the general nature of expectations for the future (CP 2.145). This vector for the future can be better understood in the light of the identification of these conclusions with mental habits.

"Habits are general patterns of action", Curley says (1969: 94), "which prepare the human organism for possible future occurrences. The generality of habits is such that it can never be fully exhausted in any given series of actual occasions". In Peirce's words, "Whatever is truly general refers to the indefinite future; for the past contains only a certain collection of such cases that have occurred. The past is actual fact. But a general (fact) cannot be fully realized. It is a potentiality; and its mode of being is *esse in futuro*" (CP 2.148).

For Curley (1969: 94), the "conscious acceptance of a habit of inference involves an expectation that the future course of experience will render that habit efficacious". Hence, "the normative distinction between good and bad inference involves not only reference to the ideal end of thinking, but also the question whether or not this expectation will be fulfilled".

Being that habit was present in nature and in the human mind, the next task was to clarify the conceptual thread which linked both. This was found in the original and complex context of Peirce's second pragmatism, which he called pragmaticism. The popularity of William James' pragmatism at the beginning of the twentieth century gave rise to a number of pragmatist versions of several different authors. Deeply unhappy with the ways his child, generated in 1878, was distorted, Peirce resumed his ideas, which included his criticism to his own first version of pragmatism, and gave new directions to his doctrine then renamed as pragmaticism (see Turrisi 1997).

This happened in the following years to 1903, when he was also deeply committed to the development of his theory of signs, especially the theory of the interpretants, and most especially the logical interpretant in which would be the touchstone for the integration of pragmatism in the theory of signs and the normative sciences (semiotics, ethics, and aesthetics), under the aegis of the notion of habit. In 1907, he would declare that the problem of the meaning of an intellectual concept, a central issue of pragmaticism, could only be solved by the study of the interpretants, or the effects properly produced by signs (CP 5. 475).

The issue of the interpretants, particularly with regard to the classification of the interpretants, is a complex and not yet fully agreed aspect of Peirce's theory of signs. Before going to the classifications, one must understand the meaning that Peirce gave to the term "interpretant". This term was invented by a juvenile Peirce. It appeared in his writings for the first time in 1866 (W 1: 464–465). A year later, in his famous study "On a new list of categories" (CP 1.545–1.567; W 2: 49–59; EP 1: 1–10), the word was already employed with aplomb. At that time, and without much future substantial changes, the interpretant was understood as the third term of the triadic relationship which constitutes the sign, as expressed in one of his classic definitions: "A REPRESENTAMEN is a subject of a triadic relation TO a second, called its OBJECT, FOR a third, called its INTERPRETANT, this triadic relation being such that the REPRESENTAMEN determines its interpretant to stand in the same triadic relation to the same object for some interpretant (CP 1.541).

This definition is quite abstract and difficult to understand at first glance. There are many other more affordable variants. In all of them, at least two constants remain important: (a) the interpretant should not be confused with the interpreter. It is a broader concept than that of an individual who interprets a given sign. The interpretant, in fact, is another sign that maintains with the object a representation relationship similar to that which the sign maintains, so that the sign functions as a mediator between the object and the interpretant, this, in turn, determining another interpretant, and so on. This leads to a second constant: (b) the object regresses to infinity and the interpretant progresses to infinity.

The progression of interpretants, endlessly generating new interpretants, gave rise to what, under the influence of some authors, especially Eco (1976, 1990), became known as infinite semiosis, that is, the action of the signs to generate new signs, or according to Eco, the interpretation of the interpretation of interpretation… in an endless process. Unfortunately, such conception of the interpretant is unaware that its definition has undergone some changes along the development of Peirce's work. Although, in fact, the changes did not leave the spirit of the notion forged in Peirce's youth, the concept became increasingly sophisticated, so that the interpretive distortions that the infinite semiosis might be subjected to were corrected by specifying the subtleties of the notion to culminate in the integration of the theory of signs to pragmaticism.

From the late 1860s to the 1890s, the concept of the interpretant underwent a long hibernation (Bergman 2003: 8). At least explicitly, the term disappeared from the writings of Peirce, worried as he was, over the years, with other issues. When the concept returned, it appeared in less abstract definitions than the original one. The interpretant appears then as a mental sign, which is produced under the influence of an excitement that the sign causes in the interpreter's mind, as can be seen in another very typical passage of this second period (1895): A sign is a thing which serves to convey knowledge to some other thing, which it is said to *stand for* or *represent*. This thing is called the object of the sign; the idea in the mind that the sign excites, which is a mental sign of the same objet, is called an *interpretant* of the sign (EP 2: 13).

It is true that, in a letter to Jourdain, in 1908 (L230a), Peirce confessed that he was forced to limit his definition of the sign because he despaired of making his abstract definition understandable. In a letter to Lady Welby, he went on to say that the simplification of his definition was a spoonful of soup he offered to Cerberus, the dog in Hell's Gate (SS: 81). For some commentators, this simplification actually damaged his notion much more than facilitated it, since it gave rise to reductionist versions, like the one, for example, which states that a sign represents something to someone.

Whatever the misunderstandings, the hypothesis with which I have worked is that only the interpretant notion that emerged after 1900, so in a third period of its generation, can help us out of the impasse, because these are definitions that put in their proper places the role of the interpreter, the psychological aspect and the collective aspect of the interpretant, its effective occurrences, and its tendency to infinity.

The Divisions of the Interpretants

Although the classification of the interpretants is still a controversial subject among commentators, I do not intend to put the controversy at issue. Rather, I will present below a version to which I came after thorough study of the scholars who seemed to me the most consistent in close comparison with Peirce's writings.

In 1903, Peirce reviewed Lady Welby's book, from which was born a personal relationship that would be of fundamental importance for the development of his theory of signs, largely conducted through correspondence between them. In his book, Welby distinguished three levels of meaning: sense, meaning, and significance. Immediately, Peirce found a correspondence not only with the three stages of thought in Hegel (see CP 8.174), but also with the three degrees of clarity in understanding the predicate of symbols that Peirce himself had established many years before, in 1878, in his essay "How to make our ideas clear" (CP 5.388–410; W 3: 257–76).

After this review, in 1904, Peirce began to develop his division of the interpretants into the immediate, the dynamical, and the final or normal interpretant. A discussion of this division can be found in Santaella (1995: 68–77), so I will limit myself here to present a very brief explanation. The sign has three interpretants, "its interpretant as represented or meant to be understood, its interpretant as it is produced, and its interpretant in itself" (CP 8.333).

The triad corresponds point by point to the three categories. The immediate interpretant is firstness; it is the potential to mean inscribed in the sign itself. "It is the range—always vaguely circumscribed—of the interpretant-generating power of the sign in a given time" (Ransdell 1983: 42). Therefore, this interpretant is inside the sign, it objectively belongs to the sign, and is independent of its encounter with any interpreter. It is only when this encounter happens that at least a portion of that potential will be put into action by the interpreter. The dynamical interpretant is secondness; the interpretant that is effectively produced in the mind of an interpreter (mind conceived of in a broad sense, as discussed in Santaella 1994). It is therefore the empirical, existential, and in the case of the human interpreter: it is a psychological fact. When the sign reaches any interpreter, an effect will be produced in that mind. This effect has always the nature of a sign or quasi sign which may find its translation into an external sign.

The final or normal interpretant, the interpretant in itself, corresponds to a final, ideal limit of interpretation, which is never actually attainable. It would be the ultimate realization of the interpretability of the sign, the ultimate realization of the potentiality to signify inscribed in the immediate interpretant. "It is the idea of the sign as it would come to be regularly and completely interpreted in an ideal long-run course of semiosis" (Ransdell 1983: 42). It is the empirical performance of the dynamical, singular interpretants that is responsible for the growth of the power of the sign to be interpreted. If it were possible to reach the ultimate limit of the sign interpretability, the final interpretant would be fully realized. As this is impossible because we are never in a position to say that such and such a dynamical interpretant is the final one, any dynamical interpretant is always *in medias res* of the final interpretant which is permanently in a state of becoming.

It is visible that this classification led Peirce to solve several problems that for some scholars still seem insoluble. There is no space here to discuss each of the problems as, for example, Peirce's statement that "it is not necessary that the Interpretant should actually exist. A being *in futuro* will suffice" for a triadic relation to exist (CP 2.92). Another example is Peirce's distinction between the interpretant and the interpreter, which has taken many commentators to the idea that the interpreter plays no role in Peirce's theory of the interpretant, plus a number of other vain discussions about false problems that the classification of interpretants can consistently lead us to solve.

However, the major controversy regarding the divisions of the interpretants is not due strictly to this first triad, but to the fact that, in 1907, in his famous MS 318, Peirce introduced a new trichotomy of interpretants also related to the three categories. They are the emotional, the energetic, and the logical interpretants. Peirce has not left any mention about the overlap of the two triads and this has puzzled many commentators.

After careful study of the subject I came to the conclusion that the trichotomy of the emotional, energetic, and logical interpretants is an internal subdivision of the dynamical interpretant, a subdivision that can be coherently extended to the immediate and to the final interpretants (see also Johansen 1985, 1993). The diagram below illustrates the idea more clearly:

INTERPRETANTS
(a) IMMEDIATE

(b) DYNAMICAL
 b.1. emotional
 b.2. energetic
 b.3. logical
(c) FINAL

or

INTERPRETANTS
(a) IMMEDIATE
 a.1. emotional
 a.2. energetic
 a.3. logical
(b) DYNAMICAL
 b.1. emotional
 b.2. energetic
 b.3. logical
(c) FINAL
 c.1. emotional
 c.2. energetic
 c.3. logical

As this is not the occasion to explore all the details of such divisions, I will only discuss the details concerning the dynamical interpretant and its subdivisions. If the dynamical interpretant is that which is experienced in every act of interpretation (SS: 111), if it is the effect actually produced in the mind of situated interpreters, then this effect can develop into three levels: emotional, energetic, and logical. These three types of interpretants concern therefore the effects produced by the sign, that is, the effect that the sign actually produces when meeting an interpreting mind. The first effect is purely emotional, namely, the feeling produced by the sign. In most cases, this is only an imperceptible sense of recognition or familiarity that gives way to a sense of effort, when some kind of physical or mental energy is involved. In most occasions, this is also so slight as not to be noticed, since it is the logical interpretant that guarantees that the sign is converted into another sign. This guarantee results from the fact that the logical interpretant acts as a rule of interpretation which is habitually actualized by the interpreter.

Sometimes, however, the hearing of a musical piece, for example, if we are sufficiently disarmed, if our mind is available, porous, then the prominent effect is just a feeling, a pure and positive inconsequential qualitative impression. So too, before something that scares us or challenges us, the energetic interpretant will be dominant, requiring active and direct response. The case of the logical interpretant, however, deserves more attention.

The development of these three types of interpretants took place in the context of the review at which Peirce was submitting his pragmatism. For him, "the problem of what the 'meaning' of an intellectual concept is"—a fundamental question to pragmaticism—"can only be solved by the study of the interpretants, or proper significate effects of signs" (CP 5.475). In this context, the concept of the logical interpretant is the most important.

The Logical Interpretant as Habit

Peirce identified the logical interpretant or mental fact with the meaning or the significant effect that is proper of an intellectual concept. In 1868, he had declared that the interpretant of a thought is another thought, and that this process is theoretically infinite. However, in 1907, in light of his second pragmatism, Peirce was looking for a logical interpretant that did not have the nature of a concept. If the logical interpretant is defined only as the intellectual apprehension of the meaning of the sign, the resulting logical interpretant will require a further logical interpretant, and so on ad infinitum. Without dismissing the existence of these logical interpretants that have the nature of signs, Peirce sought logical interpretants that would lead thought to the door of deliberate action.

According to the pragmatic maxim, the meaning of an intellectual concept is operational as long as concepts are the results of certain specifiable operations constituting this meaning. This means that "to predicate any such concept of a real or imaginary object is equivalent to declaring that a certain operation, corresponding to the concept, if performed upon that object, would (certainly, or probably, or possibly, according to the mode of predication), be followed by a result of a definite general description." (EP 1:411)

Such an analysis situates the meaning of concepts in a conditional future while maintaining that concepts have a general reference (related to effects of a general description). Now this is exactly the definition that Peirce gave of the logical interpretant, from which one is taken to identify the logical interpretant or mental fact to the meaning or significant effect that is proper of an intellectual concept. That it was some kind of mental fact Peirce had no doubt. However, what kind of mental fact could it be?

Peirce first examined conceptions, but abandoned them because although they are, in fact, logical interpretants, they cannot work as an explanation of their own nature; that they are concepts we already know. To play the role of a logical interpretant Peirce also analyzed and discarded desires and expectations, since they do not have a general application except insofar as they are tied to a concept. Desires were also rejected because they are effects of the energetic interpretant. By exclusion, Peirce arrived, then, to habit as a logical interpretant.

There are however, two distinct species of logical interpretants. As we have already seen, in his youth, Peirce postulated the constitution of semiosis as an infinite *continuum* of signs, so that there would neither be an original object nor a final interpretant. This means that the three members of the semiotic chain—sign, object, and interpretant—would have the nature of a sign. Therefore, in the interpretive process, a logical interpretant would require another interpretant of the same type, that is, another sign, and so on infinitely.

These logical interpretants of intellectual concepts would be something like quasi-habits. Similar to habits, they are conditional because they are associated with a conditional future (CP 5.483). Its conditionality also comes from the fact that they may or may not lead to action. However, they are like habits because they are general or "intimately connected to generals" (CP 5.482). They are not actions that are particular, but they are ways of acting that are generals, or rather, they are rules of action.

Later, however, Peirce replaced this vision for a more accurate notion. Besides the existence of logical interpretants that have the nature of a sign, he began to inquire about another type of logical interpretant. Quasi-habits lack something that habits should have: a future repeatability in interpretive transactions. While concepts can function as common logical interpretants, these concepts began to be accompanied by logical interpretants of another kind. Strictly speaking, there is nothing that can better fulfill the definition of a logical interpretant than habit. A rule or habit, Savan (1976: 43–44) says,

> is a pattern of actions which would, under certain appropriate conditions, be repeated indefinitely in the future. The Logical Interpretant is a Dynamic Interpretant in so far as the rule or habit is instantiated in a particular set of actions within a limited period of time. These particular sets of actions are energetic interpretants; but as exemplifying an indefinite repeatable habit, they also replicate logical interpretants. Note also that whereas emotional and energetic interpretants have a finite termination, the logical interpretant is always potentially repeatable without termination.

It is part of the logical interpretants or intellectual concepts to regulate and govern particular occurrences because they carry some implications concerning the general behavior of a conscious being, transmitting more than a feeling and more than an existential fact, that is, transmitting the "would be" and the "would do" of an habitual behavior. No set, however great, will ever fill the meaning of what "would be" (CP 5.467).

Habit is capable of such real continuity, not only because it can be exercised repeatedly but also because it regulates the events taking place under its governance. Concepts, "what would be", and laws of nature, they all share the nature of a habit. Habits can only be known to the extent that they regulate existing events, but they are irreducible to the latter. As existing phenomena, habits are discontinuous and transient, but in their continuity, habits guarantee that individual occurrences of them will be repeated in accordance with a certain regularity.

Therefore, habits precede action and not vice versa. It is due to this condition that the guiding principles of reasoning have also the nature of habits. This is also true of the logical interpretant, because without habit, there would be no rule of translation in the passage of the sign to its interpretant. Note, however, that although habit has the character of a law, this is a very *sui generis* kind of law. And here appears one more trait of Peirce's extreme originality.

> The law of habit exhibits a striking contrast to all physical laws in the character of its commands. A physical law is absolute. What it requires is an exact relation. Thus a physical force introduces into a motion a component motion to be combined with the rest by the parallelogram of forces; but the component motion must actually take place exactly as

> required by the law of force. On the other hand, no exact conformity is required by the mental law. Nay, exact conformity would be in downright conflict with the law, since it would instantly crystallize thought and prevent all further formation of habit. The law of mind only makes a given feeling *more likely* to arise. It thus resembles the "non-conservative" forces of physics, such as viscosity and the like, which are due to statistical uniformities in the chance encounters of trillions of molecules. (EP 1: 292)

Far from functioning as an inflexible force to which actions must conform, the law of habit is a guiding principle, a living force, a general guidance that leads our actions without imprisoning them in a fixed frame. That's why there is always a degree of flexibility in how actions are governed by habits. This is also why habits can be broken, with much more frequency and intensity in the human universe. That is the reason why, throughout the universe, there is nothing more plastic than the human mind, able to leave and acquire new habits. Also based on this idea of the plasticity of the human mind to acquire new habits, Peirce turned then to his attempt to characterize an ultimate logical interpretant. This also has the character of a habit, but of a very special kind:

> It can be proved that the only mental effect that can be so produced and that is not a sign but is of a general application is a ***habit-change***; meaning by a habit-change a modification of a person's tendencies toward action, resulting from previous experiences or from previous exertions of his will or acts, or from a *complexus* of both kinds of cause. (CP 5.476)

Nothing could be more apt to fulfill the function of a conditional future with a general reference than habit-change, a conditional future of a hypothetical nature. To account for the plasticity of the human mind to acquire new habits, which shows the changing nature of mind, Peirce came to the ultimate logical interpretant which tuned into the evolving nature both of the final interpretant and of pragmaticism. Furthermore, here is also a tip to link the theory of signs and pragmaticism to the logic of abduction and the process of inquiry.

Habit-Change and Evolutionary Pragmaticism

Peirce's second pragmatism was conceived of in the context of the normative sciences and in the conviction that the ideal end of thought can only be born in future experience. Therefore, the normative sciences—aesthetics, ethics, and logic —have the task of examining the conformity of things to their ends, examining what should be in a conditional future: the regulatory ideals that attract and guide feeling, conduct, and thinking, respectively. Peirce said:

> For if, as pragmatism teaches us, what we think is to be interpreted in terms of what we are prepared to do, then surely ***logic***, or the doctrine of what we ought to think, must be an application of the doctrine of what we deliberately choose to do, which is Ethics. But we cannot get any clue to the secret of Ethics—a most entrancing field of thought but soon broadcast with pitfalls—until we have first made up our formula for what it is that we are prepared to admire. (CP 5.35–36)

If logic deals with inferences and arguments that we are prepared to approve and if such approval implies self-control, then logic is a special case of ethical action: it is ethics that studies the ends that we are deliberately prepared to adopt. That would put ethics in an upward position on the normative sciences. However, Peirce came to the conclusion that ethics depends on a more basic science, aesthetics, whose task is to discern what is the ultimate goal to which our ethical commitment must turn. In the light of aesthetics, this commitment should be facing what is admirable in itself, without any ulterior reason and that, being admirable, attracts our sensitivity and captures our will. The ideal is aesthetical, the deliberate adoption of the ideal, and the commitment to achieve it are ethical. As the adoption of the ideal and the commitment to achieve it are deliberate, they give expression to our freedom at its highest degree. After facing many dilemmas, Peirce concluded that the admirable coincided with the pragmatic ideal. The highest degree of freedom of human beings is thus the aesthetic admirable that is embodied in the pragmatic ideal. But what comes to be this ideal?

The critical review of pragmatism had led Peirce to consider, first, that the pragmatic ideal should not satisfy the desires of any particular individual, but be faced to collective human purposes. To answer this demand and fulfill the requirement of being a completely satisfying goal, the ideal should be evolutionary and its full meaning should only be in the distant future, always sought, but always concretely postponed. It is an ideally conceivable future, but materially unattainable in its fullness, because it can only be approached asymptotically. Pragmaticism had discovered that in the process of evolution, that which exists more and more embodies certain classes of ideals that in the course of development prove to be reasonable. This ideal was characterized as the continued growth of the potential embodiment of the idea (Kent 1987: 158; Santaella 1999a).

While encompassing the three categories, the pragmatic ideal, which is also the aesthetic admirable, has to take into account the role of self-control on acquiring new habits as a method by which the pragmatic ideal can be achieved. Therefore, to collaborate in the growth of reasonableness, romantic feelings, and passionate voluntarisms are not enough. These are worthless without the necessary change in habits. And this, in turn, does not operate without self-criticism and self-control. Finding the essence of rationality in self-criticism which can only be born from hetero-criticism, reason is the only kind of quality that can be freely developed through human self-control.

As evolution takes its course, human intelligence plays an increasingly important role in the development of the pragmatic ideal, through its characteristic power of self-criticism and self-control. It is this power that underlies the ultimate interpretant as habit-change, since it depends on self-control, that kind of control which is exercised through the evaluation of the consequences related to habits of action. This evaluation in turn is dependent on ethics to the extent that it points to the ideal that we are deliberately prepared to adopt. This ideal, that aesthetics has the function of highlighting, is the ultimate pragmatic ideal. On the one hand, therefore, we are irresistibly attracted to what is admirable, that is, the growth of creative reason in the world. On the other hand, the power of reasonable self-criticism and self-control leads our changes of habit in order to allow ethical action to be exerted towards that ideal.

The indissoluble link of evolutionary pragmatism with the normative sciences becomes explicit. Habit-change is located in its heart, since without the change of habit, there could be no evolution. Once habit-change is characterized as the ultimate logical interpretant, the relations of evolutionary pragmatism with the theory of signs become also explicit.

Being the pragmatist ideal in constant becoming, habit-changing is what produces the constant shifting of the dynamical interpretants towards the final interpretant. This goal, as we have seen, is ideally thinkable, but concretely unattainable, since creative reason is in permanent metabolism and growth with which we may and should collaborate. When our sensibility is attracted to this ideal, our habits regenerate thanks to self-criticism and self-control, triggering our ethical commitment to make us participants, even humble, of an evolutionary process that aims to embody, always more and more, the ideals that prove to be reasonable.

Laws of Mind and Laws of Nature

Even more importantly, the dynamic process that establishes the habit of habit-change, is found in the connecting thread between the laws of the mind and the laws of nature (Santaella 1999b), as expressed in the following quotes:

> But if the laws of nature are results of evolution, this evolution must proceed according to some principle; and this principle will be itself of the nature of a law. But it must be such a law that it can evolve or develop itself. Not that if absolutely absent it would create itself perhaps, but such that it would strengthen itself, and looking back into the past we should be looking back [to] times when its strength was less than any given strength, so that at the limit of the infinitely distant part it should vanish altogether. Then the problem was to imagine any kind of a law or tendency which would thus have a tendency to strengthen itself. Evidently it must be a tendency towards generalization—a generalizing tendency. But any fundamental universal tendency ought to manifest itself in nature. Where shall we look for it? We could not expect to find it in such phenomenon as gravitation where evolution has so nearly approached its ultimate limit, that nothing even simulating irregularity can be found in it. But we must search for this generalizing tendency rather in such departments of nature where we find plasticity and evolution still at work. The most plastic of all things is the human mind, and next after that comes the organic world, the world of protoplasm. Now the generalizing tendency is the great law of mind, the law of association, the law of habit taking. We also find in all active protoplasm a tendency to take habits. Hence, I was led to the hypothesis that the laws of the universe have been formed under a universal tendency of all things toward generalization and habit-taking. (RLT: 241)

> At any rate, it is clear that nothing but a principle of habit, itself due to the growth by habit of an infinitesimal chance tendency toward habit-taking, is the only bridge that can span the chasm between the chance-medley of chaos and the cosmos of order and law. (CP 6.263)

For Peirce, the tendency of the universe to acquire new habits, a condition that has its exponent in the human mind, is what allows the continued growth of the potential of the idea. Herein lies the ultimate goal that is more consonant with pragmaticism. In its most radical vector, pragmaticism postulates that the mind (in the sense of representation) acts on matter in order to impose compliance with certain peculiar laws, called purposes.

> The way in which Mind acts upon matter is by imposing upon it conformity to certain peculiar laws called Purposes; and the manner of the reaction is that the Purposes themselves become modified and developed in being thus carried out. Logical analysis shows that it is essential to the nature of representation that it should so develop itself by imposing purposes upon matter. (MS 478: 18, second draft)

Let us consider how a law should operate in human conduct (Kent 1987: 197). The only reasonable way it could do that would be when attention to this law would create reasonable expectations for future occurrences. Everything one would expect is that expectations are not frustrated, but if they are, the resulting surprise should be a first step towards a new idea. It is, moreover, in such cases that the ultimate logical interpretant effectively comes into action having as its guide and goal the growth of concrete reasonableness. The ultimate logical interpretant, therefore, would be one that would allow the integration of aesthetics, ethics, and semiotics. This integration launches pragmaticism—Peirce's evolutionary pragmaticism in perfect harmony with synechism.

References

Cosculluela, Victor. 1992. Peirce on tychism and determinism. *Transactions of the Charles S. Peirce Society* 28(4): 741–755.

Curley, Thomas V. 1969. The relation of the normative sciences to Peirce's theory of inquiry. *Transactions of the Charles S. Peirce Society* 5(2): 91–106.

Eco, Umberto. 1976. *La struttura assente*. Milano: Bompiani.

Eco, Umberto. 1990. *The limits of interpretation*. Bloomington: Indiana University Press.

Esposito, Joseph. 1973. Synechism, socialism, and cybernetics. *Transactions of the Charles S. Peirce Society* 9(2): 64–78.

Hookway, Christopher. 1997. Design and chance: The evolution of Peirce's evolutionary cosmology. *Transactions of the Charles S. Peirce Society* 33(1): 1–34.

Johansen, Jørgen Dines. 1985. Prolegomena to a semiotic theory of text interpretation. *Semiotica* 57(3/4): 225–288.

Johansen, Jørgen Dines. 1993. *Dialogic semiosis. An essay on signs and meaning*. Bloomington: Indiana University Press.

Kent, Beverly. 1987. *Logic and the classification of the sciences*. Kingston and Montreal: McGill Queen's University Press.

Myrvold, Wayne. 1995. Peirce on Cantor's paradox and the continuum. *Transactions of the Charles S. Peirce Society* 31(3): 508–541.

Noble, Brian. 1989. Peirce's definition of continuity and the concept of possibility. *Transactions of the Charles S. Peirce Society* 25(2): 149–174.

Parker, Kelly. 1994. Peirce's semeiotic and ontology. *Transactions of the Charles S. Peirce Society* 30(1): 52–75.

Peirce, Charles Sanders. i. 1867–1913. *Collected papers of charles sanders Peirce*. Vols. 1–6, eds. Charles Hartshorne and Paul Weiss. Cambridge: Harvard University Press, 1931–1935. Vols. 7–8, ed. Arthur W. Burks. Cambridge: Harvard University Press, 1958. [References to Peirce's papers will be designated by CP, followed by volume, period, paragraph number.].

Peirce, Charles Sanders. i.1867–1913. *Writings of Charles S. Peirce: A chronological edition*. Vols. 1–6 to date, ed. the Peirce Edition Project. Bloomington: Indiana University Press. [References to these volumes will be designated by W, followed by volume number, colon, page number.].

Peirce, Charles Sanders. i. 1867–1893. *The essential Peirce: Selected philosophical writing*. Vol 1 (1867–1893), eds. Nathan Houser and Christian Kloesel. Bloomington: Indiana University Press, 1992. [References to this volume will be designated by EP 1, followed by colon, page number.].

Peirce, Charles Sanders. i. 1898. *Reasoning and the logic of things: The Cambridge conferences lectures of 1898*, ed. Kenneth Laine Ketner. Cambridge: Harvard University Press, 1992. [References to this volume will be designated by RLT, followed by lecture number, colon, page number.] Introduction, and comments, by Kenneth Laine Ketner, and Hilary Putnam: 1992: 1–102.S.

Potter, Vincent. 1997 [1967]. *Charles S. Peirce on norms and ideals*. New York: Fordham University Press (1st ed. Amherst: The University of Massachusetts Press.).

Potter, Vincent, and Paul Shields. 1977. Peirce's definitions of continuity. *Transactions of the Charles S. Peirce Society* 13(1): 20–34.

Ransdell, Joseph. 1983. Peircean semiotics. Unpublished manuscript.

Reynolds, Andrew. 1996. Peirce's cosmology and the laws of thermodynamics. *Transactions of the Charles S. Peirce Society* 32(3): 403–423.

Santaella, Lucia. 1994. Peirce's broad concept of mind. *S: European Journal for Semiotic Studies* 6(3–4): 399–411.

Santaella, Lucia. 1995. *Teoria geral dos signos*. São Paulo: Cultrix, 5th ed. 2010, São Paulo: Cengage Learning.

Santaella, Lucia. 1999a. *Estética. De Platão a Peirce*, 2nd ed. São Paulo: Experimento.

Santaella, Lucia. 1999b. A new causality for the understanding of the living. *Semiotica* 127(1/4): 497–519.

Santaella, Lucia. 2001. Esthetics, the supreme ideal of human life. *Semiotica* 135(1/4): 175–189.

Savan, David. 1976. *An Introduction to C.S. Peirce completed system of semiotics*. Toronto: Toronto Semiotic Circle Monograph 1. Toronto: Toronto University Press.

Turrisi, Patricia Ann. 1997. Introduction and commentary. In *Pragmatism as a principle and method of right thinking: The 1903 harvard lectures on pragmatism*, ed. P.A. Turrisi, 1–36. Albany: State University of New York Press.

Wells, Kelley. 1996. An evaluation of Hartshorne's critique of Peirce's synechism. *Transactions of the Charles S. Peirce Society* 32(2): 216–246.

Chapter 11
Beyond Explication: Meaning and Habit-Change in Peirce's Pragmatism

Mats Bergman

Abstract In the seminal essay "Pragmatism", Peirce discusses the end of interpretation in terms of the ultimate logical interpretant, which is varyingly characterized as habit or habit-change. While it is broadly accepted that his conception of pragmatic meaning rests on habit, the precise role of habit-change in his account of conceptual purport has not been examined in detail. In this chapter, I address this issue, which turns out to be closely linked to the pivotal question of the purpose of Peircean pragmatism itself. My primary aim is to demonstrate that Peirce's pragmatic account of the interpretant surpasses that of mere explication of habitual meaning, something that can be teased out from an embryonic account of three logical interpretants, sketched in "Pragmatism" and supported by certain suggestive references to first, second, and third pragmatistic interpretation in other writings. This investigation not only exposes the hitherto overlooked fact that Peirce recognizes a stage of conceptual clarification beyond that of the ultimate logical interpretant; it also paves the way for a reassessment of the significance of the pragmatist approach within a broader developmental-normative framework aimed at the improvements of our habits.

Keywords Pragmatism · Meaning · Habit-change · Interpretation · Ultimate interpretant · Conceptual clarification · Explication · Elucidation

Introduction

In the seminal but sprawling essay, "Pragmatism" (MS 318, 1907), C. S. Peirce depicts the goal of cognitive sign action in terms of the formation of an "ultimate logical interpretant", which he associates with the key concept of habit. This has been hailed as a genuine advance in Peirce's philosophy, a manoeuvre by which he finally succeeds in overcoming the problems of arbitrariness and endless deferral of

M. Bergman (✉)
University of Helsinki, Helsinki, Finland
e-mail: mats.bergman@helsinki.fi

D.E. West and M. Anderson (eds.), *Consensus on Peirce's Concept of Habit*, Studies in Applied Philosophy, Epistemology and Rational Ethics 31,
DOI 10.1007/978-3-319-45920-2_11

meaning that allegedly plague his earlier semiotic endeavours (cf. Gentry 1946; Short 2004, 2007a). However, it is not the overt recognition of the habit-meaning bond that constitutes the pivotal breakthrough; the connection between pragmatic import and habit dates back all the way to Peirce's first probes toward pragmatism, and plays a prominent role in "How to Make Our Ideas Clear" (1878). Rather, it is the identification of the third grade of meaning with a relatively concrete end of interpretation, conceptualised as the ultimate habit-interpretant, that completes the purported "revolution" in his later theory of signs and pragmatism (Short 2004: 228).

Yet, the link between meaning, habit of action, and the future-oriented conception of the interpretant is far from free of mist, even in Peirce's mature notions of *semiotic* and *pragmaticism.*[1] And in some ways, the viewpoint outlined in "Pragmatism" just seems to muddle matters more. There, the end of interpretation is not only portrayed in terms of habit-formation in or by the interpreting agent, but the ultimate interpretant—the "naked meaning" of the intellectual sign—is varyingly identified as *habit or habit-change*, with the latter quite plainly characterized as "a modification of a person's tendencies toward action" (CP 5.476 [1907]).

While many commentators have simply accepted that the ultimate interpretant can be broadly described as "habit or habit-change", others have argued that this is actually a blunder on Peirce's part. With characteristic sharpness, Short (1996) has pinpointed this possible error as a confusion between "the product and the event of its production" (1996: 500)—that is, a mix-up between an eventual habit and a preceding revision of extant habits. This certainly sounds plausible; habit-change would then be construed as the process that produces a firm law-like habit (the ultimate interpretant) in an interpreting agent—or, perhaps more accurately, in a "scientific intelligence", understood as "an intelligence that needs to learn and can learn (provided that there be anything for it to learn) from experience" (MS 787s: 6–7 [c. 1895–1896]; cf. CP 2.227 [c. 1897]). However, there is a second possibility, namely that Peirce's admittedly vague references to habit-change point *beyond* the methodical clarification of meaning as habit. While Short does not explicitly entertain this hypothesis, he almost suggests it when he notes that the "ultimate logical interpretant is the last interpretant, the terminus of interpretation, only when interpretation is explication. But a completed explication does not preclude other

[1]In 1905, Peirce designates his own position as "pragmaticism" in order to distinguish it from other forms of pragmatism. It is a "special and limited form of pragmatism, in which the pragmatism is restricted to the determining of the meaning of concepts (particularly of philosophic concepts)" (*Supplement to the Century Dictionary* 1909; Houser 2010: 112). Thus, this narrower type is still intended to be a part of the larger pragmatist family (CP 8.205–206 [c. 1905]; cf. Houser 2010). In several later writings (e.g., in MS 318), Peirce actually reverts to using the generic name "pragmatism" also for his own position. Although there are often good reasons to underline Peirce's narrower conception, I will mostly speak of his "pragmatism" in this article, as I feel that the themes to be discussed ultimately pertain to pragmatist thought in a broader sense that is not exclusively Peircean. When referring to Peirce's theory of signs, I will use the spelling "semiotic" rather than the more idiosyncratic version "semeiotic" (both variants occur frequently in his writings).

forms of interpretation" (Short 1996: 522). It is also worth noting that Short (2007b) later offers a far more positive assessment of the sign-theoretical role of habit-modification: "Pragmatism—in semeiotic terms, the doctrine that habit-changes are the ultimate form of intellectual interpretants—is part of Peirce's rhetorical theory" (Short 2007b: 665).

These considerations imply a fundamental issue for Peircean pragmatism: is it limited to making explicit something that lies dormant in symbolic signs, or does it also involve conceptual development in a more deliberate and creative sense? The ultimate ambition of this article is to show that Peirce's pragmatic account of the interpretant surpasses that of mere explication of habitual meaning. This can be teased out from an embryonic account of three logical interpretants, sketched in "Pragmatism" and supported by certain suggestive references to first, second, and third pragmatistic interpretation in other writings. The potential yield of this exercise is not limited to the unearthing of yet another tentative semiotic triad. Rather, the upshot of the argument will lead us toward a reassessment of the scope and goals of the pragmatist approach within a broader semiotic and normative frame.

Of Things and Effects

Peirce's original account of pragmatism is one of the most familiar parts of his philosophy, and probably the aspect that has been most vigorously analyzed, criticized, and debated. Some of the weaknesses of the first published exposition of the method are well-known, including a choice of words that can suggest narrow phenomenalism, verificationism, and even voluntarism—as well as a notorious slip into something akin to nominalism in one of the key illustrations of the method. This is the fanciful case of a diamond that materialises inside a cushion but burns up before it is actually perceived. The young Peirce controversially maintains that there is then no logical falsity in calling such a thing "soft"; the fact that we consider it to be "hard" is a matter of linguistic usage and arrangement, but not of the meaning of ideas (W 3: 267; W 3: 275 [1878]). The older Peirce is one of the harshest critics of this indifferentist standpoint, which could also be construed in terms of a denial of the objective reality of habits. Indeed, one explanation for Peirce's 1878 blunder is that he at that stage adheres to a deficient modal realism, in which the reality of possibilities and *would-bes* has not yet been fully recognised.

Much energy has been spent on excavating and dissecting the causes of Peirce's error, and the debate whether his early philosophy entails a more fundamental nominalistic strand has not been definitely settled yet. In his own mature assessment, his youthful pragmatism does fall into the nominalistic trap (cf. MS 288: 170 [1905]; CP 8.208 [c. 1905]; ILS 273 [1910]). Yet, in spite of Peirce's insistence that his later pragmaticism involves an "extreme" variant of scholastic realism—understood as an affirmation of the reality of some possibilities and other vagues as well as of some generals (cf. EP 2: 354 [1905]; cf. EP 2: 339 [1905])—it is important to see that his pragmatist scheme is meant to be non-committal with regard to the question of external metaphysical realism. Peirce is rather resolute on this point, stressing that "pragmatism is, in itself, no doctrine of metaphysics, no attempt to determine any truth of things. It is merely a method of ascertaining the meanings of hard words and of abstract concepts" (EP 2: 400 [1907]). Granted, Peirce holds that a realistic metaphysics can be derived from a logic in which pragmatism plays a rhetorical or "methodeutic" role; and perhaps a tad inconsistently, he intermittently also submits that pragmatism implies an entire system of philosophy (CP 8.191 [c. 1904]; NEM 3: 192 [1911]; cf. CP 5.64 [1903]; MS 319: 5 [1907]). But the key point here is that the primary realism associated with pragmaticism is of a distinctly conceptual kind. Put differently, it is a *habit-realism* that purportedly knocks "the pins from under every nominalistic philosophy" (MS 939: 22 [1905]), and which Peirce presumes to be one of the common denominators of the broader coalition of pragmatists (EP 2: 450 [1908]). Also, in spite of its "extremism", the realism of his scholastic credo is markedly tempered. Accordingly, Peirce emphasises that "the doctrine of scholastic realism neither is that all concepts are real (which would be the *ne plus ultra* of absurdity) nor that any concept is perfectly real; but that some concepts are real in some measure" (MS 842: 132 [c. 1905]). In view of these qualifications, perhaps Peirce should have called his position "moderate scholastic realism" after all.

Be that as it may, one of the key insights of Peircean pragmatism is that such a conceptual realism can be "cashed out" in terms of *conceivable* habits of action. Peirce maintains that intellectual concepts "essentially carry some implication concerning the general behaviour either of some conscious being or of some inanimate object, and so convey more, not merely than any feeling, but more, too, than any existential fact, namely, the '*would-acts*' of habitual behaviour" (EP 2: 401–402 [1907]). However, when it comes to the clarification of meaning that is supposedly pragmatism's proper *métier*, some complications seem to arise. In "How to Make Our Ideas Clear", Peirce on the one hand asserts that "what a thing means is simply what habits it involves", while he on the other hand contends that "the identity of a habit depends on how it might lead us to act, not merely under such circumstances as are likely to arise, but under such as might possibly occur, no matter how improbable they may be" (W 3: 265 [1878]). In the former sense, significant habit is comprehended as a state, capacity, or behavioural disposition of an object, whether this be a person, an organism, or an inanimate thing (cf. MS 673: 14–15 [c. 1911]; CP 8.380 [1913]). In the latter, habit is understood more narrowly as "a rule active in us" (W 3: 337 [1878]; cf. W 4: 249 [1881]; W 5: 162 [1885]; CP 2.170 [1902]). The presumption, it would seem, is that these two aspects of habit-meaning seamlessly cohere; but that is by no means self-evident. At any rate, one can reasonably ask whether the notion of habits *produced on us* is equivalent to that of habits *involved in things* under consideration.

It is important to keep in mind that this question pertains to ideas. As I noted above, pragmatism is primarily presented as a method of conceptual clarification that does not profess to make any substantial claims about the constitution of the external world. Thus, while a "thing" can be generically defined as "a cluster or habit of reactions" (CP 4.157 [c. 1897]), what the original pragmatic maxim actually prescribes is a scrutiny of the conceivable practical effects inherent in the objects of *our* conceptions (W 3: 266 [1878]). A later variant of this famous dictum exposes this demarcation of the clarifying habit:

> Consider what effects that might conceivably have practical bearings you conceive the object of your conception to have: then the general mental habit that consists in the production of these effects is the whole meaning of your concept. (MS 318: 22 [1907])

By emphasising that the habit implicated in the pragmatic method is mental in an inward sense, I do not mean to deny that the Peircean concept of habit is actually much more extensive than so, manifesting itself in the outer as well as in the inner world. Peirce's general idea of habit expands rapidly from a notion of nervous associations formed inductively in an organism (W 2: 232–233 [1868]; W 3: 337 [1878]) toward the suggestion that "the laws of physics [may] be habits gradually acquired by systems" (W 4: 553 [1883–1884]), culminating in an all-embracing primordial principle of habit-taking that lies behind all uniformities in the existential and logical universes (W 5: 293 [1886]; W 6: 208 [1887–1888]; W 6: 393 [1890]). At the same time, it is striking how openly *anthropomorphic* (in contrast to *anthropocentric*) Peirce's path to the hypothesis of universal habituation is (cf. RLT 241 [1898]). However, here we are dealing with a narrower context of pragmatic exposition, where "habit" is evidently to be regarded as "a state of a man in consequence of which he will on occasions of a certain description act in a certain general way" (HP 2: 912 [1901]; cf. CP 2.148 [1902]; MS 318: 34 [1907]; MS 852: 8–9 [1911]). More than that, the habit in question is quite emphatically internal to the intelligence engaged in interpretation—to the extent that the procedure prescribed by Peircean pragmatism could almost be described as introspective.

Given Peirce's well-known rejection of a special faculty of introspection (W 2: 213 [1868]), this reading may seem rather wrongheaded. Indeed, my eventual aim here is to show that Peirce does suggest ways to expand the application of his pragmatism beyond that of the strictly inner realm, which would also abrogate the strict restriction of the pragmatic method to the domain of internal ideas. Nor does this claimed inward orientation entail that habits would be merely individual possessions; in their capacities as *would-bes*, they are evidently general. Still, there is little doubt that the primary field of clarification of methodical pragmatism is the inner "theatre of consciousness" or imagination (cf. CP 8.191 [c. 1904]). True, this arena can be said to include conceptions that are relatively external (habits of acts of perception and reaction) and others that are relatively internal (habits of ideas of feelings and fancies); but in both cases we are dealing with *intellectual* signs. Peirce sometimes refers to a "double mode of association of ideas", in consequence of which "man […] makes words of two classes, words which denominate things, which things he identifies by the clustering of their reactions, and such words are proper names, and words which signify, or mean, qualities, which are composite photographs of ideas of feelings, and such words are verbs or portions of verbs, such as are adjectives, common nouns, etc." (CP 4.157 [c. 1897]). But from the perspective of pragmatic clarification, these are not necessarily hard and fast distinctions; with the exception of the purest indices, what is represented as a subject in one context can be treated as a predicate in another. Thus, given the proposition "diamonds are hard", elucidation might target "being diamond" as well as "being hard". The key point here is simply that the professed function of original pragmatism is to expound the general notions we hold—that is, such mental ideas that can plausibly be viewed as predicates—in terms of their potential consequences for conduct, whether these concepts correspond to anything externally real or not. It is not a matter of the truth of ideas, but of their *meaning*.

But even if this much is granted, one might still argue that linking the clarifying habit with both the things under consideration and the effects on the interpreting intelligence leaves too much latitude. That is, there does appear to be a germane difference that we do make between meaning as the habits of the conceived object —whether primarily regarded as external thing or internal concept—and meaning as the practical consequences that follow on the conditional acceptance of a conception of the object. Take Peirce's favourite example of "hardness". How is this to be explicated pragmatically? One way of approaching the matter is to consider what it entails for x to be hard. But does this exactly correspond to the conceivable effects on the interpreter's conduct? In reflecting upon our ideas, there is arguably still room to separate our conception of the object's potential behaviour in itself from our conceivable object-induced behaviour. Ascribing a disposition of hardness to a diamond implies a habit of the object, with which the habitual effects of this attribution on us need not correspond.

The easy way out here is to accept that a full pragmatistic explication needs to include both aspects of habit, something that indeed seems to be covered by Peirce's succinct characterization of the "kernel of pragmatism" as the thesis that "the whole meaning of an intellectual predicate is that certain kinds of events would happen, once in so often in the course of experience, under certain kinds of existential circumstances" (EP 2: 402 [1907]). An alternative formulation of the principle of pragmatism, given in a draft of "Pragmatism", renders the broader scope of clarification more explicit—in addition to underscoring the interpretational character of the procedure:

> Consider what effects that *might conceivably* have practical bearings, – especially in *modifying habits* or as *implying capacities*, – you conceive the object of your conception to have. Then your (interpretational) conception of those effects is the whole (meaning of) your conception of the object. (MS 322: 11–2 [1907]; *last two emphases added by author*)

These two facets of pragmatic elucidation are not always this clearly discernible; typically, Peirce's characterizations of the method of pragmatism stress one of the aspects, and sometimes in terms that may seem to exclude the other.[2] Overall, descriptions that place more stress on the effects on the interpreter tend to dominate.

---

[2]See the entries for "pragmatism", "pragmaticism", and "maxim of pragmatism" in *The Commens Dictionary* (http://www.commens.org/dictionary). For a more systematic attempt to articulate this dual aspect as objective disposition and agentive resolution, see the reconstructions of the pragmatic maxim in Stango (2015).

Accordingly, the gist of Peirce's original pragmatism is that "there is no distinction of meaning so fine as to consist in anything but a possible difference of practice" (W 3: 265 [1878]), a point of view elaborated in the 1903 lectures on pragmatism, where he avers that "a conception can have no logical effect or import differing from that of a second conception except so far as, taken in connection with other conceptions and intentions, it might conceivably modify our practical conduct differently from that second conception" (EP 2: 234). From this, it is a relatively short step to the far-reaching conclusion that "a *conception*, that is, the rational purport of a word or other expression, lies exclusively in its conceivable bearing upon the conduct of life" (EP 2: 332 [1905]).

Given that Peirce, in overt opposition to some other forms of pragmatism, declares that philosophy should be a strict science—"abstruse, arid, and abstract" (CP 5.537 [c. 1905])—his almost humanistic-sounding allusions to our intentions, practices, and life may feel slightly incongruous. But what he actually wishes to exclude from the purview of pragmaticism is simple quality of feeling and blind action-reaction. True, "the ultimate meaning" of a sign can consist "either in an idea predominantly of feeling or in one predominantly of acting and being acted on" (CP 5.7 [c. 1907]). In "Pragmatism" (MS 318), such upshots are characterised as *emotional* and *energetic meanings* or *interpretants*; but in a stricter sense, these are not meaningful ideas but significant effects of different kinds. In any case, pragmatic clarification is meant to be restricted to "ascertaining the meanings, not of all ideas, but only of [...] 'intellectual concepts', that is to say, of those upon the structure of which, arguments concerning objective fact may hinge" (EP 2: 401 [1907]). Hence, a "complete definition" of a concept would amount to an accurate exposition of "all the conceivable experimental phenomena which the affirmation or denial of a concept could imply" (EP 2: 332 [1905]). But equally importantly, the upshot is that the pragmatic principle "does not intend to define the phenomenal equivalents of words and general ideas, but, on the contrary, eliminates their sential element, and endeavors to define the rational purport, and this it finds in the purposive bearing of the word or proposition in question" (EP 2: 341 [1905]).

It is not accidental that such references to experimental outcomes and symbolic purposes begin to crop up more frequently in 1903 and thereafter. Two relevant developments in Peirce's philosophy occur during this period: his logic (including pragmatism as a part of the rhetorical/methodeutic branch) is recast as a full-fledged normative discipline in need of ethics and esthetics; and his long-quiescent theory of the interpretant begins to show signs of awakening. Among other things, the former manifests itself as a heightened emphasis on deliberation, self-control, and self-criticism in Peirce's pragmatism, while the latter involves a new focus on the interpretant as the primary locus of intellectual purport and growth—arguably linked to the highlighting of the significance of habit-modification. Whether these advances also imply an extension of the scope of Peircean pragmatism beyond that of narrow explication—and perhaps even of internal clarification in a broader pragmatic sense—is the question I will ponder in the rest of this essay.

Purposes of Clarification

If we take Peirce's later pragmaticistic writings as our primary guide, then it is evident that his mature account of pragmatic meaning principally boils down to the question of what form an intellectual sign, such as a concept or a proposition, should be converted into (EP 2: 340 [1905]; MS 298: 11*bis* [1906])—a viewpoint that is anticipated in *How to Reason* (aka *The Grand Logic*), where Peirce asserts that "the meaning of a sign is the sign it has to be translated into" (CP 4.132 [c. 1894]). The normative impetus of this approach already goes beyond that of mere explicatory analysis of the "A is A" type, which provides second-grade clearness by exposing the abstract constituents of the definiendum (PM 53 [c. 1895]). Accordingly, while a definitory explication of the signification that is wrapped up in a concept can bring out what Peirce labels "analytic distinctness" (MS 649: 1–2 [1910]), it is still only a description of the substance of the symbol as it is or as it has been. In contrast, elucidating "pragmatic adequacy" is meant to lay out what the meaning of a concept *ought to be* "in order that its true usefulness may be fulfilled" (MS 649: 2). This goes hand in hand with a growing emphasis on the purpose of symbols and their interpretation. Thus, Peirce asserts that "of the two implications of pragmatism that concepts are purposive, and that their meaning lies in their conceivable practical bearings, the former is the more fundamental" (CP 8.322 [1906]). This is not to suggest that he would be discarding the more methodical aspect of pragmatic explication; rather, the point is that the pursuit is now sustained by the broader telic-normative thesis that "any kind of goodness consists in the adaptation of its subject to its *end*" (EP 2: 211 [1903]).

But what is the ultimate objective of pragmatism itself? To bring clarity to our ideas, of course—but what does that entail in practice? In his 1903 Harvard lectures, Peirce distinguishes two central functions of pragmatism. In the first case, it should "give us an expeditious riddance of all ideas essentially unclear" (EP 2: 239). This refers to the *eliminative* application of the pragmatic principle, the goal of which is to bring precision to thought and discourse by streamlining our vocabularies. In a relatively innocuous form, its basic rule is laid out already in Peirce's 1871 Berkeley review: "Do things fulfil the same function practically? Then let them be signified by the same word. Do they not? Then let them be distinguished." (W 2: 483)

Generally, this milder variant of eliminative pragmatism aims at putting an end to futile philosophical disputes that are caused by words being used varyingly, inconsistently, or without definite meaning (CP 5.6 [c. 1907]). A typical application would be to seek resolutions to certain ontological questions, such as those hinging on the concept of "reality", by the use of pragmatic clarification (EP 2: 420 [1907]; cf. W 3: 271–275 [1878]). According to Peirce, the fact that such "problems" are solvable by logical means shows that they are really "epistemological" (a term that he abhors, but nonetheless uses from time to time), and not metaphysical issues in the sense of dealing with broad "positive truths of the psycho-physical universe" (EP 2: 420 [1907]).

In its harsher, *prope*-positivistic shape, the goal of eliminative pragmatism is to sweep away all "metaphysical rubbish" (CP 8.191 [c. 1904]; EP 2: 338 [1905]). In its most severe form, it entails a reduction of all substantial questions to matters of concretely sensible effects. In "How to Make Our Ideas Clear", Peirce controversially follows this rationale in an analysis of the doctrine of transubstantiation, and ends up treating the religious creed as a question of sensations merely. His conclusion is that "to talk of something as having all the sensible characters of wine, yet being in reality blood, is senseless jargon" (W 3: 266 [1878])—a rather harsh verdict, especially as he is discussing practised faith and not academic metaphysics. Here, Peirce—rather out of character—sounds like a rationalistic reformer of non-scientific language (cf. SS 20 [1904]). Moreover, his denial of "first impressions of sense" and his recognition of an interpretational element in all experience and perception render the supposition of such absolute sensationalist yardsticks of meaningfulness quite dubious (cf. CP 6.492 [c. 1896]; CP 7.538 [c. 1899]; CP 2.142–143 [1902]; NEM 4: 24 [1902]; CP 7.376 [1902]; EP 2: 223–224 [1903]).

Besides, while there are of course numerous cases of verbal dispute in which apparently simple perceptual judgments can legitimately function as benchmarks, they can hardly be said to exhaust the full range of potential consequences of a conception for habitual action. In the case of transubstantiation, Peirce focuses singularly on the capacities implied by the physical object, apparently without any consideration of the broader implications for conduct of different conceptions of that object. In any case, a solitary application of the pragmatic maxim is rarely, if ever, a sufficient reason to condemn a concept to the garbage pile.

The second task of pragmatism identified in the 1903 lectures is that of supporting and further elucidating ideas that are "essentially clear, but more or less difficult of apprehension", to which Peirce adds that this should involve "a satisfactory attitude toward the element of thirdness" (EP 2: 239). Here, we are not invited to eliminate empty concepts or to definitely settle specific disputes, but first and foremost to cultivate the symbols in question—a process that involves explication by definition, but ultimately must encompass a more consequential pragmatic clarification that fosters our concepts and other symbols in view of our goals and purposes. Accordingly, this can be dubbed the *developmental* function of pragmatism.

Although Peirce never abandons the eliminative use of pragmatism, many of his later writings display a growing appreciation for its developmental functions (cf. EP 2: 239 [1903]). This is also reflected in his view of semiotic significance. In his early and mid-period theory of signs, meaning is normally treated as something distinct from object and interpretant, and is typically identified with a grounding "conception", "form", or "idea" underlying or conveyed by the sign relation (cf. W 2: 238 [1868]; W 2: 439 [1870]; NEM 4: 309 [c. 1895]). In fact, as late as 1904, Peirce still identifies meaning with a *ground* or *representative quality* (MS 7; MS 8; see also LI 391 [1908]). And in "New Elements" (c. 1904), Peirce alludes to a confusion between "the reference of a sign to its *meaning,* the character which it attributes to its object, and its appeal to an interpretant"—to which he appends that "it is the former of these which is the more essential" (EP 2: 305 [c. 1904]). But generally, his late-period discussions of significance tend to focus on the process of interpretation, leading to the account of ultimate interpretant-meaning in "Pragmatism" (MS 318). This implies a more dynamic perspective, in which meaning is something *valued* or *desired* as an outcome (cf. MS 599: 26 [c. 1902]). It is sought-after purport, and therefore it is in a relatively future mode.

The novelty of this turn should not be exaggerated. Actually, in view of Peirce's oft-cited 1868 assertion that the "intellectual value" of a present thought "lies in what this thought may be connected with in representation by subsequent thoughts"—which renders meaning "altogether something virtual" (W 2: 227)—the claimed shift toward interpretation may not feel like much of a change at all. However, it is important to realize that by "virtual", Peirce, early and late, does not refer to an arbitrary determination postponed to the future; rather, the term implies "something, not an *X*, which has the efficiency (*virtus*) of an *X*" (CP 6.372 [1902]). In effect, this is meaning understood as a "possible habit determining how a general sign shall be applied", where the element of possibility refers to the fact that it "does not live in one mind rather than in another" (HP 2: 810 [1904]); in this respect, it is "what is in the mind, perhaps not even *habitualiter*, but only *virtualiter*, which constitutes the import" of any word (CP 5.504 [c. 1905]). Roughly, at least, this seems to accord with his intermittent specification of a habit as "nothing but a state of 'would-be' realized in any sort of subject that is itself real" (MS 671: 6–7 [c. 1911]; cf. MS 681: 22 [1913]). Viewed in this light, Peirce's claim that "meanings are inexhaustible" (CP 1.343 [1903]) does not necessarily entail that significance is endlessly deferred to future interpretations; rather, it should be understood in terms of an influence—a would-be or a possible habit—that "should never cease finally to live, as lending strength to a habit, law, or rule which is ready to produce action when occasion may arise" (MS 599: 32 [c. 1902]). As such, symbolic meaning can be said to be virtually present as a significant ground in the sign—something that in one sense precedes any semiotic incarnation (cf. W 1: 474 [1866]).

What *is* new in Peirce's later semiotic is a gradually emerging acknowledgement that meaning can be most fruitfully conceptualised in terms of interpretative outcomes and goals. Thus, Peirce not only asserts that "a purpose is precisely the interpretant of a symbol" (EP 2: 308 [c. 1904]); in the 1903 lectures on pragmatism, for the first time he introduces a "technical" definition of meaning as the *intended* interpretant of a symbol (EP 2: 218). However, it should be noted that this significant intention or purpose is portrayed as wholly internal to the sign in question. Furthermore, at this stage, Peirce accepts an essentially Kantian distinction between explicatory and ampliative judgments. Although Peirce finds Kant's perspective quite misguided because of its neglect of the logic of relations and a resulting failure to grasp that necessary reasoning is mathematical, he nonetheless uses the Kantian dictum "that necessary reasoning only explicates the meanings of the terms of the premisses to fix our ideas as to what we shall understand by the *meaning* of a term" (EP 2: 218–219 [1903]).[3] Combined, these assumptions lead to a demarcated approach to logical clarification, which singles out the sign-type *argument* as the principal meaning-bearer. Therefore, Peirce insists that "Meaning is attributed to representamens [i.e., signs] alone, and the only kind of representamen which has a definite professed purpose is an 'argument'. The professed purpose of an argument is to determine an acceptance of its conclusion" (EP 2: 218 [1903]; cf. EP 2: 308 [c. 1904]). From an analytical point of view, the other kinds of logical symbols are considered to be subservient to this objective.

> [If] by the meaning of a term, proposition, or argument, we understand the entire general intended interpretant, then the meaning of an argument is explicit. It is its conclusion; while the meaning of a proposition or term is all that proposition or term could contribute to the conclusion of a demonstrative argument. (EP 2: 220 [1903])

Putting aside the controversial suggestion that signs may thus involve an inherent reason or *telos* apart from any input of an interpreter, two things should be noted concerning this initial characterisation of meaning as interpretant. Firstly, as a conclusion, the intended interpretant is still a symbol; it is typically expressed as a proposition. Secondly, Peirce's approach entails that the meanings of terms are derivable from their contributions toward the purpose of a demonstrative argument (cf. Forster 2003). Yet, he quickly concedes that this analysis, while useful, "is by no means sufficient to cut off all nonsense or to enable us to judge of the maxim of pragmatism", for it does not provide us with "an account of the *ultimate* meaning of a term" (EP 2: 220 [1903]). Less restrictively, Peirce then argues that "in order to be of any cognitive service, it is plain that [general concepts] must enter into propositions" (EP 2: 224 [1903]). From this point of view, their function is that of a predicate, where the proper conceptual term is regarded as a *rhema* virtually containing a verb in itself (EP 2: 224 [1903]). Accordingly, a hypostatic abstraction such as "hardness" is understood as the rhematic symbol "—is hard" (other characteristic examples would be "—kills—" and "—gives—to—"). In this sense, the conceptual rhema can be approached as an incomplete proposition, judgment, or assertion (cf. CP 2.341 [c. 1895–1896]; CP 8.115 [c. 1900]).

This exposition of the formal aspects of a concept or the "chemical composition" of meaning (cf. MS 643: 3 [1909]) is a central facet of explication, bringing out actual and potential associations between ideas as well as crucial structural features of their predicative applicability. This also explains why Peirce allots the system of

---

[3]In "Reason's Rules" (c. 1902–1903), Peirce actually distinguishes *elucidations*, understood as "courses of thought calculated to awaken consciousness of beliefs that have always existed", from *arguments*, by which he means "courses of thought calculated to create beliefs" (MS 596: 22). This equates elucidation with explication as "Socratic Midwifery", a process leading to a more distinct apprehension of the conceptions we already entertain. Put differently, explicatory interpretation is a matter of establishing that terms or other symbols are equivalent "incarnations" of an original meaning (W 1: 465 [1866]). Thus, nothing new "can ever be learned by analyzing definitions"; their function is solely to provide intellectual economy by putting existing beliefs into order (W 3: 260 [1878]). Later in this article, I will outline a distinction between explication and elucidation as phases of second-stage clarification.

existential graphs such a pivotal role in many of his later attempts to expound and prove pragmaticism. While a graphical logic—no matter how iconic—can never bring out the full rhetorical and methodeutic significance implied by the pragmatic principle, explicit diagrammatisation can be construed as a vital "guide to Pragmaticism" (LI 354 [1906]), or as a first step toward a logical account of the most basic significant relations—that is, exposing the kind of rhemata we are dealing with by revealing their rudimentary properties and relational powers. In particular, the scheme of existential graphs is expected so to aid us in the explication of "the otherwise nebulous, ghostlike, dubious abstractions of metaphysics as to endue them with something of the distinctness of geometrical diagrams and with much of the convincingness of working models" (LI 353 [1906]).

However, another crucial point here is that Peirce, rejecting a traditional compositional analysis—that is, one in which "a Proposition is composed of Names, and [...] an Argument is composed of Propositions"—argues that the truly significant difference between term, proposition, and argument "does not so much consist in structure as in the services they are severally intended to perform" (CP 4.572 [1906]). With this in mind, the approach outlined above may be unnecessarily insular in its singular focus on logical arguments—an emphasis that is certainly understandable given Peirce's primary pursuit of critical logic, but one that can also lead to a neglect of the ways in which signs fulfil their purposes in other kinds of practices.

In my view, Peirce adopts an implausibly intellectualistic stance when he suggests that "in thought proper,—intellectual thought,—judgment only occurs as a part of an argument and 'concept' only as a part of judgments" (MS 298: 5 [1906]). Concepts obviously do "occur" and function meaningfully in communicative acts other than arguments or as parts thereof. Even if it be conceded that formal logic ought to be confined to ascertaining the intended interpretants as components of "thought proper"—something of which I am not wholly convinced, but will not try to dispute here—I simply see no reason to limit the application of pragmatic clarification to the more elevated forms of our semiotic activities. For that reason, I prefer Peirce's broader characterisation of pragmatism as a "method of reflexion which is guided by constantly holding in view its purpose and the purpose of the ideas it analyzes, whether these ends be of the nature and uses of action or of thought" (CP 5.13 n. 1 [c. 1902]; cf. MS 478: 5 [1903])—but perhaps adopting a more liberal attitude toward the purpose of the method than he would have allowed.

Be that as it may, Peirce's focus on argument—as well as the more plausible suggestion that terms and concepts are graspable as potential propositions or assertions—can also be viewed as part and parcel of a more distinctly normative perspective on interpretative practice, where "the essential function of a sign is to render inefficient relations efficient,—not to set them into action, but to establish a habit or general rule whereby they will act on occasion" (SS 31 [1904]). From this point of view, the ultimate aim is really the resulting habit rather than the explication of the internal purpose of the symbol as such—although these may ideally coincide. That is, the end of logical interpretation is the *generation* of a habit-interpretant, in the production of which the function of the sign can be said to be exhausted (MS 339: 287 [1906]).

> A *sign* must have an interpretation, or interpretant as I call it. This interpretant, this signification,[4] is simply a metempsychosis into another body; a translation into other language. This new version of the thought receives, in turn, an interpretation [—], and so on, until an interpretant appears which is no longer of the nature of a sign; and this I am to show to you by good evidence is, for one class of signs, a Quality, and for another, a Deed; but for intellectual concepts, is a conditional determination of the soul as to how it would conduct itself under conceivable circumstances. [...] That ultimate, definitive, and final (i.e. eventually to be reached), interpretant (final I mean, in the logical sense of attaining the purpose, is also final in the sense of bringing the series of translations [to a stop] for the obvious reason that it is not itself a sign) is to be regarded as the ultimate signification of the sign. It is the ripe fruit of thought. But this perfect fruit of thought can hardly itself be called thought, since it has no signification and does not belong to the faculty of cognition at all; but rather to the *character*. (MS 298: 11*bis*; LI 356–357 [1906])

The lesson that "Pragmatism" (MS 318) and other late writings succeed in hammering home is that the eventual aim of interpretation need not—or even cannot—take the shape of a subsequent sign. This constitutes a genuine reversal of Peirce's earlier view, according to which "a sign is not a sign unless it translates itself into another sign in which it is more fully developed" (CP 5.594 [1903]; cf. W 2: 224 [1868]; CP 8.191 [c. 1904]; CP 2.303 [1902]; but also EP 2: 388 [1906]). From this point of view, it is not feasible to maintain that symbolic meanings are in all respects inexhaustible, as ultimate meaning is something that in a pregnant sense marks the end of interpretation or sign translation. It is still debatable whether this search for the "naked or ultimate meaning" of intellectual concepts negates the previous contention that the semiotic "clothing" of meaning can never "be completely stripped off" (NEM 4: 310 [c. 1895]; cf. Short 2007a, p. 57); but it can hardly be denied that Peirce's mature account of the end of interpretation involves a crucial non-symbolic ingredient. Accordingly, he argues that "signs which should be merely parts of an endless viaduct for the transmission of idea-potentiality, without any conveyance of it into anything but symbols, namely, into action or habit of action, would not be signs at all, since they would not, little or much, fulfill the function of signs" (EP 2: 288 [1906]). Admittedly, it is not quite clear what this entails for intellectual concepts and propositions; but whether it is through their contribution to full-blown arguments or by other means, the ultimate upshot should be "embodiment in something else than symbols" (EP 2: 288 [1906]). At the very least, then, there must be a consummating result that is not itself of the nature of a conclusion as a proposition or other symbolic sign.

This more concrete aim of interpretation is further highlighted by another prominent aspect of the normativity of Peirce's later account of pragmatism, namely the contention that "conceivable practical consequences" should be understood in

[4]Here, Peirce employs "signification" indistinctly as a synonym for both "interpretant" and "meaning". In other contexts (e.g., HP 2: 810 [1904]), Peirce delimits "signification" to the image-aspect of meaning, or its "depth".

the sense of "consequences for deliberate, self-controlled conduct" (CP 8.191 [c. 1904]). Thus, in reply to the question of what kind of manifestation a propositional sign should be translated into, Peirce holds that it is, "according to the pragmaticist, that form in which the proposition becomes applicable to human conduct, not in these or those special circumstances, nor when one entertains this or that special design, but that form which is most directly applicable to self-control under every situation, and to every purpose" (EP 2: 340 [1905]). Another reason why meaning is to be regarded as substantially future-oriented is hereby revealed: it is because "the only controllable conduct is future conduct" (EP 2: 359 [1905]).

Here, we need to take note of a benign ambiguity in the notion of habit in the context of pragmatism. Generally, Peirce adopts a broad concept of habit as "any lasting state whether of a person or a thing, this state consisting in the fact that on any occasion of a certain kind that person or thing would, either certainly, or even only probably behave in a definite way" (MS 673: 14–15 [c. 1911]); in this respect, "a 'habit' is nothing but the reality of a *general fact* concerning the conduct of any subject", including the law-bound dispositions of inanimate objects (MS 671: 7 [c. 1911]). As general, logical meaning is of the nature of "a habit, in the sense in which a chemical body, or the weather, or anything else that can be said to have a 'behaviour,' or character of action, may happen to have more or less settled habits" (MS 321: 19d [1907]). Evidently, it is this generic notion of habit that is at stake when we expound the import of "hardness" in terms of how a hard object would act and react under different circumstances.

However, Peirce also recognises a relevant distinction between disposition and habit in the "proper" sense of an acquired law (CP 5.538 [c. 1902]; CP 2.292 [c. 1902]; CP 5.476 [1907]). While it is true that this division is not hard and fast in his metaphysics, where even natural law is a product of evolution, it is of great concern in the context of pragmatic clarification. Accordingly, Peirce specifies that the habits under consideration in "Pragmatism" are to be regarded as "voluntary habits, i.e. such as are subject in some measure to self-control" (MS 318: 48 n. [1907]). From this narrower viewpoint, habits are explicitly *not* dispositions; they are outcomes that are at least to some extent consciously governable.

> Habits differ from dispositions in having been acquired as consequences of the principle, virtually well-known even to those whose powers of reflexion are insufficient to its formulation, that multiple reiterated behaviour of the same kind, under similar combinations of percepts and fancies, produces a tendency, – the *habit*, – actually to behave in a similar way under similar circumstances in the future. Moreover, – *here is the point*, – every man exercises more or less control over himself by means of modifying his own habits; and the way in which he goes to work to bring this effect about in those cases in which circumstances will not permit him to practice reiterations of the desired kind of conduct in the outer world shows that he is virtually well-acquainted with the important principle that *reiterations in the inner world, – fancied reiterations, – if well-intensified by direct effort, produce habits*, just as do reiterations in the outer world; *and these habits will have power to influence actual behaviour in the outer world*; especially, if each reiteration be accompanied by a peculiar strong effort that is usually likened to issuing a command to one's future self. (EP 2: 413 [1907])

Thus, when Peirce speaks of the goal of interpretation as habit, what he has in mind is not merely the relatively passive grasping of the behavioral implications of an object and whatever behavioral dispositions it might bring about in the interpreter, but more fully "voluntary action that is self-controlled, i.e. controlled by adequate deliberation" (CP 8.322 [1906]; cf. EP 2: 348 [1905]). An intellectual concept is a sign on which controlled reasoning can turn; but perhaps more pointedly, its "Eventual Interpretant (and in their measures, the Initial and Middle interpretants also,) must be habits of self-controlled fact" (MS S46). Put differently, the "deliberately formed, self-analyzing habit,—self-analyzing because formed by the aid of analysis of the exercises that nourished it,—is the living definition, the veritable and final [i.e., ultimate—*MB*] logical interpretant" (EP 2: 418 [1907]).

But if this is accepted, then the mere explication of the virtual capacity of a sign cannot cover the entire aim of the process of pragmatism; rather, its animating goal is the controlled improvement of our symbolic habits. Even more boldly, it may be suggested that pragmatism, as a part of a broader normative course of self-criticism and self-control, is not restricted to the development of conceptual habits in a narrow sense; using a term suggested by the quote given above, its ethical objective is to contribute to the formation of character (cf. CP 4.611 [1908]).

This is admittedly quite a leap—one that seems to fly in the face of Peirce's oft-repeated contention that pragmatism is just "a method of reflexion having for its purpose to render ideas clear" (CP 5.13 n. 1 [c. 1902]). However, while what I am suggesting may extend pragmaticism beyond its modest role as "a mere rule of methodeutic" (MS 322: 12 [c. 1907]), I would argue that Peirce frequently does the same—often implicitly, by allowing broader ethical reflexions (that is, deliberations that reach beyond reasoning in a narrower sense) to seep into applications of the pragmatic principle, but sometimes also quite overtly. This is perhaps most unmistakably expressed in a late-period letter to J. H. Kehler, where Peirce discusses the acquisition of habits by imaginary practice, and adds that out "of such considerations, which turn, as if upon a pivot, about the idea that a thought is nothing but a habit connected with a sign, one can build up quite a little philosophy which is what I meant by 'pragmatism'" (NEM 3: 192 [1911]).

All of this points toward an understanding of the purpose of pragmatic interpretation in terms of deliberate formation of habit of action or habit-change, the latter professedly being "the only mental effect […] that is not a sign but is of a general application" (CP 5.476 [1907]). This perspective does not exclude the explication of meaning in terms of the capacities of objects or a narrower clarification of conceptual import in terms of its contribution to logical argument; but the stress on self-control turns the focus toward the question of how adjustment of habit can and should be achieved. This control is not mere negative self-discipline, "proportionate to the intensity of the passion that is held in check"; in its positive aspect, it can be construed as development of intellectual virtue (MS 499 s [1906]). Pragmatic clarification is not only a question of eliminative control of language and discourse; it can also have a more constructive function in habit-change and the growth of character.

Stages of Interpretation

As I noted at the outset, the designation of habit-modification as an end of interpretation is contentious, as it seems to involve a confusion of the process with the product. Is it not more plausible to identify the end at which pragmatic interpretation aims—the veritable meaning—as habit rather than as habit-change? In one respect yes; but what may have prodded Peirce to emphasise modification in this manner is a broader focus on interpretation as a potential improvement of habits rather than as mere explication of a habit. At any rate, it is evident that his most detailed treatment of the matter, which develops into an embryonic theory of three logical interpretants or interpretative stages,[5] is primarily oriented toward questions of how alterations of habit occur and how they can be deliberately controlled. At the beginning of this discussion, Peirce actually submits that the notion of habit-change is in a sense broader than that of habit as such: the former "excludes natural dispositions, as the term 'habit' does, when it is accurately used; but it includes beside associations, what may be called 'transsociations', or alterations of association, and even includes *dissociation*" (CP 5.476 [1907]). This suggests that one reason for underscoring habit-modification is that there are courses of interpretation the main outcome of which may be the deprecation or elimination of habits rather than habit-sustaining or habit-forming.

In his account of the stages of interpretation, Peirce at first remarks that there are two causes of habit-change that are strictly speaking non-cognitive, namely experiences forced upon the mind and purely muscular efforts. However, although such reactions and actions may indeed produce habit-modification, he insists that involuntary experiences are incapable of creating new mental associations and argues that "nothing like a concept can be acquired by muscular practice alone" (CP 5.478–479 [1907]). On the other hand, such causes do play vital roles in the initial formation of intellectual signs; every "concept, doubtless, first arises when upon a

[5]This aspect of Peirce's theory on interpretants, most fully developed in MS 318, has received hardly any attention at all. One reason for this neglect is the fact that the relevant text has been badly cut up in the *Collected Papers* (in 5.481, to be exact), in effect joining completely different portions of the manuscript—and without letting the reader know that Peirce's discussion of second and third logical interpretant has been omitted. It may also have been taken as just another variant of the three grades of clarity—familiarity, distinctness, and pragmatistic exposition—outlined in many of Peirce's writings (e.g., W 3: 258–266 [1878]; MS 835 [c. 1895], CP 3.457 [1897]; EP 2: 496–7 [1909]; MS 649: 1–2 [1910]). However, while the 1907 description of the logical interpretants and the better-known account of clearness are closely related, they are not precisely equivalent. The most substantial difference is that Peirce's identification of three logical interpretants primarily describes the process of deliberation involved in the adoption and modification of habits, rather than degrees of clarity of meaning *per se*. It should be noted that he explicitly rejects the notion that the grades of clearness would be stages in the sense that a higher grade supersedes a lower one (MS 649: 1–2). Still, when looking at the matter from the point of view of habit-change, it does make sense to speak of phases or stages of an unfolding course of elucidation, as long as it is understood as a cumulative process, where the earlier moments do not lose their relevance along the way.

strong, but more or less vague, sense of need is superinduced some involuntary experience of a suggestive nature" (CP 5.480 [1907]). Even so, for an idea to be formed, there needs to be at least some "accompanying inward efforts" (CP 5.479 [1907]). In this respect, all concepts, no matter how directly perceptual or immediately intuitive they may appear, are mental "habits formed by exercise of the imagination" (MS 318: 44 [1907]).

In human cognition, such "first concepts (first in the order of development, but emerging at all stages of mental life) take the form of conjectures, though they are by no means always recognized as such"; these "ideas are the *first logical interpretants* of the phenomena that suggest them, and which, as suggesting them, are signs, of which they are the (really conjectural) interpretants" (CP 5.480 [1907]). All new concepts enter the mind by means of such nascent judgments (CP 5.546 [c. 1908]). Thus, the initial logical interpretant could be roughly characterised as an abductive judgment, which is typically of a perceptual nature. In the developmental account delineated in "Pragmatism", Peirce highlights the habitual character of such a seemingly instant interpretation; even the first logical interpretant "is equivalent to, or is expressive of, such a habit that having a certain desire one might accomplish it if one could perform a certain act" (CP 5.480 [1907]). This habit is basically an unanalysed belief, which in spite of its apparent simplicity encompasses the rudiments of deliberate self-control. As such, it might also be conceptualised as immediate interpretation—or immediate object and immediate interpretant *in tandem*—understood as "so much of a Sign that would enable a person to say whether or not the Sign was applicable to anything concerning which that person had sufficient acquaintance" (SS 110 [1909]). Put differently, a first-degree logical interpretant enables an informative attribution of a significant character or ground to an object by a proposition (or, more generally, "dicisign"), something that Peirce designates the *first pragmatistic interpretation* in "The Bed-Rock Beneath Pragmaticism" (1908). This first interpretation is "pragmatistic" because the service that the symbol performs "consists in the fact that when its interpreter meets with an object to which the subject applies, he is informed by the proposition that it either has (if it be universal) or may have (if it be particular), the character signified by the predicate" (LI 392 [1908]). In their immediacy, such interpretations are acritical; but apart from rare (possibly inexistent) exceptions, they leave a latitude that can be reflectively explored.

In the next stage of interpretation, the first interpretant-habits rouse different voluntary actions in the inner world: we "imagine ourselves in various situations and animated by various motives; and we proceed to trace out the alternative lines of conduct which the conjectures would leave open to us" (CP 5.481 [1907]). Through comparisons emerging from this inward processing, possibilities of slightly modifying the conjectures are noted; and such observations constitute the first steps toward control and modification of conduct in view of conceivable future consequences. In this way, the initial conjectures come to be delineated more clearly, leading to definitions that Peirce designates the *lower second logical interpretants*. These enable us to observe relations between the revised conjectures, and take note of delimited but apparently stable characteristics of the modifications themselves; "and thus we are led to generalizations and to abstracting the forms of conjectures which (with much else) will constitute the *higher second logical interpretant*" (MS 318: 45 [1907]).

This procedure facilitates more complex abstractions, in which our processes of symbolisation become objects of scrutiny in themselves; it is therefore the basis of higher-order logical thought. But more broadly, what Peirce is outlining here is the course of internal experimentation. In a sense, it is only through some such procedure that we really become aware of our habits; we "can only know that we have formed a habit by some experiment, although it may be an involuntary experiment or may be an experiment in the imagination" (CN 3: 188 [1904]). When deliberately and methodically pursued, this can be construed as an application of the method of the successful sciences to mental signs (EP 2: 400–401 [1907]; MS 322: 7 [1907])—or a laboratory approach to conceptual analysis, advancing by means of tracing out conceivable consequences, forming abstractions, and even feigning hesitancy in a way not possible in the outer world (cf. Colapietro 1988: 73). It is in this arena that pragmatic clarification primarily takes place.

> The second logical interpretants constitute the ultimate normal and proper mental effect of the sign taken by itself (I do not mean removed from its context but considered apart from the effects of its context and circumstance of utterance). They must, therefore, be identified with that "meaning" which we have all along been seeking. In that capacity, they are habits of internal or imaginary action, abstracted from all reference to the individual mind in which they might happen to be implanted, and whose future actions they would guide. For it must never be forgotten that habits called internal, as having been produced by internal exercise[,] take effect in external actions, unless a particular inhibition has been laid upon such action. (MS 318: 46 [1907])

Two things bear emphasizing here. Firstly, in the passage above, Peirce clearly states that the ultimate logical interpretant is constituted by the second logical interpretants. Given that a numerically higher interpretant awaits, this may seem surprising; but the outlook is actually in line with his oft-repeated contention that that the method of pragmatism should be limited to the inward exposition of mental habits. Secondly, however, it should also be noted that Peirce stresses that the conceivable consequences of internal habits refer to external actions; in this regard, it is not a matter of introspective musing or even of reasoning merely. Experimentation in the imagination can lead to the kind of *habituation* that really would guide our behaviour, were the circumstances to arise (CN 3: 278 [1906]). In "Bed-Rock", such regulation of the interpreter's conduct is characterised as *second* or *statically pragmaticistic* interpretation (LI 392 [1908]). In a relevant respect, this already moves beyond explication understood as unpacking what is implicitly embodied in a symbol. We might therefore designate this mode of interpretation as *pragmatic elucidation* in contrast to the kind of *analytic explication* that is aimed at making conceptions distinct by means of abstract definition.

As demarcated here, "analytic explication" is nearly equivalent to that which Peirce designates "distinctness" or the second grade of clearness (cf. W 3: 258–61 [1878]; MS 649: 1–2 [1910]; CP 8.214 [c. 1910]). As a stage in pragmatic clarification, elucidation can be taken to involve explication, understood as a process leading to a more distinct apprehension of the conceptions we already entertain but not yet to a revision or formation of habits. That is, such explicatory scrutiny can produce a recognition of implicitly held conceptions and beliefs, which is "the half-way house on the road to Doubt; and doubt is the usual prerequisite to improvement of one's beliefs" (MS 596: 22 [c. 1902–1903]). This implies a vital role for explication as an initial stage in the breaking up of old conceptual habits; but more than that, it also suggests that one of the broader aims of this process as a whole is to seek out doubts—not counterfeit "paper doubts", for sure, but genuine misgivings produced by deliberate testing. This, in turn, calls into question the notion that pragmatism would be essentially passive, concerned with nothing but rendering extant concepts and beliefs clearer and more stable. Thus, although Peirce sometimes (for example in a letter to Josiah Royce dated 30 June 1913) submits that pragmatism has more to do with the security than the "uberty" (fruitfulness) of reasoning, his writings also suggest a more active conception of pragmatism as conceptual critique.

For its part, elucidation moves beyond exposing that which is implicitly given in our conceptions; in a sense, this phase of clarification can be construed as a response to doubts—or cracks in our conceptual habits—caused by deliberate explications or unplanned difficulties in application. In elucidation, conceptions are basically approached as potential schemas for action, or as rudimentary assertions; and as such, they inevitably include experiential considerations related to our current state of knowledge. The procedure involves the formation of a kind of diagram of the associations involved in the concept. However, when it comes to clarifying potential consequences, we must appeal not only to previous information, but also to conceivable future experience and signification—and place ourselves into the imagined action, so to speak. This procedure is schematic, but also in a pertinent sense experiential and normative, moving beyond what one would normally expect of conceptual analysis. Peirce characteristically describes this as abstractive observation, analogous to an ordinary mental performance taking place whenever a person reflects on the outcomes of different courses of conduct, and therefore "makes in his imagination a sort of skeleton diagram, or outline sketch, of himself, considers what modifications the hypothetical state of things would require to be made in that picture, and then examines it, that is, observes what he has imagined, to see whether the same ardent desire is there to be discerned" (CP 2.227 [c. 1897]). This can also be construed as a form of internal dialogue, where the reasoner debates the merits of possible upshots with him- or herself; it is in such a communicative process that a conception matures (CP 5.546 [c. 1908]). By making this dialogical setting explicit we bring out thought in operation, manifested as experiments on diagrammatic signs; and this (borrowing a line from Peirce's justification of his system of existential graphs) is of critical normative import because "nothing can be controlled that cannot be observed while it is in action" (MS 280: 32 [c. 1905]).

Internal experimentation, conceived along the lines sketched above, is creative in the sense of introducing and operating on a number of hypothetical images and associations not given at the outset of the process. In this, we may note a suggestive parallel to Peirce's discussion of *corollarial* and *theorematic* deduction. In the former "the premisses act as stimulus to a suggestion according to general logical associations"; while in the latter, "associations should be introduced of which the premisses afford not the slightest hint" (MS 318: 55 [1907]). Put differently, the "peculiarity of theorematic reasoning is that it considers something not implied at all in the conceptions so far gained, which neither the definition of the object of research nor anything yet known about could of themselves suggest, although they give room for it" (NEM 4: 49 [1902]). While this variety of deduction is not straightforwardly equivalent to the course of internal pragmatic elucidation, there is a relevant similarity between the two.[6] Perhaps most importantly, the theorematic procedure involves the imaginative introduction of "may-bes" into our diagrams (cf. EP 2: 502 [1909])—which, extended to the arena of habit-modification, corresponds to the use of imagination to conceive of possible circumstances and outcomes in view of aims and purposes. This is the procedure through which we can go beyond our current habits of action by means afforded by our cerebral habits. However, in distinction from the ideal freedom of pure mathematical or "schematoscopic" reasoning, the applied theorematic reasoning of pragmatic elucidation will always be doubly circumscribed. As Peirce puts it in an oft-cited maxim, the "elements of every concept enter into logical thought at the gate of perception and make their exit at the gate of purposive action; and whatever cannot show its passports at both those two gates is to be arrested as unauthorized by reason" (EP 2: 241 [1903]).

Thus framed, elucidation essentially becomes a case of ethical deliberation, in which general determinations or "virtual habits" are formed and tested by experimentation on imagined diagrams (cf. MS 620: 24 [1909]). This limited power of self-control "consists (to mention only the leading constituents) first, in comparing one's past deeds with standards, second, in rational deliberation concerning how one will act in the future, in itself a highly complicated operation, third, in the formation of a resolve, fourth, in the creation, on the basis of the resolve, of a strong determination, or modification of habit" (CP 8.320). Here, self-control—"the principal ingredient of efficiency, *alias* virtue" (MS 620: 27 [1909])—is conceived to be operative in logical as well as in moral criticism. In both kinds of process, the formation of habits under imaginary action is an essential ingredient; the principal difference is that the former allows much greater leeway for the imagination (EP 2: 347 [1905]). But both follow a broadly schematoscopic model, which commences by "forming an image of the conditions of the problem, associated with which are

[6]As far as I know, Peirce does not make this connection. It should also be noted that the way I employ the terms "explication" and "elucidation" here does not accord with all of Peirce's uses. In "A Neglected Argument for the Reality of God" (1908), he distinguishes explication as a primary part of deduction from deductive demonstration or argumentation, which is then divided into corollarial and theorematic reasoning; the first step explicates the hypothesis by logical analysis "to render it as perfectly distinct as possible" (EP 2: 441). This should be compared to his earlier qualified acceptance of "Kant's dictum that necessary reasoning is merely explicatory of the meaning of the terms of the premisses" (EP 2: 218 [1903]).

certain general permissions to modify the image, as well as certain general assumptions that certain things are impossible", after which "certain experiments are performed upon the image, and the assumed impossibilities involve their always resulting in the same general way" (CP 5.8 [c. 1907]).

However, there seems to be at least one significant discrepancy between Peirce's construal of internal ethical deliberation and his pursuits of conceptual clarification. Namely, whereas the former by default contextualizes the process of reflexion to practices and circumstances related to the self, the latter rarely involve overt considerations of what we (using a term Peirce borrows from Augustus De Morgan) might designate their specific "universes of discourse". This possible neglect is partly explained by Peirce's contention that the primary arena of application of pragmatic clarification is ontology. Thus, his elucidation of "reality" is presumably meant to be valid in any philosophical universe of discourse whatsoever; and the pursuit of the clarification in question is justified by nominalistic aberrations and other conceptual confusions.

But another favourite example—that of "hardness"—may reveal some shortcomings of this approach. To begin with, Peirce does not spell out what practical or theoretical challenges supposedly prompts us to pursue such an elucidation in the first place. And to make things worse, it is far from evident what is really gained by defining the ordinary concept of "being hard" in terms of conditional statements such as "would not be scratched by many other substances" (W 3: 266 [1878]; CP 1.615 [1903]). To put it bluntly: do we really have a clearer idea of "hardness" after this exercise in elucidation? Of course, one might retort that this is just an illustration of the workings of a method; but why would Peirce choose to pay so much attention to a concept of little consequence—and one that additionally lures him into nominalistic snares? Putting the case into context may provide a clue. What Peirce most likely has in mind here is a specific technical sense of hardness, namely the conception of *scratch-hardness* that has been used to establish scales in mineralogy since the 1820s. Such a delimitation to the universe of discourse of a particular discipline renders the rationale of Peirce's clarification more fathomable; it is modelled on the procedure of a "successful science". Still, it should be noted that the results are hardly generalizable so as to elucidate other uses of the concept, such as "a hard problem", "a hard man", or "a hard concept". In truth, Peirce does not provide any convincing reason to adopt scratch-hardness as our primary exemplar; and I suspect that such an archetypical conception, divorced from relevant universes of discourse, is neither desirable nor practically attainable in this case. And even if we stick to the narrower scientific context, the question of what added value an application of the pragmatic maxim brings to the proceedings still remains—especially as a formulation such as "many other substances" can hardly be described as an epitome of clearness. This simply highlights the question of what kinds of conceptual habits actually call for elucidation. More promising candidates for pragmatic clarification than "hardness" can be found in some contexts that Peirce does not really consider, for instance in disputes concerning genuinely contested concepts, such as "objectivity" or "democracy". Consider, for example, what difference it could conceivably make—to social inquiry and to social life—

whether we adopt and develop an Athenian, republican, or Deweyan-pluralist conception of democracy.

But whatever its proper context of application may be, any pragmatic elucidation of conceptual meaning will eventually come down to some conceivable habitual effects on us—or more profoundly, on our norm-governed social practices. Significantly, this involves ends-in-view, which are themselves of a habitual nature; although "esthetic valuation" need not be actually involved "in every intellectual purport", "it is a virtual factor of a duly rationalized purport" (CP 5.535 [c. 1905]). More fully, pragmatic elucidation will therefore involve questions of how to ameliorate our habits of action in view of ethical principles and esthetic ideals—including higher-level criticism of those principles and ideals themselves (cf. CP 1.574 [1906]). Where such a process of evaluation may come to rest arguably depends on the purposes of our practices and inquiries (cf. Fitzgerald 1966: 163); but it should be an *ultimate aim* in the sense of being an objective that can be actually adopted and consistently pursued (EP 2: 202 [1903]). However, for a "conditional idealist" such as Peirce, any partial actual purpose will not suffice (EP 2: 345 [1905]). According to him, "an ultimate end of action *deliberately* adopted, —that is to say, *reasonably* adopted,—must be a state of things that reasonably recommends itself in itself aside from any ulterior consideration" (EP 2: 201 [1903]). Put differently, the improvement of our concepts and habits should ultimately be pursued in view of the *summum bonum*, understood as a process of rational evolution or "the development of concrete reasonableness" (CP 5.3 [1902]; CP 1.590 [c. 1903]; EP 2: 343 [1905]).

While pragmatic elucidation is always conditioned by the contextual ground of preceding signification and information, a meaning cultivated in the above manner —whether our aim is set at the highest possible good or at more modest ends-in-view—differs distinctly from the kinds of embodied meanings from which we embark. This developmental pragmatism may seem to end up uncomfortably close to a Humpty-Dumpty view of language, where meaning would be arbitrarily assigned by individuals clarifying their concepts—something that would of course make ordinary communication practically impossible. However, the elucidation of pragmatic meaning is not only always socially grounded; it is also primarily a question of practice-oriented *conceptions* rather than of linguistic concepts per se. Thus, there is no inconsistency in accepting that elucidation may little affect the conventional meanings of a language, while still maintaining that the procedure can bring out or develop the meanings of the habitual conceptions involved in or implied by the terms and propositions in question. Here, Peirce's term *purport*, with its inherent evocation of purpose, can be employed in order to mark meaning in the explicitly pragmatic, developmental sense. As part of a wider normative practice, where the "continual amelioration of our own habits [...] is the only alternative to a continual deterioration of them" (MS 674: 1 [c. 1911]), purport can be said to imply much more than stable habit. From this angle, it is not all that surprising that Peirce links the ultimate logical interpretant to the broad notion of habit-change, with its dual comprehension of both process and result.

Still, Peirce contends that a clarification of conceptions by means of generating second logical interpretants does not bring the action of intelligent elucidation and habit-scrutiny to completion. This is professedly the field of the third logical interpretant.[7] According to Peirce, such interpretants are called into action when the activity is for some reason turned "from the theatre of internal to that of the external experience" (MS 318: 46 [1907]). Put differently, it entails that the experience which is consequent upon the production of the second (or, sometimes, directly upon that of the first) logical interpretant is sought and found by a deliberate, self-controlled, purposive, muscular effort (MS 318: 47 [1907]). This is external experimentation, which like internal experiments can lead to habit-modifications,[8] but likely with more definite confirmatory or disruptive force.

> From this experimentation result deliberate self-controlled habit-changes, which may be dissociative of the second interpretants or of parts of them, or associative or confirmatory, or may involve slight modifications of them (as, for example, by substituting one approximate value or valuation for another.) In the third logical interpretant, or interpretants, the work of the intellect comes to a demicadence, a provisional and partial consummation; so that it is of supreme logical importance. (MS 318: 46–47)

If the first logical interpretant is primarily abductive and the second broadly deductive, then the third can be characterised as substantially inductive. However, none of these phases of interpretation is reducible to a pure form of reasoning as such. Furthermore, these are not hard and fast divisions. In particular, the second and third stage must not be understood as preparatory ratiocination followed by energetic testing; according to Peirce, our notion of reasoning should not exclude functions where physical effort is a contributing factor (MS 634: 3 [1909]). Nor does all this entail that change per se would be the final goal of the process, but in contrast to Peirce's original account of the fixation of belief, his incomplete theory of logical interpretants suggests a more proactive and outward-oriented conception of pragmatistic interpretation and habit-modification. The proper significate outcomes of a particular course of conceptual inquiry does not necessarily need to be the confirmation of a previous habit or the formation of a new habit; it can also be dissociation. This is backed up by Peirce's stress on the importance of habits of changing habits (NEM 4: 142 [1897–8]); not only "the tendency to the formation of

[7]Unfortunately, only two short scraps of Peirce's treatment of the third logical interpretant have survived in MS 318. I do not know whether he developed the idea elsewhere; but he does suggest that his discussion would eventually lead to a differentiation of three subtypes of this interpretant. It is possible that he thought about this in analogy with the three kinds of induction (e.g., CP 2.756–759 [c. 1908]).

[8]In an earlier reading, I connected habit-change primarily with the third logical interpretant (cf. Bergman 2012). This was motivated by the worry (noted at the beginning of this article) that Peirce had confused product and process; and my tentative solution was to associate habit and habit-change with different logical interpretants. This proposal was based on a too-narrow understanding of Peirce's conception of habit-change, for such modifications are purportedly achieved on the level of second as well as third pragmatistic interpretation. Also, I did not consider that the notion of habit-change could be construed as broader than that of habit from a developmental point of view.

habits, but also the liability of habits to get broken up, [...] is certainly a most important characteristic of the intellectual man" (CN 3: 188 [1904]; cf. CP 6.86 [1898]).

As it has been depicted here, the second logical interpretant—even in its highest form as ultimate interpretant—is an affair of internal examination. Again, this should not be understood too narrowly, for imaginary experimentation can produce surprises and doubts just like outer experience (CP 5.524 [c. 1905]). It is also imperative to keep in mind that the ideas thus developed ought to be outwardly effective in the sense of potentially affecting conduct. But the interpretative stage delimited as the second logical interpretant does stop short of an actual engagement with the external world of existence. In addition to directly confronting this inevitable clash of expectation and hard experiential reality, the third logical interpretant moves beyond pragmatic elucidation and inner habit-change in two crucial respects. Firstly, when conceptions are brought out into an arena of public experimentation, they are liable to affect a broader social community of habit than internal self-control can. This is not to say that products of the latter could not be publicly consequential, such as through exemplary actions and publication in books; but such outward efforts are already in effect moving toward the third logical interpretant, virtually opening up the ideas to external testing. Of course, in actual practice, these stages are hardly ever cleanly separated, as external experimentation overlaps with internal elucidation, each informing the other. Nonetheless, there is a further aspect of meaning-development—which we might, following Peirce and evoking John Dewey, call *consummation*—that is only captured by the notion of the third logical interpretant.

Secondly, the aim of *third* or *evolutionarily pragmaticistic interpretation* can be characterised as bringing about "a reconciliation or interadjustment between reason and the facts of experience" (LI 392 [1908]). Mostly, for sure, this is a matter of the interpreter adapting his or her habits to those of the world (cf. Altshuler 1981). It might be described in terms of "the consequences of surprise at the unexpected and the counter expected, where an apparent rupture in Nature's habit produces a real rupture of our associations" (CN 3: 188 [1904]). More broadly, this is doubt produced by dumbfounding and recalcitrant external experiences—not only as the initial stimulus to inquiry, but also as something deliberately risked in the conduct of experimentation itself. The world may not accord with the practical implications of the conceptions we have painstakingly developed, in spite of them functioning ever so flawlessly in our imagination. However, the admittedly vague notion of interadjustment suggests something more than mere adjustment. It involves at least a degree of modification of the conditions of experience itself.

This almost sounds like one of those "seeds of death" that Peirce detects in such un-pragmaticistic notions as "the mutability of truth" (EP 2: 450 [1908]). To this possible worry one might counter that we are not dealing with truth here, but with meaning and habit. But this will not do, for with the third logical interpretant we have definitely moved out from the confines of the internal world. It must be admitted that the principle of pragmatism is being stretched beyond what Peirce usually allows, if it in some respects can be said to encompass changes produced in the outer world of experience. Yet, that is the direction toward which Peirce's developmental approach to interpretation seems to be heading. It is not just a matter of acknowledging that the informational meanings embodied in symbols grow (EP 2: 10 [c. 1894]).

While Peirce "grants that the continual increase of the embodiment of idea-potentiality is the *summum bonum*", he insists that the "growth of idea-potentiality" itself requires non-symbolic incarnation (EP 2: 388 [1906]). And although he underscores that symbols possess real powers over sign-users (CP 2.149 [1902]), they do not mature unaided. In human cognition, the relationship between mind and symbol is better viewed as one of mutual amelioration in which the deliberate development of internal *and* external habits plays a crucial, albeit not necessarily dominant, role (cf. W 2: 241 [1868]). This amounts to a qualified denial of the strongest version of the semiotic autonomy thesis, according to which symbols "grow alone" (Nöth 2014: 176; cf. Ransdell 1992).

Admittedly, these claims need to be tempered in view of the constraints Peirce places on human freedom (cf. CP 8.320; CP 4.611 [1908]; but see also CP 5.536 [c. 1905]; EP 2: 459–460 [c. 1911]). Also, Peirce stresses that real habits are not something that can be easily cast off; and he warns against a sweeping application of the pragmatist method of clarification to familiar everyday conceptions, which often are "far more trustworthy than the exacter concepts of science" (EP 2: 433 [1907]). Put differently, not all conceptual clarifications are of genuine interest, and some might even become detrimental—something that Peirce exemplifies by futile doubts brought on by attempts to set down a perfectly definite notion of "the order of nature" (CP 5.508 [c. 1905]). While "the theory that all vagueness is due to a defect of cogitation or cognition […] is a natural kind of nominalism" (MS 70: 6 [1905–1906]), vagueness can be real and indispensable in many communicative contexts, and is "no more to be done away with in the world of logic than friction in mechanics" (CP 5.512 [c. 1905]). To this list of caveats we might add Peirce's conservative view of linguistic reform: "the past cannot be reformed; and consequently, its memory and records subsisting still, no prevalent mode of expression can be annihilated. The most you can do is to introduce an *additional* way of expressing the same meaning." (SS 20 [1904])

This does not mean that habits of communication cannot be changed, but only that it is a gradual process, more likely to advance by powers outside of individual control than by calculated decisions (cf. HP 2: 811 [1904]). Perchance, Peirce is a bit too complacent in his "sentimental conservatism": conscious linguistic reforms are often attempted, and they do sometimes succeed, possibly altering meanings along the way. As a counterbalance to Peirce's linguistic conservatism, it should be noted that he suggests that the highest grade of clearness of them all is linked to "the development of concrete reasonableness" (CP 5.2 [1902]), and that evolution, in its higher stages, "takes place more and more largely through self-control" (EP 2: 344 [1905]). From this perspective, deliberate experimentation can be seen as a part of a developmental process in which generals become embodied in existents (EP 2: 343 [1905])—or, somewhat more concretely, as the incarnation and development "of general ideas in art-creations, in utilities, and above all in theoretical cognition"

(EP 2: 443 [1908]). In a significant respect, this involves both conceptual elucidation and consummation as well as the kind of higher-order ethical habit-change in which semiotic ideals are revised "under the influence of a course of self-criticisms and of heterocriticisms" (EP 2: 377 [1906]). Such a transformation, "by modifying the rules of self-control modifies action, and so experience too—both the man's own and that of others" (CP 5.402 n. 3 [1905]). In Peirce's view, this is the only way we can rationally affect the future, limited as our capacity to do so might be. By extension, this entails an enrichment of symbolic meaning by means of adjusting conditions of experience. But it all begins with the effort to cultivate, control, and change our habits.

References

Altshuler, Bruce. 1981. Peirce on progress and meaning. In *Proceedings of the C. S. Peirce bicentennial congress*, eds. K.L. Ketner et al., 63–69. Lubbock: Texas Tech Press.

Bergman, Mats. 2012. Improving our habits: Peirce and meliorism. In *The normative thought of Charles S. Peirce*, eds. C. de Waal and K.P. Skowronski, 125–148. New York: Fordham University Press.

Bergman, Mats, and Sami Paavola, eds. 2014. *The commens dictionary: Peirce's terms in his own words.* New Edition. http://www.commens.org/dictionary.

Colapietro, Vincent M. 1988. Dreams: Such stuff as meanings are made on. *VS* 49: 65–79.

Fitzgerald, John J. 1966. *Peirce's theory of signs as foundation for pragmatism.* The Hague: Mouton.

Forster, Paul. 2003. The logic of pragmatism: A neglected argument for Peirce's pragmatic maxim. *Transactions of the Charles S. Peirce Society* 39(4): 525–554.

Gentry, G. 1946. Peirce's early and later theory of cognition and meaning: Some critical comments. *The Philosophical Review* 55(6): 634–650.

Houser, Nathan. 2010. The church of pragmatism. *Semiotica* 178(1/4): 105–114.

Nöth, Winfried. 2014. The growth of signs. *Sign Systems Studies* 42(2/3): 172–192.

Peirce, Charles Sanders. i. 1867–1913. *Writings of Charles S. Peirce: A chronological edition.* Vols. 1–6 to date, ed. the Peirce Edition Project. Bloomington: Indiana University Press. [References to these volumes will be designated by W, followed by volume number, colon, page number.].

Peirce, Charles Sanders. i. 1867–1913. *Collected papers of Charles Sanders Peirce.* Vols. 1–6, eds. Charles Hartshorne and Paul Weiss. Cambridge: Harvard University Press, 1931–1935. Vols. 7–8, ed. Arthur W. Burks. Cambridge: Harvard University Press, 1958. [References to Peirce's papers will be designated by CP, followed by volume, period, paragraph number.].

Peirce, C.S. 1865–1913. *Historical perspectives on Peirce's logic of science: A history of science.* 2 vols, ed. C. Eisele. Berlin: Mouton Publishers. [Cited as HP v:p].

Peirce, C.S. 1865–1914. *The new elements of mathematics.* 4 vols, ed. C. Eisele. The Hague: Mouton Publishers. [Cited as NEM v:p].

Peirce, C.S. i. 1867–1893. *The essential Peirce: Selected philosophical writing.* Vol 1 (1867–1893), eds. Nathan Houser and Christian Kloesel. Bloomington: Indiana University Press, 1992. [References to this volume will be designated by EP 1, followed by colon, page number.].

Peirce, C.S. i. 1893–1913. *The essential Peirce: Selected philosophical writing.* Vol 2 (1893–1913), ed. the Peirce Edition Project. Bloomington: Indiana University Press, 1998. [References to this volume will be designated by EP 2, followed by colon, page number.].

Peirce, C.S. 1975. *Charles Sanders Peirce: Contributions to "the Nation": Part I: 1869–1893*, eds. Kenneth Laine Ketner and James Edward Cook, Lubbock: Texas Tech University Press.

Peirce, C.S. 1978. *Charles Sanders Peirce: Contributions to "the Nation": Part II: 1894–1900*, eds. Kenneth Laine Ketner and James Edward Cook. Lubbock: Texas Tech University Press.

Peirce, C.S. 1979. *Charles Sanders Peirce: Contributions to "the Nation": Part III: 1901–1908*, eds. Kenneth Laine Ketner and James Edward Cook. Lubbock: Texas Tech University Press.

Peirce, C.S. 1987. *Charles Sanders Peirce: Contributions to "the Nation": Part IV: Index*, eds. Kenneth Laine Ketner and James Edward Cook. Lubbock: Texas Tech University Press.

Peirce, C.S. 1877–1910. *Illustrations of the logic of science*, ed. C. de Waal. Chicago: Open Court [Cited as ILS p].

Peirce, C.S. 2009. *The logic of interdisciplinarity: The monist-series*, ed. Elize Bisanz. Berlin: Akademie Verlag GmbH.

Peirce, C.S. 1895–1908. *Philosophy of mathematics: Selected writings*, ed. M.E. Moore. Bloomington: Indiana University Press [Cited as PM p].

Peirce, C.S. i. 1898. *Reasoning and the logic of things: The Cambridge conferences lectures of 1898*, ed. Kenneth Laine Ketner. Cambridge: Harvard University Press, 1992. [References to this volume will be designated by RLT, followed by lecture number, colon, page number.] Introduction, and comments, by Kenneth Laine Ketner and Hilary Putnam: 1992: 1–102.S.

Peirce, Charles Sanders and Victoria Lady Welby. 2001. *Semiotic and Significs: The Correspondence between Charles S. Peirce and Victoria, Lady Welby*. Charles S. Hardwick, ed. with the assistance of James Cook, 2nd ed. The Press of Arisbe Associates.

Ransdell, Joseph. 1992. Teleology and the autonomy of the semiosis process. In *Signs of humanity/L'homme et ses signes,* Vol 1, eds. Michel Balat, Janice Deledalle-Rhodes, and Gérard Deledalle, 239–258. Berlin: Mouton de Gruyter.

Short, Thomas. L. 1996. Interpreting Peirce's interpretant: A response to Lalor, Liszka, and Meyers. *Transactions of the Charles S. Peirce Society* 32(4): 488–541.

Short, Thomas. L. 2004. The development of Peirce's theory of signs. In *The Cambridge companion to Peirce*, ed. Cheryl Misak, 214–240. Cambridge: Cambridge University Press.

Short, Thomas. L. 2007a. *Peirce's theory of signs.* Cambridge: Cambridge University Press.

Short, Thomas. L. 2007b. Response. *Transactions of the Charles S. Peirce Society* 43(4): 663–693.

Stango, Marco. 2015. The pragmatic maxim and the normative sciences: Peirce's problematic "fourth" grade of clarity. *Transactions of the Charles S. Peirce Society* 51(1): 34–56.

Chapter 12
In What Sense Exactly Is Peirce's Habit-Concept Revolutionary?

Erkki Kilpinen

Abstract This article argues that Peirce uses the term "habit" in a meaning that differs radically from its meaning in mainstream philosophy, both before and after him. However, Peirce's use is not only different but is from a completely fresh viewpoint, one that turns the received meaning of "habit" upside down, colloquially speaking. Peirce's revolutionarily new meaning, however, is due to considerable development in his philosophy, from nominalism toward realism. In Peirce's final usage, the habit-term does not refer to the routine character, but rather to the *process* character, of human action. A process, however, can be known only through its instantiations, and in Peirce's final meaning individual "actions", the traditional objects of interest, are instantiations of more comprehensive phenomena, "habits". For Peirce, habitual action is not outside the control of the acting subject's consciousness. The consciousness is present even in so radical sense that the habitual character of action is supposed to correlate positively with its logicality and rationality.

Keywords Action · Routine · Realism · Nominalism · Pragmatism

Introduction: Peirce's Marriage of Habituality with Intentionality

Even a modest perusal of Charles Peirce's writings suffices to tell that he frequently uses the concept of "habit", a seemingly ordinary term. If one peruses a bit more, it will turn out that he does not use it quite in its ordinary sense. However, the further ensuing questions: (i) In what sense exactly does he use it then? ... (ii) how does his usage differ from that of others, philosophers, or lay people? ... and (iii) might all this have also further implications? ... have remained without satisfactory answers so far. In this paper I outline some such answers, but before that I state explicitly my

E. Kilpinen (✉)
University of Helsinki, Helsinki, Finland
e-mail: erkki.kilpinen@helsinki.fi

D.E. West and M. Anderson (eds.), *Consensus on Peirce's Concept of Habit*,
Studies in Applied Philosophy, Epistemology and Rational Ethics 31,
DOI 10.1007/978-3-319-45920-2_12

opinion about Peirce's contribution. My general thesis is that his seemingly strange use of the term "habit" signals an entire revolution in the psychological and philosophical theory of action. His position, expressed, e.g., in the *prima facie* enigmatic assertion that "knowledge is habit" (CP 4.531; 1906), puts the relation of thought and action into a completely new perspective, such that means advancement beyond previous treatments of the thought/action relation.

The received view about that relation comes out in authoritative dictionary definitions, which explain that habit is "a thing a person does often and almost without thinking, especially something that is hard to stop doing", as per *The Oxford English Dictionary*. *The Random House Dictionary* adds that a habit is "an acquired behavior pattern regularly followed until it has become almost involuntary". These definitions echo old ideas, such that "habit rules the unreflecting herd", as the poet William Wordsworth once put it. Peirce's understanding of the matter suggests that it does not have to be so. His position can be expressed in a maxim that has the classic Kantian form, but changes the order of things.[1] It would be that *Intentionality without habituality is empty; habituality without intentionality is blind.*

Peirce on "Habit": Ambiguities on the First Acquaintance

The best-known part of Peirce's philosophy are the two articles that begin the six-piece series, "Illustrations of the Logic of Science" (1877–1878), and one of his best known sayings appears in the latter of them, "How to Make Our Ideas Clear" (1878). It says that "the whole function of thought is to produce habits of action" (EP 1: 131). This expression may have appeared rather innocuous to most readers. It might easily be taken to mean that thought produces a singular independent action, then another, and eventually a number of such, one at a time, so that they through time, by the effect of repetition, assume a habitual form. However, if applied to Peirce's position, this idea would be problematic, because he goes on to say, in the same paragraph, that "the identity of a habit depends on how it might lead us to act, not merely under such circumstances as are likely to arise, but under such as might possibly occur, no matter how improbable they may be" (*ibid.*). This gives one of the first hints about his later *realist* (as opposed to nominalist) position about modalities, the one where "might-be's and would-be's are as real as is's and have-been's," as he colloquially put it in his later philosophy (cf. CP 8.216). If we apply this interpretation of modalities to our present problem, the relation between habit and individual actions, then the realist interpretation will suggest that the

[1]Funke (1958) and Camic (1986) provide extensive historical discussions about how "habit" has been understood in the history of philosophy and the human sciences. Neither of them pays attention to Peirce's use. I have provided some preliminary comparisons between his use and that of classical philosophy in Kilpinen (2009, 2012, 2015).

reality of habit does not depend on its empirical occurrences, as is the case when *repetition* is taken as the defining characteristic.

However, it apparently would be hasty to suggest that Peirce takes "habit" in a realist sense right from the beginning. The truth rather is that, in the period that we are now dealing with, his usage of the term teeters between positions. He is striving toward the final, realist interpretation, but is not there yet. There are namely passages where Peirce appears to follow the ordinary understanding, relate the phenomenon of "habit" to repetition without further ado, and to equate it also with the notion of "routine". In other words, he often still falls into the ordinary, *nominalist* interpretation of "habit".

Let me give some examples. Peirce's article in *The American Journal of Mathematics*, "On the Algebra of Logic" (1880), throws light on his general development in more than one sense. The ordinary picture of Peirce as a logician is that he is a very consistent "anti-psychologist" in logic, in other words, he maintains that the conclusiveness of logical truths is completely independent of their reliability and validity as psychological explanations. This is indeed his position about logical truth, but there are also other issues involved that pertain to the relation of logic and psychology. Some of them come out in Peirce's 1880 paper. Namely, he begins it with the statement: "In order to gain a clear understanding of the origin of the various signs used in logical algebra ... we ought to begin by considering how logic itself arises" (EP 1: 200). Logic itself arises, he goes on, in "cerebration", which is "subject to the general laws of nervous action" (*ibid.*). These general laws, in turn, are *habits* of nervous action, but on this occasion Peirce does not have anything specific to say about their status; he is content to take them in the traditional, nominalist sense:

> Now, all vital processes tend to become easier on repetition. ... Accordingly, when an irritation of the nerves is repeated, all the various actions which have taken place on previous similar occasions are the more likely to take place now, and those are most likely to take place which have most frequently taken place on those previous occasions. ... Hence, a strong habit of responding to the given irritation in this particular way must quickly be established. (EP 1: 201; 1880)

After this passage, Peirce's discussion of "cerebration" goes on nearly three pages, and during its course he derives from the operations of cerebration the basic concepts in his logical analysis, "belief", "thought", "judgment", and "inference", after which he is ready to take up logical relations, his main subject in the article. Such an approach puts his general reputation as a logical anti-psychologist in a somewhat problematic perspective. Logic does not rest on psychological truths, (as Peirce maintains consistently), but from this it does not follow that logic and psychology have nothing to do with each other.

There are also other examples where Peirce relates habit to repetitive action. One occurs in his well-known article of 1893, "Evolutionary Love", which closes his five-piece series in *The Monist* (1891–1893). There he uses the term "habit", again in the ordinary sense, by saying that it is "mere inertia, a resting of one's oars, not a propulsion", as he says while refuting the Lamarckian suggestion that evolution might take place "by the force of habit" (EP 1: 360; 1893). A mere page later, he goes on in the same vein that "Everybody knows that the long continuance of a routine of habit makes us lethargic, while a succession of surprises wonderfully brightens the ideas" (EP 1:361). The vexing term "habit" appears here concomitantly with "routine", and Peirce's observation about its psychological effects might be plausible. Now, judging exclusively by these passages, one might conclude that in this period, in the early 1890s, he is still in the snares of the received view about "habit", though later on transcends it. However, I think that such an interpretation would not be satisfactory, and that Peirce's development—on the issue that now vexes us—is not so straightforward. I rather suggest that there is a *tension* in his view on the issue from early on, such a tension where the realist reading of "habit" is present from the beginning—though not prominently at first—and that it finally supersedes the nominalist one.

To consider this question, we need to turn to Peirce's earliest published writings, and specifically to the middle one of his three "anti-Cartesian" papers (1868–1869), "Some Consequences of Four Incapacities" (1869). There he writes, while explaining the role of attention in the modification of consciousness, that a "habit arises, when, having had the sensation of performing a certain act, *m*, on several occasions, *a*, *b*, *c*, we come to do it upon every occurrence of the general event, l, of which a, b, and c are special cases. … Thus the formation of habit is an induction and is therefore necessarily connected with attention or abstraction" (EP 1: 46–47; 1869).

Here Peirce draws an analogy between habit-formation and inductive inference, an analogy that might serve a pedagogic purpose. I think, however, that it remains a mere analogy this time, but more serious issues get involved when Peirce returns to it. That return occurs some 10 years later, in the "Illustrations"-series, already cited. In the first piece of that series, "The Fixation of Belief" (1877), Peirce again compares habit with inductive inference, but now habit does not any longer refer just to the outcome or conclusion of an induction. Now it also refers to the conscious act of *drawing* the inference:

> That which determines us, from given premises, to draw one inference rather than another, is some habit of mind, whether it be constitutional or acquired. The habit is good or otherwise, according as it produces true conclusions from true premises or not; … The particular habit of mind which governs this or that inference may be formulated in a proposition whose truth depends on the validity of the inferences which the habit determines; and such a formula is called a *guiding principle* of inference. (EP 1: 112; 1877; original emphasis)

I am tempted to take the formulation, "guiding principle of inference", as Peirce's original suggestion, and also as important, because such a principle presumably is meant as a *normative* principle. Accordingly, it is to be followed *consciously* in one's course of reasoning, even though Peirce equates it with "a particular habit of mind". In other words, he *is* now asserting that a phenomenon can be habitual and conscious at the same time. This accordingly gives a first suggestion that conscious thinking and habitual action need not be taken as mutually exclusive.

A possible critical rejoinder might be that Peirce talks here about a habit of *mind*, rather than corporeal doing, so that bodily habits may remain closed to the acting subject's reflection, whatever the case with mental ones. That rejoinder's presupposition, however, would be acceptance of the Cartesian mind/body dualism, and it is widely known that Peirce rejected it already in his earliest writings. Accordingly, I find reason to stick to my above suggestion about an essential tension in Peirce's early understanding of "habit". On some occasions, he relates it to repetition and routine, in other cases he takes it as open for the acting subject's reflection—even during its occurrence, not just retrospectively. This is because he sees normative, and therefore reflective, aspects in it. My hypothesis thus is that this latter, realist tendency eventually carries the day in Peirce's consideration of the phenomenon, along with his more thorough-going acceptance of ontological realism, the realist interpretation of modalities in particular. However, before we can move to observe that development, there are still a couple of preliminary issues to be dealt with. The first of them concerns philological Peirce scholarship, the question about his outside influences, whereas the other one is about how justifiable an explicitly action-theoretic interpretation of his habit-concept may be.

The Role of Joseph J. Murphy

For the first issue, we need to return to Peirce's 1880 article, "On the Algebra of Logic", already cited. I quoted a passage where he relates habit-formation to repeated physiological irritation. The editors and commentators of *The Essential Peirce*, volume 1, Nathan Houser and Christian Kloesel, have here added a note where they refer to the Irish philosopher and physician Joseph J. Murphy (1827–1894) (see Houser and Kloesel 1992: 380; note 1 to Chap. 13). They quote Murphy's book *Habit and Intelligence* (first published in 1869) to the effect that "The definition of habit, and its primary law, is that all vital actions tend to repeat themselves; or [at least] tend to become easier on repetition." Houser and Kloesel suggest this to be the source for the formulation that Peirce used above (EP 1: 200–201; 1880). It is known that Peirce owned a copy of Murphy's book and drew on it in his own work.[2] However, it is questionable whether he needed to turn to it for the idea that a habit might be due to effects of repetition. *That* idea was old hat, the only new ingredient that Murphy brought to this discussion was to relate it explicitly to physiological psychology. What instead was original in Murphy's contribution was expressed already in the title of the work, *Habit and Intelligence*. He was suggesting that habit and intelligence might go together and correlate positively, instead of being at war with each other.

---

[2]Regarding Peirce's relation to Murphy, I am much indebted to suggestions by my colleague and friend, Ahti-Veikko Pietarinen.

In order to establish such a correlation, Murphy needed to redefine both of these concepts to some extent. As he said (1879: 87), "we generally use the word 'habit' with special reference to the mysterious border-land between the conscious and the unconscious function." Murphy widened this "general use of habit" so that he could admit some *overlap* between it and the "conscious function" (or reflective thinking, if you like). What is perhaps even more important, he put the notions of intelligence and consciousness into a new, reversed order. Traditionally, intelligence had been taken as an attribute of the mind, particularly of the conscious human mind. Murphy, for his part, took "intelligence" as the foundational phenomenon, so that its denotation became "almost coextensive with life", as the pragmatist Mead (1938: 68) later on came to say about his own intelligence-concept.[3] Mind, or consciousness, in turn, came to be a particular sub-case of intelligence.

For Peirce, this offered new support to the idea that he already had held, namely that all mental operations are of inferential character, but not necessarily conscious operations, the idea that he originally credited to the German psychologist Wilhelm Wundt (CN 1: 37; 1869). The locus of reasoning now came to be within intelligence, so that we can see the point in Peirce's macabre metaphor, "A decapitated frog almost reasons" (CP 6.286; c. 1893). The poor frog's physiological reactions can be reconstructed according to the syllogistic formula of elementary logic, so that the only thing that he lacks "is the power of preparatory meditation" (*ibid.*).

Regarding the human use of reason, in turn, Wundt's and Murphy's conclusion enabled Peirce to hold the view that his explicator, Murray G. Murphey, half a century ago paraphrased as follows: "We reason as we breath, involuntarily, and whether these inferences are recognized as such is immaterial" (Murphey 1961: 359). Our consciousness does play a role as well, but it does not initiate the process of reasoning. According to Peirce, it rather steps into *criticize* the putative mental associations, to see whether they are to be taken as logical conclusions or not. Mental associations, in turn, take place by themselves, "in a sort of Bacchic train", as Peirce once illustriously puts it (W5: 326; 1886). This Bacchic train provides the raw material out of which genuine inferences are made by means of critical self-control. "Logic is the criticism of conscious thought", Peirce avers in his 1903 Harvard lectures (EP 2: 169), the point being that the task of conscious thought is to review critically such associations (candidates for putative conclusions) that have emerged out of unconscious thought. In order to perform its critical function, conscious thought first needs to have something to be criticized. That emerges from our sub-consciousness, as we twenty-first century people might say.

---

[3]By calling *habit* "at least coëxtensive with life" (EP 1: 223; 1884), Peirce establishes an interesting parallel between himself and his later fellow-pragmatist, Mead. They share also other important affinities, like the triadic theory of meaning, which makes an interesting research theme, in view of the fact that Mead, who died early in 1931, cannot have learned those ideas straight from Peirce's texts. I have broached this theme in Kilpinen (2002) but much more remains to be said about it.

But Is It Indeed About Action?

I have referred to ambiguities in Peirce's earlier treatment of "habit" and similar ambiguities turn up when one attempts to contextualize his discussions. I proposed that he does have a theory of action and that it develops concomitantly with the general development of his philosophy. However, there are text passages that might suggest that this interpretation is untenable, whereas there are also places, in Peirce's writings, that seem to speak in favor of it.

One document that states rather definitely that Peirce has had in mind a psychological theory of action (among his great variety of interests) is his plan for a projected series of books, with the awesome title, *Principles of Philosophy: Or Logic, Physics, and Psychics, Considered as a Unity, In the Light of the Nineteenth Century*. It was intended to consist of twelve volumes, and Peirce printed and circulated an advertising prospectus describing its planned contents, in 1893. The prospectus contained a summary of each of the intended volumes, and the list of topics to be treated in the eighth volume, tentatively entitled *Continuity in the Psychological and Moral Sciences*, ran as follows:

> Mathematical economics. Precisely similar considerations supposed by utilitarians to determine individual action. But, this being granted, Marshall and Walras's theorem leads to a mathematical demonstration of free will. Refutation of the theory of motives. The true psychology of action expounded. (RLT: 14; 1893).

In a previous publication (Kilpinen 2010), I have suggested that though Peirce has never quite expounded the true psychology of action, such a conception can be reconstructed by collating some of his texts from the periods both preceding and following the above date. I submitted then that Peirce's psychology of action is mainly a unification of two doctrines that he has brought together, namely Wilhelm Wundt's physiological psychology and Peirce's own "doubt/belief model of inquiry" that he has situated within the former conception, to provide a kind of logical skeleton for it. However, regarding the doubt/belief model as a constituent in a psychological theory of action, it turns out that though such a reading may have answered Peirce's original purpose, there is also the embarrassing fact that he himself, later on, seems to have rejected it. We need to take a closer look.

The embarrassing fact is a well-known passage in the first paragraphs of the fifth volume of the *Collected Papers* (originally written for *Baldwin's Dictionary*), where Peirce exercises self-criticism in regard to his above earlier position: "The doctrine appears to assume that the end of man is action—a stoical axiom which, to the present writer at the age of sixty, does not recommend itself so forcibly as it did at thirty" (CP 5.3; 1902). This has prompted some commentators to conclude that action is of no importance for the mature period Peirce. Such interpretation is enhanced as Peirce later on explains what he actually had meant by the idea of pragmatism, whose prototype had appeared in the 1870s writings. In the article "What Pragmatism Is" (1905), he renames his theory of meaning (*not* his entire doctrine) "pragmaticism", and states emphatically that "if pragmaticism really made Doing to be the Be-all and the End-all of life, that would be its death" (EP 2: 341;

1905). These dramatic words suggest at first that Peirce wanted to drop "action" (or doing) as a philosophical subject, but I do not take them as his final words. I think it is more correct to say that he warns against *reducing* life to mere doing, but doing or material action never vanishes from his agenda. Peirce is very critical toward the position that he calls Stoical nominalism, "the germinal conviction … that the only end of man lies in action, and that knowledge, as such, is an idle accomplishment" (CN 3: 26; 1901). But to say this is not the same thing as to say that knowledge and action have nothing to do with each other. The conclusion that they have to do with each other is suggested by Peirce's well-known humorous phrase that "a court cannot be imagined without a sheriff" (CP 1.213; 1902). Or, as he elsewhere puts it in closer detail but only slightly less metaphorically:

> We forget that thinking implies existential action [that is: corporeal action in and against the outer world – E.K.] though it does not consist in that; or if we remember that thought implies the action of forces upon a brain or something like it, we still more perversely regard that as lowering the dignity of thought, and as making it a "mere" existential event; whereas the truth is just the opposite. In that thought requires existential acting, and further requires something else beside that, it ought to be plain enough that it exceeds the existential acting. The ruler of a nation depends upon his cook and his secretaries. That does not place him lower than they, but higher. Thought is higher than brute fact in much the same way that a statesman is higher than his secretaries: namely, it needs the existential facts, but regulates them. (CP 6.324; 1909)

Accordingly, I think that we are justified to consider Peirce as a contributor to the theory of action. But now it is high time to state explicitly what role the idea of habit plays in his theory, as I have promised to do above.

Habituality and Fallibilism as Constituents in Human Action

Above I have suggested that one of the welcome clarifications that Peirce brings to the understanding of "habit" is ridding it of what I am tempted to call "the illusion of repetition". Instead, habitual action is to be taken as an already ongoing process. His development of this idea may be reconstructed in a three-phase logical order, which does not necessarily coincide with the temporal order of the development. The phases are, in an ascending order of importance, as follows: (i) Peirce dissociates the idea of "habit" from repetitive action (which has been the received view). (ii) He admits, if not insists, on the presence of consciousness in habitual action. (iii) He maintains, furthermore, that logicality and rationality have their locus precisely *within* the habitual mode of action, not anywhere outside it in singular instantaneous actions. Let me now attempt to prove these assertions.

(i) Regarding the question whether repetitive action suffices to define a habit, I have mentioned that Peirce's first scattered remarks about it are equivocal. Later on, however, he appears explicitly to discard the idea about repetition as a defining characteristic. In some first drafts for what became the series, *Reasoning and the*

Logic of Things, he gives also a new, rather strictly worded explication about what is to be meant by habit:

> Habits are not for the most part formed by the mere slothful repetition of what has been done, mere tendency to repeat any action you happened to perform, but by the logical development of the potential germinal nature of the man, generally by an effort, the accident of having done this or that merely having an adjuvant effect. (NEM 4: 143; c. 1898).

On the preceding page, Peirce made a statement that a *change* in one's habits is due to the subject's interaction with the objective outside world, and to the surprises and failures that are encountered in that interaction. In the late period of Peirce's philosophy, his habit-concept is closely related to his principle of *fallibilism*, to the idea that action can also fail and falter, and that sooner or later this will happen. In Peirce's own words,

> It is the catastrophe, accident, reaction which brings habit into an active condition and creates a habit of changing habits. To learn is to acquire a habit. What makes men learn? Not merely the sight of what they are accustomed to, but perpetual new experiences which throw them into a habit of tossing aside old ideas and forming new ones. (NEM 4: 142; c. 1898).

In Peirce's final, delivered version of the lectures *Reasoning and the Logic of Things*, the concept of habit plays a prominent role, it is the theme of an entire lecture (the 7th), but then in the meaning of "cosmic habits", by which he means mediation between strict mechanical causation on the one hand, and objective chance, on the other. That interesting theme does not directly bear on the present discussion about human action, so that I return instead to the former of the above quoted passages and pay attention to what exactly Peirce says about the role of repetition, or "the accident of having done this or that" as his own words went. He says that it has a "merely adjuvant effect"; he does not say that it has no effect at all. This observation is notable from at least two viewpoints. In the first place, it serves as a prophylactic to such interpretations that might try to equate Peirce's "habit" to Wittgenstein's notion of rule-following, which has spawned a wide philosophical literature. Such an equation would be spurious precisely because the aspect of material doing is present in Peirce's notion (though only as an aspect, not as the whole story). Philosophers (e.g., Rorty 1961) have seen similarities in Peirce and Wittgenstein, and there may be some, but they are mostly of superficial nature. This is for the reason that in final analysis the two philosophers belonged to opposite meaning-theoretical camps. These camps are sustained, in the phrase of the late Jaakko Hintikka (1929–2015), by the conception of language as a universal medium, that Wittgenstein held; versus language taken as a calculus, which was Peirce's position (for closer details of this distinction, see Hintikka 1997).

But, to return to Peirce's treatment of "habit", in the first decade of the 20th century he sticks to his above idea about habit's independence of repetition. As he puts it in explicit words, in 1905: "I need not repeat that I do not say that it is the single deeds that constitute the habit. It is the single 'ways,' which are conditional propositions, each general,—that constitute the habit" (CP 5.510; 1905). Peirce's dissociation of habit from repetitive action is thus in accordance with his advancing acceptance of ontological realism, where a principle (here: "ways") is taken as real, regardless the frequency of its empirical instantiations.

Toward the end of this paper I shall have one more brief word about repetition, about how Peirce takes it as an asset rather than liability in the ongoing process of action. At the moment, I move to consider his interpretation about the role of consciousness in habitual action.

The prevailing idea has been and still is that the former has no role at all in such action. Peirce leaves this idea behind. Already in the 1898 lectures *Reasoning and the Logic of Things* (published in 1992), he had suggested, while explaining how reasoning takes place, that "there is a direct consciousness of habit-forming or learning" (RLT: 191). This suggestion is not yet very radical, as it refers to *habit-forming*, and just a couple of sentences further Peirce explains that the question is now about a "mental habit". Most people accept, I think, that we can be directly conscious about our learning, call it habit-forming or not. But can it be so also about *established* habits, and can it be so about habits other than mental? In his late philosophy, Peirce gives affirmative, though not perfectly explicit answers to both of these questions, so that some teasing out of his meaning remains as the reader's task.

He opens the habit-notion for conscious reflection quite outspokenly, though in a place that might easily escape commentators' attention. Central among Peirce's late period manuscripts is MS 318 of 1907, entitled by posthumous editors "Pragmatism", because it is the place where he goes to most pains to explain what he had originally meant by that doctrine, and what he means by it at the time of writing. There are several variants of that manuscript, which has not been published in its entirety, and at one place Peirce explains definitely that he now means by "habit", a *voluntary* habit:

> Involuntary habits are not meant, but voluntary habits, i.e., such as are subject in some measure to self-control. Now under what conditions is a habit subject to self-control? Only if what has been done in one instance with the character, its consequents, and other circumstances, can have a triadic influence in strengthening or weakening the disposition to do the like on a new occasion. This is as much as to say that voluntary habit is conscious habit. (EP 2: 549, n49 to chapter 28).

The idea of conscious habit is a novelty in Peirce's treatment of action, an idea to which he sticks henceforth, for example while explaining his intended meaning in a letter to his friend and co-founder of pragmatism, William James: "Consciousness of habit is a consciousness at once of the substance of the habit, the special case of application, and the union of the two" (CP 8.304; 1909). Here Peirce is marrying habit and consciousness happily together. He is not content to speak merely about consciousness regarding the substance of the habit; that might enable a recalcitrant reader to think that the question is merely about retrospective consciousness about the habit. Instead, Peirce refers explicitly to the special case of application, suggesting that he does not take habit as an autonomous motor routine—though it certainly is a continuous phenomenon in his treatment. Consciousness of the case of application reveals that the subject is supposed to be self-consciously aware when and how to apply his habitual disposition to act. And, to clinch the argument, Peirce talks about the union between the general substance of the habit and its case of application. I think that this justifies my interpretation about the presence of consciousness in habit, in Peirce's late-period understanding, that is, I have defended the above thesis (ii). However, the further important question still remains to be treated, namely what then is the logical status of that consciousness, (the theme of my thesis, iii).

The mere fact that habit is taken to be open for the acting subject's consciousness, of course alters its status as an action-theoretic concept. The traditional understanding, which geared habit together with repetition had been backward-looking, or past-oriented. Habit had then been, in a sense, a sum or synthesis of past activities. In Peirce's hands, the newly-interpreted habit rather assumes a forward-looking, future-oriented character:

> Intellectual concepts, … the only sign-burdens that are properly denominated "concepts," … convey more, not merely than any feeling, but more, too, than any existential fact, namely, the "would-acts" of habitual behavior, and no agglomeration of actual happenings [read: individual instantaneous actions] can ever completely fill up the meaning of a "would be." (EP 2: 401–402; 1907).

I inserted into the above passage my own suggestion that "actual happenings" are here to be understood to mean individual actions, those which have received the pride of place in philosophy, today in what is known as "philosophy of action". My insertion was not preposterous, because later on in the same writing Peirce argues to the same effect. This happens as he joins his semiotic concepts together with his treatment of habitual action. He explains both his rather well-known tripartite characterization of signs—index, icon, symbol—as well as what he means by the division, sign-object-interpretant. About interpretants, furthermore, he takes up their possible division into emotional, energetic and logical ones. For those not well-versed in Peirce's semiotic, it could be pointed out that these latter terms are not to be taken as kinds of things, but as possible modes of behavior. More strictly speaking, they are modes of interpretive behavior, but for Peirce all conscious behavior is interpretive. The radical point that has not often received treatment, even by Peirce scholars, is that (a) singular action(s) cannot make a logical interpretant, they make only an energetic interpretant, use of energy in effort, whereas, logical interpretant refers precisely to (a) habitual mode(s) of behavior.

> In every case, after some preliminaries, the activity takes the form of experimentation in the inner world; and the conclusion (if it comes to a definite conclusion) is that under given conditions, the interpreter will have formed the habit of acting in a given way, … The habit conjoined with the motive and the conditions has the action for its energetic interpretant; but action cannot be a logical interpretant because it lacks generality. … The deliberately formed, self-analyzing habit, – self-analyzing because formed by the aid of analysis of the exercises that nourished it, – is the living definition, the veritable and final logical interpretant. (EP 2: 418).

A brief moment earlier in the same text, Peirce had considered the possibility whether desire and expectation might come to question as criteria of logical interpretant, and he concluded that they cannot serve in that capacity: "there remains only habit, as the essence of the logical interpretant" (EP 2: 412; 1907).

It perhaps is apt to note one more time that Peirce's conception is only superficially akin to Wittgenstein's notion of rule-following. I iterate this point, because suggestions to that effect come up regularly should one attempt to explain Peirce's position to analytic philosophers. Of these two philosophers, it is actually Wittgenstein who has remained closer to the traditional understanding of human behavior. He said that, "When I obey a rule, I do not choose. I obey the rule *blindly*" (1968[1953]: §219; Wittgenstein's emphasis). This idea, of following blindly a rule or a habit, Peirce precisely rules out. Moreover, in his conception of internal rehearsal (call it rule-following if you like) and outer performance of the activity are mutually constituted:

> Habits differ from dispositions in having been acquired as consequences of the principle, virtually well known even to those whose powers of reflection are insufficient to its formulation, that multiply reiterated behavior of the same kind, under similar combinations of percepts and fancies, produce a tendency, – the *habit*, – actually to behave in a similar way under similar circumstances in the future. Moreover, – *here is the point*, – every man exercises more or less control over himself by means of modifying his own habits; … *reiterations in the inner world, – fancied reiterations, – if well-intensified by direct effort, produce habits, just as do reiterations in the outer world; and these habits will have power to influence actual behavior in the outer world* (EP 2: 413; 1907; Peirce's emphases).

This passage shows that Peirce has achieved a synthetic view that includes elements of previous conceptions, but also puts them into their proper places. Wittgenstein's idea of rule-following is not entirely foreign to this conception, insofar as Peirce talks about reiterations in the inner world. However, these reiterations do not remain mere matters of the inner world, as he also points out that they eventually have to prove their mettle in doings in the outer world—and then perhaps meet a shock or surprise, as were Peirce's colorful terms for failing action. The idea that the possibility of failure needs to be included in the action-conception right from the beginning has remained surprisingly foreign to the "philosophy of action" influenced by Wittgenstein (though the latter himself did take it into account, but not prominently enough, it now turns out).

Regarding the prevailing received meaning about "habit", Peirce's words about "multiply reiterated behavior" show that he has not thrown the idea of repetition completely onto the wayside. However, he has made an asset out of this phenomenon that traditionally has been taken as a hindrance. Multiply reiterated behavior is now subordinate to the logical and intellectual side in habit-formation and in the application of habit. As Peirce once humorously puts the matter, "It is bad economy to employ the brain in doing what can be accomplished mechanically, just as it would have been bad economy for Napoleon to write his own dispatches" (NEM 4: 71; c. 1902). Peirce's friend William James captured this point as well, as he explained that "the more details of our daily life we can hand over to the effortless custody of automatism, the more our higher powers will be set free for their own proper work" (James 1890/1950: 1.122). Later on in that work, James introduced a technical term to refer to this principle by calling it "the principle of parsimony" in human reason and conduct. Peirce, for his part, explained about those higher powers of ours, that even their operation is all the higher if it makes use of habit and its applications.

So, what might such action look like, on closer descriptive inspection? We have come across various definitions by Peirce, but not so much across detailed descriptions. However, he satisfied even this demand. In the penultimate article of his publishing career, "Some Amazing Mazes" (1908), he gives the following illustrious example:

> Every action of Napoleon was such as a treatise on physiology ought to describe. He walked, ate, slept, worked in his study, rode his horse, talked to his fellows, just as every other man does. But he combined those elements into shapes that have not been matched in modern times. Those who dispute about Free-Will and Necessity commit a similar oversight [as do those who treat Napoleon one action at a time]. ... Our power of self-control certainly does not reside in the smallest bits of our conduct, but is an effect of building up a character. All supremacy of mind [over matter] is of the nature of form. (CP 4.611; 1908)

The traditional mistake in philosophy has been the attempt to search for self-control (or character formation, or any other exemplification of rationality) in the smallest bits of our conduct. Peirce's message is that the picture needs to be turned upside down, and Napoleon's (or any other man's) rationality is to be searched from top down, so to speak. His actions, or smallest bits of conduct are but exemplifications of his general habits, or of his character, which is but the same thing, seen from the normative side. Regarding a general theory of action, it perhaps is more incumbent to talk about the intentionality and/or rationality of action, rather than about character per se. An interesting conclusion arises here. This action-conception, which gives a foundational role to habit, does not, after all inflict any harm on those notions that traditionally have been taken as the most burning issues in the treatment of action. Instead, it begins to dawn on us that their job-descriptions are considerably enlarged, when action is understood as a habitual process. They now have to see the entire performance through, over its vicissitudes, not just send the acting subject on his or her way, as was the understanding in the traditional conception which set out from the "one intention-one action" premise. This is what gives us the right to call Peirce's habit-oriented upheaval in action theory "revolutionary" in the full sense of the term.

It deserves this name also for another reason, in that like a real revolutionary, Peirce is at the head of an entire movement, not just a solitary voice in the wilderness. That movement is, of course, the pragmatic movement in philosophy, whose founding father Peirce is recognized to be. However, I do not agree with those Peirce scholars (or pragmatism scholars) who tend to belittle—sometimes even write off—the contributions of the other members of the movement. They are not to be neglected as minor figures who just misunderstood Peirce, though it of course is true that they did not agree with him on each and every point, and that this sometimes was for their own detriment. I think that the best way to form an opinion here is to let Peirce himself pass the verdict. What appears to be his final considered opinion about his relation to the other pragmatists is expressed in another article from the year of 1908, entitled "A Neglected Argument for the Reality of God". In its closing paragraph, he enumerates strengths and weaknesses of the other pragmatists and brings out his dissatisfaction with their "angry hatred of strict logic". Nevertheless, there are also important points of agreement, which Peirce expressed as follows:

> At the same time, it seems to me clear that their approximate acceptance of the Pragmaticist principle, and even that very casting aside of difficult distinctions (although I cannot approve of it), has helped them to a mightily clear discernment of some fundamental truths that other philosophers have seen but through a mist, and most of them not at all. Among such truths, – all of them old, of course, yet acknowledged by few, – I reckon their denial of necessitarianism; their rejection of any "consciousness" different from a visceral or other external sensation; their acknowledgment that there are, in a Pragmaticistical sense, Real habits (which Really *would* produce effects, under circumstances that may not happen to get actualized, and are thus real generals); and their insistence upon interpreting all hypostatic abstractions in terms of what they *would* or *might* (not actually *will*) come to in the concrete. (EP 2: 450; 1908; Peirce's emphases).

Other philosophers have seen through a mist, or not at all, that habits are to be taken as real generals, in other words, in ontologically realist terms, but pragmatists have seen this more clearly, according to Peirce's testimony. We need not get entangled with the question whether his realist interpretation of other pragmatists' habit-concepts is philologically perfectly correct or not. The point to be taken is that they apparently can be reconstructed along those lines, and such reconstruction may enhance even the notion that Peirce already achieved. The pragmatist habit-concept is not to be treated with awe as Peirce's magnificent achievement, but to be taken as a challenge for future research.

References

Camic, Charles. 1986. The matter of habit. *American Journal of Sociology* 91(5): 1039–1087.

Funke, Gerhard. 1958. Gewohnheit. *Archiv für Begriffsgeschichte* 3: 9–606.

Hintikka, Jaakko. 1997. *Lingua universalis vs. calculus ratiocinator: An ultimate presupposition of twentieth-century philosophy.* (*Jaakko Hintikka, Selected Papers* vol. 2). Dordrecht/Boston/London: Kluwer.

Houser, Nathan, and Christian Kloesel. 1992. Editors' comments on the edition, *The essential Peirce: Selected philosophical writings*, vol. 1 (1867–1893). Bloomington and Indianapolis: Indiana University Press.

James, William. 1950[1890]. *The principles of psychology*, I–II. New York: Dover.

Ketner, Kenneth, and Hilary Putnam. 1992. Introduction: The consequences of mathematics. Editors' introduction to *Reasoning and the logic of things*, by C.S. Peirce. Cambridge: Harvard University Press.

Kilpinen, Erkki. 2002. A neglected classic vindicated: The place of George Herbert Mead in the general tradition of semiotics. *Semiotica* 142.1/4: 1–30.

Kilpinen, Erkki. 2009. The habitual conception of action and social theory. *Semiotica* 173.1/4: 99–128.

Kilpinen, Erkki. 2010. Problems in applying Peirce in social sciences. In *Applying Peirce: Proceedings of the applying Peirce conference*, ed. M. Bergman et al., 86–104. Helsinki: Nordic Pragmatism Network. An electronic book publication, ISSN 1799-3954.

Kilpinen, Erkki. 2012. Human beings as creatures of habit. In *The habits of consumption*, ed. A. Warde and D. Southerton, 45–69. Helsinki: Helsinki Collegium for Advanced Study.

Kilpinen, Erkki. 2015. Habit, action, and knowledge, from the pragmatist perspective. In *Action, belief and inquiry: Pragmatist perspectives on science, society and religion*, ed. U. Zackariasson, 157–173. (Vol. 3, *Nordic studies in pragmatism*. [An electronic volume, ISSN 1799-3954; ISBN 978-952-67497-2-3].) Helsinki: Nordic Pragmatism Network.

Mead, George H. 1938. *The philosophy of the act*, ed. C.W. Morris et al. Chicago: University of Chicago Press.

Murphey, Murray G. 1961. *The development of Peirce's philosophy*. Cambridge: Harvard University Press.

Murphy, Joseph J. 1879[1869]. *Habit and intelligence*. 2nd impression. London: Macmillan and Company.

Peirce, Charles Sanders. i. 1867–1913. *Collected papers of Charles Sanders Peirce*. Vols. 1–6, ed. Charles Hartshorne and Paul Weiss. Cambridge: Harvard University Press, 1931–1935. Vols. 7–8, ed. Arthur W. Burks. Cambridge: Harvard University Press, 1958. [References to Peirce's papers will be designated by CP, followed by volume, period, paragraph number].

Peirce, Charles Sanders. i. 1987. *Charles Sanders Peirce: Contributions to "The Nation": Part IV: Index*, ed. Kenneth Laine Ketner and James Edward Cook. Lubbock: Texas Tech University Press.

Peirce, Charles Sanders. i. 1866–1913. *New elements of mathematics, by Charles S. Peirce*, vol. 4 (in 5 volumes), ed. C. Eisele. The Hague, 1976: Mouton. Cited as NEM.

Peirce, Charles Sanders. i. 1867–1913. *Writings of Charles S. Peirce: A chronological edition*. Vols. 1–6 to date, ed. the Peirce Edition Project. Bloomington: Indiana University Press. [References to these volumes will be designated by W, followed by volume number, colon, page number].

Peirce, Charles Sanders. i. 1898. *Reasoning and the logic of things: The Cambridge conferences lectures of 1898*, ed. Kenneth Laine Ketner. Cambridge: Harvard University Press, 1992. [References to this volume will be designated by RLT, followed by lecture number, colon, page number.] Introduction, and comments, by Kenneth Laine Ketner and Hilary Putnam: 1992: 1–102.S.

Peirce, Charles Sanders. i. 1867–1893. *The essential Peirce: Selected philosophical writing*. Vol. 1 (1867–1893), ed. Nathan Houser and Christian Kloesel. Bloomington: Indiana University Press, 1992. [References to this volume will be designated by EP 1, followed by colon, page number.].

Peirce, Charles Sanders. i. 1893–1913. *The essential Peirce: Selected philosophical writing*. Vol. 2 (1893–1913), ed. the Peirce Edition Project. Bloomington: Indiana University Press, 1998. [References to this volume will be designated by EP 2, followed by colon, page number].

Rorty, Richard. 1961. Pragmatism, categories, and language. *The Philosophical Review* 70.2:197–223.

Wittgenstein, Ludwig. 1968[1953]. *Philosophical investigations*. (Translated from the German by G.E.M. Anscombe.) Oxford: Blackwell.

Chapter 13
Indexical Scaffolds to Habit-Formation

Donna E. West

Abstract This inquiry advances the claim that Indexes in Secondness constitute the earliest and firmest foundation for the establishment of habit, undergirded by habit-change and logical interpretants. Convincing evidence for the primacy of Index as an implicit source to determine regularities in lived experience and in objective logic is proffered. The case is made that Indexical regularities form directional templates early in development which prime the semiosis of more objective logic-based regularities (played out in revisions of logical interpretants). While a predisposition may preempt salience of indexical templates (gaze trajectories, motion and force toward objects), more complex indexical regularities (perspective-taking), and focus with the mind's-eye on potentialities (virtual habit) constitute acquired, more consciousness-based habits. Ultimately, Peirce's use of habit transcends conformity to compulsory participation in events—occasional non-conformity to a pattern is essential to what he means by "habituescence," or, the conscious awareness of taking a habit (c. 1913: MS 930).

Keywords Development of index · Dialogue · Attention · Logical interpretant · Virtual habit · Action habit

Introduction

The influence of Peircean habit to direct mind and matter is unparalleled. In it resides the power to discern regularities, while implementing divergence from them—whether by deliberate and/or conscious decisions of mind, or by free-flowing physical changes in the natural world. Although Peirce refers to consciousness as "one of the most mendacious witnesses that ever was questioned" (1902–1903: 1.580), he explicitly allows for the operation of habit flowing from less intentional, less delib-

D.E. West (✉)
Department of Modern Languages, State University of New York at Cortland, Cortland, NY, USA
e-mail: westsimon@twcny.rr.com

D.E. West and M. Anderson (eds.), *Consensus on Peirce's Concept of Habit*, Studies in Applied Philosophy, Epistemology and Rational Ethics 31, DOI 10.1007/978-3-319-45920-2_13

erate processes, but favors more conscious modes of belief and conduct, in which habit-change can flourish. Accordingly, Peirce's concept of habit encompasses both a coalescence of similar ways of thinking and behaving, as well as habit-change.

Recognition of patterns in nature (since most organisms are non-sentient) is not, *prima facie*, conscious but still qualifies as habit; and because habit encompasses consciousness to foster habit-change in sentient beings, it can surface as revisions in behavior or in belief structure. In fact, Peirce privileges the latter (in terms of beliefs and concepts) as underlying habit, especially consequent to purposes beyond subjective modes of self-serving action. In MS 313: 32 (1903), Peirce demonstrates that habit indeed possesses this objective character: "I think it [practical effects of our conception of the object] has the fault of seeming to exaggerate the element of Doing, of Secondness, and of not clearly marking that our conception of all the possible practical effects of a conception must be a broadly general conception, of which Thirdness is the very life and soul" (cf. Houser, this volume, for further consideration of this latter issue). Furthermore, a focus on doing and Secondness alone deprives habit of its effect, its logical interpretant, without which habit would be stripped of its revisionary character toward habit-change.

In either case, Peircean habit requires indexical properties to highlight relational phenomena (be they physical entities or mental constructs), and to discern whether to conform to established patterns of belief and action. In fact, index's status as informational index applies to both the practical and the logical—moderating any tendency toward blind conformity to norms. Index stands at the threshold—highlighting event pattern-change in the physical world, while beckoning new directions for sound inferencing and modal matrices in living systems.

Habit and Habit Change

Peirce is adamant that habits must include some element of regularity "…the person or thing that has the habit, *would* behave (or usually behave) in a certain way *whenever* a certain occasion should arise" (1913: L477). Nonetheless, he guards against casting habit as a mechanistic, non-deliberate paradigm in which responses constitute mere verbatim reactions. Instead, habit must entail some element of progression, revision, or change; but, instigating each paradigm change/shift is a Logical Interpretant uniting all of the instantiations. As such, some objective, law-like meaning regulates the conduct/natural process. Although Peirce's use of habit springs from a Logical Interpretant (not an Emotional or Energetic one), it may or may not require the presence of an Ultimate Logical Interpretant: "It is now necessary to point out that there are three kinds of interpretants. Our categories suggest them, and the suggestion is confirmed by careful examination. I term them the Emotional, the Energetic, and Logical interpretants. They consist respectively in feelings, in efforts, and in habit-changes" (1907: MS 318: 43). A habit must consist in more than either Emotional or Energetic interpretants, consequent to its dependence on more than a single turn of event. Because Energetic Interpretants arise from one experience, they do not furnish the regularity in Thirdness necessary to qualify as habit. Habits, in contrast, are characterized by a meaning or effect (actual or potential) that spans a set of circumstances, but is, nevertheless, open to semiosis. Some common context is foundational to the belief or action tendency which ordinarily materializes; consequently, habit requires the presence of a Logical Interpretant. At the same time, these Logical Interpretants must not truncate inquiry or preclude reaching the Final Interpretant—they must be open to change and chance, both as Logical, Ultimate Logical, or as Final Interpretants (cf. Short 2007: 57–58 for a more extensive treatment of this issue).

Put in more pragmatic terms, habit is either what an individual/set of individuals do in similar circumstances (more passive form of Logical Interpretant), and/or a common projection of mind perpetuated by a kind of feeling, event/experience (a more active form of Logical Interpretant—an Ultimate form approaching Final Interpretant status. Peirce makes this active-passive distinction plain in his 1909 letter to William James: "When the captain of infantry gives the word 'Ground arms!' the Dynamic Interpretant [which can surface as a Logical Interpretant (cf. Short 2007: 178)][1] is in the thump of the muskets on the ground, or rather it is the Act of their Minds. In its more {Active/Passive} forms, the Dynamical Interpretant indefinitely approaches the character of the {Final/Immediate} Interpretant; and yet the distinction is absolute" (EP 2: 499).

By describing the distinction as "absolute," Peirce frees the interpretant from a particular subjectivity and context. In the call to arms, the Active Dynamical Interpretant resides in a Logical Interpretant, with the common conception on the part of the soldiers at large that the enemy is at the threshold—underlying the response for immediate offensive and defensive action. Such is not a response of a single party, nor is it a comprehensive reaction of a group on a single occasion; it is a habit, an Active Logical Interpretant—a general, somewhat durative belief/action that is expected to operate, but not entirely without fail. This more active form of logical interpretant can shade into virtual habit, if conjecture transcends actualized consequences to incorporate possible ones. In this way, only possible habits can serve as habits, although virtual habit (could-bes or would-bes) in Firstness and Thirdness, respectively are not excluded altogether. The potential incorporation of virtual habit into habit proper emphasizes the central place for conjectural interpretants in Peirce's semiotic. The Active Logical Interpretant is an Ultimate one, since it approaches the character of a Final Interpretant in suggesting a would-be.

For Peirce, habits do not express a perfect approach or set of affairs; consequently, they are stable, but awaiting revision. An irritant/surprising consequence, or anomalous event is often perceived to be the catalyst for a habit-change (cf. Aliseda, this volume; Colapietro, this volume). Unrecognized/under-addressed,

[1]Short (2007: 178) claims that the Immediate, Dynamic, and Final Interpretants comprise one trichotomy; the Emotional, Energetic, and Logical Interpretants apply to Peirce's phaneroscopy, allowing each latter member to be enjoined to any member of the former triad.

however, is the reason for the irritant, namely, the emergence of a logical Interpretant which is incongruent with the Logical Interpretant already in place.

Habit as Logical Interpretant

Peirce indicates that Logical Interpretants are of three major types: Firsts, grounded in concepts/propositions, Seconds, having their foundation in action/reaction recommendations, and Thirds, advocating conjectural conditions: "...when its interpreter meets with an object to which the subject applies, he is informed by the proposition that it either has (if it be universal) or may have (if it be particular), the character signified by the predicate; but *first* only, because if the interpretation is to stop at that point, the information neither contributes toward regulating the interpreter's conduct, which would be the second, or statically pragmatistic interpretation, nor does it tend toward bringing about a reconciliation or inter-adjustment between reason and the facts of experience, which would be a *third* and evolutionary pragmatistic interpretation" (1908: LI 392, MS 300).

The first logical interpretant approaches habit, in that classifications are determined which can qualify as the foundation for belief. This materializes, in that the intrinsic properties of an object (perceptual, functional) are settled upon, such that they underlie the nature of interactions with them. While they themselves are not sufficient for habit, because they do not recommend course of action modifications to resolve an unexplained consequence, they do stabilize simple coordinated sensorimotor schemes. Second logical interpretants unquestionably give rise to recommended courses of action; but, as Peirce indicates, the actions are static sequences more tightly bound as behavioral chains (cf. Baddeley 2007: 95–96), and consequently are less subject to being severed from particular goal aggregates. Conversely, the effect of each conduct within action sequences of third logical interpretants are more discernable; hence their contribution, if any, to the over-all performance is more likely to be realized. Moreover, their status as habit is increased by the fact that they are more subject to chance and habit-change, in that they can be displaced from a conventional conduct scheme and imported to different goals.

An irritant (conflictual Logical Interpretant) can materialize on the level of concepts, assertions, or arguments (First Logical Interpretant alterations) to resolve belief conflicts within or across kinds of Logical Interpretants. Conceptual or propositional conflicts are resolved when adjustments are made to subject-predicate associations, e.g., when the meaning of "I" is extended from self-reference to any speaker. The Logical interpretant (habit) of the sign-object relation is changed from self as agent to speaker as agent. In the case of assertions (Second Logical Interpretants), a modification in action recommendations could entail sharing agent role in turn-taking exchanges; whereas changes in Thirdness-based Logical Interpretants might take the form of altering argument structure, e.g., accepting the premise that turn-taking enhances the quality of inquiry, as well as human dignity.

In sum, Habit change is equivocal to arriving at different Logical Interpretants at any of the three category levels.

Whether the change results in getting rid of modes of operation, or simply reconciling to a more veridical premise, it is nonetheless a crucial precondition. The irritant responsible for doubt regarding earlier Logical Interpretants though, does not qualify as habit because it, itself, is not the change, but rather the precondition for the change (for further discussion, cf. Bergman, this volume). The irritant is merely the catalyst and precursor for habit. However insufficient it is to qualify, it (an irritant) is a necessary precondition for habit—often it constitutes the trigger, bringing automatic, unconscious habits to conscious consideration, and ultimately bringing them to deliberate and intentional reflection. The irritant can surface as a puzzling event (actual or cognitive, probable/improbable) that becomes salient to the mind after exposure to or acceptance of a different already conceived belief/action pattern; and the two pose a conflict.

To resolve the conflict, a decision must be settled upon; otherwise dissonance will not allow either to qualify as habit, since waiver/indecision does not permit the Logical Interpretant to flourish; consequently, habit formation would be truncated. Habits may exist as originary insights (implicit knowledge), or may undergo several revisions and may be consequent to learned knowledge (either implicitly or explicitly derived). Nonetheless, neither the irritant, nor the decision to change the earlier assumption, can ever qualify as habit in itself, because habit is the resultative pattern from the decision to the change (the new Logical Interpretant). In other words, the over-all effect after many additional diverse instantiations of similar circumstances (iterations of Energetic Interpretants) need to be manufactured in the mind (drawing upon both Firstness and Secondness) to genuinely qualify as habit (Thirdness). In fact, the will to believe or a deliberate act still falls short of habit, in that it fails to embrace the result of a process—the conscious taking (c. 1913: MS 930) of new: concepts, assertions, or arguments. Essentially, "conscious taking" of a habit entails accepting an argument as one's own, which requires a discrete commitment to the meaning/effects of the argument:

> But hell is paved with good resolutions; and therefore to this promise must be attached good security, or, in other words, the resolve, which is compared to…thinking, must be baked into the hard brickbat of a real determination of the habit-machinery of his organism, which shall have force to govern his actions. A determination is a virtual habit. (1909: MS 620: 24)

In other words, at minimum, determinations are the belief-imperative to act; without such imperative—recommendations for new courses of action for self—would be truncated; and taking habits would lack "force to govern his actions." Absent determinations, the thinking would not "be baked into the hard brickbat…of the habit machinery of his organism." In short, determinations are the impetus for habit-change, where promises to act are kept; whereas resolutions constitute but scant overtures toward change.

Bergman (this volume) likewise notes that habit is not the change, but the result of the change. Habit requires some form of alteration, going beyond mere disposition or resolution. The upshot is that mere automatic conformity to previous modes of thought or action are rescued from monotonous routine, and hence (according to Peirce) oblivion:

> This is the central principle of habit: and the striking contrast of its modality to that of any mechanical law is most significant. The laws of physics know nothing of tendencies or probabilities: whatever they require at all they require absolutely and without fail, and they are never disobeyed. Were the tendency to take habits replaced by an absolute requirement that the cell should discharge itself always in the same way, or according to any rigidly fixed condition whatever, all possibility of habit developing into intelligence would be cut off at the outset; the virtue of Thirdness would be absent. It is essential that there should be an element of chance in some sense as to how the cell shall discharge itself; and then that this chance or uncertainty shall not be entirely obliterated by the principle of habit, but only somewhat affected. (c. 1890: 1.390)

It is for this reason that Peirce employs the terminology: the "conscious taking of a habit," because "taking" demarcates not merely the onset of the new regularity or Logical Interpretant but the fact that it is adopted as one's own.

Adopting a measure as one's own entails either incorporating new modes of conduct or abolishing the conduct: "The third power consists of powers of taking habits, which, by the meaning of the word includes getting rid of them, since…in my nomenclature a 'habit is nothing but a state of 'would be' realized in any sort of subject that is itself real…" (1911: MS 670). Emphasis on "taking" strengthens Peirce's claim that habit is not the change, nor what impelled the change, but the new pattern itself, which for Peirce is a conscious act: "[The] Third mode of consciousness may be briefly denominated 'The consciousness of taking a habit,' or in one word, 'habitucscence'" (1913: MS 930: 31).[2]

As such, habit change breaks through the barrier of blind adherence to previous relational trajectories (causal, and otherwise), suggesting alternative physical and logical connections. This process of meaning and structural change is a quintessential illustration of how newly conceived diagrams infiltrate physical and mental sign systems to innovate modes of representation, and how they eventually acquire status as established habit. As Magnani, Arfini, and Bertolotti (this volume) make plain, responses to newly perceived affordances represent a primary impetus for taking habits; and as Coletta (this volume) illustrates, non-living systems are equally able to change habit. The upshot is that both sentient, conscious organisms, as well as non-sentient systems exercise habit-change, making convincingly clear Peirce's central and necessary tenet—the establishment and re-establishment of courses of conduct compelled by disposition (1909: MS 637: 12). As such, all members of Peirce's categories are alive and well in habit: internal feeling as disposition in Firstness, patterns of events (action/reaction, simple routing of inanimate happenings as existential Secondness, and the continuity/coherence of either of the former processes as Logical Interpretants in Thirdness).[3]

[2]While Robin's (1967) catalogue determines the manuscript to be undated, subsequent investigation by the Peirce Edition Project has revealed that Peirce composed the manuscript in 1913.

[3]Peirce defines Logical Interpretant in MS 318 (1907) as follows: "meaning of a general concept."

In fact, Peirce explicitly states that the categories suggest the existence of three distinctive interpretants/meanings intrinsic to signs, with the latter especially relevant to habit change: "Our categories suggest them [interpretants], and the suggestion is confirmed by careful examination. I term them the Emotional, the Energetic, and the Logical Interpretants. They consist respectively in feelings, in efforts, and in habit changes" (MS 318: 43: 1907). The habit change necessarily consists in a Logical Interpretant; and it is the habit change which differentiates fleeting feelings/false action-starts, from determinations. As such, for humans, habit change requires an element of consciousness, not merely a deliberate diversion from a previous order of things. It entails some recognition, however implicit, that some previous order of things is found to be inferior, as well as a more adequate paradigm to explain and to re-govern feelings and efforts, as when attaining "the supreme art" (c. 1911: MS 674: 14–15).

In this way, Peirce's categories make their mark—habit change representing an alteration in the Logical Interpretants that govern/ground the emotive associations and actions of the event. Here novel Thirdness, as adoption of a different hypothesis for states of affairs controls; in turn, affect and effort are modified. Habit change is tantamount to a new law or abduction from which recommendations for new courses of action materialize. This is what Peirce means by "the consciousness of taking a habit" or "habituescence" (c. 1913: MS 930: 31)—the conscious exercise of self-control gives rise to a revised course of action. Essentially, by virtue of a belief shift, novel avenues to effect an outcome are implemented. Peirce's term, "habituescence", features (in a single legisign) the very nature of intellectual advancement—seeking more expedient remedies for compelling consequences. "Everything being explicable, everything has been brought about; and consequently everything is subject to change and subject to chance. Now everything that can happen by chance, sometime or another will happen by chance. Chance will sometime bring about a change in every action; or at least, this is near a correct statement of the matter as readily be drawn up…" (1894: EP 1: 220).

The issue of change and chance particularly control in diagrammatic systems where indexical relations are primary. Determinations whose inferences are grounded in event-relational components, e.g., causation, are pivotal to habit construction, because of their reliance upon indexical components, which have the advantage of inscribing themselves upon living and non-living systems alike. By virtue of its means to integrate temporal and spatial qualities of events (uniting time and space through its "universal concept…of the present" and referring to individuals (W2: 49)), index establishes, maintains, and draws new visionary event-paths—uniting unforeseen bedfellows, and reuniting/strengthening old ones. Index features motion; thus it establishes the most primary relations (Mandler 2010: 25, 2012: 431)—in some cases even without the benefit of interpretants. As such, index has the means to measure event relations, e.g., before, after and the like, prior to the assignment of identity or conventional meanings. Hence, index measures changes in event structure from early stages of semiosis (cf. West 2014 for elaboration). Through its motion toward concurrent objects/happenings, index can move observers' attention from one event to another, suggesting (or not) potential relations between them. This forms the foundation for predicting whether and to what degree precedent/unorthodox events actually trigger particular outcomes (whether expected or unforeseen). Absent index, this process (ordinarily abductive) would be compromised. As such, recognition of Index's role in determining which kind of chance is operating (Ordinary or Absolute) is pivotal (cf. Dearmont 1995).

Dearmont contends that habit permeates both ordinary and absolute chance, differentiating regularity of practice (habit as ordinary chance) from revisionary predictive processes (habit-change as absolute chance). In fact, the productivity which holds between precedent/contributory events and particular outcomes (the frequency and quality of their associations), is but one factor in determining whether Ordinary or Absolute Chance defines the habit: "Habit-taking can take one of two forms, either a tendency to reinforce previous outcomes or a tendency to counteract previous outcomes" (Dearmont 1995: 186–187). Dearmont's claims are in line with Peirce's concept of habit: Ordinary chance defines habit when subsequent events are assumed to conform to past practices; in this case little, if any, conjecture is required, only affixing different events to quite predictable outcomes. Conversely, affixing novel contributory events (especially those which are latent) to unexpected consequences characterizes absolute chance.

According to Dearmont, Peirce favors habit defined by ordinary chance, the kind of chance that reinforces adherence to previous regularities. Dearmont's assumption is precipitous; he fashions his claim on Peirce's early work and the frequency with which Peirce uses ordinary chance to illustrate instances of habit, without attending sufficiently to Peircean explanations of how ordinary chance alone (chance absent re-interpretive possibilities) can ever qualify as habit. Even as early as 1885, Peirce emphasizes the influence of Index and action upon his semiotic—that observation and implementation of action schemes tend to reinforce not merely their likelihood to resurface, but may likewise hasten apprehension of unfounded event connections/hasty errors. The character of index to compel event diagrams (physically and mentally) is the most graphic of Peirce's signs, furnishing the raw material for inferences about logical relations—the essence of Logical Interpretants. In fact, the importance that Peirce accorded to indexical features and to enactment of such as event constituents within episodes, supports Peirce's commitment to habit as absolute chance.

The present account of Peircean habit accentuates potential would-be relations across events: "…the formation of habits could be accounted for by the principles of probability, and I have shown by experiment how a certain regularity of arrangement can be impressed upon a pack of cards by imitating the action of habits. The main element of habit is the tendency to repeat any action which has been performed before… Imagine a large number of systems in some of which there is a decided tendency toward doing again what has once been done, in others a tendency against doing again what has once been done…" (EP 1: 223). While Peirce alludes to ordinary chance when he refers to the likelihood to reenact past practice (to 'repeat what has already been done'), he concludes the passage recognizing the power of non-conformity to such practices—reversing such patterns/conduct. For Peirce, Absolute Chance represents perhaps a more viable force in "taking a habit," in that consideration of how the entire trajectory of related successive practice actually produces success toward an outcome (Peirce's "supreme art") is more instrumental in habit formation than are past practices themselves. In fact, novel action pathways deriving from the presence of absolute chance ensure establishment of the most plausible assumptions about event connections, which in turn hasten more revisionary hypotheses. In short, predictions that explain the predictive adequacy of event relations are, by their very nature, more prospective in character and represent absolute more than ordinary chance. While Ordinary Chance highlights significant co-occurrences between events and their degree of influence upon one another, Absolute Chance checks the validity of established action associations in light of subsequent relevant facts and predictions for subsequent ones, and integrates collateral cognitions to preclude unfounded presumptions.

Index—A Tool to Facilitate Habit Formation

The claim advanced here is that index constitutes the single most important tool hastening habit change. The fact that index can be grounded in its object or in its interpretant, and that its sign is so locally determined (temporally and spatially), makes it a prime candidate to promote ascribing novel logical interpretants to diverse object sets. Objects of index can control the state of modal logic by virtue of whether they are static physical existent things, whether they are dynamic aggregates of conduct of a single individual (as in motion paradigms), or ultimately of interactive communicative events (whose shifts are often verbal and symbolic). Because index is so influenced by its Object, it occupies a strategic place in the semiosis of habit change. Given its dependence upon its Object, its sign becomes rather incidental, taking the form of: physical gestures, event structures illustrating some logical syntax, and/or linguistic directives with modal significance. In fact, it is attention to the position of its Objects in relation to each other that establishes the representamen of index. This relational character is responsible for the fact that index's primary function is to promote notice of locations, physical movement, and episodic trajectories, all of which constitute Objects of index. As such, the relational character of index involves implicit inferences about how it is that parts of relations work together to form a habit. In short, because Peirce's concept of index encompasses non-living, natural sign relations, e.g., smoke-fire, the weathercock, gestures, footnote, "The index is a sign that is not a thought, not conceptual, not general, not descriptive; it can therefore occur outside of thought, e.g., a symptom of a disease or other natural sign" (Short 1996: 489).

After 1885, Peirce significantly augmented his definition of index from reference to present objects to could-bes and would-bes—widening the kinds of interpretants surfacing in indexical triads (cf. West in press). Essentially, the habit associated with indexical signs later gives rise to a far more amplified modal operation involving changes in its Logical Interpretant. Short (1996: 488–490) likewise notes the sweeping alteration in index's scope, and posits that such modification of index (the universally present to its determination of the individual) not merely broadened Peirce's semiotic, but commanded far-reaching effects on his metaphysic: "In addition to broadening the scope of semiotic, the discovery of the index had other revolutionary implications" (Short 1996: 490). In 1885 (W5: 164) Peirce alludes to the expansion of index as follows: "The actual world cannot be distinguished from a world of imagination by any description. Hence the need of pronouns and indices..." Peirce develops this trajectory and its significant semiotic and metaphysical effects with his development of interpretants in his ten-fold division of signs (1904, 1906, and 1908). Peirce employed index as the sign most irreducible and consequently the most capable of illustrating interpretant augmentation. Interestingly, the trajectory with which Peirce supplies us, mirrors children's universal representational path. Initially Index is restricted to a gesture (pointing) into a co-present physical objects; and afterward it extends to physical event actions. Still later, it serves as an assertion or an argument, having would-be interpretants.

Visual indexes (e.g., gestures) result in attention to and focus on co-present objects which are likewise visually accessible; hence their interpretants are often Emotional—steeped in hypostatic abstraction in Firstness. Alternatively, when indexes force attention to the structure of an action with respect to others, (those having event motility), interpretants are likely to be Energetic in nature. Finally, when index relates two or more events into an episode classification (bearing the conventional wrappings of legisigns), its interpretants favor the logical kind. In either of the three functions, index represents a habit change, in that a habit change is proffered: the first suggesting an object qualia-related proposition, the second demonstrating a novel way for participants to interact, and the third recommending conditions for resultant states of affairs (cf. NEM IV: 248). In all cases, index impels others to take notice of implied propositions and arguments about object, action, and event utility; consequently, they qualify as indexes which convey information (implicitly or explicitly) (Bellucci 2014: 536).

These informational Indexes constitute the earliest and firmest foundation for the establishment of habit for Peirce; and habit, by its very nature, consists in a Logical Interpretant, even early on in ontogeny via implementation of physical gestures in attentional paradigms. Habit, as Logical Interpretant, later takes the form of implicit well-founded inferences regarding event participant expectations and ultimately surfaces in assumptions of event contributory effects. In these cases, the viable regulatory construct underlying the inference is law-like and open to revision, and hence constitutes habit. The law-like but regulatable paradigm gives rise to inferential structures (habit) that in ontogeny is especially subject to semiosis—advancing to higher levels of regenerative representational facility. In this way, indexical gestures implying imperatives (gaze, pointing) to another become indexical legisigns ("that", "this") that command attention; and afterward, modal operators ("may", "might") control how events are cast ultimately to influence/control another's conduct. Accordingly, Index serves as an imperative to recommend courses of action and belief, illustrating Peirce's concept of habit on increasingly higher levels of representation; it forms a "real and living logical conclusion" (1907: 5.491). In short, Index, especially given its informational character, is an implicit source to trace and monitor progress in living action schemas and in scientific genres.

While indexical signs are ordinarily associated with Secondness, in view of their dependence upon object-sign coexistence (1902: 2.304)[4]—drawing attentional templates in the physical context, they compel the operation of the other two categories. The relevance of Firstness and Thirdness become obviated when index takes on an informational character (either implicitly or explicitly), advancing the recognition of regularities and changes in regularities in the physical and mental worlds (habit). In fact, the most primary expression of habit is Secondness-based spatial and temporal sequences of the state of things in the physical world, even prior to the brute-force experiences of sentient beings where consciousness becomes relevant: "One of these kinds is the *index*, which like a pointing finger exercises a real physiological *force* over the attention, like the power of a mesmerizer, and directs it to a particular object of sense" (1885: 8.41). Peirce continues down this path by arguing "…B. *Designatives* (or *Denotatives*), or Indicatives, *Denominatives*, which like a Demonstrative pronoun, or a pointing finger, brutally direct the mental eyeballs of the interpreter to the object in question, which in this case cannot be given by independent reasoning" (1908: 8.350). Gradually, regularities in Secondness insinuate themselves upon the mind (whether consciously or unconsciously), such that they constitute foundational (perhaps universal) tracing devices, ultimately utilized to generate plausible hypotheses in Thirdness.

This model draws attention to the primacy of directional signs in meaning-making, particularly in the hypothesis-making/inferencing process—because it is via this process that habit reaches its destiny. Index finds its mark in animate and inanimate venues alike—it sets parameters that objects/persons inhabit —determining their within-space; as such, it establishes figure-ground imaging. In establishing within-space for events, Index foregrounds entities, places and actions, and preserves them within a context to be memorialized when the inferencing process draws upon past observances. Intellectual Indexes likewise can promote apprehension of novel regularities in which forward-thinking inferencing elevates habit to a prospective endeavor (akin to claims of both Kilpinen and Bergman, this volume). Index facilitates the existence and recognition of habit, in that it serves as the instrument to situate inhabited events in their fullest sense, pinpointing the location of participants, and determining the breadth of each event's potential range of application. As situator, Index localizes objects and event participants, measuring near and far distances from implicitly established landmarks. By directing gaze toward landmarks, then toward objects inside the contained space, Index establishes a visual and semantic frame to establish and guide propositions. This is so given the pivotal influence produced by anticipated object interaction (expected effect/s that are often latent) inside of contained spaces (cf. West under review). Hence, perceptual judgments that draw upon perceived distance and orientation with respect to

[4]"An *index* is a sign which would, at once, lose the character which makes it a sign if its object were removed, but would not lose that character if there were no interpretant."

the landmark/s or from the parameter of the contained space qualifies as habit quite early in ontogeny, clearly illustrating the inception of abductive rationality.

These foundational perceptual judgments constitute the earliest abductions and proto-habit; they cannot qualify as full-fledged habit. The fact that self-control is still in need of refinement. Although at 0;3.5 infants utilize landmarks to locate and identify other objects to which they ultimately wish to attend (Quinn 1994: 66–67), what is absent at this early developmental stage is both regularity of conduct and a semblance of self-control. Until children intentionally (not merely deliberately) employ particular landmarks toward a similar purpose, and come to substitute originary perceptual judgments consequent to revisionary judgments, they do not clearly display self-control. Self-control is obviated when an earlier hypothesis is either supplanted or is significantly reformed by a more fitting one, such that habit change liberates activity from blind adherence to convention. In fact, habit change demonstrates that the underlying judgment is impelled by a Logical Interpretant—a newly conceived effect whose conditions regularly surface under similar conditions.[5] At this later stage, revised hypotheses are associated with a kind of precedent condition, not a single, isolated object. This qualifies as habit, in that it is a deliberate and conscious consideration of Logical Interpretants (usual effects if an object is within a particular distance from another), are not derivable without some reference to Energetic Interpretants of retrospective and prospective observations of states of things and in their changes. In this capacity, index draws attention to objects' stationary arrays and to their patterns of motility.

This path-finding function that Peirce readily attributes to index, is, in fact, more primary than that of iconic imaging. Two rationale support this line of reasoning: the first emanates from the ontogeny of infants' mental files, while the second is derived from state of the art findings about the foundational components necessary for making salient visual or tactual trajectories. Without demarcations supplied by indexes (either in elementary Firstnesses or existent Secondnesses), images would lack form, and would lack crispness to make them attentionally ripe (cf. West 2016 for more a more in-depth analysis of visual index). Emblazoned Indexes form directional templates early in development, priming the semiosis of more objective logic-based regularities. While a predisposition may preempt salience of indexical templates (gaze trajectories, motion, and force toward objects), more complex indexical regularities, e.g., attention to landmarks, perspective-taking skills, constitute acquired, more consciousness-based habits.

In CP 2.297 (1894), Peirce makes plain that habits can either be acquired or can exist as a predisposition. In either case, Peirce's concept of habit must exhibit a degree of continuity, vis-à-vis its regular actualization in indexical templates. The earliest indexical templates include: tacit recognition of the effect of motion and force on particular objects, together with appreciation of the actor's likelihood for

[5]"[Logical interpretants] must, therefore, be identified with that 'meaning' which we have all along been seeking. In that capacity, they are habits of internal or imaginary action, abstracted from all reference to the individual mind in which they might happen to be implanted, and those whose future actions they would guide."

independent means to change course during the trajectory of motion. Additional indexical templates consist in: noticing participation slots, apprehending event characteristics, and recognizing relations between events, e.g., cause-effect. The latter entails appreciation for the effect of neighboring events—those which constitute simultaneous and sequential actions and states of affairs. Later assuming the Origo of another within indexical event templates illustrates how Index continues to hasten apprehension of regularities within and across event roles, such that participant roles determine certain orientations and distances. This habit relies upon mastery of a rather complex system of deictic competencies.

Mastery of these deictic competencies validates Peirce's claim that regularity does not culminate in extreme mechanistic conformity, because these indexical competencies involve different actors, different events, different loci of events, and different orientations to objects and other actors. In fact, should some degree of nonconformity fail to materialize, all that would exist is an absolute without any access to semiosis, since alteration and truth-seeking via the Final Interpretant would be truncated (c. 1890: CP 1.390). Ultimately, Peirce's use of habit transcends conformity to compulsory participation in events—occasional non-conformity to a pattern is essential to what he means by "habituescence," or, the conscious awareness of taking a habit (c. 1913: MS 930).

Developmental Considerations

Index traces infants' knowledge and expectations regarding graphical arrays, foundational to the establishment and promotion of diagrammatic conceptualizations and reasoning. The significance of this developmental account is to capture graphical knowledge from its emergence, and to show the twists and turns of the logical interpretants associated with each graphical advance. To adequately trace the meanings/effects accorded to graphical arrays across development, the evolution of index's use must be examined at a micro level; and habit changes from one kind of logical interpretant to another can shed particular light on the distinctive complexions of habit.

Initially, habit takes the form of index in production of tongue movement, typically in response to observation of another's movement of same. This regularized behavior has been documented at 0;1[6] (Meltzoff and Moore 1977). Although at first it may appear to constitute habit, with a closer analysis it is but an involuntary conduct, and appears not to qualify as a sign, given that it often lacks an interpretant consequent to its unintentional character. It is, in fact, an illustration of proto-index since it does not demonstrate voluntary behavior where interpretants flourish consonant with habit-change, but rather involuntary behavior. Because this

[6]This notation signifies age in years followed by months; this is the convention employed in developmental science literature.

conduct emanates from involuntary directional movement (Mandler 2004: 31), it does not rise to the level of habit—since it does not illustrate a first logical interpretant. The latter must constitute, at least, deliberate, and arguably conscious, intentional physical directional movement to qualify as habit.

Between 0;1 and 0;2, infants engage in pre-reaching conduct—ballistic arm movement and fist swiping toward objects that they notice in the near surround (Bruner and Koslowski 1972; Trevarthen 1977), demonstrating ascendance to more volitional, directional behavior, approaching first logical interpretants. Following the path of another's eye gaze at 0;2 (Scaife and Bruner 1975: 265), likewise supersedes involuntary graphical demonstrations. At 0;2.5, infants notice (via gaze-following) the entry and exit of a person from a room (Hespos and Baillargeon 2001: 145). Using indexical gestures (gaze) to follow the movement of an object/person is likewise proto-indexical at this stage (at approximately three months). Even these more goal-oriented behaviors fall short of carrying logical interpretants, since none of the behaviors entail sources or goals, only formulaic paths—hence, referent points are un-established. As such, they do not gain sign status as indexes, but constitute proto signs, without logical interpretants. In some cases, Energetic Interpretants are connected with early indexical use, given notice of directional tracing without a beginning or end point. Additionally, there is no discernable evidence that infants have, even implicitly, constructed perceptual judgments from simple percepts—a qualification of first logical interpretants; neither following with the arm/hand, nor with the eyes reveals an implicit proposition applying to the subject, which underlies perceptual judgments: "Every concept, every general proposition of the great edifice of science, first came to us as a conjecture. These ideas are the first logical interpretants of the phenomena that suggest them…" (1907: 5.480).

Soon thereafter, infants demonstrate the onset of more full-fledged habit, with performance of some degree of intentionality in planned sensorimotor goals for ego. Infants likewise demonstrate inferential reasoning, which indicates that they are at the threshold of habit change, consequent to reasoning in the form of conjecture. Inferential reasoning is present when they become startled after observing results different from what was expected. The following findings support inferential reasoning and intentional conduct. At 0;3.5 infants' direct attention longer toward the location of surprising consequences, e.g., upon the unexpected occlusion of an object behind a screen (Baillargeon 1987: 662). Gaze following and reaching for objects in the light at 0;4 demonstrates still more solidified apprehension of graphic relations and habit-change, in that inferences are made prior to reaching for and grasping objects. Accordingly, how to fashion the hand and determine arm length are prior considerations to be successful at grasping. Realization of different distances and object shape/location are indexical considerations pivotal to targeted reach. In other words, targeted reach is successful upon several viable inferences; and these inferences are drawn specifically from indexical features of like objects which, as observed, have formed habits of operation (perceptual and functional) within their typical contexts. Habits which materialize consequent to first logical interpretants include: lengthening arm extension, pulling the object toward ego, and the like. These indexical behavior changes are derived from the location of objects with respect to spatial reference points which are often apart from infants' own bodies. This constitutes a habit-change, since beforehand, the source and only zero point for finding objects and for determining distances is ego. In other words, at this age, infants perceive more than a single reference point to find and obtain sought-after objects. They likewise are increasingly cognizant of the different sources, different paths, and more than a single goal/consequence. For example, infants at 0;4 determine that they can attain the sought-after object by dragging it with their feet, biting it, or by grasping it either with two hands, or a single hand. Although these constitute physical habits by way of habit-change, underlying the change is an accommodated belief structure (likewise a habit) foundational to how the objects ordinarily appear or how they function.

Piaget and Inhelder (1969 [1966]: 10) bolster the present claim—implying the necessity of logical interpretants for habit and habit-change. They evidence the fact that targeted reach represents intentional guided grasping such that the grasp is not a consequence of accidental or random object attainment, but of purposive, measured extension of the arm, consequent to planning guided by some inference, however rudimentary. Here targeted reach obviously demonstrates implementation of graphical indexes which, although conforming to a template, are, nonetheless, revisionary, demonstrating habit-change.

Perceptual judgments appear to materialize, however tacitly, given infants' awareness of the effect of one action upon another, e.g., looking briefly at expected consequences, while looking longer at surprising ones, such as a toy emerging or not from behind a screen. Looking longer upon re-emergence indicates a judgment that the toy's re-appearance was not expected. The infants' conduct illustrates a primitive form of conjecture—that is present in Peirce's first logical interpretants:

> These ideas [conjectures] are the first logical interpretants of the phenomena that suggest them, and which, as suggesting them, are signs of which they are the (really conjectural) interpretants.... Meanwhile, do not forget that every conjecture is equivalent to, or is expressive of, such a habit that having a certain desire one might accomplish it if one could perform a certain act. (1907: 5.480)

As such, direction and length of gaze (looking time) represent in graphical form, infants' conjectures regarding what happens to objects once they are hidden—do they simply cease to exist? Here, children do "accomplish it" when they "perform a certain act," namely, looking/reaching toward the object and its previous location—where index plays a primary role.

The influence of mental index—Visual Working Memory (VWM), internalizes the indexical action-habit. It evidences Index's early role in facilitating object-finding and recognition. The first appearance of index files is documented at 0;5 (Leslie et al. 1998: 11; Leslie and Káldy 2007: 117), when it situates objects, associating them first with their physical placement. Leslie et al.'s (1998) and Leslie and Káldy's (2007) findings that indexical information constitutes the first to be noted/remembered in memory files demonstrates that, universally children use index (object's location) first to identify objects, not object attributes. In fact, "At five months, infants will index-by-location without binding features [of the object] ..." (1998: 17). Oakes et al. (2007: 89) indicate that non-indexical attributionary features, e.g., color, are not bound to location with a single object until 0;8; and not until 0;10, do infants mentally bind the same color to multiple objects (Oakes et al. 2007: 85). Nonetheless, it is not until 1;1 that color, as a feature of an object file, is bound to the object index of that file. Essentially, infants rely upon indexical features (location), rather than qualia to think and act, as demonstrated by these early index files. As such, indexical signs frame infants' earliest meanings and constitute the primary device to regulate how often particular responses are likely to materialize.

Indexical signs whose representamen constitute directional gaze are utilized to find objects, as well as to indicate surprise upon seeing their re-appearance. In fact, attention via gaze to neighboring objects was utilized to measure success at object-finding—visual notice of a landmark with respect to the object determined success at locating the sought-after object. Quinn (1994: 66–67) reports that at 0;3, infants recognized (looking longer) that objects had been displaced from their previous location, i.e., above or below a particular point of reference, when they visually fixated on the point (landmark) prior to the displacement. Moreover, Bremner and Bryant (1977) found that the more salient the landmark, the more likely infants were to notice them and utilize them for object recovery. They reported that at 0;6 infants exploited landmarks (table edges) to recover objects when the table's sides were covered with distinctive cloths (cloths displaying particular geometric shapes). At 0;6, infants demonstrate the means to associate their own actions in movement scenarios with landmarks (Acredolo 1988: 369); and at 0;9 they begin to "demonstrate an appreciation for the spatial designation function of landmarks… that one object or feature can 'stand for' another in one's memory…" (Acredolo 1988: 370). The fact that attention to the landmark leaves a trace in memory, and as such hastens object-finding illustrates, among other things, the effectiveness of index in suggesting object relations and ultimately in promoting habits of thought and action.

It is the presence of a logical interpretant underpinning the indexical use (another near object means that the sought-after object is likewise located in a certain proximate position) that affords infants the raw material to utilize index so effectively. The presence of a logical interpretant (in this case the meaning of the landmark in picking out the object of choice) energizes infants to reach and obtain such objects. Infants are impelled to act (object-find) by virtue of logical interpretants, not merely energetic ones, given their application of a meaning that fits all behavioral attempts—acting upon their belief that finding one object will show them another. It is obvious then, that the logical interpretant, bordering on the lower second level, is crucial in developing and revising habits of belief and action at this juncture. It is the lower second logical interpretant that motivates specific voluntary modes of action, which begin to surface at this stage: "In the next step of thought [beyond the first logical type], those first logical interpretants stimulate us to various voluntary performances in the inner world. We imagine ourselves in various situations and animated by various motives; and we proceed to trace out the alternative lines of conduct which the conjectures would leave open to us" (1907: 5.481). Second lower logical interpretants do not unequivocally underlie habit altogether at this point, since infants do not yet possess the requisite deictic skills to "imagine" themselves "in various situations and animated by various motives"; nor can they "trace out the alternative lines of conduct which the conjectures would leave open; ...", but indexical competencies operating at this stage do mark the onset of their influence. What is orchestratable at this stage is are indexical memory processes, which is equivocal to exercising "various voluntary performances in the inner world," because in the inner world of memory (via index files and traces derived from physical landmarks) children select objects, identify them, and code their precise location.

Indexes determined in the inner world are transformed into action-habits (cf. Stjernfelt 2014: 118) in an effort to show the self or others the points of reference which have been internally graphicalized. Many external indexes surface as visual devices, e.g., pointing, simply in response to the virtually universal compulsion to individuate objects, which is a primary function of index. In fact, pointing, whether to establish and maintain object-focus for self or to share it with another, constitutes a precursor to naming (cf. West 2015a, b)—a pre-linguistic indexical marker for topic formation. Although unidirectional gestures (noticing objects for self only) begin with gaze-following and prehension, they become further refined in hand shape and in extent of reach. With index, infants can select objects for focus which are increasingly further from them and other observers, to the degree that eventually objects of physical indexes are absent from the spatio-temporal context (cf. West 2011a). To this end, index modifies its representamen to pointing with the index finger at 8 months of age (Bates 1976: 61; West 2011b: 92; West 2013), which develops in social contexts as arm extension in giving and receiving exchanges at 9 months and thereafter (Carpenter et al. 1998: 681; Volterra et al. 2005: 9). When extending the arm, infants often need to alter their hand shape to accommodate the object to be received—taking the form of an open hand, cupped hand, et cetera; and the co-occurrence of eye gaze to facilitate and coordinate motor indexes (hand and arm) is still indispensible. Nonetheless, indexical gestures at this stage are for ego alone (ego as the only origo), such that they are not concurrent with eye gaze toward another, nor mutual eye gaze exchange (Carpenter et al. 1998: 153).

Index as Perpetuator of Image into Action

In the face of relatively undeveloped cognitive and pragmatic systems, even at 0;9, infants rely heavily on affect (less on effort or logic) to drive action. To illustrate, feelings/preferences are affective components which motivate the use of early indexes. Directional gaze and gaze coordinated with reach and/or pointing, are elicited by a Firstness-based agent, namely, subjective preference for an object and/or care-giver's encouragement. However, in perfecting what for Peirce is the "supreme art," effort becomes an indispensable agent for implementing novel action-habits. In MS 674: 14–15 (c. 1911), Peirce models how effort increasingly influences affect in ontogeny (and the reverse); and, in MS 620: 22–23 (1909), he discusses how modes of conduct become implemented by virtue not merely of the will to act, but consequent to a discrete determination: "Namely, it [ratiocination and moral virtue] begins with running over in one's mind how one has conducted one's self on each occasion, and in asking one's self whether one's actual conduct was in all respects such as was best calculated to effect the achievement of one's heart's desire." A culminating feeling/affect ("one's heart's desire") is accomplished only with the expenditure of significant effort, in performing novel action-habits which are directed by determinations. Peirce continues (later in MS 620: 22–23) illustrating the process of arriving at determinations toward habit-change: "The first step toward this is to recognize, and get a lively image of, what one's heart's desire really is;...One's very realist desire is not whatever one may happen to desire, but is what he would desire upon sufficient information and a sufficiently detailed imagination." For Peirce, imaginations (particularly those for which self-control is exercised) are indispensible for implementing determinations, especially those developed from creative hallucinations in the inner world, and hence for habit-change.

Nonetheless, later Peirce demonstrates how effort, if gradually harnessed in children prior to 5;0, impels action to ascertain Peirce's supreme art. The imperative to act out one's imaginations proceeds as follows:

> 1st, how to make an effort; 2nd, how to make a great effort, preceded by that mysterious action, or brief voluntary process, (it should be deliberate) by which the various elementary powers that are to be simultaneously put forth on the 'great effort' may be coordinated. This power of 'gathering one's forces' is an art (and as such a habit); and is to be cultivated on the same general principles as any other; 3rd, how a performance is to be facilitated by repetition; 4th, how it is still more facilitated by intense attention to the precise modes of effort and precise feeling of effort at each stage of the performance. (c. 1911: MS 674)

While feeling constitutes a primary impetus, it is effort that energizes novel action-habits, supplying the direct link for habit-change.

Effort, which implements determinations, continues to produce far-reaching higher order developmental benchmarks. Between 1;0 and 1;2 search (gaze as index) for hidden referents upon other's request materializes (Baldwin and Saylor 2005); and at 1;4 joint gaze (both parties in an exchange looking toward one another) emerges (Saylor 2004: 608). The emotive nature of this joint index triggers another's compliance with children's implied imperative (their effort) to "look at what I am looking at"; it tacitly expresses an intent to influence another's line of focus and to confirm the legitimacy of noticing that particular object (Bates 1976: 61). Bates contends that joint gaze is, in fact, triangular at this stage—from children to care-giver, to object, then again to care-giver, confirming the universal function of index to itself confirm the soundness of object salience. Children are obviously exercising determinations—employing attentional indexes to gain "their heart's desire;" as such, their effort is to align another's focus to their own.

Nevertheless, with efforts to implement joint index, deictic composites are mastered—accommodating for the shifting locations, orientations, and perspectives of external referent points (Clark 2009: 166). In other words, when children make index joint, they augment its logical interpretant—heralding its use as role administrator. This surfaces especially when index is used as a legisign to identify conversational and narrative roles, e.g., "I," "this" "here" indexing locations accorded to speaker who is likewise a participant in the narrated event. These kinds of indexes qualify as informational ones (cf. Stjernfelt 2014: 60), in that they are akin to indexes containing icons and accompanied by legends (Dicisigns); their use implies the same level of information that the picture with a legend makes explicit (c. 1902–1903: 5.543). Like pictures, these pronouns incorporate iconic features, i.e., speaker's visual likeness and voice; but, their interpretants are far more conventional and objective. Unlike pictures bearing legends, the meaning of these pronouns need not be augmented with an explicit title, because their codification as conversational and event participants is sufficiently implied to communicate the logical interpretant. In fact, despite the shifting nature of particular objects associated with pronominal indexes, their meaning is, nonetheless, obviated by concurrent auditory indexes—speaker's voice marking speaker's location/orientation. Moreover, the degree of practice implementing conversational indexes is extensive, even at young ages, since speaker-addressee turn-taking is negotiated in virtually every verbal exchange.

The productive use of these indexical legisigns, beginning at 3;0 (West 1986: 142, 158; West 2011b: 95), approximates meanings accorded to higher second logical interpretants, since indexical use is normative and traces different contextualized lines of focus that represent worlds far beyond children's own localized, parochial ones: "The second logical interpretants constitute the ultimate normal and proper mental effect of the sign taken by itself (I do not mean removed from its context but considered apart from the effects of its context and circumstance of utterance). They must, therefore, be identified with that "meaning" that we have all along been seeking. In that capacity, they are habits of internal or imaginary action, abstracted from all reference to the individual mind in which they might happen to be implanted, and whose future actions they would guide. For it must never be forgotten that habits called internal, as having been produced by internal exercise[,] take effect in external actions, unless a particular inhibition has been laid upon such action" (MS 318:46 [1907]).

Deictic uses of pronouns carry higher second logical interpretants consequent to supreme dependence upon their context; and because the general, codified meaning is objective, their meaning is "abstracted from individual minds", in that the objective meaning via the logical interpretant (speaker and/or speaker's location) is implied by the sign, and as such can be applied indiscriminately to every use. These kinds of indexes qualify as habits, produced "by internal exercise" (by virtue of their status as implied argument) and realized in "external action". The implied argument is equivocal to: this particular individual who is narrating establishes the point of orientation (physically and psychologically), unless otherwise specified. Furthermore, these lines of perspective come into being through "internal or imaginary action", in that the location of objects is seen in the mind's eye as if the imaginer were the speaker. In short, deictic pronouns are quintessentially "habits of… imaginary action", because they draw a perspectival path often not directly experienced at that moment, which illustrates the source (speaker) and the goal (an external entity under focus or an internal point of view). In fact, in identifying the source and goal (beginning and endpoint), the path is established anew. In the productive use of "this", what constitutes internal habits are imaginings either of geometric relations of near, as opposed to far objects from that particular speaker's vantage point, or the speaker's likely idiosyncratic likes, dislikes, and logical capacities given inferences drawn from the respective knowledge-base. What "take [s] effect in external action" is the reciprocal give-and-take field of negotiations within conversational speaker-listener roles, and/or the affective and cognitive perpetual re-orientation inherent in agent, receivership, and benefactor role shifts in narrated events. It is indisputable that the advance from use of index as gesture to its use as linguistic placeholder and path-tracer illustrates a formative shift from lower second to higher second logical interpretants.

This shift is hastened by attainment of the ultimate stage in Peirce's Supreme Art, in which effort and feelings are informed by logical interpretants. In the seventh stage (c. 1911: MS 674:14–15), Peirce illustrates how determination fosters self-control; and as such, new logical interpretants are realized in the concentrated enactment of conduct:

> …7th, how the facilitation asserted in the 4th point, where it is caused by attention to feelings, where the attention of the nature of an inward exertion of power, is perhaps even greater when a different kind of exertion is substituted for the attention to feeling, this different kind of exertion being describable to a person who has experienced it as an act of giving a compulsive command to oneself. Some books call it 'self-hypnotization,' whatever that may signify. This [is] effective whether there be any 'disposition,' i.e., any imperfectly developed or otherwise imperfect habit, or not. (MS 674: 11–14 c. 1911)

The level of self-control characteristic of this seventh stage supersedes the art of inner focus by icon or envisionment alone. It relies more heavily upon index in Secondness—the self's performance of an action or set of actions, which, when orchestrated, defines the future direction of the goal. In other words, enacting the feeling from Firstness supplies a more particularized imperative for the individual when it is concretized in Secondness. In short, the action itself, or (better said) the process of taking that action, confirms the plausibility of a new hypothesis, and propels a suitable course of remediative action. Essentially, an image/icon conceived of in Firstness by "self-hypnosis" (although often vital for establishing habit-change) can easily fade in the absence of a course of directed action (action habits with indexical templates) in Secondness.

The Habit of Inter- and Intrapersonal Dialogue

Peirce makes the argument that determinations, as the epistemological mechanism of virtual habit and the force behind abduction, are rooted in dialogue. The quintessential characteristic of dialogue is its commitment to illustrating more advanced indexical templates. It can do so inter/intrasubjectively. The former develops first, and has the advantage of informing children of a perspective apart from their own; despite the obvious benefits of this kind of dialogue, it fails to inform absent shared interpretants between conversational partners: "No man can communicate the smallest item of information to his brother-man unless they have που στωσι [a place to stand] of common familiar knowledge; where the word 'familiar' refers less to how well the object is known than to the manner of knowing" (MS 614: 1908). "How the object is known" is what supplies the common place to stand/shared knowledge for interlocutors, according to Peirce. Knowing the object principally entails the meanings that have been associated with the sign and its object, namely, the interpretant. This is so particularly since knowledge of the object is derived from either direct or indirect experience—from tangible interaction, or from linguistic reference to it by interlocutors. In short, it is not fastening upon meanings/effects implied in the conversational partner's arguments that precludes communication, nor inferring others' logical interpretants that constitutes the greater challenge, but discerning their emotional or energetic ones.

Intrasubjective dialogue, on the other hand, need not meet the requirement of a shared "place to stand", because although the interpretant may be distinct between different but simultaneously considered perspectives, they, nonetheless, exist within the same mind, and hence meanings are not hidden but latent. With dependence upon indexical charting of argument structure, intrasubjective dialogue frames and traces inner conversation, incorporating both present means to convince the speaker-listener self, together with contingencies arising subsequently; hence it underlies actual inference-construction, as well as hypotheses that might surface later. In 1908, Peirce posits to Lady Welby that, "all thinking necessarily is a sort of dialogue, it is an appeal from the momentary self to the better considered self from the immediate self to the self that is general and future" (SS: 195). Premises in progress are developed by the "momentary self"; while those that are potential derive from "the better considered self". The latter approximate habits—they are but virtual habits, in that they are mere potentialities. Nonetheless, because these potential habits provide opportunities for habit-change in deciphering new event relations, they are primary to habit proper. Talking to the self is a form of action albeit internal; it demonstrates how one can transitionalize from either a conventional/idiosyncratically derived argument that is flawed, to intermediate and final belief structures; hence Peirce refers to the "intermediate…future self". Necessary to following arguments generated internally is index, with its power to establish and maintain attention to propositions/assertions that are topicalized from those which are simply old information and de-emphasized. Advanced mental indexes trace the principle components of the argument, and integrate minor elements.

The kind of dialogue to which Peirce refers is orchestrated within—it uses the self as the source, path, and goal to conceive of insights in the form of creative hallucinations in Firstness, or to deliberate on the relative validity of one argument over others in Thirdness. To determine the sufficiency of each argument, it must be analyzed and synthesized; and the comparative strength of arguments must be considered to select which is most viable. It is then when new habits have enough specificity to surface. Accordingly, habit through dialogue requires sufficient memory resources to adequately representationalize and organize the many factual ingredients, and arrange them in logical sequences.

For Peirce, inner dialogue constitutes a viable habit to fuel action, and at the same time, it fosters mastery of the supreme art (c. 1911: MS 674: 11–14)—not merely as inter-subjective communication to recommend a course of action (1909: MS 637: 12), but as internal dialogue to command the self to take up a new action habit. This qualifies as forerunner of Vygotskiian intra-subjectivity; this inner dialogue carries a command or imperative to behave in a certain way under certain circumstances, thus directing and shaping subsequent inferences (cf. West 2015c).[7] The process of commanding the self to take up a course of action solidifies what once were fleeting beliefs—making them determinations. Through self-talk or inner speech children specify the integral structure of how events logically serve one another; then they enact such—and find greater success targeting action to goals/consequences.

Once egocentric speech makes explicit arguments that were implicit, the process reverses itself, such that explicit arguments become implicit in inner speech (though the arguments are conventional, not diverging from paths already trodden). Only in inner speech do arguments rise to the level of virtual, obviating which inferences qualify as plausible and abductive. It is obvious then, that inner speech (articulating each step of novel action-habits) is a useful remediator; it establishes which inferences are viable that, in turn, liberates the working memory system by making the habit more automatic, and by providing resources for operations of inferential reasoning (cf. Baddeley 2007: 198–203 for a more detailed account of the functionality of the working memory system).

At this juncture, when thought and language become one and when index becomes more mental, working memory has increased means to administer semiotic and logical relations. This paradigm of giving silent voice proceeds according to Vygotsky's paradigm of—articulated arguments, to whispered arguments, and finally to a form of inner speech—arguments that mature in the mind (Vygotskii 1962[1934]: 16–17, 149). This shift in the functionality of working memory resources permits children to more effectively reflect upon a greater number of arguments simultaneously, establishing new means to govern facts, would-bes, and possibilities. The internalization of language actuates a higher and more efficient course of mental action—integrating assumptions and determining which hunches are reasonable about states of affairs. This higher mental capacity introduces new habits (habit change), which ultimately facilitates abductive reasoning. It does so by introducing a novel logical interpretant—one that approaches the third logical kind. The process of self-talk introduces an ultimate interpretant to indexical signs, because transcending from the has-been and the is, to the may-be and the would-be

[7]"[To] believe the concept in question is applicable to anything is to be prepared under certain circumstances, and when actuated by given motives, to act in a certain way" (1907: EP 2: 432).

(in the case that particular deictic shifts are met) elevates meanings (emotional, energetic, logical) to possible worlds. Hence, habit enters the realm of epistemic habit-change—creating virtual percepts, virtual judgments, and virtual worlds.

Index's means to discern and follow the complexion of implied logical and modality relations through dialogue is reminiscent of Peirce's third logical interpretant: "In the third logical interpretant, or interpretants, the work of the intellect comes to a demicadence, a provisional and partial consummation; so that it is of supreme logical importance" (MS 318:46–7 [1907]). Index measures the "demicadence" of event relations, showing the particular wheres and whens of focus-shifts from action to action among event partners. Index's means to represent the individual in changing contexts supplies the deictic character necessary to capture event progression; it brings the work of the "intellect to a demicadence" by showcasing new sources for actions, and by capitalizing upon new intents/speaker-goals. In this way, the "demicadence" of intellectual matters is incorporated into the composition of argument structure—by drawing an attentional connect between events/persons which once were disconnected.

In dialogue, index integrates logic with experience, the ultimate form of the third logical interpretant: "a reconciliation or interadjustment between reason and the facts of experience" (LI 392 [1908]) is the substance of the third logical interpretant. "Interadjustment between reason and the facts of experience" is accomplished when gaze from one party to an event exchange invites/commands another party to exercise a particular function in that conversational or narrated event (e.g., suggesting that the new-comer inflate a ball needed in a game), ordinarily an event in which the agent is likewise participating. An interadjustment is made when facts of experience (e.g., need to assume the role of inflator) are initially incongruent with the current state of logic, e.g., balls are not inflatable by human agency. Reconciling the two (e.g., the agent who received the implied imperative) can orchestrate the inflation, given the different characteristics of this ball—that it is hollow. Ultimately, the "work of the intellect comes to…a provisional and partial consummation; so that it is of supreme logical importance" when the child agent actually changes belief and action-habits (to fill the ball with air/water), demonstrating the depth of the reason-experience-adjustment. Inner/intersubjective dialogue provides a new forum to repackage action icons into more informational signs whose interpretants factor in objective meanings, hence to convert energetic Interpretants into Logical ones.

Conclusion

Novel logical interpretants are, without question, exemplary of habit, since they incorporate chance, and change—ultimately resulting in non-mechanistic regularity in which some element of self-control resides. Children ascertain new Logical Interpretants by using increasingly more refined semiotic instruments—namely, indexes clothed with intellectual properties (cf. West in press). During the course of development, alterations in the objects and interpretants of index give rise to new semiotic instruments; and learning to use these instruments to advance the state of logic toward reaching the "supreme art" is Peirce's primary objective. Engaging in intra/intersubjective dialogue demonstrates how index is responsible for advances to higher levels of logical and semiotic functioning—from physical signs to physical objects virtually devoid of an interpretant, to linguistic indexes whose objects and interpretants trace perspectival-shifts. In the latter case, index shapes either intra-subjective event constructions within a single mind, or inter-subjective conversational alternations, suggesting new logical foci and locality templates in speaker-listener episodes. Thus, the attentional shifts which index affords both illuminate logical relations from event to event within a single mind, and trace variability in modality—highlighting which event partner issues the imperative and which receives it—to look somewhere for informational focus. That "somewhere" necessarily involves taking a habit by accepting as one's very own a habit-change —a new logical interpretant.

References

Acredolo, Linda. 1988. From signal to "symbol": The development of landmark knowledge from 9 to 13 months. *British Journal of Developmental Psychology* 6: 369–393.

Aliseda, Atocha. this volume. Belief as habit. In *Consensus on Peirce's concept of habit: Before and beyond consciousness*, ed. Donna E. West and Myrdene Anderson. (Studies in Applied Philosophy, Epistemology and Rational Ethics [SAPERE]) New York: Springer.

Baddeley, Alan. 2007. *Working memory, thought, and action*. Oxford: Oxford University Press.

Baillargeon, Renee. 1987. Object permanence in 3½ and 4½-month-old infants. *Developmental Psychology* 23(5): 655–664.

Baldwin, Dare A., and Megan Saylor. 2005. Language promotes structured alignment in the acquisition of mentalistic concepts. In *Why language matters for theory of mind*, ed. J. Wilde, and J.A. Baird, 123–143. Oxford: Oxford University Press.

Bates, Elizabeth. 1976. *Language and context: The acquisition of pragmatics*. New York: Academic Press.

Bellucci, Francesco. 2014. "Logic, considered as Semeiotic": On Peirce's philosophy of logic. *Transactions of the Charles S. Peirce Society* 50(4):523–547.

Bergman, Mats. this volume. Habit-change as ultimate interpretant. In *Consensus on Peirce's concept of habit: Before and beyond consciousness*, ed. Donna E. West and Myrdene Anderson. (Studies in Applied Philosophy, Epistemology and Rational Ethics [SAPERE]) New York: Springer.

Bremner, J.Gavin, and Peter Bryant. 1977. Place versus response as the basis of spatial errors made by young infants. *Journal of Experimental Child Psychology* 23: 162–171.

Bruner, Jerome S., and Barbara Koslowski. 1972. Visually preadapted constituents of manipulatory action. *Perception* 1: 3–14.

Carpenter, Malinda, Katherine Nagell, & Michael Tomasello. 1998. Social cognition, joint attention, and communicative competence from 9 to 15 months of age. *Monographs of the Society for Research in Child Development,* Serial 255, 63(4).

Clark, Eve. 2009. *First language acquisition*, 2nd ed. New York: Cambridge University Press.

Colapietro, Vincent. this volume. Habits, awareness, and autonomy. In *Consensus on Peirce's concept of habit: Before and beyond consciousness*, ed. Donna E. West and Myrdene

Anderson. (Studies in Applied Philosophy, Epistemology and Rational Ethics [SAPERE]) New York: Springer.

Coletta, W. John. this volume. The "Irrealevance" of habit formation: Peirce, Hofstadter, and the rocky paradoxes of physiosemiosis. In *Consensus on Peirce's concept of habit: Before and beyond consciousness*, ed. Donna E. West and Myrdene Anderson. (Studies in Applied Philosophy, Epistemology and Rational Ethics [SAPERE]) New York: Springer.

Dearmont, David. 1995. A hint at Peirce's empirical evidence for tychism. *Transactions of the Charles S. Peirce Society* 31(1):185–204.

Hespos, Susan, and Renee Baillargeon. 2001. Infants' knowledge about occlusion and containment events: A surprising discrepancy. *Psychological Science* 12(2): 141–147.

Houser, Nathan. this volume. Social minds and the fixation of belief. In *Consensus on Peirce's concept of habit: Before and beyond consciousness*, ed. Donna E. West and Myrdene Anderson. (Studies in Applied Philosophy, Epistemology and Rational Ethics [SAPERE]) New York: Springer.

Kilpinen, Erkki. this volume. In what sense exactly is Peirce's habit-concept revolutionary? *Consensus on Peirce's concept of habit: Before and beyond consciousness*, ed. Donna E. West and Myrdene Anderson. (Studies in Applied Philosophy, Epistemology and Rational Ethics [SAPERE]) New York: Springer.

Leslie, Alan, and Zsuzsa Káldy. 2007. Things to remember: Limits, codes, and the development of object working memory in the first year. In *Short- and long term memory in infancy and early childhood*, ed. L. Oakes, and P.J. Bauer, 103–125. Oxford: Oxford University Press.

Leslie, Alan, Xu Fei, Patrice Tremoulet, and Brian Scholl. 1998. Indexing and the object concept: Developing "what" and "where" systems. *Trends in Cognitive Sciences* 2: 10–18.

Magnani, Lorenzo, Selene Arfini, and Tommaso Bertolotti. this volume. Of habit and abduction: Preserving ignorance or attaining knowledge? In *Consensus on Peirce's concept of habit: Before and beyond consciousness*, eds. Donna E. West and Myrdene Anderson. (Studies in Applied Philosophy, Epistemology and Rational Ethics [SAPERE]) New York: Springer.

Mandler, Jean. 2004. *The foundations of mind: Origins of conceptual thought*. Oxford: Oxford University Press.

Mandler, Jean. 2010. The spatial foundations of the conceptual system. *Language and Cognition* 2 (1): 21–44.

Mandler, Jean. 2012. On the spatial foundations of the conceptual system and its enrichment. *Cognitive Science* 36(3): 421–451.

Meltzoff, Andrew N., and M. Keith Moore. 1977. Imitation of facial and manual gestures by human neonates. *Science* 198(4312): 75–78.

Oakes, Lisa, Shannon Ross-Sheehy, and Steven Luck. 2007. The development of visual short-term memory in infancy. In *Short- and long-term memory in infancy and early childhood*, ed. L. Oakes, and P. Bauer, 75–102. Oxford: Oxford University Press.

Peirce, Charles S. i. 1866–1913. *The New Elements of Mathematics*, Vol IV, ed. Carolyn Eisele. The Hague: Mouton Press, 1976. Cited as NEM.

Peirce, Charles Sanders. i. 1867–1913. *Collected papers of Charles Sanders Peirce*. Vols. 1–6, ed. Charles Hartshorne and Paul Weiss. Cambridge: Harvard University Press, 1931–1935. Vols. 7–8, ed. Arthur W. Burks. Cambridge: Harvard University Press, 1958. [References to Peirce's papers will be designated by CP, followed by volume, period, paragraph number].

Peirce, Charles Sanders. i. 1867–1893. *The essential Peirce: Selected philosophical writing*. Vol. 1 (1867–1893), ed. Nathan Houser and Christian Kloesel. Bloomington: Indiana University Press, 1992. [References to this volume will be designated by EP 1, followed by colon, page number.].

Peirce, Charles Sanders. i. 1893–1913. *The essential Peirce: Selected philosophical writing*. Vol. 2 (1893–1913), ed. the Peirce Edition Project. Bloomington: Indiana University Press, 1998. [References to this volume will be designated by EP 2, followed by colon, page number].

Peirce, Charles Sanders. 2009. The logic of Interdisciplinarity: The Monist-Series, ed. Elize Bisanz (Berlin: Akademie Verlag GmbH) Peirce, Charles S. (i. 1866–1913). *The new elements of mathematics*, vol. IV, ed. Carolyn Eisele. The Hague: Mouton, 1976.

Peirce manuscripts in Texas Tech University Library at Texas Tech University, Institute of Studies of Pragmaticism, beginning with MS—or L for letter—and followed by a number, refer to the system of identification established by Richard R. Robin in Annotated Catalogue of the Papers of Charles S. Peirce (Amherst: University of Massachusetts Press, 1967), or in Richard R. Robin, "The Peirce Papers: A Supplementary Catalogue," Transactions of the Charles S. Peirce Society.

Peirce, Charles Sanders. i. 1867–1913. *Writings of Charles S. Peirce: A chronological edition*. Vol. 1–6 to date, ed. the Peirce Edition Project. Bloomington: Indiana University Press. [References to these volumes will be designated by W, followed by volume number, colon, page number].

Peirce, Charles S. & Victoria, Lady Welby (i. 1898–1912). *Semiotic and Significs: The Correspondence Between Charles S. Peirce and Victoria, Lady Welby*. Charles Hardwick and James Cook, (eds.). Bloomington: University of Indiana Press, 1977. Cited as SS.

Piaget, Jean and Bärbel Inhelder. 1969[1966]. *The psychology of the child*. H. Weaver (Trans.). New York: Basic Books.

Quinn, Paul C. 1994. The categorization of above and below spatial relations by young infants. *Child Development* 65: 58–69.

Saylor, Megan M. 2004. Twelve- and 16-month-old infants recognize properties of mentioned absent things. *Developmental Science* 7(5): 599–611.

Scaife, Michael, and Jerome Bruner. 1975. The capacity for joint visual attention in the infant. *Nature* 253: 265–266.

Short, Thomas L. 1996. Interpreting Peirce's Interpretant: A response to Lalor, Liszka, and Meyers. *Transactions of the Charles S. Peirce Society* 32(4):488–541.

Short, Thomas L. 2007. *Peirce's logic of signs*. Cambridge: Cambridge University Press.

Stjernfelt, Frederik. 2014. *Natural propositions: The actuality of Peirce's doctrine of dicisigns*. Boston: Docent Press.

Trevarthen, Colwyn. 1977. Descriptive analyses of infant communicative behaviour. In *Studies in mother-infant interaction*, ed. H.R. Schaffer. London: Academic.

Volterra, Virginia, Maria Caselli, Olga Capirci, and Elena Pizzuto. 2005. Gesture and the emergence and development of language. In *Beyond nature-nurture: Essays in honor of Elizabeth Bates*, ed. M. Tomasello, and D. Slobin, 3–40. New Jersey: Lawrence Erlbaum Associates.

Vygotskii, Lev S. 1962[1934]. In *Thought and language*, ed. E. Hanfmann and G. Vakar. Cambridge, MA: MIT Press.

West, Donna. 1986. *The acquisition of person and space deictics: A comparison between blind and sighted children*. Unpublished Doctoral Dissertation, Cornell University.

West, Donna. 2011a. Indexical reference to absent objects: Extensions of the Peircean notion of index. In *Semiotics 2010*, ed. Karen Haworth, Jason Hogue, and Leonard Sbrocchi, 149–161. Toronto: Legas Press.

West, Donna. 2011b. Deixis as a symbolic phenomenon. *Linguistik Online* 50(6): 89–100.

West, Donna. 2013. *Deictic imaginings: Semiosis at work and at play*. Heidelberg: Springer.

West, Donna. 2014. Perspective switching as event affordance: The ontogeny of abductive reasoning. *Cognitive Semiotics* 7(2): 149–175.

West, Donna. 2015a. The primacy of index in naming paradigms. *Part I. Respectus Philologicus* 27(32): 23–32.

West, Donna. 2015b. The work of secondness as habit in the development of early schemes. *Public Journal of Semiotics* 6(2): 1–13.

West, Donna. 2015c. Dialogue as habit-taking in Peirce's continuum: The call to absolute chance. *Dialogue (Canadian Review of Philosophy)* 54(4): 685–702.

West, Donna. 2016. Individuating in the dark: Diagrammatic reasoning and attentional shifts. *Semiotica* 210: 35–56.

West, Donna. in press. Index as scaffold to logical and final interpretants: Compulsive urges and modal submissions, *Semiotica*. Special Issue on the Ten-Fold Division of Sign.

West, Donna. Under Review. Individuating in undifferentiated space: Trends within normative versus blind populations. *Spatial Cognition and Computation*.

Chapter 14
Dicisigns and Habits: Implicit Propositions and Habit-Taking in Peirce's Pragmatism

Frederik Stjernfelt

Abstract Peirce's notion of "habit" is famously wide, including also natural dispositions. Another Peircean notion generalized from its normal use is his term for propositions, "Dicisigns". What is the connection between the two? It goes via the pragmatist notion of belief: "A *belief* in a proposition is a controlled and contented habit of acting in ways that will be productive of desired results only if the proposition is true" (Kaina Stoicheia 1904). This paper charts the important connection between habits, beliefs and Dicisigns.

Keywords Dicisigns · Diagrams · Proposition · Pragmatic maxim · Laws · Continuity · Realism · Self-control · Action · Pragmatism

Background

"Dicisigns" is the concept developed by Peirce, in the context of his post-1900 generalized semiotics, in order to cover his vast generalization of standard conceptions of propositions. In his mature semiotic architectonic taking its beginnings from the *Syllabus* (1903), Peirce generalized the basic trichotomies of *term-proposition-argument* and *icon-index-symbol* to become, each of them, exhaustive, so that all signs will be either a term, a proposition, or an argument, as well as an icon, an index, or a symbol.[1] During the composition of the *Syllabus*, yet another trichotomy, that of *qualisign-sinsign-legisign* was added (as the first one). This gave rise to the possibility of combining the three trichotomies to give the *Syllabus* table of ten combined sign types.

[1]The "or" in the claim was not an exclusive-or—a sign may be *both* an icon, an index, and a symbol, but no sign may belong to a fourth category.

F. Stjernfelt (✉)
University of Aalborg at Copenhagen, Copenhagen, Denmark
e-mail: stjern@hum.aau.dk

D.E. West and M. Anderson (eds.), *Consensus on Peirce's Concept of Habit*, Studies in Applied Philosophy, Epistemology and Rational Ethics 31,
DOI 10.1007/978-3-319-45920-2_14

The later extensions of Peirce's semiotics, particularly in the Lady Welby letters, in terms of further trichotomies, up to a total of at least ten trichotomies, were established with the same claim for exhaustivity in order to fit the same combinatorial pattern, famously giving a total of 66 combined signs. As to the conception of propositions in particular, the generalization indicated by the neologism "Dicisigns" (also "Dicent Sign", "Pheme", and such) vastly extended its range from linguistically expressed truth claims to include propositions using diagrams, pictures, gestures, and more, as well as a vast swathe of "quasi-propositions" covering more or less natural signs such as weathercocks, fossils, and the like.

The rationale behind this generalization was the interconnected definitions of Dicisigns: (1) by means of their ability to take a truth value, and (2) by their functionally interpreted predicate-subject structure, according to which they function by means of simultaneously *indicating* an object by means of a subject and *describing* that same object by means of a predicate. In *Natural Propositions* (Stjernfelt 2014), I provided a reconstruction of Peirce's elaborated theory of propositions as well an overview over actual interpretation possibilities of that theory. In this paper, I investigate the relations between the Dicisign doctrine and the central conception of "habit" in Peirce's logic, semiotics, and metaphysics. An immediate connection is indicated by the fact that most non-quasi propositions are symbols, and Peircean symbols are defined by their object connection relying on a habit: "A Symbol incorporates a habit, and is indispensable to the application of any intellectual habit, at least." ("Prolegomena to an Apology for Pragmaticism", 1906: CP 4.531)

The implication here is that such propositions, often referred to by Peirce as "beliefs", hold not only for the moment, but rely upon thought habits holding also for an indefinite future. So, beliefs are propositions as well as habits and thus function as a basic, important connection between the two concepts: *beliefs are those habits which are also propositions* (although they may be arguments, not merely propositions). Further investigation, however, reveals a series of complications to this simple scheme, taking us deep into fundamental structures and issues of pragmatism.

Aspects of Habits

A major issue is that Peirce's conception of habit, central as it is to pragmatism and semiotics alike, appears as somewhat less well-defined than most of the other central concepts of that edifice. Even if habit is central already in his early 1860s papers, Peirce's conception of it changes considerably over the years. Let us run through some of the ambiguities or tensions involved.

A first important complication is that symbols involve *two* set of habits; those of the sign itself, as a rule-bound legisign capable of repetition, and those of the purported behavior of the object referred to by the symbol "The word and its meaning are both general rules" (*Syllabus*, 1903: CP 2.292f, see also Nöth 2010:

85; Pietarinen and Bellucci, this volume). One thing is the habit, which governs the production of still new replicas of the symbol sign itself; another is the habit claimed to govern the behavior of the object referred to by that symbol. The former belongs to symbol expression in semiotics, the latter belong to the meaning expressed—and, if the symbol is true, to the (type of) objects referred to. Thus, the propositional symbol accounts for some of the habits of the object indicated: In contrast to the icon and the index, intellectual conceptions convey more about their object "… than any feeling, but more, too, than any existential fact, namely, the 'would-acts', 'would-dos' of habitual behaviour; and no agglomeration of actual happenings can ever completely fill up the meaning of a 'would-be.'" ("Pragmatism", MS 318, 1907; EP 2: 402; CP 5.467).

Habits in the Pragmatic Maxim

These, however, are results of Peirce's mature semiotics. As early as in Peirce's 1866 writings, habits appear as one of three basic elements of the mind, to be introduced as categories in "A New List" the year after. He claims that there are "… three kinds of inference: 1st, Intellectual inference with its three varieties Hypothesis, Induction and Deduction; 2nd, Judgments of sensation, emotions, and instinctive motions which are hypotheses whose predicates are unanalyzed in comprehension; and 3rd, Habits, which are Inductions whose subjects are unanalyzed in extension. This division leads us to three elements of consciousness: 1st, Feelings or Elements of comprehension; 2nd, Efforts or Elements of extension; and 3rd, Notions or Elements of information, which is the union of extension and comprehension." ("Consciousness and Language", 1866: CP 7.580).

Here, the three categories are thus closely connected to the extension and intension of propositions serving as conclusions to inferences. Already here, habits are thus allied to propositions: (a) they are inferred from vague inductions; (b) their information is what is provided by propositions. Extension and intension being independent in propositions, the product of the two is taken to form the information they carry. Importantly, habit constitutes a structural element of mind which is not actually present at all times and whose type and degree of consciousness shall continue to form a matter of contention for years to come; see below. More generally, habit shall continue, during the development of Peirce's thought, to appear as one of the major, regular means of characterization of the category of Thirdness, along with Continuity, Generality, Law, and so on, of which it is sometimes a synonym, other times a subtype.

A particularly central role is played by habit in the articulation of the Pragmatic Maxim,[2] allegedly taking its beginnings in the early 1870s and appearing in its classic formulation in the 1878 papers. Inspired by Alexander Bain's definition of

[2]As observed by Colapietro (Colapietro 2009: 368).

Belief as "… that upon which a man is prepared to act", introduced in the Metaphysical Club by Nicholas St. John Green and much discussed there, the pragmatic maxim forms an analysis of belief in terms of possible action habits.[3] Here, Belief is established as a particular subtype of habit in human thought: "And what, then, is belief? It is the demi-cadence which closes a musical phrase in the symphony of our intellectual life. We have seen that it has just three properties: First, it is something that we are aware of; second, it appeases the irritation of doubt; and, third, it involves the establishment in our nature of a rule of action, or, say for short, a *habit*." ("How to Make our Ideas Clear", 1878: EP 1: 129; CP 5.397) Here, beliefs are those habits that we are aware of and which mitigate doubt.

Uncontroversially, Peirce assumes that familiarity with the use of a notion to form the first standard step of clearness, the ability explicitly to define the notion to form the next step. Deeming these insufficient, he famously adds the third and final step of clearness to be that expressed in the pragmatic maxim: "It appears, then, that the rule for attaining the third grade of clearness of apprehension is as follows: Consider what effects, that might conceivably have practical bearings, we conceive the object of our conception to have. Then, our conception of these effects is the whole of our conception of the object." (EP 1: 132; CP 5.402). Here, habit is involved no fewer than on two occasions. The very conception of possible effects of the object forms a habit of thought—thought is taken to be a particular type of action: "the action of thinking" whose purpose is the removal of doubt. This habit of thought, in turn, establishes a further habit of action, relating to the effects of the object, in itself transcending thought: "The *final* upshot of thinking is the exercise of volition, and of this thought no longer forms a part; but belief is only a stadium of mental action, an effect upon our nature due to thought, which will influence future thinking." (EP 1: 129, CP 5.397)

This idea of the final meaning of a concept as consisting in a habit of non-mental action continues to absorb Peirce in his attempts to construct a proof of pragmatism in the years after the turn of the century. Here, the meaning of a proposition—a belief—is reducible to a claim about the conceivable effects of its object, while it is not addressed whether the resulting, final volitional action beyond thought but still governed by a general principle, also has, in itself, propositional structure. Habit being general, it possesses a schematic structure, as Rosenthal insists (1982: 231)—a diagram, as Peirce would say, incarnating the possibility of drawing particular action inferences from it.[4]

[3]Even late in life, Peirce continued to refer to this idea in his discussions of habit: "For our present purpose it is sufficient to say that the inferential process involves the formation of a habit. For it produces a belief, or opinion; and a genuine belief, or opinion, is something on which a man is prepared to act, and is therefore, in a general sense, a habit." ("Minute Logic", 1902: CP 2.148).

[4]"Thus, when you say that you have faith in reasoning, what you mean is that the belief habit formed in the imagination will determine your actions in the real case. This is looking upon the matter from the psychological point of view. Under a logical aspect your opinion in question is that general cognitions of potentialities *in futuro*, if duly constructed, will under imaginary conditions determine *schemata* or imaginary skeleton diagrams with which percepts will accord when the real

Habit, Continuity, and Realism

A constant theme in Peirce's further development of the habit concept is its generality. A habit not only involves more than one occurrence of the relevant action, it also transcends any finite number of such instances (Letter to Lady Welby, 24 December 1908: EP 2: 487). Even if each single such occurrence constitutes an individual event, the structure permitting the indefinite extension of such occurrences is, in itself, general and thus forms a prime example of Peirce's description of generality in terms of continuity. A habit transcends any number of actualizations, just like the continuum transcends any number of individual points, even infinite numbers. For that reason, habits form a central example of general patterns referred to by Peirce's realism of universals: habits are not themselves sums of individual existents or events, rather, they constitute patterns which possess the real power to make such existences incarnate—even in the extreme case of never once becoming so actualized.

Again, this structure, connecting some general rule with its possible instantiations in single cases mirrors that of propositions—consisting of indices pointing out objects referred to, on the one hand, and of general predicates on the other hand. Another way of expressing said realism is that some of those general predicates describe real patterns—habits—of reality; and their presence in the mind can never exhaust them but must, by the same token, be one of a habitual disposition, different from any here-and-now content of the mind. This also becomes evident the many times Peirce recognizes that the only way of presenting a habit is by predicatively describing the general behavior sequence common to each of its instantiations:

> To get back, then, to the die and its habit—its "would-be"—I really know no other way of defining a habit than by describing the kind of behavior in which the habit becomes actualized. (*Syllabus*, 1903: CP 2.666)

Habits thus share the predicate/subject structure with propositions—*general* propositions due to the inherent generality of habits. The particular occasion that calls into action the general habit acts like the object of the proposition, the ensuing volitional act appearing as an inference from that proposition, as it is described in this long and pretty early quote locating this logical habit structure in neuropsychology with a frog as an example:

> The cognition of a rule is not necessarily conscious, but is of the nature of a habit, acquired or congenital. The cognition of a case is of the general nature of a sensation; that is to say, it is something which comes up into present consciousness. The cognition of a result is of the nature of a decision to act in a particular way on a given occasion. In point of fact, a syllogism in Barbara virtually takes place when we irritate the foot of a decapitated frog. The connection between the afferent and efferent nerve, whatever it may be, constitutes a nervous habit, a rule of action, which is the physiological analogue of the major premiss.

(Footnote 4 continued)

conditions accord with those imaginary conditions." ("Minute Logic", 1902: CP 2.148). Cf. Stjernfelt (2007: Chap. 4).

> The disturbance of the ganglionic equilibrium, owing to the irritation, is the physiological form of that which, psychologically considered, is a sensation; and, logically considered, is the occurrence of a case. The explosion through the efferent nerve is the physiological form of that which psychologically is a volition, and logically the inference of a result. When we pass from the lowest to the highest forms of innervation, the physiological equivalents escape our observation; but, psychologically, we still have, first, habit—which in its highest form is understanding, and which corresponds to the major premiss of Barbara; we have, second, feeling, or present consciousness, corresponding to the minor premiss of Barbara; and we have, third, volition, corresponding to the conclusion of the same mode of syllogism. Although these analogies, like all very broad generalizations, may seem very fanciful at first sight, yet the more the reader reflects upon them the more profoundly true I am confident they will appear. They give a significance to the ancient system of formal logic which no other can at all share. ("A Theory of Probable Inference", 1883: CP 2.711)

Here, logical habit leading from the general habit proposition (the major premise), occasioned by the appearance of the relevant particular information in a perceptual judgment proposition (the case, the minor premise) the action conclusion, is instantiated in the neurophysiological system—propositional habit thereby extending also to cover inherited behavior structure: (1) Habit: "In case of A, do B"; (2) Occasion: A; (3) Action: B. The habit proposition—the major premise conditional—is later described as: "Real Habit—its subject would under certain conditions behave in a certain way, even if those conditions never actually do get fulfilled" ("A Sketch of Logical Critics", 1909: EP 2: 457)—the "certain conditions" given in the minor premise activate the habit conclusion. Habits in this very general sense, involving inherited biological instinct, thus form general, conditional propositions. So, not only explicit, consciously adapted beliefs, among habits, are propositional. Indeed, it seems that habit is propositional all the way down to biology. In short, habit in this sense is a general, conditional proposition urging a type of action, generally described, to occur on given conditions. Some of those habits, of course, may be intellectual, so that the resulting action is a thought; in that case the relevant habits themselves are rules of inference.

But if all habits have a propositional structure, forming the major premise of action arguments, beliefs are no longer those habits that are propositions. What then distinguishes beliefs? A mature version of the pragmatic maxim: "A belief in a proposition is a controlled and contented habit of acting in ways that will be productive of desired results only if the proposition is true." ("New Elements" (Kaina Stoicheia) 1904: EP 2: 312). The subtype of habits that are beliefs are now those subject to *control* (see below). This is obviously a different criterion from that of the 1878 pragmatic maxim where the defining feature of beliefs as habits were awareness and assuaging of doubt.

Importantly, these structures give rise to a couple of corollaries. One is the mirror definition of doubt as something which is only real if it actually breaks an already existing belief. Already from the Metaphysical Club period, Peirce refuses "parade" doubt[5] that may be expressed explicitly but which is not evidenced by

[5]"... for it is the belief men *betray* and not that which they *parade* which has to be studied." ("Issues of Pragmaticism", EP 2: 349n).

hesitation or changed behavior, that is, without effects upon habit. The pragmatic maxim is therefore also a means to distinguish real doubt from parade doubt: "A true doubt is accordingly a doubt which really interferes with the smooth working of the belief-habit. Every natural or inbred belief manifests itself in natural or inbred ways of acting, which in fact constitute it a belief-habit. (I need not repeat that I do not say that it is the single deeds that constitute the habit. It is the single "ways," which are conditional propositions, each general)." ("Consequences of critical Common-Sensism", 1905: CP 5.510) We remark in passing that belief-habits may be inbred and are thus *not* subject to explicit control, unlike beliefs in the 1904 quote above.

Another corollary is the realization that the existence of conscious habits necessitates that the mind has direct access to general objects, that is, not fully determined objects—not unlike Husserl's notion of "categorical intuition": "We can understand one habit by likening it to another habit. But to understand what any habit is, there must be some habit of which we are directly conscious in its generality. That is to say, we must have a certain generality in our direct consciousness. Bishop Berkeley and a great many clear thinkers laugh at the idea of our being able to imagine a triangle that is neither equilateral, isosceles, nor scalene. They seem to think the object of imagination must be precisely determinate in every respect. But it seems certain that something general we must imagine. (…) At any rate, I can see no way of escaping the proposition that to attach any general significance to a sign and to know that we do attach a general significance to it, we must have a direct imagination of something not in all respects determinate." (CP 5.371 Footnote 1, 1893).[6] Habit was introduced in order to understand structures of the mind which transcend immediate consciousness. But the fact that the mind is able to make conscious (some of) those habit structures has important consequences for the contents also of immediate consciousness fragment. The fact that it is indeed

[6]An important variant idea occurring several times in Peirce is that beliefs, pragmatically defined, are *at odds* with propositions of science (despite the fact that the maxim was originally conceived of as a meaning clarification procedure in scientific terms) fragment. Thus, in 1898, he says ("Cambridge Lectures on Reasoning and the Logic of Things: Philosophy and the Conduct of Life", CP 1.635) the following. "… I hold that what is properly and usually called *belief*, that is, the adoption of a proposition as a {ktéma es aei} to use the energetic phrase of Doctor Carus, has no place in science at all. We *believe* the proposition we are ready to act upon. *Full belief* is willingness to act upon the proposition in vital crises, *opinion* is willingness to act upon it in relatively insignificant affairs. But pure science has nothing at all to do with *action*. The propositions it accepts, it merely writes in the list of premisses it proposes to use. Nothing is *vital* for science; nothing can be. Its accepted propositions, therefore, are but opinions at most; and the whole list is provisional. The scientific man is not in the least wedded to his conclusions." A similar argument is repeated five years later (CP 7.606). The idea seems to distinguish the definition of conceptions by the sum of conceived effects of their objects, on the one hand, and the action consequence inferred from (some of) those effects—so that the latter is reserved for "vital" issues only, remote from the cool, detached relation of the scientist to his results. In this variant idea, then, "beliefs" differ from scientific propositions, radically narrowing the explicitly broad definition of "belief" so as to cover any assent, of some endurance, to a proposition.

possible to be aware of a habit is thus of central importance: this necessitates the controversial existence of not-fully determined, general, representations.

To sum up, habit is a conditional, general proposition, realist in the sense that it covers an indefinite amount of possible instantiations, which, given the appearance of a particular occasion of a certain general description, leads to action, generally described. Explicit beliefs, as a subset of belief-habits, are the subjects of awareness and of control.

Acquired Habits, Innate Habits, Laws

Until now, we have implicitly assumed that habit is something generalized from the human mind to cover other types of biological cognition, such as in the frog example. But as so often with Peirce, generalization must be driven as far as possible. A controversial and pretty consistent implication of the habit doctrine, is that habit not only extends to animals but also spans across the received innate/acquired distinction, coming out of the principle of using "If I may be allowed to use the word "habit," without any implication as to the time or manner in which it took birth, so as to be equivalent to the corrected phrase "habit or disposition", that is, as some general principle working in a man's nature to determine how he will act, then an instinct, in the proper sense of the word, is an inherited habit, or in more accurate language, an inherited disposition. But since it is difficult to make sure whether a habit is inherited or is due to infantile training and tradition, I shall ask leave to employ the word "instinct" to cover both cases." ("Minute Logic", 1902: CP 2.170) This, however, is not only a *façon-de-parler*, rather it is an ontological claim which insists that there is no principal difference between habits acquired during the phylogenetic course of evolution and habits acquired in the ontogenetic development of the individual: "The old writers call [them] dispositions, but I do not think there was any advantage in calling them by a separate name, but rather the reverse. Some call them 'hereditary habits'. If they are that, they are innate." ("Materials for Monist article", 1905: MS 288: 65–67)

The basic idea that one of the essential elements of every possible mind is habit excludes the possibility that habits as such could be accidental developments during individual lifetime only: "… every animal must have habits. Consequently, it must have innate habits. In so far as it has cognitive powers, it must have *in posse* innate cognitive habits, which is all that anybody but John Locke ever meant by innate ideas. To say that I hold this for true is implied in my confession of the doctrine of Common-Sense—not quite that of the old Scotch School, but a critical philosophy of common-sense. It is impossible rightly to apprehend the pragmaticist's position without fully understanding that nowhere would he be less at home than in the ranks of individualists, whether metaphysical (and so denying scholastic realism), or epistemological (and so denying innate ideas)." ("Consequences of critical Common-Sensism", 1905: CP 5.540)

Thus, Peircean habit does not comprise only patterns of behavior acquired in the ontogenetic timescale of individual organisms, but also patterns of behavior acquired in phylogenetic timescale of species' lineages.[7] Despite the fact that the "narrow" interpretation of habit to cover only the former is widespread, even to the degree that it forms a prejudice of our time, the biosemiotic idea that there is no deep ontological distinction between the two is supported by Peirce's argument. Inherited habits, thus, form implicit conditional propositions ready to give inference to action if perceptual occasion adds the relevant minor premiss needed.

Given that Peircean habits thus pervade biology, the next issue called for by tentative generalization is whether they extend into the pre-biological, purely physical universe as well. Immediately, there is a tendency to the exact opposite, to strongly contrast habits to physical laws. In Peirce's first major outline of a cosmology, the "Guess at the Riddle" (1887), he describes habits in terms of neurophysiology, generalizing the frog example and anticipating Hebb's law that connections used are connections strengthened:

> Fourth, if the same cell which was once excited, and which by some chance had happened to discharge itself along a certain path or paths, comes to get excited a second time, it is more likely to discharge itself the second time along some or all of those paths along which it had previously discharged itself than it would have been had it not so discharged itself before. This is the central principle of habit; and the striking contrast of its modality to that of any mechanical law is most significant. The laws of physics know nothing of tendencies or probabilities; whatever they require at all they require absolutely and without fail, and they are never disobeyed. Were the tendency to take habits replaced by an absolute requirement that the cell should discharge itself always in the same way, or according to any rigidly fixed condition whatever, all possibility of habit developing into intelligence would be cut off at the outset; the virtue of Thirdness would be absent. (CP 1.390)

The "Law of Mind" cosmology of the first series of *Monist* papers around 1892 sophisticates that point: "The law of habit exhibits a striking contrast to all physical laws in the character of its commands. A physical law is absolute. What it requires is an exact relation. Thus, a physical force introduces into a motion a component motion to be combined with the rest by the parallelogram of forces; but the component motion must actually take place exactly as required by the law of force. On the other hand, no exact conformity is required by the mental law. Nay, exact conformity would be in downright conflict with the law; since it would instantly crystallize thought and prevent all further formation of habit. The law of mind only makes a given feeling more likely to arise. It thus resembles the "non-conservative" forces of physics, such as viscosity and the like, which are due to statistical uniformities in the chance encounters of trillions of molecules." ("The Architecture of Theories", 1891: CP 6.23)

[7]Add to this idea the actual realization that the sharp distinction between phylogeny and ontogeny holds for higher animals with gendered reproduction only; for bacteria which comprise the vast majority of the biosphere, DNA exchange is not confined to meiosis but takes place continuously even across species so that phylogeny and ontogeny are rather aspects of the same process.

Here, the bottom-line contrast, however, is more precisely that between conservative and non-conservative physical laws. The former are defined by dealing with those forces, like gravity, whose work on an object between two points is independent of the trajectory taken; the latter comprising particularly cases involving friction, thus the statistical laws of thermodynamics. This argument is allied to Peirce's simultaneous idea of the objective existence of chance—the absence of "exact conformity" being responsible for merely statistical tendencies on the one hand as well as the possibility of development of novelty on the other.

But already in the same paper series, Peirce famously continues the generalization of the habit concept in the famous exclamation that "… what we call matter is not completely dead, but is merely mind hidebound with habits." ("The Law of Mind", 1892: EP 1: 331, CP 6.158). But then physical laws, even pertaining to conservative forces, are *also* habits, only very stiff habits. Similar ontological ideas stabilize after the 1897 adoption of the idea of the objective reality of "real possibilities" or "would-bes", e.g., in the "Minute Logic": "For every habit has, or is, a general law." (1902: CP 2.148). Thus, a mere physical probability, such as that of a die, is now "quite analogous" to human habits, the difference being only one of degrees of simplicity: "… the "would-be" of the die is presumably as much simpler and more definite than the man's habit as the die's homogeneous composition and cubical shape is simpler than the nature of the man's nervous system and soul; and just as it would be necessary, in order to define a man's habit, to describe how it would lead him to behave and upon what sort of occasion—albeit this statement would by no means imply that the habit consists in that action—so to define the die's "would-be," it is necessary to say how it would lead the die to behave on an occasion that would bring out the full consequence of the "would-be"; and this statement will not of itself imply that the "would-be" of the die consists in such behavior." (Notes on "Doctrine of Chances"; 1910: CP 2.664)

The crude oppositions of habits vs. laws of the period around 1890 thus seems to give way to a more continuous conception according to which natural laws and human habits are but ends of one large generalized continuum of "would-be's", only differing in complexity and plasticity. Thus, we seem to have a habit continuum along the lines of:

> conservative physical laws -> non-conservative physical laws -> innate biological patterns of behavior -> acquired biological patterns of behavior -> deliberately acquired human patterns of behavior -> deliberately acquired human patterns of thought (beliefs)[8]

with increasing plasticity, and where the former influence the latter but do not determine them fully. Oftentimes, however, more "narrow" habit concepts in Peirce may still be used to single out only later phases of this series. Still, a seminal difference seems to prevail between the physical and the biological phases of the continuum depicted. Biological habits serve a semiotic function because they

[8]Of course, "human" should be construed with caution here; we only know human realizations of such processes, but it is in no way precluded that other organisms or automata could satisfy the relevant criteria.

describe certain possible environmental conditions, the actualization of which will release organism action with the local purpose of survival. Purely physical habits hardly could be said to serve such functions (if we do not subscribe to teleological theories of the whole of cosmos). Even if it is possible to render the gravitational pull of an object as a conditional proposition: "Object A is heavy, and if another heavy Object B appears, there will be a gravitational force between them proportional to the product of their masses", this is hardly in itself a quasi-proposition except when appearing in the Umwelt of some organism.[9]

An important idea based on the plasticity increase along the habit continuum above is that of the *variation of habits*, becoming more and more crucial to Peirce. Within biology, this gives rise to the idea that human reason is more plastic than the reasoning of lower animals—making it more prone to error than reasoning in simpler species, but at the same time functioning as a precondition of intellectual growth: "It is a truth well worthy of rumination that all the intellectual development of man rests upon the circumstance that all our action is subject to error. *Errare est humanum* is of all commonplaces the most familiar. Inanimate things do not err at all; and the lower animals very little. Instinct is all but unerring; but reason in all vitally important matters is a treacherous guide. This tendency to error, when you put it under the microscope of reflection, is seen to consist of fortuitous variations of our actions in time. But it is apt to escape our attention that on such fortuitous variation our intellect is nourished and grows. For without such fortuitous variation, habit-taking would be impossible; and intellect consists in a plasticity of habit." ("Detached Ideas, Causation and Force", 1898: CP 6.86). Habit-taking thus considered along Darwinian lines as the combination of variation and selection places an increasing emphasis not only on the initial *establishment* of habits, but also on the subsequent *variation, selection, development*, and *changes* of them. This is also connected to an important development in Peirce's logic, namely the sophistication of the concept of deduction which would lead, ultimately, to the important corollarial/theorematic distinction after the turn of the century.

A basic idea here is that deduction has been erroneously simplified, generalizing from syllogisms where there is but one deductive conclusion to be inferred—giving the received Kantian impression that there is nothing in the conclusion which was not already clearly there in the premises, and that deduction is thus algorithmically automatizable.[10] But as Peirce realizes, in axiomatic systems, there is nothing like

[9]Following Peirce's definition of a fact: "What we call a 'fact' is something having the structure of a proposition, but supposed to be an element of the very universe itself. The purpose of every sign is to express 'fact,' …" ("New Elements", 1904: EP 2: 304), it is evident that physical laws are general facts and thus have the structure of propositions—but that is not the same thing as saying that they *are* themselves propositions, as only their semiotic or scientific representations are.

[10]As was the case as late as in 1878 when Peirce wrote: "As for deduction, which adds nothing to the premisses, but only out of the various facts represented in the premisses selects one and brings the attention down to it, this may be considered as the logical formula for paying attention, which is the volitional element of thought, and corresponds to nervous discharge in the sphere of physiology." ("Deduction, Induction, and Hypothesis", 1878: CP 2.643).

"*the* conclusion": "There is but one conclusion of any consequence to be drawn by ordinary syllogism from given premisses. Hence, it is that we fall into the habit of talking of *the* conclusion. But in the logic of relatives there are conclusions of different orders, depending upon how much iteration takes place. What is the conclusion deducible from the very simple first principles of number? It is ridiculous to speak of *the* conclusion. The conclusion is no less than the aggregate of all the theorems of higher arithmetic that have been discovered or that ever will be discovered." ("Detached Ideas, The First Rule of Logic", 1898: CP 5.579). Consequently, even deductive inferences imply the need for the variation of inference habits—seeking by trial-and-error the comparison and selection between a variety of different possible proof trajectories. This also considerably complicates the pragmatic core idea that the meaning of a conception is the set of conceiveable effects and correlated action habits—for the sum of those effects may, for a given conception, such as "the first principles of number" be far from simple and fully realized only in an idealized future.[11]

Habit Straddling the Unconscious/Conscious Distinction

The most complicated and open issue, however, in Peirce's lifelong habit discussion, concerns the degree to which habits, in the narrower biological and human senses of the word, are subject to awareness, consciousness, deliberation, and self-control. We have already explored the idea that beliefs are not habits that are propositions; they are, rather, habit propositions subjected to control. As to belief in particular, Peirce's standard conception was that it is "something that we are aware of" as we saw in "How to Make our Ideas Clear" (1878). This, however, is subject to many qualifications and even contradictions. This seems to have to do with the Scotist roots of Peirce's conception of habit. In his famous early articulation of his "scholastic realism", Peirce wrote in 1871, addressing Scotus' solution to the problem of universals:

> … it may be asked, first, is it necessary to its [the universal's] existence that it should be in the mind; and, second, does it exist in re? There are two ways in which a thing may be in the mind, – *habitualiter* and *actualiter*. A notion is in the mind *actualiter* when it is actually conceived; it is in the mind *habitualiter* when it can directly produce a conception. It is by virtue of mental association (we moderns should say), that things are in the mind *habitualiter*. In the Aristotelian philosophy, the intellect is regarded as being to the soul what the eye is to the body. The mind *perceives* likenesses and other relations in the objects of sense, and thus just as sense affords sensible images of things, so the intellect affords intelligible images of them. It is as such a *species intelligibilis* that Scotus supposes that a conception

[11]Of course, this is equivalent to Hilbert's *Entscheidungs*-problem, which was, famously, proved to be undecidable by Gödel thirty years later. The implication of this for pragmatism is that for certain conceptions, not only the sum of conceivable effects may be practically unattainable but in some cases not even *principally* attainable. Consequently, the same holds for the related action habits. In most cases, however, this makes the concept of number no less pragmatically clear.

exists which is in the mind *habitualiter*, not *actualiter*. (Review of Fraser's Works of Berkeley, 1871: EP 1: 92; CP 8.18)

Thus, Peirce's conception of how a habit inhabits the mind is derived from the Scotic theory of universals: the habit simply *is* the way that a universal is in the mind, for the universal, just like its counterpart in reality, is not exhausted by any actual occurrence in the mind of conscious tokens of it.[12] So it forms part of the mind's structure, also when it is not present to the mind. But this implies the surprising consequence that this habitual existence does *not* depend upon consciousness: "This *species* is in the mind, in the sense of being the immediate object of knowledge, but its existence in the mind is independent of *consciousness*." (ibid.) This holds important consequences for Peirce's realism, but also for our actual interest in the mode of existence of habits in the mind: "… to say that an object is in the mind is only a metaphorical way of saying that it stands to the intellect in the relation of known to knower." (ibid.) But as the existence of habits is independent of consciousness, this knowledge of habits must be unconscious or potential.

Already in 1867, Peirce had insisted on the threefold character of Scotus' distinction: "I adopt the admirable distinction of Scotus between actual, habitual, and virtual cognition." (CP 2.398 fn)

Virtual cognition comprises the whole universe of possible forms that the mind may possibly address; an example lies in the fact discussed above that certain implications in the ultimate meaning of a conception may be logically possible but never reached, neither actually nor by the (use of the) existing habits concerning the meaning "… I do not think that the import of any word (except perhaps a pronoun) is limited to what is in the utterer's mind actualiter, so that when I mention the Greek language my meaning should be limited to such Greek words as I happen to be thinking of at the moment. It is, on the contrary, according to me, what is in the mind, perhaps not even habitualiter, but only virtualiter, which constitutes the import. To say that I hold that the import, or adequate ultimate interpretation, of a concept is contained, not in any deed or deeds that will ever be done, but in a habit of conduct, or general moral determination of whatever procedure there may come to be, is no more than to say that I am a pragmaticist." ("Consequences of critical Common-Sensism", 1905: CP 5.504) The triad of actual, habitual, and virtual may be resumed as follows: *Actualiter* are the Greek words or sentences I may be processing at any moment; *habitualiter* is my general knowledge of Greek and *virtualiter* is the whole of the Greek language, including those parts I never learnt.

So, the Scotist distinction between *virtualiter, actualiter*, and *habitualiter* cognition—to resume the three in the order of Peirce's categories—may be explained using, again, the logical example of inference from habit: *Habitualiter*: an empirical thought habit may be: 'If there is lightning, then there is thunder' *Actualiter*: Any existing occurrence of 'lightning' to the mind is an actual cognition. *Virtualiter*: the

[12] "The Scotistic form or essence functions in precisely the same manner that Peirce's habit does; it determines how a thing "would be" disposed to behave under certain specifiable conditions." (Raposa 1984: 157).

combination of the two *may* lead to the conclusion: 'it thunders'. But even if the mind in question holds the habit mentioned and actually has the experience cited, there is no guarantee that the relevant conclusion will be drawn—it thus may remain virtual. In the pragmatic maxim meaning definition, we may surmise that many among the sum of the conceivable effects of a given conception will, at any point of time, remain virtual only. And to say that virtual cognitions, even if logically implied by a presently conscious cognition, are in any sense "in the mind", may be to stretch the point beyond normal usage—which may be why Peirce sometimes mentions two out of the Scotist trichotomy only.

Actual cognitions are thus taken to be conscious, at least in general, while virtual cognitions are not. Habitual cognitions are more than their actual, conscious instantiations and thus have an unconscious basis; but they may, on the other hand, become conscious as objects of deliberate consideration. Thus, as to the definition of belief as a thought-habit, Peirce is bewilderingly inconsequent as to its deliberate, conscious, self-controlled character—which seemed clear in the 1878 pragmatic maxim version. Suffice it to compare the following later quotes:

> A belief is a habit; but it is a habit of which we are conscious. The actual calling to mind of the substance of a belief, not as personal to ourselves, but as holding good, or true, is a judgment. An inference is a passage from one belief to another; but not every such passage is an inference. ("How to Reason, Essence of Reasoning, Chap. 6", 1893: CP 4.53)

> A belief need not be conscious. When it is recognized, the act of recognition is called by logicians a judgment, although this is properly a term of psychology. A man may become aware of any habit, and may describe to himself the general way in which it will act. For every habit has, or is, a general law. (…) What particularly distinguishes a general belief, or opinion, such as is an inferential conclusion, from other habits, is that it is active in the imagination. (…) … a belief habit formed in the imagination simply, as when I consider how I ought to act under imaginary circumstances, will equally affect my real action should those circumstances be realized. (*Minute Logic*, 1902: CP 2.148)

> The purpose of reasoning is to proceed from the recognition of the truth we already know to the knowledge of novel truth. This we may do by instinct or by a habit of which we are hardly conscious. But the operation is not worthy to be called reasoning unless it be deliberate, critical, self-controlled. In such genuine reasoning we are always conscious of proceeding according to a general rule which we approve. It may not be precisely formulated, but still we do think that all reasoning of that perhaps rather vaguely characterized kind will be safe. This is a doctrine of logic. We never can really reason without entertaining a logical theory. That is called our *logica utens*. ("Logical Tracts no. 2", 1903: CP 4.476)

> Belief is not a momentary mode of consciousness; it is a habit of mind essentially enduring for some time, and mostly (at least) unconscious; and like other habits, it is (until it meets with some surprise that begins its dissolution) perfectly self-satisfied. (…) a process of self-preparation will tend to impart to action (when the occasion for it shall arise), one fixed character, which is indicated and perhaps roughly measured by the absence (or slightness) of the feeling of self-reproach, which subsequent reflection will induce. Now, this subsequent reflection is part of the self-preparation for action on the next occasion. Consequently, there is a tendency, as action is repeated again and again, for the action to approximate indefinitely toward the perfection of that fixed character, which would be marked by entire absence of self-reproach. The more closely this is approached, the less

> room for self-control there will be; and where no self-control is possible there will be no self-reproach. ("What Pragmatism Is" 1905: CP 5.417)

> … habit is by no means exclusively a mental fact. Empirically, we find that some plants take habits. The stream of water that wears a bed for itself is forming a habit. Every ditcher so thinks of it. Turning to the rational side of the question, the excellent current definition of habit, due, I suppose, to some physiologist (if I can remember my bye-reading for nearly half a century unglanced at, Brown-Sequard much insisted on it in his book on the spinal cord), says not one word about the mind. Why should it, when habits in themselves are entirely unconscious, though feelings may be symptoms of them, and when consciousness alone – i.e., feeling – is the only distinctive attribute of mind? ("Pragmatism", MS 318, 1907; EP 2: 418; CP 5.492)

The chronological organization of these quotes may give the idea that the emphasis on the basically unconscious status of habits (including beliefs) is growing over Peirce's mature period. There are, however, also counterexamples during that period ("[Readiness] to act in a certain way under given circumstances and when actuated by a given motive is a habit; and a deliberate, or self-controlled, habit is precisely a belief.", "Pragmatism", MS 318, 1907; CP 5.480), but the tendency goes in the direction of the doctrine that habits as well as beliefs are basically unconscious, even if they give rise to conscious feelings when instantiated. Habits themselves may, however, become the object of consciousness—as in the important case of the deliberate adoption of habits, in thought as in action.

Adoption of habits may take place inductively, the habit being established by the repetition of similar acts, which are not necessarily deliberate and conscious—or, it may take place by deliberate, conscious, imaginative experimentation in the mind, ensued by deliberate decision, akin to addressing an order to the future self which is, necessarily conscious ("Pragmatism", MS 318, 1907; EP 2: 413; CP 5.487). Only the latter, the deliberate and conscious inference of one proposition from another, qualifies as *reasoning*, as Peirce repeatedly insists. The automatized drawing of inferences in a mechanical logic machine, however refined, will never be but quasi-inferences because they lack the quality of deliberate, conscious self-control. Thus, the very role of consciousness in mind, is to make possible that increased level of self-control which characterizes real reasoning: "… I am far from holding consciousness to be an "epiphenomenon", though the doctrine that it is so has aided the development of science. To my apprehension, the function of consciousness is to render self-control possible and efficient" (MS 318: 74–76, 1907). But human beings do lots of things which are not characterized by deliberate, conscious self-control. To repeat the nested series of processes from the above section:

> conservative physical laws -> non-conservative physical laws -> innate biological patterns of behavior -> acquired biological patterns of behavior -> deliberately acquired human patterns of behavior -> deliberately acquired human patterns of thought (beliefs)

Human beings, of course, partake in all of them, and only the latter small subset qualify as reasonings. This would explain Peirce's seeming vacillation as to the conscious status of habits and beliefs: the crucial subset of reasonings require conscious deliberation but human beings constantly acquire, follow, and change many habits and beliefs without that underpinning by explicit reasoning.

Peirce's criterion for reasoning, that it is the subject of deliberate, conscious self-control, is only really developed from around 1902 ("Minute Logic"), figuring centrally in the 1903 Pragmatism lectures and the last *Monist* papers series. In 1905, Peirce dates, a bit hesitant, the appearance of the idea of ethical constraints on logic, based on the self-control criterion of reasoning, as late as to the 1903 Lowell Lectures ("Consequences of critical Common-Sensism", 1905: CP 5.533), and it certainly takes center stage only from around 1902. The discussions around that criterion opens a bundle of questions which we shall approach as a conclusion:

(1) What is the relation between consciousness and self-control? … the former as a special tool serving the latter.
(2) How is self-control developed ? … the hierarchy of levels of self-controls.
(3) What is the relation between self-control more generally and self-control of logical thought in particular?
(4) What is the status of the final action habit claimed by the pragmatic maxim as the ultimate meaning of conceptions?

Self-control and Consciousness

We saw how Peirce's mature theory of consciousness simply makes it a tool for efficient-self-control. Self-control thus is a wider phenomenon which has consciousness as one of its higher-level instruments. Connected to this is the realization that conscious control can only encompass certain highlighted steps of inference, not all of their preconditions all the way to the bottom. Consciousness, as a mark of deliberate reasoning, could never require full perspicuity as to all levels, preconditions, and implications of reasoning. This is indicated, of course, by the *logica utens/logica docens* distinction: only the scientific logician makes of logic an explicit doctrine; the normal reasoner, including scientists and mathematicians, makes do with his *logica utens* that does not necessarily include access to the explicit formulation of logical rules and principles. So deliberate, conscious self-criticism is taken to be possible with less than *logica docens*. Thus, reasoning normally makes do with a triple fundament of "perceptual judgments, original (being indubitable because uncriticized) beliefs of a general and recurrent kind, as well as indubitable acritical inferences." ("Issues of Pragmaticism", 1905, EP 2: 348)—none of them subject to conscious self-control. The latter two are only in the mind *habitualiter*—that is, not necessarily in the conscious present, and the former, perceptual judgments, appear in the mind unconditionally, beyond criticism and must be taken at face value (of course, certain perceptual judgments may be criticized, but only on the basis of other such judgments which then lie beyond conscious control):

… to say that an operation of the mind is controlled is to say that it is, in a special sense, a conscious operation; and this no doubt is the consciousness of reasoning. For this theory requires that in reasoning we should be conscious, not only of the conclusion, and of our deliberate approval of it, but also of its being the result of the premiss from which it does result, and furthermore that the inference is one of a possible class of inferences which conform to one guiding principle. Now in fact we find a well-marked class of mental operations, clearly of a different nature from any others which do possess just these properties. They alone deserve to be called reasonings; and if the reasoner is conscious, even vaguely, of what his guiding principle is, his reasoning should be called a logical argumentation. There are, however, cases in which we are conscious that a belief has been determined by another given belief, but are not conscious that it proceeds on any general principle. Such is St. Augustine's "cogito, ergo sum." Such a process should be called, not a reasoning, but an acritical inference. Again, there are cases in which one belief is determined by another, without our being at all aware of it. These should be called *associational suggestions of belief.* (Issues of Pragmaticism, 1905: CP 5.441)

So there are vast amounts of inference habits which are not subjected to conscious control (but may, of course, later be so subjected) and thus not proper logical reasoning, and even within such reasoning, the guiding principle may be the subject of vague acceptance only. Peirce obviously does not want normal scientists, those with little or no acquaintance with explicit formal logic, to be bereft of reasoning abilities, so he admits sufficient conscious control to remain just vague.

On the other hand, self-control is a far wider subject than pertaining to logical inference habits only; it potentially addresses all other types habits, particularly moral habits of which logic is taken to be a special example only: "… while I hold all logical, or intellectual, interpretants to be habits, I by no means say that all habits are such interpretants. It is only self-controlled habits that are so, and not all of them, either." ("Pragmatism", MS 318, 1907; EP 2: 431)

Thus, logical reasoning is but a subtype of the last category of the continuum above:

deliberately acquired human patterns of thought (beliefs) -> logical reasonings

It comes as no surprise that the structure of deliberately establishing new habits by means of logical reasoning in the broad sense is co-extensive with scientific epistemology:

In the process of inference, or the self-controlled formation of new belief on the basis of Knowledge already possessed, I remark three chief steps. They are, first, the putting together of facts which it had not occurred to us to consider in their bearings upon one another, second, experimentation, observation, and experimental analysis, which is substantially the same process whether it be performed with physical apparatus such as the chemist uses or with an apparatus of diagrams of our own creation, such as the mathematician employs, and third, the generalization of experimental results, that is, the recognition of the general conditions governing the experiment, and the formation of a habit of thought under the influence of it. (CP 7.276, n.d. but late, as Peirce speaks about having studied logic for 40 years).

Hierarchies of Self-control

Thus self-control is a far wider subject than consciousness and is assumed to be under development already in biology, and the relevant self needs not be a single organism, rather, the ongoing adaptation of a lineage to its environment after Darwinist principles should probably be classed as an early degree of self-control. Growth in self control is not, however, interpreted to be a gradual increase of the same capability, rather taking the shape of iteration of controls over controls, habits controlling other habits (see also Shapiro 1973: 38; Stjernfelt 2014: Chap. 6), constructing a hierarchy of nested habit controls:

> To return to self-control, (…) of course there are inhibitions and coördinations that entirely escape consciousness. There are, in the next place, modes of self-control which seem quite instinctive. Next, there is a kind of self-control which results from training. Next, a man can be his own training-master and thus control his self-control. When this point is reached much or all the training may be conducted in imagination. When a man trains himself, thus controlling control, he must have some moral rule in view, however special and irrational it may be. But next he may undertake to improve this rule; that is, to exercise a control over his control of control. To do this he must have in view something higher than an irrational rule. He must have some sort of moral principle. This, in turn, may be controlled by reference to an esthetic ideal of what is fine. There are certainly more grades than I have enumerated. Perhaps their number is indefinite. The brutes are certainly capable of more than one grade of control; but it seems to me that our superiority to them is more due to our greater number of grades of self-control than it is to our versatility. ("Pragmaticism", 1905, CP 5.533).

Self-control, thus, is a hierarchy ultimately aiming at some esthetic norm (in Peirce's very broad sense of esthetics as that which charts all purposes valuable of pursuit): "… it is by the indefinite replication of self-control upon self-control that the *vir* is begotten, …" ("Consequences of Pragmaticism", 1906: CP 5.402fn). In human beings, a particular device is deemed central, namely that of hypostatic abstraction, of making a subject of thought out of something which has already been thought:

> … thinking is a kind of conduct, and is itself controllable, as everybody knows. Now the intellectual control of thinking takes place by thinking about thought. (…) One extremely important grade of thinking about thought, which my logical analyses have shown to be one of chief, if not the chief, explanation of the power of mathematical reasoning (…) This operation is performed when something, that one has thought about any subject, is itself made a subject of thought. ("Consequences of critical Common-Sensism", 1905: CP 5.533)

This process, of course, may be iterated, so thinking about thinking of thought, and so on, gives rise to a number of habit levels which simultaneously constitute what is referred to as human freedom:

> … a man is a machine with automatic controls, one over another, for five or six grades, at least. I, for my part, am very dubious as to man's having more freedom than that, nor do I see what pragmatic meaning there is in saying that he has more. The power of self-control is certainly not a power over what one is doing at the very instant the operation of self-control is commenced. It consists (to mention only the leading constituents) first, in

> comparing one's past deeds with standards, second, in rational deliberation concerning how one will act in the future, in itself a highly complicated operation, third, in the formation of a resolve, fourth, in the creation, on the basis of the resolve, of a strong determination, or modification of habit. This operation of self-control is a process in which logical sequence is converted into mechanical sequence or something of the sort. (CP 8.320, undated letter)

The final resolve—typically, again expressed in a conditional proposition with the intention of regulating future behavior—is converted into a mechanical procedure so as to ease future use and avoid repeating the complicated reasoning process over and over again at each potential occasion. Thus, the idea of the "deliberate, conscious" quality of reasoning is fleshed out in two doctrines: (a) that of a hierarchy of self-controls, taking their beginnings deep in biology; (b) that of the higher levels of such controls taking the shape of thinking about thought, using the semiotic tool of hypostatic abstraction, permitting a higher level of thought to control the next lower level.

The Status of the Final Action Habit—A Proposition or Not?

Even if an explicit conclusion in the shape of a conscious, deliberately assented proposition is one type only among self-controlled habits, it takes a special position in Peirce's works because it is the condition for science and is a central subject for logical and epistemological investigation. But here a particular conundrum appears. The pragmatic maxim, in its different guises, identifies the final meaning of any conception with the bundle of action habits which the truth of that conception would ultimately give rise to. But is that set of action habits, in itself, a sign? Peirce tends to answer "no". Being a fact, of course, such an action habit will still possess the structure of a proposition, but it will not, in itself, be a sign. The idea seems to be that ever so long chains, hierarchies, and diagrams of logical inferences serve, in the last resort, the arc taking us from perceptual judgment to action, two ends not in themselves logical.

> In every case, after some preliminaries, the activity takes the form of experimentation in the inner world; and the conclusion (if it comes to a definite conclusion), is that under given conditions, the interpreter will have formed the habit of acting in a given way whenever he may desire a given kind of result. The real and living logical conclusion is that habit; the verbal formulation merely expresses it. I do not deny that a concept, proposition, or argument may be a logical interpretant. I only insist that it cannot be the final logical interpretant, for the reason that it is itself a sign of that very kind that has itself a logical interpretant. The habit alone, which though it may be a sign in some other way, is not a sign in that way in which that sign of which it is the logical interpretant is the sign. The habit conjoined with the motive and the conditions has the action for its energetic interpretant; but action cannot be a logical interpretant, because it lacks generality. The concept which is

> a logical interpretant is only imperfectly so. It somewhat partakes of the nature of a verbal definition, and is as inferior to the habit, and much in the same way, as a verbal definition is inferior to the real definition. The deliberately formed, self-analyzing habit—self-analyzing because formed by the aid of analysis of the exercises that nourished it—is the living definition, the veritable and final logical interpretant. Consequently, the most perfect account of a concept that words can convey will consist in a description of the habit which that concept is calculated to produce. But how otherwise can a habit be described than by a description of the kind of action to which it gives rise, with the specification of the conditions and of the motive? ("Pragmatism", 1907: EP 2: 418; CP 5.491)

But, as Hookway says (2009: 26–27): the very verbal description of the habit implied by the pragmatic maxim is, in itself, conceptual and thus a sort of further logical interpretant of it; in that sense the action habit could not be the final logical interpretant. The same tension is present in the Peirce quote just given: the central argument that "action cannot be a logic interpretant, because it lacks generality." The conclusion of the habit inference, however, as discussed above, is no single action event. It is a general structure surpassing any number of such events. And that is exactly what the end of the quote addresses: the *kind* of action. Action is the dynamic interpretant, the conception of action is the logical interpretant. So it must be in another sense that the action habit is the final logical interpretant.

The issue is not made simpler, obviously, by the fact that those ultimate action habits defining meaning also comprise cognitive action habits. Pietarinen and Bellucci (this volume) rightly argue that in the special but central case of the leading principle of an inference, the explicit expression of that (taking it from *logica utens* to *docens*, so to speak) is, in a certain sense superfluous: "To regard a habit as a sign is analogous to regard the leading principle of reasoning as a premise of reasoning: by some sort of deduction theorem one can always express the leading principle (logical rule) as a conditional proposition (logical proposition), but as he had shown already in 1867, nothing is gained by so doing when the leading principle is *logical*, for the very same principle will be governing the new argument thereby obtained. In semeiotical terms, in reasoning no final logical interpretant can be considered as a sign without employing in this reasoning that very same logical interpretant." This might give us the clue also to the final logical interpretant of thought signs in general.[13] The finality of the set of action habits should rather be seen from the perspective that they furnish the final arbiter of truth of the proposition of which they serve as interpretants. If any of those habits fail in some respect, it will be a sign there is something to be investigated about the relevant proposition.

[13]But probably not more than a clue—the generalization to the non-sign, action habit character of the final interpretants of *all* concepts is probably equivalent to the much sought-after "proof of pragmatism".

Habits and Dicisigns Revisited

This long development finally paves the way for our conclusion as to the relation between habits and propositions. Habits, in general, are articulated with the structure of conditional propositions—this includes even non-biological, conservative, physical laws. This does not make them propositions: consider Peirce's definition of fact as a part of reality having the structure of a proposition (but not in itself *being* a proposition). As to the biological/human habit types of innate habits, acquired habits (specifications of more general innate habits), deliberately acquired habits, and such, these will also be structured as conditional propositions. Animal and human behaviors, according to this analysis, have propositional structure, and a significant part of them, whether conscious or not, qualify as beliefs—the decisive criterion here being whether they articulate the double habit structure so that they themselves constitute rule-bound signs, in turn, referring to other rule-bound habits in their objects.

This result probably, from the point of view of the parsimonious ontologist, gives us a vastly overpopulated universe. Why couldn't we just do with individual objects, while human minds take care of the analyses of those objects in terms of similarities, habits, laws, signs, propositions, inferences, and such? The basic reason is that such a universe is ineradicably dualist, and what is more, strange. Throwing most of the complexity of the cosmos into the human mind, leaving only individuals out there, gives us two radically different and incommensurable parts of that cosmos, that is, the human mind and the rest. This makes it difficult to envisage how the human mind could ever have evolved out of that naked universe.

Thus, Peirce's conception is deep down motivated by his naturalist monism. As to habits, it is monism that make him construct the chain of being, connecting regularities, spanning from conservative physical laws in one end of the chain and conscious human habit-taking in the other. As to propositions, it is monism that takes us from facts—structured like propositions—in one end of the chain, to fully-fledged, deliberately adopted, explicitly expressed proposition signs in the other. So, the basic motivation is naturalism. Thus, Peirce may serve as a possible inspiration for actual naturalism attempts, reminding us that naturalism may not necessarily give us a very simple ontology, rather the opposite. But why should it not? Chemistry progressed only when it realized that the number of elements exceeded four.

Thanks to Francesco Bellucci for comments and discussions.

References

Colapietro, Vincent. 2009. Habit, competence, and purpose: How to make the grades of clarity clearer. *Transactions of the Charles S. Peirce Society* 45.3: 348–377.

Hookway, Christopher. 2009. Habits and interpretation: Defending the pragmatist maxim. Paper presented at the conference, Peirce and early analytic philosophy, Helsinki, Finland,

19–20 May 2009. (http://www.nordprag.org/papers/Hookway%20-%20Habits%20and%20 Interpretants.pdf. Accessed 11 July 2015).
Nöth, Winfried. 2010. The criterion of habit in Peirce's definitions of the symbol. *Transactions of the Charles S. Peirce Society* 46.1: 82–93.
Peirce, C. 1966. *Selected writings*, ed. P. Wiener. New York: Dover Publications.
Peirce, C. i. 1867–1913. *Writings of Charles S. Peirce: A chronological edition*. Vols. 1–6 to date, the Peirce Edition Project. Bloomington: Indiana University Press. [References to these volumes will be designated by W, followed by volume number, colon, page number].
Peirce, C. i. 1867–1893. *The essential Peirce: Selected philosophical writing*. Vol. 1 (1867–1893), ed. Nathan Houser and Christian Kloesel. Bloomington: Indiana University Press, 1992. [References to this volume will be designated by EP 1, followed by colon, page number].
Peirce, C. i. 1867–1913. *Collected papers of Charles Sanders Peirce*. Vols. 1–6, ed. Charles Hartshorne and Paul Weiss. Cambridge: Harvard University Press, 1931–1935. Vols. 7–8, ed. Arthur W. Burks. Cambridge: Harvard University Press, 1958. [References to Peirce's papers will be designated by CP, followed by volume, period, paragraph number].
Peirce, C. i. 1893–1913. *The essential Peirce: Selected philosophical writing*. Vol. 2 (1893–1913), the Peirce Edition Project. Bloomington: Indiana University Press, 1998. [References to this volume will be designated by EP 2, followed by colon, page number].
Pietarinen, Ahti-Veikko, and Francesco Bellucci. (this volume). Habits of reasoning: On the grammar and critics of logical habits. *Consensus on Peirce's concept of habit: before and beyond consciousness*, ed. Donna E. West and Myrdene Anderson. (Studies in Applied Philosophy, Epistemology and Rational Ethics [SAPERE]) New York: Springer.
Raposa, Michael L. 1984. Habits and essences. *Transactions of the Charles S. Peirce Society* 20.2: 147–167.
Rosenthal, Sandra B. 1982. Meaning as habit: Some systematic Implications of Peirce's pragmatism. *The Monist* 65(2): 230–245.
Shapiro, Gary. 1973. Habit and meaning in Peirce's pragmatism. *Transactions of the Charles S. Peirce Society* 9.1(Winter 1973): 24–40.
Stjernfelt, Frederik. 2007. *Diagrammatology: An investigation on the borderlines of phenomenology, ontology, and semiotics*. Dordrecht: Springer Verlag.
Stjernfelt, Frederik. 2014. *Natural propositions: the actuality of Peirce's doctrine of dicisigns*. Boston: Docent Press.

Part III
Mental Complexions of Habit

Chapter 15
Habits of Reasoning: On the Grammar and Critics of Logical Habits

Ahti-Veikko Pietarinen and Francesco Bellucci

> "The world hates sound reasoning as a child hates medicine" (Peirce, MS 650, 1910).

Abstract We explain the grammar and the critics of the habits of reasoning, using Peirce's 1903 Lowell Lectures and the related *Syllabus* as the key textual source. We establish what Peirce took sound reasoning to be, and derive a major soundness result concerning his logic as semeiotic: an argument is valid if for any object that the premises represent, the conclusion represents it as well, which in semeiotic terms translates to a sign being a valid argument if for any object that the sign represents, the interpretant sign represents it as well. The perfect adherence of the grammar to the critics is evidenced by the un-eliminability of leading principles. Just as a logical leading principle is an un-eliminable element of reasoning, because any attempt to use it as a premise engenders an infinite regress, so a logical representative interpretant is a habit that cannot be rendered a sign. The logical representative interpretant is a principle not itself a premise, a rule not itself subject to rules, a habit not itself a sign.

Keywords Speculative grammar · Logical critics · Semeiotic · Sign · Interpretant · Leading principle · Reasoning · Justification of deduction

---

The second author is the main author of the second section, the first author of the third section. Other sections are jointly written.

A.-V. Pietarinen (✉) · F. Bellucci
Tallinn University of Technology, Tallinn, Estonia
e-mail: ahti-veikko.pietarinen@ttu.ee

F. Bellucci
e-mail: bellucci.francesco@gmail.com

A.-V. Pietarinen
Xiamen University, Xiamen, China

D.E. West and M. Anderson (eds.), *Consensus on Peirce's Concept of Habit*, Studies in Applied Philosophy, Epistemology and Rational Ethics 31,
DOI 10.1007/978-3-319-45920-2_15

Introduction

According to Peirce, ethics is the science that distinguishes good from bad conduct. Ethical conduct is *deliberate* and *self-controlled* conduct, because only of a deliberate conduct is it possible to say whether it is good or bad. Likewise, reasoning is self-controlled *thinking*. Drawing its principles from ethics, logic studies a special kind of deliberate conduct, *thought*, and distinguishes good from bad thinking. First of all, then, the purpose of logic is to distinguish valid from invalid reasoning (CP 5.120–150; EP 2: 196–207, 1903).

But to say that reasoning is self-controlled thinking amounts to saying that whenever one reasons, one has to reason according to some general principle of reasoning. We cannot say that an inference is valid unless we first admit that there is in fact *some* principle according to which the inference is drawn. Reasoning does not consist simply in deriving a proposition (the conclusion) from other propositions (the premises). Rather, it involves the approval of the rationality of the derivation, or the judgment that the reasoning is valid because it is an instance of *some* class of valid arguments.[1] One can reason, Peirce maintains, without knowing the principles of reasoning, just as one can play pool without understanding the laws of analytical mechanics (CP 5.319, 1869). Reasoning is *controlled* thinking, that is, it consists in passing from one proposition to another on the basis of *some* principle.

We propose, following Peirce on this point, that the principles of reasoning are habits of reasoning. Thus, habits of reasoning are of the utmost interest in logic and in philosophy of logic.[2] In this paper we examine Peirce's mature views on the nature and the functioning of these habits of reasoning. Our focus is on his 1903 Lowell Lectures that he delivered in Boston in November and December and which

[1]"[A] man cannot truly reason without having some notions about the classification of arguments. But the classification of arguments is the chief business of the science of logic; so that every man who reasons (in the above sense) has necessarily a rudimentary science of logic, good or bad" (MS 692: 5, 1901); "no argument can exist without being referred to some special class of arguments. The act of inference consists in the thought that the inferred conclusion is true because *in any analogous case* an analogous conclusion *would be true*" (EP 2: 200, 1903); "to accept the conclusion without any criticism or supporting argument is not what I call reasoning" (MS 293: 7, 1906).

[2]That this was Peirce's position has long been recognized, for example by Boler: "the leading principle fits Peirce's description of habit" (1964: 388). Pietarinen (2005a) studies habits of reasoning in the context of the *logica utens* and *logica docens* distinction. Contemporary logic and cognitive sciences, which have partially revived the old idea of the possibility of theorizing about the nature of the instinctive faculty of logic, the *logica utens*, have set a new agenda in which leading principles could be identified in novel cognitive and behavioral contexts such as unconscious processes of thinking and acting or abductive and imaginative reasoning in the sciences of discovery.

have to date remained unpublished, as well as on the *Syllabus* which he wrote to accompany these lectures.[3]

Peirce took logic to have three branches. The first branch, Speculative Grammar, is the "physiological" department of logic, and consists in the definition and classification of signs. The second branch, Logical Critics, classifies a specific kind of signs, arguments, which it divides into valid and invalid kinds. The third branch, Methodeutic, contains the doctrine of method (*Syllabus*; EP 2: 272). Although the "habits of reasoning" should receive a treatment in each of logic's branches, it is evident that Logical Critics is the branch most concerned with logical habits. The task of the second department of Peirce's logic is to submit principles of reasoning to scrutiny and to classify them according to the validity and strength of the arguments they govern. Logic, in its most general sense, becomes an analysis and classification of habits of reasoning. But Logical Critics is preceded by the analysis and classification of all signs, because Peirce was persuaded that the broadening of the scope of logic through the classification of signs was a valuable methodological generalization contributing to a better definition and analysis of logic's objects and boundaries (MS 675, 676, 1911; MS 12, 1912). Thus, the logical habits that are under scrutiny in Logical Critics must have received a semeiotic definition and categorization at the level of Speculative Grammar—the classification of arguments, the principal purpose of Logical Critics, must be preceded by a preliminary classification of signs, of which the classification of arguments becomes a part. Likewise, once Critics has pronounced upon the validity of certain logical principles, it is up to Methodeutic to investigate not whether a principle is *valid,* but whether and to what extent it is *advantageous* for scientific inquiry (MS L 75, 1902). In this paper, we explain the grammar and the critics of the habits of reasoning, leaving the discussion of the methodeutic point of view for a future occasion.[4]

Grammar of Logical Habits

Logical principles are habits of reasoning, and such habits are not signs. But reasoning is a sign, namely the sign that the premises are a sign of the conclusion. In Peirce's terms, an argument is a sign that is represented as a sign by its interpretant. There are thus different and highly sophisticated semeiotic relations involved in reasoning, and since as we know reasoning is based on habits, these latter should figure somewhere in the semeiotic description of inference.

[3]*A Syllabus of Certain Topics of Logic*. The main text is MS 478, with drafts, variants and additions found in MSS 2–3, MSS 508–512 and MSS 538–542. There are at least 350 draft pages related to the Syllabus altogether, The total number of surviving manuscript pages for his 1903 Lowell Lectures is at least 1300 (Pietarinen 2015b).

[4]On Peirce's later views on the interconnection between the different kinds of reasoning and on the methodeutic of reasoning, especially the abductive, see Pietarinen and Bellucci (2014).

That the premises are a sign of the conclusion was the Aristotelian and Stoic doctrine which Peirce adopted already in 1867. In hypothesis (later, abduction or retrodiction), the premises are an *icon* of the conclusion, in induction an *index*, and in deduction a *symbol* of the conclusion (*New List*, W 2: 58). In an 1870 letter to Jevons, Peirce writes that "all inference proceeds by the substitution of one sign for another on the principle that a sign of a sign is a sign, and that reasoning differs according to the different kinds of signs with which it deals, and that signs are of three kinds" (W 2: 446–447). More than thirty years later, we find him still experimenting with a similar, though not identical, semeiotic classification of the three basic forms of reasoning:

> If the premiss of induction is a symbol of the conclusion, then it should be a term, proposition, or syllogism.
>
> If the premiss of deduction is an index of the conclusion, it should be a term or a proposition.
>
> But if the premiss of abduction is an icon of the conclusion it can only be a term.
>
> (*Logic Notebook*, MS 339: 196r, 1901)
>
> Tersigns are arguments [...] or inferences. The conclusion is the interpretant sign which is intended to be determined. But this is represented by the fact which is itself represented in the premiss. The premiss is usually a copulative proposition. The praemitted fact may be a sign of the conclusion in various ways. (*Logical Tracts II*, MS 492: 41, October 1903)

The *Syllabus* of the Lowell Lectures—a document closely connected to the *Logical Tracts*—registers the first of a series of systematic applications, experimentations, and variations of this idea. All signs represent an object, although only some of them are also represented as representing their object as they in fact do. These signs are what Peirce calls "symbols":

> A Symbol is a Representamen whose Representative character consists precisely in its being a rule that will determine its Interpretant. All words, sentences, books, and other conventional signs are Symbols. We speak of writing or pronouncing the word "man"; but it is only a replica, or embodiment of the word, that is pronounced or written. The word itself has no existence although it has a real being, consisting in the fact that existents will conform to it. It is a general mode of succession of three sounds or representamens of sounds, which becomes a sign only in the fact that a habit, or acquired law, will cause replicas of it to be interpreted as meaning a man or men. (*Sundry Logical Conceptions*, MS 478, EP 2: 274, 1903)

This passage comes from the first version of speculative grammar that Peirce drafted for the *Syllabus* (*Sundry Logical Conceptions*, MS 478: 43–105). Here he presents two trichotomies of signs, one into *icon*, *index* and *symbol* and the second into *rheme*, *dicisign* and *argument*. The second trichotomy was usually considered by Peirce as a division of the last term of the first trichotomy (symbols, cf. e.g., the *Logical Tracts I*, MS 491, 1903). In *Sundry Logical Conceptions*, on the contrary, the two trichotomies are *not* presented as so related, so that one should expect, for example, that rhemes and dicisigns exist which are not symbols. Remarkably, in *Sundry Logical Conceptions* we find Peirce spilling much ink to discuss the dicisign as the "sign whose interpretant represents it as an index" (MS 478, EP 2: 275–79). But MS 478 lacks fine enough analysis of arguments, and their semiotic characterization remains in the dark.

Soon enough, however, he realizes that the two trichotomies are insufficient for certain important logical distinctions. Therefore, in a revised version of the *Syllabus* (MS 540, *Nomenclature and Division of Triadic Relations*), he adds to the above a third trichotomy, that into *qualisign, sinsign,* and *legisign.* To interpret a sequence of written signs as the word "man", I first have to consider it as *something significant*, as a legisign, "a law that is a sign". It is "not a single object, but a general type which, it has been agreed, shall be significant" (EP 2: 291). The legisign professes to be a legisign, namely a general sign that occurs in replicas. This is the true innovation of MS 540 over MS 478: the separation of the material aspect (*suppositio materialis*) from the formal aspect (*suppositio formalis*) of the signification of symbols and other signs.

First, the sign has to be regarded as significant (recognized as legisign). Only then it can be considered as conveying a certain general idea (recognized as symbol). The habit of interpretation is different in these two cases. Peirce coupled and confused these two habits in the pre-MS 540 treatments of symbols; but in MS 540 he perceives that it is one thing to recognize a legisign as such and a different thing to recognize that which the legisign signifies. One can recognize a word as a word, that is, as a symbol, without knowing what the word means, namely what idea the symbol is supposed to excite. This happens when we read a language we do not know. As Peirce puts it in the *Logical Tracts I*, symbols are not only general signs *materialiter*, they are also general signs *formaliter* (MS 491, 1903). Not all signs can represent a general object. Icons represent possible objects, indices individual objects, but only symbols can represent, or convey the idea of, a general object. The symbol "man" excites, conveys or determines in the mind of the interpreter an *idea*, the idea of a man, together with all that is known of men. But it does this only in so far as that symbol is recognized as a sign of men. The symbol "man" itself does not represent the form of any man, nor is it connected with any man; but it does represent men because it is *represented* as representing men.

Symbols may be of three basic kinds—terms, propositions, and arguments—and thus each kind must differ from the others *according to the way each is represented by its interpretant*. This is the doctrine exposed in MS 540: "A *Rheme* is a sign which, for its Interpretant, is a sign of qualitative possibility [...] A *Dicent Sign* is a sign which, for its Interpretant, is a sign of actual existence [...] An *Argument* is a sign which, for its Interpretant, is a sign of law" (EP 2: 292). A qualisign is a law, a symbol refers to its object by means of a law, and an argument is a sign of a law. This reflects the fact that the argument not only represents something according to a law, but it also represents the very law according to which it represents what it does.

An argument "urges" its law, it presents the law as a justification of that which it represents:

> The Interpretant of the Argument represents it as an instance of a general class of arguments, which class on the whole will always tend to the truth. It is this law, in some shape, which the argument urges; and this "urging" is the mode of representation proper to arguments. The Argument must, therefore, be a Symbol, or Sign whose Object is a General Law or Type. It must involve a Dicent Symbol, or Proposition, which is termed its *Premiss*; for the Argument can only urge the law by urging it in an instance. This Premiss is, however, quite different in force (i.e., in its relation to its interpretant) from a similar proposition merely asserted; and besides, this is far from being the whole Argument. As for another proposition, called the Conclusion, often stated and perhaps required to complete the Argument, it plainly represents the Interpretant, and likewise has a peculiar force, or relation to the Interpretant. (MS 540, EP 2: 293)

This doctrine arguably[5] requires a division of interpretants, the first hint of which is, not surprisingly, to be found in *Sundry Logical Conceptions*, where Peirce distinguishes the fact that the sign determines an interpretant from the fact that the sign is represented by its interpretant as being related to its object (MS 478, EP 2: 272–273). This distinction is developed in subsequent writings. In the *Logic Notebook*, we find Peirce registering a division among immediate, dynamic, and representative interpretants. The dynamical interpretant is the sign's actual effect upon an actual interpreter, or a "determination of a field of representation exterior to the sign" (MS 339: 253r, October 1905). The immediate interpretant "is the Interpretant as represented in the sign as determination of the sign. To which the sign appeals" (*ibid.*). The representative interpretant is "the interpretant that truly represents that the sign represents its object as it does" (*ibid.*), or that which "correctly represents the Sign to be a Sign of its Object" (MS 339: 255r). In the 1906 *Prolegomena to an Apology for Pragmaticism* we find again the famous triad of interpretants. The representative interpretant, here termed the Final, is defined as "the manner in which the Sign tends to represent itself to be related to its Object" (CP 4.536, 1906).

Some such division of interpretants becomes of the utmost importance for the grammatical analysis of arguments. For according to the *Syllabus* doctrine, a proposition is represented by its interpretant as an *index* of its object, while an argument is represented as a *symbol* of its object:

> It is easy to see that the proposition purports to intend to compel its Interpretant to refer to its real Object, that is, represents itself as an index, while the argument purports to intend not compulsion, but action by means of comprehensive generals, that is, represents its character to be specially symbolic. (EP 2: 183, 1903)

> [T]he Final (or quasi-intended) Interpretant of an Argument represents it as representing its Object after the manner of a Symbol. (CP 4.572, 1906)

> An argument, on the other hand, intends its representative interpretant to represent its immediate object, not now as a brute effect, but as a *sign* of its external object. (MS 145: 29–30, n.d.)

[5]See Bellucci (2014) for more details.

If we accept Peirce's distinction between the immediate and the final or representative interpretant of a sign, and if we accept that an argument is a sign, then we are forced to conclude that the most thorough grammatical analysis of the argument reveals that an argument is a sign *whose representative interpretant represents it as representing the immediate interpretant according to a logical habit.*

The next step in the classification is therefore a division as to the different ways in which arguments are represented to represent their immediate interpretants:

C. According to the Nature of the Sign as represented in the Representative Interpretant as determining its Interpretant

Professes to be exclamatory Abductive

" " " imperative Deductive

" " " enlightening Inductive

(*Logic Notebook*, 9 October 1905, MS 339: 255r)

C is a division of arguments only, because *only arguments explicitly represent interpretants.* In the *Logic Notebook* entry of 30 August 1906, Peirce gives the following division: "Sign by common nature, Sign by diagram, Sign by Experiment" (MS 339: 284r, 1906). In letters to Lady Welby we find the following divisions: "I *think* that one [division of signs] must be into Sign assuring their Interpretants by Instinct, Experience, Form" (EP 2: 481, 1908); "X. As to the Nature of the Assurance of the Utterance: *Assurance of Instinct, Assurance of Experience, Assurance of Form*" (EP 2: 490, 1908). The grammatical division of arguments seeks to answer the question: in which different semeiotic ways arguments are represented as representing their interpretants? That is, in which logical ways do arguments determine or represent their conclusions? According to Peirce, in three ways: abductively, deductively, and inductively. That is, arguments are represented by their representative interpretant as representing their immediate interpretants according to an abductive, a deductive, or an inductive leading principle or rule of inference. As it is evident, this is a noteworthy refinement of the *New List* doctrine, in which it was simply stated that arguments are icons, indices, or symbols of conclusions.

Habits of reasoning are therefore representative interpretants. In the first place, all signs must determine an interpretant sign if they are to function as signs and refer properly to an object. In the second place, symbols alone among signs can refer to an object and convey some idea about it only if they are represented as so doing by their (representative) interpretant. In the third place, while all symbols are represented as representing an object, *not all symbols are also represented as representing an interpretant*: only arguments are represented as representing an (immediate) interpretant. A proposition represents a state of affairs and is represented by its representative interpretant as an index of that state of affairs; that is, the proposition professes to be true, or implicitly asserts its own truth, and there its function is exhausted. In still other words, propositions do represent immediate objects but do not represent immediate interpretants.[6] An argument, by contrast, represents a state of affairs that is in its turn the representation of another state of affairs, which is true when the former is true. In other words, arguments represent an object as a means for representing an immediate interpretant. In order to do so, an argument must profess to be representative of a general method of procedure, or to represent what it does according to a general principle (*Lowell Lecture III*, MS 465: 37, 1903). The argument not only represents a certain conclusion. It is also represented as representing that conclusion according to a valid logical principle, a principle whose validity transcends the validity of the argument at hand, and is thus general. The argument, while it represents its own conclusion, also evidences the general rule according to which it represents that conclusion. The habit according to which an inference is drawn is, in semeiotic terms, the representative interpretant that represents *how* the argument represents the immediate interpretant. The representative interpretant is the manner in which arguments indicate or reveal a conclusion.

Critics of Logical Habits

Speculative grammar can at most describe the physiology of arguments and classify them according to their structure. It cannot pronounce on the conditions of the *validity* of each. This investigation—the examination of the conditions of validity of reasoning and of its different varieties—is the task of Logical Critics. Critics and Grammar meet in the classification of arguments. But while Grammar regards the classification as a product of a combinatorial approach, Critics regards it as the outcome of a critical analysis.

In a sense, to be explicated later, leading principles, when they are *logical*, govern reasoning without themselves being premises or parts of reasoning. Reasoning is sound when its leading principle is true. The validity of reasoning is thus first and foremost a question concerning logical habits. But what makes a principle true and a reasoning sound in each different class of reasoning? What is the peculiar form of reasonableness that characterizes each forms of inference? The answer to this question is the key upon the door of Logical Critics.

As regards *deductive* or *necessary* reasoning—the reasoning of mathematics—Peirce was prepared to give an answer right in the beginning of his 1903 Lowell Lectures. "The gist of the reasoning", he tells his audience, "is to state in the most general terms that relation of the state of things expressed in the conclusion of each

[6]See Peirce: "every proposition asserts its own truth" (CP 5.340, 1869); "[e]very proposition besides what it explicitly asserts, tacitly implies its own truth" (CP 3.446, 1896); cf. MS 292: 16, 1906. On Peirce's "redundancy theory of truth", see Chauviré (1995: 152–153) and Thibaud (1997: 290). On Peirce's doctrine of propositions see also Hilpinen (1982, 1992), Stjernfelt (2014, Chap. 3; 2015). For an argument why the object of a proposition is, properly speaking, its *immediate* object see Bellucci (2015).

inferential step to the state of things expressed in the premises" (MS 450). This remark is a pertinent one. In spelling out that general character of the relation between the states of things expressed in premises and conclusion, Peirce stands out as a lone early figure in the model-theoretic and semantic tradition in logic. His place in the model-theoretic tradition is well confirmed in numerous other places of his writings, too, both published and unpublished, as well as in the earlier literature.[7] Peirce's answer could in fact have led to the semantic definition of the logical consequence relation and a development of the model-theoretic approach much earlier than what actually happened, even a third or half a century ahead, if only his audience could have understood the impact of the remark well, or if only G.H. Putnam had agreed to publish his Lowell Lectures according to Peirce's own requests (GHP to CSP, 21 December 1903). As of 2015 CE, Peirce's lectures still remain unpublished.

At any event, Peirce had worked out the essence of the model-theoretic idea in the first draft lecture which he wrote around June 1903 (MS 454). Later in that fall, he rewrote the lecture into an informal exposition, which he delivered as the famous opening lecture of the series, entitled "What makes a reasoning sound?" (23 November 1903, MS 448–449; EP 2: 242–257). The delivered lecture aimed not at a positive answer but at a definite refutation of the earlier logicians' fallacious attempts at the answer. Those attempts had, according to Peirce, failed to analyze in a logical manner the required changes in the forms of expression. The earlier logicians were attempting to describe the workings of the mind or consciousness and failed to develop the theory of logic on its own terms.[8]

But what is the *positive* answer to Peirce's celebrated question? His solution is in fact deceptively simple. But it certainly was not a straightforward one at that time when the idea was altogether novel. The reason why the nature of this question has puzzled later commentators is that his answer does not appear in the delivered lecture in any direct manner. But the answer is found in the early draft of that lecture, which for expository and pedagogical reasons, following James's advice, Peirce had decided to supersede by the more accessible, *negative* answer to the question (namely what does *not* make the reasoning sound?). Yet even in the delivered lecture, he did remark on where to look for the positive answer: "[S]ound reasoning is such reasoning that in every conceivable state of the universe in which the facts stated in the premisses are true, the fact stated in the conclusion will thereby and therein be true" (EP 2: 250).

So what precisely, is Peirce's solution? Look at that early draft lecture in MS 454. The planned lecture, whose notebook cover states "Early D. Lowell Lecture I (used for II)", directly begins with the presentation of the basic principles and transformation rules for the Alpha and Beta parts of his theory of Existential Graphs

---

[7]See especially Tiercelin (1991), Hintikka (1997) and Pietarinen (2006, Chap. 6).

[8]Peirce's reading list for his Lowell Lectures, which has been recovered only recently, reveals that his criticism was targeted at works such as Bosanquet (1892), Bradley (1883), Erdmann (1896), Sigwart (1895), Trendelenburg (1840), Ueberweg (1871) and Wundt (1893–1895); see Pietarinen (2015b).

(hereafter EGs). This non-delivered draft is thus a fairly technical exposition of some novel facts concerning the logic of EGs, which Peirce later decided to avoid discussing in the opening lecture. But the draft is quite striking in that it actually provides an answer to his initial question. What makes a reasoning sound is *the existence of the proof of the soundness of the rules of transformation* of his newly-founded diagrammatic logic of EGs.

The naturalness and simplicity of the idea lies in this. What makes a reasoning sound must be a property that we can find and isolate when analyzing the processes of reasoning. In a system of logic such as EGs, that property is a common invariant of those rules of transformation upon which its inferential system is built. Those transformations change one token of a graph into another, modified token. What is preserved in all these transformations is the fact that they never change tokens of graphs true in the state of things that they describe, into a modified token that would not be true in that same state of things. If such a general relation indeed holds between these two occurrences of the state of things, it is impossible, as Peirce recounts in MS 450, that the premises are true without the conclusion being true.

It took a while until logicians routinely began to express such a relation in terms of statements of logic being "true-in-a-model". But the idea is just the same. If the premises are true (or satisfied) in some state of things, and under a variable assignment in case there are free variables (which the original theory of EGs did not have), then the conclusion is also true (or satisfied) in that same state of things, under the same variable assignment. In brief, the security of the truth of the conclusion is encapsulated in the truth of the premises. This is the semantic definition of inference, and it is Peirce's definition—the seminal definition of the semantic consequence relation, one might say.

The notion of semantic consequence is needed in order then to state the related, fundamental property of the rules of reasoning in a deductive system such as that of the graph transformations of EGs: If one has a derivation of a conclusion from the premises (that is, if there is a proof for that argument), then that conclusion must be a semantic consequence of those premises (that is, the argument in question is valid). This of course is the statement of the *soundness* of the system of proofs. It appeals to the semantically characterized notion of the validity of reasoning.

In the draft lecture (MS 454), Peirce indeed begins proving the soundness of his four "basic rights". These four rights are the basic rules of transformation of the Alpha part of the EGs. He had worked on such rules of transformation ever since the invention of the EGs in fall 1896. By the summer of 1898 he would already have the final (sound and complete) forms of those rules at hand. They are the (irreversible) rules of *erasure and insertion* (Right I), the (reversible) rules of *deiteration and iteration* (Right II), the rule of *double cut* (Right III), and the rule for *the pseudograph* (Right IV). The 1903 manuscript breaks off after the proof of the soundness of the first right (erasure and insertion), but the surviving material suffices to understand what the general character and tendency of the argumentation is that Peirce had in mind.

Although in this draft lecture he does not complete his proof of soundness, it demonstrates beyond reasonable doubt that he managed to describe the semantic notions of validity and logical consequence, and that he used these to characterize a general property (soundness) of his system of EGs. All these notions are needed to gather a positive answer to the principal question. Once he had described these notions and the general property of soundness covering all the basic rights, the main job was done. Peirce had delivered a modern answer to the question that had puzzled his audience, his contemporaries and the early logicians, but also the generations that followed.

It is this proof of the soundness of the rules of transformation that, in the actually delivered lecture, came to be replaced by an informal and negative, and what he must have thought a more agreeable, exposition of the principal question. In the delivered lecture, he reaches the point in which the audience was expected to see how the previous attempts at explaining the phenomenon of reasoning actually compromise soundness. Peirce does not attempt in the delivered lecture to proceed from that informal exposition of the negative answer to the exposition of the positive answer. Luckily, the latter is preserved in the early draft of MS 454.

Since Peirce had proved the soundness of the system of the rules of transformation of his newly-developed diagrammatic logic of EGs, his answer to the question of what makes a reasoning sound now becomes crystal clear: *it is the existence of the general property of the soundness of the reasoning that that system exhibits*. Although the manuscript breaks off before completing the inductive proof of the soundness of the basic rights of EGs, proofs of the soundness of the rules of transformation are found in several other places in his works (Pietarinen 2015b; MS 484, 1898; MS 491, 1903). It is merely his intention to present such demonstration that suffices for our purposes. His intention alone shows what the "thorough and formal refutation of the fallacy" was supposed to be that he asserted to have in his possession (MS 448, see EP 2: 234). The editors of EP 2 claimed that such refutation "has not been found" (EP 2: 553), but here we have every reason to believe that that refutation consists in the proof of the soundness of the rules of transformation for EGs as found in MS 454: that proof positively replaces any earlier attempt to establish what sound reasoning would or was thought to consist of, such as Sigwart's "feeling of logicality" or other appeals to psychological notions. It was only the fear of presenting material that would be too technical and to the public that prohibited him to present that "formal refutation" in the actually delivered lecture.

A further, related point is worth mentioning. Deductive reasoning is carried out in Peirce's system in terms of "four basic rights of transformation". These rules were published for the first time in the *Dictionary of Philosophy and Psychology* in 1901 (in the entry "Symbolic Logic", Rules I–IV for the Alpha part). In that entry, one also finds different, and one is almost tempted to say, "symbolic", versions of the graphs, rewritten using parentheses, brackets, and braces instead of ovals. Since Peirce had previously had very few opportunities to explain his system in public, by 1903 he obviously must have been eager to communicate his major discoveries to the live audience. But this is not all. A remarkable observation follows the description of these basic rights in the lecture, namely that "by using these rights we can draw from any premisses any inferences that they justify" (MS 454). This comes close to stating the semantic *completeness* of the system of transformations.

In the draft, Peirce first begins demonstrating the other direction of such statement, namely that everything that is so deducible by these rights is a semantic consequence of the premises. This soundness proof is a semantic proof in which "a right to transform an entire graph" from M to N, that is, the transformability, or the existence of a derivation sequence that leads from a graph M to a graph N, is taken to mean that "every state of things in which M is true is a state of things in which N is true". That is, we actually do have these rights to transform the graphs in the manner they prescribe. The draft does not contain the proof of completeness, as it does not show the adequacy (if the relationship between the premises and the conclusion is stated in terms of a semantic consequence relation, then there is an inference based on these basic rights), but that proof is easy to complete (Roberts 1973). Peirce also makes similar statements on completeness of the system in other places of the Lowell Lectures and elsewhere (MS 491).

Peirce manages to articulate in the Lowell Lectures a wealth of surprisingly modern logical ideas, such as semantic and model-theoretic treatment of reasoning, including the notions of semantic consequence, quantification theory, the notion of truth-in-a-model, the essence of the semantic tree method, the soundness and completeness of a logical system, systems of modal logic and possible-worlds semantics (Pietarinen 2005a, b, 2011). The year 1903 is not the first time for some of them (Pietarinen 2006, 2015, In press), but what makes his method of EGs of his *annus mirabilis* stand out from previous expositions is that the presentation of the theory invariably begins with the description of the elaborate system of *conventions* upon which the theory is erected. The presentation of permissive rules and rights *follows* the description of the conventions. The conventions, which there were altogether fourteen, provide the theoretical context in which what we nowadays recognize as the semantic and model-theoretic approach begins to blossom.[9] Most of what he had to say of these semantic conceptions underlying the development of modern logic and reasoning he had never published before, however, and so the Lowell Lectures were the first and largely the only sufficiently elaborate attempt to communicate those pioneering ideas in public and to the posterity.

Justifying Habits of Reasoning

The set of notes Peirce drafted for his second Lowell lecture is no longer concerned on setting out the conventions of the system of EGs (MSS 455, 455(s), 456, S-29, S-32-34). Instead, it takes up the issue of the *justification* of the principles of reasoning. What, after all, is the nature of the relationship between premises and

[9]The conventions rely heavily on the idea of reasoning that is "nothing but the discourse of the mind to its future self" (MS 450), that is, to dialogic, game-theoretic and interactive patterns. Peirce takes diagrams to be the best form that gives the precepts according to which the interaction between various phases of mind operates. Much of what is found in his conventions are attempts to capture the workings of such interactive operations.

conclusions which the semantic notion of validity formally characterizes? How to ascertain ourselves of the soundness and validity of a piece of reasoning that is altogether of a new kind, never seen and never thought of before? And third, what makes a reasoning not just sound but *evident* to reason? These are questions that pertain to the philosophy of logic; questions that formal logic with its notions of truth-in-a-model, validity, and soundness alone does not answer.

In the second lecture Peirce addresses and refutes what has in the subsequent literature been termed the "paradox of deduction". The problem stems at least from Carroll's enigmatic article (Carroll 1895). According to that "paradox", still much debated in the philosophy of logic, deductive reasoning that rests upon logical principles must not be used to justify those principles.

Peirce's solution in the Lowell Lectures is quite clever.

> Shall I give you, then, a strict demonstration of each principle? There is a logical objection to that, too. I do not mean the objection which has sometimes been urged, that since reasoning is to rest upon logical principles, therefore logical principles must not be made to rest upon reasoning. For this objection arises from a pedantic confusion of thought. I call it pedantic because it is not the outcome of any unsophisticated natural reasoning. It is essentially a logician's fallacy. That is, it comes about, as many other fallacies of professional logicians come about, from the mistaken application of a logical rule. The person who first raised the objection was thinking of the form of fallacy which logicians call "begging the question". That fallacy really consists in assuming as a premiss some thing you wish to prove not admitted to be true. (MS 455s: 5–6, Lowell Lecture II (B), 1903)

Peirce then rightly recalls that the *statement* of the principle of reasoning and *reasoning according to* that principle do not amount to the same thing. When one states the principle according to which an inference is draw one uses a conditional proposition: "if such and such premises are true, then such and such conclusion is also true". Given this difference, what logicians call "begging the question" is not a real objection to the project of demonstrating the principles of reasoning. For it is true that the proof of a principle has to contain this principle in its premises. But a principle of reasoning can be a premise only in a very special sense. According to Peirce, any proof of a leading principle can only consist in showing that the leading principle to be proved is already admitted to be true in another form in the argument that is supposed to prove it. The only proof of a principle is the proof of its un-eliminability.

The essentials of this doctrine had been published by Peirce in 1867 (W 2: 24) and 1880 (W 4: 167), and subsequently discussed in other writings (MS 411: 184–185; MS 594: 60, c. 1894; NEM 4: 174–176, 1898). In one place he connects this point to what Aristotle says at *An. Post.* 77a10-12: "No demonstration assumes that it is not possible to assert and deny at the same time". That is, no argument assumes the principle of contradiction as an axiom. Peirce says: "Aristotle remarks respecting one of the most important logical principles, that it is never taken as a premise but only used in drawing the conclusion. This seems quite right. *Nothing is gained by laying down a true logical principle as a premise, since nothing can be deduced from it without virtually assuming that very same principle*" (MS 411: 184–185, our emphasis). In the *Cambridge Conferences* (1898), he again returns to this point, and claims that a complete argument is one whose leading principle has reached a *maximum degree of abstractedness* (NEM 4: 174–176). The most abstract leading principle is a logical principle. A logical principle is a purely formal principle, containing nothing material. Peirce's "proof" of a logical principle consists in showing that the principle in question is already admitted in another form in the reasoning that is supposed to prove it.[10] In the second Lowell lecture Peirce returns to this idea, and declares that the standard objection of begging the question in the justification of reasoning is a confusion of thought. A proper logical principle cannot, in Peirce's terms, be laid down as an additional premise of the argument of which it is the leading principle, without producing another argument whose own logical leading principle is identical to the leading principle at hand.

A logical principle is a logical habit, and a logical habit corresponds, as we showed in Sect. 2, to the representative interpretant discovered by means of the grammatical taxonomy. The question thus arises whether the representative interpretant is un-eliminable in the same sense. If logical habits are found to manifest a special logical behavior (that is, they are unanalyzable), then this fact must be reflected at the level of speculative grammar. Peirce's answer to this question is contained in *A Survey of Pragmaticism* (1907):

> The real and living logical conclusion is that habit; the verbal formulation merely expresses it. I do not deny that a concept, proposition, or argument may be a logical interpretant. I only insist that it cannot be the final logical interpretant, for the reason that it is itself a sign of that very kind that has itself a logical interpretant. The habit alone, which though it may be a sign in some other way, is not a sign in that way in which that sign of which it is the logical interpretant is the sign. (CP 5.491)

The representative interpretant is the habit-governing reasoning but is not itself an element of reasoning. To regard a habit as a sign is analogous to regard the leading principle of reasoning as a premise of reasoning: by some deduction theorem one can always express the leading principle (a logical rule) as a conditional proposition (a logical proposition).[11] But as he had shown already in 1867, nothing is gained by so doing when the leading principle is *logical*, for the very same principle will be governing the new argument thereby obtained. In semeiotic terms, in reasoning no final logical interpretant can be considered as a sign without

[10]On Peirce's doctrine of the un-eliminability of logical leading principle see the classic Thompson (1953: 6–8). Boler (1964: 390) gives the following explanation: "while the argument finds its completion in the leading principle, the function of the leading principle can never be made explicit within the argument itself. The progression that results from stating the leading principle is trivial enough, but the point Peirce wants to make is not". See also Bellucci (2013: 192–194).

[11]In fact Peirce states many further logical results in the Lowell Lectures, including the deduction theorem. Namely, he observes that if a graph Y is transformable from the graph X, then X *scrolls* Y (that is, X implies Y) is a theorem. Following his statement of the deduction theorem, the permission to add a double cut around any graph follows as a corollary.

employing in this reasoning that very same logical interpretant. A final logical interpretant is thus un-eliminable: no matter how many times you transform it into a sign, it continues to act as a habit just the same.[12]

Conclusion

It appears from our reconstruction that the results of the critical analysis of logical principles—the idea underlying model theory—closely correspond to the results of the grammatical analysis—the semiotic physiology of arguments—, so much so as to be virtually identical to it.[13] The answer to the question "What makes a reasoning sound" was that a reasoning is sound, as he demonstrated in the context of EGs, when it is based on a system of transformations that never transforms a graph that is true in a certain state of things into a modified graph that is not true in that same state of things.

This can be stated, and Peirce often did state it, semeiotically. In 1865 he had proposed that logic should be considered the science of the relations of symbols in general to their objects. His reason was that "that inference is logical which conforms to the general conditions of passing from premises which have a real object to a conclusion which has a real object" (W 1: 309, 1865). In an unpublished fragment

[12]In this perspective, then, Short errs in claiming that Peirce in 1907 "broke out of the hermetic circle of words interpreting words and thoughts interpreting thoughts" (2007: 59). According to Short, before 1907 Peirce had maintained that the interpretant of a sign is always a sign. This, according to Short, immediately engenders the paradox of unlimited semiosis, so much praised by Eco and Derrida, but would ultimately fail to explain what the meaning of a sign actually is: by saying that the meaning of a sign is the sign it is translated into Peirce, into, Peirce makes significance dependent on further significance, *ad infinitum*. In 1907, says Short, Peirce makes a "final decisive change in his semeiotic" (2007: 56): the claim that the "habit alone, which though it may be a sign in some other way, is not a sign in that way in which that sign of which it is the logical interpretant is the sign" (CP 5.491, 1906–1907) is the required necessary correction to his earlier, flawed semeiotical theory. Now, if one admits that arguments are signs, and that in 1867 Peirce had already elaborated and subsequently published his doctrine of the un-eliminability of logical leading principles, then there is good reason to maintain that already in 1867—that is, at the dawn of Peirce's semeiotic investigations—what would later be called the final, ultimate, or representative interpretant is considered in some sense not of the nature of a sign but of the nature of a habit. In fact, one may say that in those early papers Peirce gave a demonstration—that while representative interpretants can be considered as signs, *logical* representative interpretants cannot on pain of redundancy. So if one admits that arguments are signs for Peirce, then there is no reason to believe that Short's "third flaw" in Peirce's theory needed correction in 1907. For, the idea needed for its correction—"that it is the habit itself, and not a concept of it, that is the ultimate interpretant of a concept" (2007: 58)—was born together with the theory itself.

[13]To show that they really are identical we would need a *semeiotic completeness* result: everything that is a conclusion in Logical Critics is an interpretant in Speculative Grammar. But are, for instance, emotional interpretants conclusions? We believe that after the 1903 declaration of soundness, Peirce spent rest of his life trying to demonstrate the semeiotic completeness of his architectonic. We will leave this important issue for a future paper.

from c. 1893 we read that "the merely formal study of the laws of symbols [...] show[s] the general symbolical laws to which symbols in general must conform *in order to retain, in transformations of them, their correspondence with their objects*" (MS 810: 1, our emphasis). In a 1908 letter to Jourdain, Peirce writes that logic is "the theory of the control of signs in respect to their relation to their objects" (NEM 3: 886). In other words, logic controls that one is passing from a sign with an object to another sign with the *same object*, that is, with the *same truth-value*. An inference is sound if whatever the premises represent, the conclusion represents it as well.

The semeiotic version of the 1903 model-theoretic idea, and thus the ultimate answer to the question in what the *soundness* of reasoning consists of, is therefore an establishment of a passage from the clause that an argument is valid *if for any object that the premises represent, the conclusion represents it as well*, to the clause that an argument is valid *if for any object that the signs represent, the interpretant represents it as well.* We get the latter clause from the facts that the premises are a sign of the conclusion and that the conclusion is the (immediate) interpretant of the sign.

The latter clause of course states something that is quite familiar to the attentive reader:

> A *Sign*, or *Representamen*, is anything which so stands in relation to a second, called its *Object*, as to be capable of determining a third, called its *Interpretant*, to be in the same triadic relation to that object in which it stands itself. (*Sundry Logical Conceptions*, MS 478: 43, 1903)

The sign determines the interpretant to stand in the same relation to the object in which it itself stands; that is, the sign determines the interpretant to be *a true representation of the object* if the sign is one. This is why the relation should be genuine: the interpretant must represent the object *as the sign does*, and because it is so determined to represent it by the sign. The relation of illation is the paramount semiotic relation.[14]

The perfect adherence of the grammatical to the critical side is also evidenced by Peirce's theory of the un-eliminability of leading principles. Just as a *logical* leading principle is an un-eliminable element of reasoning, because any attempt to use it as a premise engenders an infinite regress, so a *logical* representative interpretant is a habit that cannot be rendered a sign. The classification of signs of the late years mirrors the discovery made in 1867: the logical representative interpretant is a principle not itself a premise, a rule not itself subject to rules, a habit not itself a sign.

---

[14]Murphey was the first to see that the "reduction of illation to the sign relation" (1961: 63) was an essential move in Peirce's early logical investigations, although he was mostly interested in the "reduction" as a decisive move in his early *categorial* investigations, of which he gave an unsurpassed reconstruction (1961: 65–91). However, in the light of the results of the present paper, speaking of a "reduction" of illation to the sign relation is not quite right, as soundness is not a reduction but a central metalogical result of the relationship between the two levels of the logical system, grammar and critics.

Acknowledgments Research supported by the *Estonian Research Council*, Project PUT267, and the Academy of Finland, project 1270335, "Diagrammatic Mind: Logical and Communicative Aspects of Iconicity," Principal Investigator Prof. Ahti-Veikko Pietarinen.

References

Bellucci, Francesco. 2013. Peirce's continuous predicates. *Transactions of the Charles S. Peirce Society* 49: 178–202.

Bellucci, Francesco. 2014. 'Logic, considered as semeiotic'. On Peirce's philosophy of logic. *Transactions of the Charles S. Peirce Society* 50: 523–547.

Bellucci, Francesco. 2015. Exploring Peirce's speculative grammar. The immediate object of a sign. *Sign Systems Studies 43*(4): 399–418.

Boler, John. 1964. Habits of thought. *Studies in the philosophy of Charles Sanders Peirce*, ed. E. Moore and R. Robin, 382–400. Amherst: The University of Massachusetts Press.

Bosanquet, Bernard. 1892. *Knowledge and reality*. London: Kegan Paul, Trench, & Co.

Bradley, Francis H. 1883. *The principles of logic*. London: Kegan Paul, Trench, & Co.

Carroll, Lewis (Charles L. Dodgson). 1895. What the Tortoise said to Achilles. *Mind* 1: 278–280.

Chauviré, Christiane. 1995. *Peirce et la signification*. Paris: Puf.

Erdmann, Johann E. 1896. *Outlines of logic and metaphysics*. Translated from the German by B.C. Burt. London: S. Sonnenschein & Co.

Hilpinen, Risto. 1982. On C.S. Peirce's theory of the proposition: Peirce as a precursor of game theoretical semantics. *The Monist* 65: 182–188.

Hilpinen, Risto. 1992. On Peirce's philosophical logic: Propositions and their objects. *Transactions of the Charles S. Peirce Society* 28: 467–488.

Hintikka, Jaakko. 1997. The Place of C.S. Peirce in the history of logic. *The rule of reason*, ed. J. Brunning and P. Forster, 13–33. Toronto: University of Toronto Press.

Murphey, Murray G. 1961/1991. *The development of Peirce's philosophy*. Cambridge: Harvard University Press (new 1991 edition, Hackett).

Peirce, Charles Sanders. i. 1867–1913. *Collected papers of Charles Sanders Peirce*. Vols. 1–6, ed. Charles Hartshorne and Paul Weiss. Cambridge: Harvard University Press, 1931–1935. Vols. 7–8, ed. Arthur W. Burks. Cambridge: Harvard University Press, 1958. [References to Peirce's papers will be designated by CP, followed by volume, period, paragraph number].

Peirce manuscripts in Texas Tech University Library at Texas Tech University, Institute of Studies of Pragmaticism, beginning with MS—or L for letter—and followed by a number, refer to the system of identification established by Richard R. Robin in Annotated Catalogue of the Papers of Charles S. Peirce (Amherst: University of Massachusetts Press, 1967), or in Richard R. Robin, "The Peirce Papers: A Supplementary Catalogue," Transactions of the Charles S. Peirce Society.

Peirce, Charles Sanders. i. 1976. *The new elements of mathematics by Charles S. Peirce*, vol. 4, ed. by Eisele, C. The Hague.

Peirce, Charles Sanders. i. 1867–1913. *Writings of Charles S. Peirce: A chronological edition*. Vols. 1–6 to date, ed. the Peirce Edition Project. Bloomington: Indiana University Press. [References to these volumes will be designated by W, followed by volume number, colon, page number].

Peirce, Charles Sanders. i. 1893–1913. *The essential Peirce: Selected philosophical writing*. Vol. 2 (1893–1913), ed. the Peirce Edition Project. Bloomington: Indiana University Press, 1998. [References to this volume will be designated by EP 2, followed by colon, page number].

Pietarinen, Ahti-Veikko. 2005a. Compositionality, relevance, and Peirce's logic of existential graphs. *Axiomathes* 15: 513–540.

Pietarinen, Ahti-Veikko. 2005b. Cultivating habits of reason: Peirce and the *logica utens* versus *logica docens* distinction. *History of Philosophy Quarterly* 22: 357–372.

Pietarinen, Ahti-Veikko. 2006. *Signs of logic: Peircean themes on the philosophy of langauge, games, and communication*. Dordrecht: Springer.
Pietarinen, Ahti-Veikko. 2011. Existential graphs: What a diagrammatic logic of cognition might look like. *History and Philosophy of Logic* 32: 265–281.
Pietarinen, Ahti-Veikko. 2015. Exploring the beta quadrant. *Synthese* 192: 941–970.
Pietarinen, Ahti-Veikko, ed. In press. *Logic of the future: Peirce's writings on existential graphs*
Pietarinen, Ahti-Veikko, and Francesco Bellucci. 2014. New light on Peirce's conceptions of retroduction, deduction, and scientific reasoning. *International Studies in the Philosophy of Science* 28: 353–373.
Roberts, Don D. 1973. *The existential graphs of Charles S. Peirce*. The Hague: Mouton.
Short, Thomas L. 2007. *Peirce's theory of signs*. Cambridge: Cambridge University Press.
Sigwart, Christoph. 1895. *Logic*. Vol. 2, 2nd edition. Translated from the German by Helen Dendy. London: Macmillan & Co.
Stjernfelt, Frederik. 2014. *Natural propositions. The actuality of Peirce's doctrine of dicisigns*, Boston: Docent Press.
Stjernfelt, Frederik. 2015. Dicisigns: Peirce's semiotic doctrine of propositions. *Synthese* 192: 1019–1054.
Thibaud, Pierre. 1997. Between saying and doing: Peirce's propositional space. *Transactions of the Charles S. Peirce Society* 33: 271–327.
Thompson, Manley. 1953. *The pragmatic philosophy of C.S. Peirce*. Chicago: The University of Chicago Press.
Tiercelin, Claudine. 1991. "Peirce's semiotic version of the semantic tradition in formal logic", *New Inquiries into Meaning and Truth*, N. Cooper & P. Engel (eds.), Harvester Press,187–213.
Trendelenburg, Friedrich A. 1840. *Logische Untersuchungen*. Berlin: Bethge Verlag.
Ueberweg, Friedrich. 1871. *System of logic, and history of logical doctrine*. Translated from the German by T.M. Lindsay. London: Longmans, Green, & Co.
Wundt, Wilhelm M. 1893–1895. *Logik*. Vol. 2, 2nd edition. Stuttgart: Ferdinand Enke.

Chapter 16
Thirdness as the Observer Observed: From Habit to Law by Way of *Habitus*

Göran Sonesson

Abstract In order to study the notion of habit as an instance of Thirdness in Peirce's work, it is necessary to go back to the intuitions at the basis of Peirce's categories, trying to spell out concretely, as I think this has not been done before, the meaning of the three categories. This involves entangling the notions of fallibilism and of the collaborative work of the community of scholars, which may not have been taken seriously by most scholars pursuing the Peircean tradition. It is suggested that Peirce's phenomenology is a version of Husserl's phenomenology imposing a lot of constraints on the variation in imagination. In order to make sense of habit as Thirdness, we have to extend Peircean phenomenology into Husserlean phenomenology, abandoning the language of degeneracy, which is not very enlightening. Important contributions to the study of habit has also been made by several sociologists and psychologists.

Keywords Phenomenology · Phaneroscopy · Thirdness · Mediation · Husserl

Preamble: Scholars, Make a Little Effort to Become Peirceans

As compared to Husserlean phenomenology, the phenomenology (or, as he later called it, the Phaneroscopy) of Peirce seems to be frozen in time. Husserl's phenomenological discoveries have been reviewed, revised, and elaborated many times over, first by himself, and then by his most devout followers, such as Gurwitsch, Schütz, Merleau-Ponty, Sokolowski, Marbach, Drummond, and Zahavi, let alone by such revisionists as Heidegger and Derrida. Meanwhile, most students of Peirce seem to be satisfied with only trying to find out what Peirce really wanted to say. Yet, Husserl and Peirce similarly staked their hope on the community of scholars which were going to pursue the task that they had initiated.

G. Sonesson (✉)
Lund University, Lund, Sweden
e-mail: goran.sonesson@semiotik.lu.se

D.E. West and M. Anderson (eds.), *Consensus on Peirce's Concept of Habit*,
Studies in Applied Philosophy, Epistemology and Rational Ethics 31,
DOI 10.1007/978-3-319-45920-2_16

Late in life, Peirce penned some observations on his on work that are seldom quoted by scholars styling themselves as Peirceans. At one point, looking back on his own earlier work, he noted: "All my notions are too narrow. Instead of 'sign', ought I not to say *Medium*" (1906: MS 339: 526). Even in the admittedly chaotic first collection of Peirce's work, the *Collected Papers*, there is this observation from 1898: "I did not then [in 1867] know enough about language to see that to attempt to make the word *representation* serve for an idea so much more general than any it habitually carried, was injurious. The word *mediation* would be better. Quality, reaction, mediation will do." (1898: CP 4.3) This has not impeded latter-day followers of Peirce from continuing to use terms such as "sign", "representation" and other similarly "injurious" terms for Peirce's much broader ideas.

These remarks should be viewed on the background of Peirce's idea of an "ethics of terminology", an idea that I think he has pioneered and which has not found any appreciable following so far. Basically, the ethics of terminology states that not just any term may be used for any idea, because it is impossible to ignore the weight of history, of culture, of language in general and of scholarly language in particular. Space won't allow me to quote all of the precepts forming part of the ethics of terminology, so I will just give an extract corresponding to point 6 and 7:

> For philosophical conceptions which vary by a hair's breadth from those for which suitable terms exist, to invent terms with a due regard for the usages of philosophical terminology and those of the English language but yet with a distinctly technical appearance. Before proposing a term, notation, or other symbol, to consider maturely whether it perfectly suits the conception and will lend itself to every occasion, whether it interferes with any existing term, and whether it may not create an inconvenience by interfering with the expression of some conception that may hereafter be introduced into philosophy. To regard it as needful to introduce new systems of expression when new connections of importance between conceptions come to be made out, or when such systems can, in any way, positively subserve the purposes of philosophical study. (1903: CP 2.226)

It is all too easy to point out that Peirce himself hardly followed these wise precepts, since he tended to change his basic terminology all the time, but then it must be taken into account that most of Peirce's posthumous papers merely were scribbles never prepared for publication.

There is, however, even a more general point to be made, anticipated above: Peirce's notion of fallibilism is not obviously different from Husserl's notion of *Evidenz*, although the terminology may suggest otherwise, for, in both cases, the idea is that any scientific results (also in philosophy/phenomenology) are subject to revision. No conclusion is final, and all findings have to be reviewed by a community of scholars. No doubt, Husserl would have needed to heed Peirce's call for an ethics of terminology. Also, Husserl and Peirce both felt that the task they had taken on themselves was too big for them to accomplish alone, so they put their trust in a "community of scholars", to use Peirce's term, to pursue their work, which means that they both anticipated their own conclusions being called into doubt.

Why, then, was the posterity of Husserl and Peirce so different? Husserl certainly held a stable position at the university most of his life, while Peirce only had passing university employment. Husserl also published several books during his lifetime (although these are dwarfed by all the posthumous manuscripts), whereas Peirce only published some articles, most of his work being published posthumously. Husserl had some followers already during his lifetime, though those he trusted most, such as Heidegger and Fink, fundamentally disowned him, as did, less radically, the Göttingen school, while those who now seem to be his truest disciples at the time, Gurwitsch and Schütz, do not seem to have had much direct contact with him. Since he published very little, Peirce, on the other hand, could hardly have much contemporary influence, and yet his ideas made a tangible impact on William James (who also basically kept him alive by his financial assistance), who was (and is) an important figure in the history of psychology, but as soon as James took up some of his ideas, Peirce felt betrayed and marked his distances. This is, as the French say, "la petite histoire". I don't know to what extent this explains that the posterity of Husserl has been very fecund in new discoveries, whereas the followers of Peirce have been fundamentally limited to doing the hermeneutics of the Peircean texts. I do not propose that we should abandon Peircean hermeneutics. I just suggest that we should give priority to the task Peirce left us in inheritance: to pursue the process of inquiry, always fallible, within the community of scholars.

A book that undoubtedly must be taken as following such a tendency is that by Donna West (2013), who tries to incorporate the findings of, among others, Bühler and Piaget, into a Peircean framework. Much of what I have written could also be taken in this sense. An earlier work that might be understood to take Peircean ideas further is that of Dines Johansen (1993) and of course the recent contributions by Stjernfelt (2007, 2014). There is, however, as we shall see, a problem of finding a transitional language, which stays within the Peircean framework, without sacrificing the subtleties of thinkers such as Piaget and Bühler.

The Trichotomy and Its Degeneracy

Elsewhere, I have tried to show that the Peircean and Husserlean phenomenologies are largely identical when it comes to their operations, though Peirce imposes more constraints on these operations. Briefly put, Peirce posits an a priori triadic framework instead of "going to the things themselves", and his phenomenology is thus not independent of "presuppositions" in Husserl's sense (cf. Sonesson 2013). Together with the (originally mathematical) notion of degeneracy, the three categories may seem to offer a way out of the Peirce's own self-critical remarks quoted above. Before we can get at the bottom of this issue, however, it is necessary to get a little deeper into the meaning of the three categories.

The Peircean trichotomies involve two kinds of presuppositions. The most obvious one is numerical, in at least a triple sense: there are three overall categories; all categories of Thirdness subdivide into three categories, while the category of Secondness only has two subdivisions, and that of Firstness none; and, while Firstness is made up of a single term, Secondness involves two, and Thirdness three (which is without counting the relations or predicates, as we shall see). Thus we get the following characterization:

> Firstness is the mode of being of that which is such as it is, positively and without any reference to anything else. Secondness is the mode of being of that which is such as it is, with respect to a second but regardless of any third. Thirdness is the mode of being of that which is such as it is, in bringing a second and third into relation to each other" (1904: CP 8.328).

Firstness and Secondness could here almost be understood as approximations to Husserl's (1913, 2:1, 225ff) distinctions between independent and dependent parts, with the exception that there is no proviso for the difference between mutual and one-sided dependence. This then raises the question of what the business of Thirdness is. If it involves a relation between two terms, instead of only one term and a relation, as Secondness could perhaps be understood to be, or a relation between relations, why then should we not go on defining Fourthness, and so on? Of course, Peirce claimed that all relations beyond Thirdness could be dissolved into several relations, but Thirdness itself could not be so resolved. It is not clear whether this is indeed a phenomenological fact. For the present, however, we will accept is as such.

The definitions of the categories are not exhausted by numbers. Peirce supplies a lot of content-full definitions of the three categories. The trouble is that, in different passages, these characterizations are rather different. Elsewhere, I have tried to extract the common denominator of (most of) these characterizations (Sonesson 2009a, 2013). Summarizing all of Peirce' s different attempts at pinning down the nature of *Firstness*, we could probably say that it is something that appears without connection to anything else. It is thus prior to all relationship. *Secondness* is not only the second term that comes into play, but it is also made up of two parts, one of which is a property and the other a relation. It is something the function of which is to hook up with something already given. In this sense, it is a reaction, in the most general sense, to Firstness, where the first part is the connection to the property independently appearing and the second part describes the nature of this relationship. *Thirdness* is not only the third term which is ushered in, but it consists of three parts, two of which are relational; one which is hooked up to the term of Firstness and another which is connected to the relation of Secondness, together with which we find a third term describing the relationship between these two terms. It is thus an observation of the reaction.

If we consider the numerous and varied descriptions that Peirce gave to his categories, it might be suggested that everything said about Firstness boils down to a meaning roughly paraphrased as "something there", that those phrases describing Secondness are equivalent to "reaction to the appearance of something", and that Thirdness can be reduced to "observing the appearance as well as the reaction to the appearance" (see Fig. 16.1 and Sonesson 2009a, 2013). A few glosses on these conclusions remain to be spelled out. First of all, nothing can appear without appearing to somebody (even if that somebody is only a "quasi-mind", whatever that is), so even if Firstness, exemplified by iconicity, only appears for "a fleeting

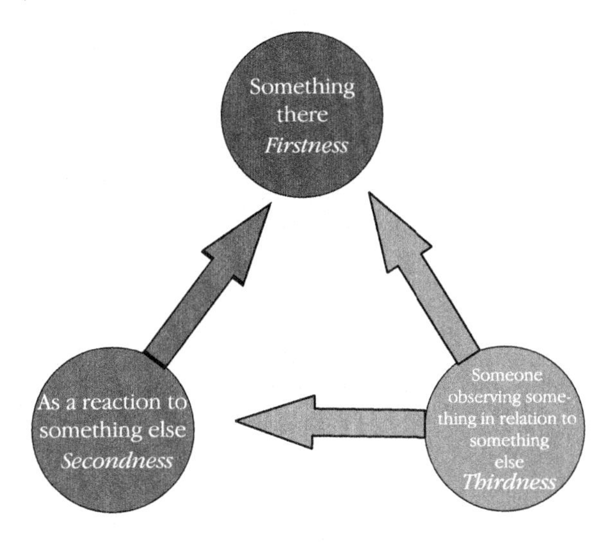

Fig. 16.1 The interpretation of the three Peircean categories, according to Sonesson (2013)

moment", as Peirce observes, it is still a relation, in spite of Peirce's insistence that it is not, or else it cannot even be an appearance. Still, we can recognise in Secondness a reaction in a fuller sense, something that may be an action, or also an awareness of the phenomenon. Thirdness may then either be the acknowledgment of the action or of the percept ascribed to Secondness. From this point of view, it is easier to understand why Peirce argues that Thirdness is different from Secondness, but that any higher relation is reducible to Thirdness: the observation of a reaction is different from a reaction, but the observation of an observation is just another observation.

Appearance is monadic, reaction is dyadic, and observation is triadic. I do not think it is sufficient to say that Firstness, Secondness, and Thirdness correspond to a one-place predicate, a two-place predicate, and a three-place predicate, respectively, as Ransdell (1989) maintains. This cannot explain the workings of the categories. Rather, Firstness must be a one-place predicate with one term in the slot, Secondness a second-place predicate having two terms, and Thirdness a three-place predicate including three terms. According to Peirce, "a fork in the road is a third, it supposes three ways: a straight road, considered merely as a connection between two places is second, but so far as it implies passing through intermediate places it is third" (c. 1875: CP 1.337).

If a sign or representation is made up of a Firstness, combined with a Secondness and a Thirdness in this very general sense, then, of course, it is quite understandable that biosemioticians can find signs everywhere, even within the cell (cf. Sonesson 2006), and others, such as Deely (2010) and Caivano (2015), claim to discover signs even in inanimate nature. On the other hand, there are many more concrete descriptions of the sign, with or without "a sop to Cerberus", in Peirce's writings.

One may wonder, at this point, whether the notion of degeneracy can be used to bridge these two extremes. In mathematics, to which Peirce refers, a degenerate case is a limiting case in which an element of a class of objects is qualitatively different from the rest of the class and hence belongs to another, usually simpler, class. Thus, a degenerate case has special features, which depart from the properties that are generic in the wider class, and which would be lost under an appropriate small perturbation. A point, for example, is a degenerate circle, namely one with radius 0. Or, to quote Peirce himself: "Consider, first, the thirds degenerate in the first degree. A pin fastens two things together by sticking through one and also through the other: either might be annihilated, and the pin may continue to stick through the one which remained" (c. 1890, CP 1.366). If we start from the more concrete definition of the sign, the combination of Firstness, Secondness, and Thirdness may appear to be a degenerative case. In quite another vocabulary, we may say that the concrete definition of the sign is the prototype of the category (cf. Rosch 1975), and that all other cases of combining the three categories are situated at more or less great distances from this prototype. If we start from the abstract definition, on the other hand, we get the opposite of a prototype. However useful the notion of degeneracy may be in mathematics, and in spite of prototype concepts being abundant in ordinary language, it may seem that, from the point of view of phenomenology, this way of defining concepts is not very enlightening. Perhaps this is the insight which dawned on Peirce in his later years.

The Grounding of It All

An icon is a sign, in which the "thing" serving as expression is similar in one respect or another to (or has properties in common with) the "thing" that serves as its content; but the sign is an icon only if the similarity between these relata (the elements which are related) obtains independently of the sign relation and independently of possible relations between the relata as such. In other words, similarity is not a result of the sign relation; and the iconicity of the two things related taken separately cannot explain the similarity before the sign. In the same way, an index is a sign in which the "thing" serving as expression is connected in one respect or another to the "thing" that serves as its content; but the sign is an index only if the connection between these relata (the elements which are related) obtains independently of the sign relation. In other words: it is not the sign relation which makes these two items into an index. In a symbol (a conventional sign), in contrast, there is nothing, apart from the sign relation, which relates expression to content.[1]

To go from the concept of iconicity to the iconic sign, as well as from indexicality to the indexical sign, we have to ponder the meaning of a notion,

[1]As I have argued elsewhere (Sonesson 1996, 2009b, 2010), iconicity and indexicality are not always independent of the sign character, but we will accept Peirce's position here, for the sake of the argument. In another context, Peirce (c. 1897: CP 4.157) does indeed argue that "it is not the resemblance that causes the association, but the association that constitutes the resemblance", anticipating the similarity critique of Nelson Goodman (see Sonesson 1989: 226ff).

sporadically, but often significantly, used by Peirce, i.e., the notion of *ground.* As applied to signs, I here suppose, iconicity is one of the three relationships in which a representamen (expression) may stand to its object (content or referent) and which can be taken as the "ground" for their forming a sign: more precisely, it is the first kind of these relationships, termed Firstness, "the idea of that which is such as it is regardless of anything else" (1903: CP 5.66), as it applies to the relation in question. In one of his well-known definitions of the sign, a term which he here, as so often, appears to use to mean the sign-vehicle, Peirce (1897: CP 2:228) describes it as something which "stands for that object not in all respects, but in reference to a sort of idea, which I have sometimes called the *ground* of the representamen".

The operation in question, I submit, must be *abstraction* or, as I would prefer to say, *typification.* In one passage, Peirce himself identifies "ground" with "abstraction" exemplifying it with the blackness of two black things (c. 1894: CP 1.293).[2] It therefore seems that the term *ground* could stand for those properties of the two things entering into the sign function by means of which they get connected, i.e., both some properties of the thing serving as expression and some properties of the thing serving as content. In case of the weathercock, for instance, which serves to indicate the direction of the wind, the content ground merely consists in this direction, to the exclusion of all other properties of the wind, and its expression ground is only those properties that makes it turn in the direction of the wind, not, for instance, the fact of its being made of iron and resembling a cock (the latter is a property by means of which it enters an iconic ground, different from the indexical ground making it signify the wind). If so, the ground is really a *principle of relevance*, or, as a Saussurean would say, the "form" connecting expression and content: that which must necessarily be present in the expression for it to be related to a particular content rather than another, and vice versa. This phenomenon is well-known from linguistics, where often conventional rules serve to pick out some properties of the physical continuum, differently in different languages, which have the property of separating meanings, i.e., of isolating features of the expression on the basis of the content, and vice versa. The difference is, of course, that, in the iconic ground, the relation that determines one object from the point of view of the other is basically non-conventional (cf. Sonesson 1989: 202ff).

The ground explains the relation that exists between the two items forming the sign, but it does not of itself guarantee that a sign is present. For this to happen, it is necessary to add Thirdness, and, in fact, a very peculiar version of Thirdness, corresponding to what I have elsewhere defined as the sign, relying on criteria suggested by Edmund Husserl and Jean Piaget (see Sonesson 1989, 2010). At the level of grounds, it will be necessary to further generalize the description of the categories: if Firstness may still be seen as the phenomenon appearing (to someone), Secondness has to be viewed simply as one phenomenon appearing together

[2]For the argument that "ground" should be used for both expression and content, or rather for the relation between them, cf. Sonesson (2009b, 2010).

Table 16.1 The different kinds of Peircean grounds as related to the three sign types

	Firstness	Secondness	Thirdness
Firstness	Iconicity	–	–
Secondness	Iconic ground	Indexicality = indexical ground	–
Thirdness	Iconic sign (icon)	Indexical sign (index)	Symbolicity = symbolic ground = symbolic sign (symbol)

with another, that is, as contiguity,[3] but Thirdness, since it is supposed to shift the level, must be seen as the coming together of several phenomena into a new whole. As suggested above, it entails some kind of meta-cognition (See Table 16.1).

Habit as Thirdness

If Thirdness, in the Peircean sense, can be interpreted, as I have suggested above, as "the observer being observed", or, more generally, as "the reaction being observed", then it is located at the level of meta-cognition, or at least it involves the coming into awareness of an action or event, and, if so, it seems rather natural to treat rules and laws as being instances of Thirdness. However, it is more difficult to see how habits, which are also Thirdness according to Peirce, let alone instincts, innate releasing mechanism, or the like, which Biosemioticans assimilate to this Peircean notion, can be considered as meta-cognition. Peirce clearly thinks of habits as a kind of meta-cognition, because he says they are "either habits about ideas of feelings or habits about acts of reaction" (1892: CP 6.145), that is, they refer either to Firstness or Secondness: according to my interpretation, illustrated in Fig. 16.1, we should expect them to be ideas of both Firstness and Secondness at the same time. On the other hand, not unlike contemporary biosemioticians, Peirce (cf. c. 1890: CP 1.390; c. 1897: 4.157; c. 1893: 6.277) not only treats the cell as something able to acquire habits, but, not unlike Deely (2010), he even thinks that natural laws are a kind of habits. Then again, he claims that for Thirdness to exist, it is essential "that there should be an element of chance" and that "this element of chance or uncertainly should not be entirely obliterated by the principle of habit, but only somewhat affected" (CP 1.390). Indeed,

> The most plastic of all things is the human mind, and next after that comes the organic world, the world of protoplasms. Now the generalizing tendency is the great law of mind, the law of association, the law of habit taking. We also find in all active protoplasm a

[3]Cf. Peirce (c. 1897: CP 4.157): "For to be contiguous means to be near in space at a time; and nothing can crowd a space for itself but an act of reaction". Still, if all reactions are indeed contiguities, the opposite does not seem to be necessarily the case.

tendency to take habits. Hence I was led to the hypothesis that the laws of the universe have been formed under a universal tendency of all things toward generalization and habit-taking. (c. 1898: CP 7.515)

In this plasticity, it is easy to recognize, first, consciousness, and second, life, the two emergent properties in nature, as described by Thompson (2007). Although Thompson claims these properties are emergent, he goes on to say that how life is produced from non-animate nature is easily explained by chemistry, and the same he expects to hold for consciousness. But even if life is easily explained chemically, it does not follow that it has been explained phenomenologically. Even if consciousness could also be chemically explained, this does not take away the phenomenological conundrum. Being by profession a chemist, Peirce may have thought the same, but it is not clear in what way chemistry can account for the kind of plasticity that Peirce talks about here. These considerations obviously presuppose, contrary to what Peirce is wont to say, that the mind plays a more important part in phenomenology than simply as a "sop to Cerberus". We simply need more distinctions than what are offered by the Peircean framework.

If we admit that the sign, in the narrow sense implicitly recognized in Peirce's late remarks, is a kind of Thirdness, it is certainly a Thirdness of another kind than habits and laws. There clearly are habits and conventions that are not signs, or which pre-exist to signs. For instance, if the iconic sign for woman in the manual sign language of the North American Indians is a gesture showing the extension of the braids on both sides of the head that was the typical hairdo of the Indian woman, the sign merely depicts a convention that must have existed before the sign. It is easy to think of rules that are not signs, such as traffic regulations, the rules for playing chess, social rules (such as the *habitus* of Elias and Bourdieu), and aesthetic norms (such as the canon according to Mukařovský 1978). I conclude that Table 16.1 must be amended in the way suggested by Table 16.2.

In other traditions of semiotics, such as the Prague school, it has been argued that there is a continuous scale going from habit to laws by way of norms (see Mukařovský 1978), and, within the Greimas School, at least Hammad (2002) has suggested that regularities can be transformed into rules, such as, for instance, the places where the leaders of the conference are used to sit turning from habitual into

Table 16.2 An amended version of the Peircean grounds, including a kind of symbolic ground which is not a sign

	Firstness Impression	*Secondness* Relation	*Thirdness* Habituation/Rule
Firstness Principle	Iconicity	–	–
Secondness Ground	Iconic ground	Indexicality = indexical ground	Symbolicity = symbolic ground
Thirdness Sign	Iconic sign (icon)	Indexical sign (index)	Symbolic sign (symbol)

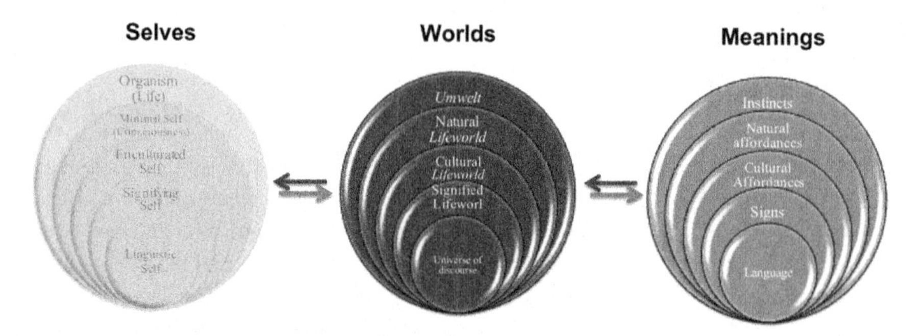

Fig. 16.2 Five kinds of selves, worlds, and meaning (from Zlatev 2009), with the addition of a distinction between the cultural Lifeworld and the signified Lifeworld (Sonesson 2015a) and the level of meanings (Sonesson, in press)

implicitly mandated. This seems to correspond to Peirce's notion, which however adds more continuity on the lower, less "plastic", end of the scale. Still, between normalcy and normativity, there clearly is a qualitative difference. To Mukařovský, habits, norms, and laws are not identical, even if they are akin to each other, and to Hammad a regularity only becomes a rule once it has been repeated over and over again, and has been observed to be so repeated. There would therefore seems to be an increasing degree of "plasticity", which here must mean the involvement of mind (or the self), in Thirdness going from dead matter to language and other signs via affordances as given to perception. On a very general level, there are a number of qualitative differences within a semiotic hierarchy, as first suggested by Zlatev (2009) and extended by Sonesson (2015a; in press), with reference to selves, worlds, and meanings (see Fig. 16.2).

Instincts, reflexes and innate releasing mechanisms would be on the level of the *Umwelt*. Another circle would be required for the "habits" of dead matter, and between them perhaps another circle for processes taking place in the cells. The latter, according to Peirce (c. 1890: CP 1.390) only retain "the virtue of Thirdness" because there is no "requirement that the cell should discharge itself always in the same way". At the end of the *Umwelt*, we get to the natural and cultural lifeworlds, basically inhabited by natural and cultural affordances, respectively. I here refer to James Gibson's notion of affordances that I have adopted in earlier works (e.g. Sonesson 2010), arguing for the distinction between affordances that are naturally given (such as graspability) and those which only become possible in a cultural Lifeworld (such as seeing the difference between a post-box and a litter-bin and acting on it), without becoming signs, in the sense of double-layered structures. Interestingly, West (2013, 2014: 118f) has also pointed to a parallel between Gibson's affordances and Peirce's habits.

The really interesting distinctions may actually be localized in the thin layer (in Fig. 16.2) of the cultural Lifeworld. Here we find Mukarovský's (1978) distinction between habit, norm, and law. On a more general level, here is the distinction between normalcy (that which we expect to happen) and normativity (that which we think should happen; see Sonesson 2001). In the first case, it is sufficient to observe the observer (or the reaction), but in the second case, when it is a question of emitting mandatory judgments, it clearly is necessary to occupy a meta-meta-level. Somewhere on the border between normalcy and normativity is found the notion of *habitus*, an originally scholastic term (translating an Aristotelian notion) made famous by Pierre Bourdieu, but used before him by, among others, Norbert Elias, Marcel Mauss, and, not least, Edmund Husserl (Mennell and Goudsblom 1998; Moran 2011). Roughly the term seems to have the same meaning for all these thinkers: a structure of the mind, acquired through the activities and experiences of everyday life, and characterized by a set of acquired schemata, sensibilities, dispositions, and tastes.

From a theoretical point of view, however, no other thinker has taken this idea further than Husserl, as is apparent mostly from his posthumous publications (cf. Welton 2000; Steinbock 1995; Sonesson 2015b). To Husserl, every object in our experience has a *genetic* dimension: it results from the layering, or *sedimentation*, of the different acts that connects it with its origin, which gives its validity, in the way in which geometry, as Husserl's (1954:378ff) observes, derives from the praxis of land-surveying. There is also the further dimension of *generativity*, pertaining to all objects, and which results from the layering, or *sedimentation*, of the different acts in which they have become known—whether acts of perception, memory, anticipation, imagination, or otherwise. The term generativity is meant to evoke the idea of generations following upon each other, as well as the trajectory accomplished by each individual from the cradle to the grave. In this formulation, the properties in question are attributed to the object of attention, but they could just as well be ascribed to the subject attending, in which case they are different varieties of habits, sometimes not so far from being matter, in the Peircean sense of "indurated habits /…/ with a peculiarly high degree of mechanical regularity, or routine." (c. 1893: CP 6.277), but still with a certain amount of "chance or uncertainly" (c. 1890: CP. 1.390) being retained.

Conclusion

These remarks only serve to introduce a productive domain of study. In order to go further, the best choice is, suggested elsewhere, to treat Peircean phenomenology as a special case of Husserlean phenomenology, making Thirdness at least a case of double intentionality, in the Husserlean sense of directedness. Whatever the value of the Peircean meta-language, including Thirdness of Thirdness, and so on, it seems to be insufficiently precise for accomplishing the task of phenomenology. While Peirce's observations of the life of the mind (or quasi-mind) are precious, we need to extend his framework in the Husserlean sense in order to make sense of them. Peirce's claim for the continuity of being is fundamental, in particular from the point of view of evolution and child development, but, as Peirce's own observations show, continuity cannot go without a certain amount of discontinuity.

There clearly are qualitative differences, as for instance, those between dead matter and life, and between different kinds of consciousness, as Peirce recognizes when talking about "plasticity". The meta-language of degeneracy is not comprehensive enough to account for these differences.

References

Caivano, J. L. 2015. Cognición y procesamiento semiósico de estimulos luminoses en varios órdenes del mundo cultural y natural. Lecture at the XIth conference of the International Association for Visual Semiotics, Liège, Belgium, 8–11 September 2015.

Deely, John N. 2010. *Semiotic animal: A postmodern definition of "human being" transcending patriarchy and feminism*. South Bend, Indiana: St. Augustine's Press.

Dines Johansen, Jørgen. 1993. *Dialogical semiosis*. Blomington: Indiana University Press.

Hammad, M. 2002. *The privatisation of space*. Lund: Department of Theoretical and Applied Aesthetics, School of Architecture, Lund University.

Husserl, E. 1913. *Logische Untersuchungen*. Tübingen: Max Niemeyer.

Husserl, E. 1954. *Husserliana: gesammelte Werke. Bd 6, Die Krisis der europäischen Wissenschaften und die transzendentale Phänomenologie : eine Einleitung in die phänomenologische Philosophie*. Haag: Nijhoff.

Mennell, S., and J. Goudsblom. 1998. Introduction. In *On civilization, power, and knowledge: Selected writings, by Norbert Elias*, ed. S. Mennell, and J. Goudsblom, 1–48. Chicago: University of Chicago Press.

Moran, D. 2011. Edmund Husserl's phenomenology of habituality and habitus. *Journal of the British Society for Phenomenology* 42(1): 53–77.

Mukařovský, J. 1978. *Structure, sign, and function: Selected essays*. New Haven: Yale University Press.

Parmentier, Richard J. 1985. Signs' place *in medias res*: Peirce's concept of semiotic mediation. In *Semiotic mediation: Sociocultural and psychological perspectives*, ed. E. Mertz, and R. J. Parmentier, 23–48. Orlando, Florida: Academic Press.

Peirce, Charles Sanders. i. 1867–1913. *Collected papers of Charles Sanders Peirce*. Vols. 1–6, eds. Charles Hartshorne and Paul Weiss. Cambridge: Harvard University Press, 1931–1935. Vols. 7–8, ed. Arthur W. Burks. Cambridge: Harvard University Press, 1958. [References to Peirce's papers will be designated by CP, followed by volume, period, paragraph number.].

Peirce, Charles Sanders. i. 1867–1893. *The Essential Peirce: Selected Philosophical Writing*. Volume 1 (1867–1893), eds. Nathan Houser and Christian Kloesel. Bloomington: Indiana University Press, 1992. [References to this volume will be designated by EP 1, followed by colon, page number.].

Peirce, Charles Sanders. i. 1893–1913. *The Essential Peirce: Selected Philosophical Writing*. Volume 2 (1893–1913), ed. the Peirce Edition Project. Bloomington: Indiana University Press, 1998. [References to this volume will be designated by EP 2, followed by colon, page number.].

Peirce manuscripts in Texas Tech University Library at Texas Tech University, Institute of Studies of Pragmaticism, beginning with MS—or L for letter—and followed by a number, refer to the system of identification established by Richard R. Robin in Annotated Catalogue of the Papers of Charles S. Peirce (Amherst: University of Massachusetts Press, 1967), or in Richard R. Robin, "The Peirce Papers: A Supplementary Catalogue," Transactions of the Charles S. Peirce Society.

Ransdell, Joseph. 1989. Peirce est-il un phénoménologue? *Ètudes Phénoménologiques* 9–10, 51 75. http://www.cspeirce.com/menu/library/aboutcsp/ransdell/phenom.htm.

Rosch, E. 1975. Cognitive reference points. *Cognitive Psychology* 7(4): 532–547.

Sonesson, G. 1989. *Pictorial concepts. Inquiries into the semiotic heritage and its relevance for the analysis of the visual world.* Lund, Sweden: Aris/Lund University Press.
Sonesson, G. 1996. An essay concerning images. From rhetoric to semiotics by way of ecological physics. *Semiotica* 109.1/2 (March 1996): 41–140.
Sonesson, G. 2001. The pencils of nature and culture. *Semiotica* 136.1/4: 27–53. —. 2006. The meaning of meaning in biology and cognitive science. A semiotic reconstruction. *Sign Systems Studies* 34.1: 135-214.
Sonesson, G. 2009a. The view from Husserl's lectern. Considerations on the role of phenomenology in Cognitive Semiotics. *Cybernetics and Human Knowing* 16.3–4: 107-148.
Sonesson, G. 2009b. Prologomena to a general theory of iconicity. Considerations on language, gesture, and pictures. In *Naturalness and iconicity in language,* eds. K. Willems and L. De Cuypere, 47 72. Amsterdam: Benjamins.
Sonesson, G. 2010. Semiosis and the elusive final interpretant of understanding. *Semiotica* 179.1/4: 145 258.
Sonesson, G. 2013. The natural history of branching: Approaches to the phenomenology of firstness, secondness, and thirdness. *Signs and Society* 1.2: 297–326.
Sonesson, G. 2015a. Le jeu de la valeur et du sens. In Dans *Valeur. Aux fondements de la sémiotique*, ed. Amir Biglari, 85–100. Paris: L'Harmattan.
Sonesson, G. 2015b. Phenomenology meets semiotics: Two not so very strange bedfellows at the end of their cinderella sleep. *Metodo* 3.1: 41–62.
Sonesson, G. (in press) *Kognitiv Semiotik. Ett Program.* Stockholm: Riksbankens Jubileumsfond.
Steinbock, A. 1995. *Home and Beyond: Generative Phenomenology after Husserl.* Evanston, Illinois: Northwestern University Press.
Stjernfelt, F. 2007. *Diagrammatology: An investigation on the borderlines of phenomenology, ontology, and semiotics.* Dordrecht: Springer.
Stjernfelt, F. 2014. *Natural propositions.* The Actuality of Peirce' Doctrine of Dicisign. Boston: Docent Press.
Thompson, E. 2007. *Mind in life: Biology, phenomenology, and the sciences of mind.* Cambridge: Belknap Press of Harvard University Press.
Welton, D. 2000. *The other husserl: The horizons of transcendental phenomenology.* Bloomington: Indiana University Press.
West, Donna E. 2013. *Deictic imaginings: Semiosis at work and at play.* Heidelberg: Springer.
West, Donna E. 2014. From habit to habituescence: Peirce's continuum of ideas. In *Semiotics 2013*, ed. Jamin Pelkey and Leonard G. Sbrocchi, 117–126. Ottawa: Legas Press.
Zlatev, J. 2009. The semiotic hierarchy: Life, consciousness, signs, and language. *Cognitive Semiotics* 4: 169–200.

Chapter 17
Habits, Awareness, and Autonomy

Vincent Colapietro

Abstract The word *consciousness* is, if anything, even more ambiguous than habit and more or less closely allied terms or expressions (e.g., disposition, practice, routine, ritual, convention, and pattern of action). The pragmatist consensus regarding habit *change* (and it is the change of habits, not simply habits, that is at the center of this consensus) encompasses an account of consciousness or awareness in one or more of its most central senses. According to the pragmatists, the arrest of habits intensifies or heightens awareness; and such an alteration of consciousness aids agents in exercising control over both environing circumstances and their somatically rooted habits. That is, consciousness is not a mere epiphenomenon: "it seems to me," Peirce claims, "that it exercises a real function in self-control" (c. 1906, CP 5.493). In addition to the intimate connection between arrested action and heightened awareness, however, we must also consider the integration of habits and the emergent, precarious, and yet quite effective form of autonomy characteristic of human agents. While the consensus in question involves a shift in perspective, one from consciousness to habituation and hence the alteration of habits, it does not simply jettison the concept of consciousness. It rather tries to explain consciousness in reference to the operation, dissolution, and modification of habits.

Keywords Consciousness · Habit change · Self-control · Doubt · Belief · Virtues

---

V. Colapietro (✉)
Pennsylvania State University, State College, USA
e-mail: vxc5@psu.edu

D.E. West and M. Anderson (eds.), *Consensus on Peirce's Concept of Habit*,
Studies in Applied Philosophy, Epistemology and Rational Ethics 31,
DOI 10.1007/978-3-319-45920-2_17

Introduction

Life is, John Dewey suggests, a series of interruptions and recoveries (*MW* 14 [1922]: 125).[1] In *Art as Experience*, he claims, the "live being [hence, the active and engaged one] recurrently loses and reestablishes equilibrium with his surroundings. The moment of passage from disturbance into harmony is that of intensest life" (*LW* 10 [1932]: 22). Insofar as this is a cognitive or epistemic drama, the disturbance or disruption is identifiable with doubt, the recovery of harmony with the reestablishment of belief. The passage is when consciousness is most heightened, life most intense.

C.S. Peirce's doubt-belief theory of inquiry is, hence, a detailed account of a specific instance of a pervasive phenomenon. The pragmatist consensus regarding habit *change* is rooted in this theory (see Nöth 2016). On Peirce's account, the eruption of doubt is an occasion of interruption, while the establishment of belief in the teeth of doubt is an occurrence of recovery. But for Peirce no less than Wittgenstein (see, e.g., 1972[1969], #115, #141, #150, and especially #160), doubt presupposes belief. Since beliefs are at bottom forms of habit, that doubt is a disruption of, or impediment to, the operation of a habit or cluster of habits. Peirce's doubt-belief theory of inquiry, then, is one in which inquiry is seen, first and foremost, as a process inaugurated by the interruption of doubt and sustained as a struggle to recovery the agential fluidity of our habitual comportment. In this chapter, however, I want to focus not on this specific instance, but more generally on the arrest of habits and, intimately connected to such arrests, the intensification of consciousness (Colapietro 1989). More precisely, I want to look at this in relationship to the heightening and operation of awareness or consciousness. While the operation of habits is, in the majority of cases, below the threshold of awareness, the interruption of them is intimately tied to states and processes crossing that threshold. That is, habits characteristically operate behind and beyond consciousness, but consciousness itself can only be adequately understood when it is conceived in reference to the breakdown and recovery of habit. Much goes on, as Hegel puts it, behind the back of consciousness (Hegel 1807[1977]: 56).[2] But consciousness provides invaluable clues for what is taking place behind and indeed

[1]The critical edition of John Dewey's writings are divided into The Early Works of John Dewey, The Middle Works, and The Later Works. The standard way in which these works are cited is by *EW* for *Early Works*, *MW* for *Middle Works*, and *LW* for *Later Works*, followed by volume number and then page number. For example, "*MW* 14: 125" refers to page 125 of volume 14 of the *The Middle Works*, as the following citation "*LW* 10: 22" refers to page 22 of volume 10 of *The Later Works*. In this essay, however, I will modify this standard form of citation by inserting the *original* date of publication (not the date of the critical edition but that of the publication of the work when it first appeared).

[2]When I am citing a translation of a work, such as in this case of Hegel's *Phenomenology of Spirit*, I will include outside of brackets the original date of publication of the original work and, then, within brackets the date of the translation being cited.

beyond it, not least of all clues regarding the operation and, thus, the nature of our evolved and evolving habits.

Human consciousness is far from a reliable witness, a trustworthy observer of even its own defining operations. Indeed, Peirce goes so far as to assert our consciousness "may be set down as one of the most mendacious witnesses that ever was questioned" (1902–1903, CP 1.580). He is however quick to point out, it "is the only witness there is." So "all we can do is to put it in the sweat-box and torture the truth out of it, with such judgment [that is, practical wisdom] as we can command" (ibid.). What consciousness takes itself to be, is one thing; what consciousness, for the most part unwittingly, reveals about itself and, moreover, what is going on in the deeper recesses of the human psyche, is quite another. Consciousness encourages the illusion that it is transparent to itself and the world is immediately available to it. What reflection reveals however is that such self-transparency and such cognitive immediacy are illusory. Even so, consciousness is neither a mere epiphenomenon nor a surface show in no way *indicative* of the more hidden workings of the human psyche (see, e.g., 1903, CP 5.493). Consciousness is, among other things, *symptomatic* of turmoil, not least of all the turmoil ensuing upon disruption of the integrated dispositions at the heart of effective agency. This means, as it turns out, that, among its other meanings, the term *consciousness* signifies "a sense of taking a habit, or disposition to respond to a given kind of stimulus in a given kind of way" (c. 1905, CP 5.440). While the word might legitimately be taken to designate "simple feeling" (cf. Dewey 1925, *LW* 1: 226ff.), consciousness "is in its ultimate nature (meaning in that characteristic element of it that is not reducible to anything simpler)" a felt sense of undergoing a specific kind of agential modification (in other words—the ones just quoted—"a sense of taking a habit") (Peirce 1905, CP 5.440).

It is just this that I want to examine here; and I want to do so in light of Peirce's writings on this phenomenon. To repeat, the pragmatist consensus regarding human habituation, with its defining emphasis on *habit change*, consolidates and develops insights offered by Peirce. But it truly is a working consensus at or, at least, near the center of pragmatist tradition.[3] As an aid in carrying out this task, therefore, I will draw heavily on the writings of John Dewey as well (see Colapietro 2004). More than any of the other pragmatists after Peirce, he consolidates and develops his predecessor's insights. In *Human Nature and Conduct* and other texts,[4] indeed,

[3]"The history of a tradition … is," John E. Smith notes, "an indispensable resource for philosophical understanding" (1992: 86). This is nowhere more evident than in reference to the working consensus regarding habitual change under consideration in this essay and indeed all of the essays in this volume. It is also not amiss here to recall Peirce's remark about consensus in science: "Professor Baldwin … distinctly places himself upon the platform that philosophical and scientific questions ought to be settled by majorities. 'We are many: you are one', he says. But in the history of science majorities short of unanimity have more often been wrong than right. Majorities do not form their opinions rationally" (1902, CP 7.367n7).

[4]Dewey's *Unmodern Philosophy and Modern Philosophy* (2012) is a rich but very much unappreciated resource for elaborating a pragmatist account of human habituation.

Dewey is filling in some of the most important details of what is demonstrably a Peircean account of human habituation. While my focus is primarily on Peirce, Dewey will prove at various junctures to be an instructive guide in bringing into clear focus salient features of a complex terrain.

Awareness and the Breakdown of Habits

At this point, it is imperative to appreciate a defining feature of human habits. Such habits are *propulsive*: quite apart from consciousness or intention, they propel us forward (see, e.g., Dewey 1922, *MW* 22: 21–22). They are what we go on. At any moment, we are, whether we realize it or not, actually going on them. It is indeed characteristic of our habits to be going on them *without* realizing this. Moreover, there is no need to wait for them to initiate a course of conduct, for such a course is always already underway (even sleeping is an activity in which the operation of diverse habits is demonstrable!). Stimuli do not prompt an inert organism to act, but inform and likely re-direct the course of an activity always already in process (Dewey's "The Reflex Arc in Psychology" [1896] in *EW* 5, 96–109; see also Backe). Activity is coextensive with life: to be alive is to be active, to be engaged in some process or other. Seen in this lights, we are warranted in seeing habits as providing the background for the interruptions and recoveries constitutive of our lives. Indeed, they more than anything else provide the impetus for the activities susceptible to interruption or arrest, also the drive to recover the fluency of exertion. In large measure, the propulsive force of innumerable habits accounts for the dynamism of our lives.

In any event, habits are in effect "demands for certain kinds of action" (Dewey 1922, *MW* 22: 21).[5] For example, our linguistic habits prompt us to seize or create occasions for utterance, our prehensile ones occasions for grasping and handling objects. Organisms by their very constitution appear to delight in the exercise of their capacities and skills quite apart from anything realized by their activities (e.g., some work of handicraft[6]). As such, they constitute nothing less than the self, but always the self-in-the-world (Dewey 1922, *MW* 14: 20). As far as the actual identity of any agent goes, habits are hardly external, adventitious garments to be put on and taken off at will; they are, for the most part, constitutive, integral features of this identity (see Colapietro 2006). Our inability to shed bad habits at will is, as Dewey notes, telling. We tend to think of habits "as passive tools waiting to be called into

[5]Immediately after asserting that habits "are demands for certain kinds of activity," Dewey claims, "*they constitute the self*" (1922, *MW* 22: 21; emphasis added). This is a central thesis in the pragmatist consensus and, accordingly, it will be considered in this essay.

[6]It is certainly significant that the quality of a piece of work is so closely tied to the delight in which its production or performance is carried out. While this claim unquestionably needs to be qualified, the quality of the work and that of the experience of the agent in the production of the work are, in the majority of cases, intimately tied.

action from without" (1922, *MW* 14: 20). A bad habit however exposes the fatal flaw in this commonplace conception.

> A bad habit suggests an inherent tendency to action and also a hold, a command over us. It makes us do things we are ashamed of, things which we tell ourselves we prefer not to do. It overrides our formal resolutions, our conscious decisions. When we are honest with ourselves we acknowledge that a habit has this power because it is so intimately a part of ourselves. It has this hold upon us because *we are the habit.* (Ibid.; emphasis added)

Conscious intentions are more often than not thwarted by dispositional inertia. But, then, the impeded operation of deeply rooted habits occasions a heightening of awareness, an intensification of consciousness. As already indicated, the topic of this paper is just this phenomenon—the arrest of habitual action and the consequent heightening of awareness—as two pragmatists (Dewey but especially C.S. Peirce) highlight and delineate this phenomenon.

Time and again, our habitual comportment is interrupted. In response to such interruptions, we almost always strive and sometimes strenuously struggle to recover the fluency of action. The self as an integration of habits takes the place of the soul as the principle of life and even consciousness as a dying echo of the traditional idea of the substantial soul (cf. James).[7] The continuity, integration, and efficacy of habits usurp the place of the unity of such a soul or consciousness. No conception of a simple, indissoluble seat or locus of agency, identity, or consciousness is any longer creditable. All that we need is a conception of the integration of habits (however, see Allport 1939). The pragmatists replace antecedently given unity with historically emergent integration. "Concrete habits", Dewey insists, "do all the perceiving, recognizing, imagining, recalling, judging, conceiving, and reasoning that is done" (*MW* 22: 124). It is more accurate to say that the organism, by virtue of its innate and (to a far greater extent) acquired habits, performs these activities. But the living organism, equipped with its concrete habits, displaces the classical concept of the soul and the modern ones of consciousness, mind, and self (or "I"). "'Consciousness' ... expresses functions of habits, phenomena of their formation, operation, their interruption and reorganization" (*MW*

---

[7]In "Does Consciousness Exist?" (1904), James announced, "I believe that 'consciousness,' when once it has evaporated to this estate of pure diaphaneity, is on the point of disappearing altogether. It is the name of a nonentity, and has no right to a place among first principles. Those who still cling to it are clinging to mere echo, the faint rumor left behind by the disappearing 'soul' upon the air of philosophy" (1904[1976]: 3–4). But his conclusion is far less radical than it might appear, since James is denying that *consciousness* names an entity but affirming the word designates a function. Peirce's response in a letter to his friend's thesis is worth recalling, especially since it bears directly on the topic of consciousness: "your paper floors me at the very opening and I wish you would do me the favor (I suppose it to be a simple matter) of explaining what you mean by saying that 'consciousness' is often regarded as an entity. I do not think you capable of setting up a straw man. ... But this word, in modern philosophy, has never conveyed to my mind any idea except that it is a sign the writer is setting up some man of straw whom he imagines to entertain opinions too absurd for definite statement" (28 September 1904; *CP* 8.279). As we have seen, consciousness *is*, for Peirce at least, a function bound up with the operation of habits, including their alteration and dissolution. Functions are as real as anything else.

22: 124). While presenting this position in his own name, Dewey is in effect also presenting the views of C. S. Peirce on the centrality and ubiquity of habit in the constitution and exercise of our agency (see, e.g., Peirce c. 1898, CP 7.512ff.). Concretely, somatic agents such as the human animal *are* agents primarily by virtue of the incredibly complex integration of truly innumerable dispositions. Consciousness is a function of the operation of habits, including especially their breakdown or break up. Peirce could not be more explicit about this: when habits are broken up, the mind is at its liveliest, consciousness is at its highest ("Evolutionary Love" [1893], EP 1: 361). "Everybody knows that the long continuance of a routine of habit makes us lethargic, while a succession of surprises [of unexpected events] wonderfully brightens the ideas" (ibid.).

Where "history is a-making", new forms of subjectivity and identity, linked to new genres of practice and hence modes of agency, are *coming into being*, For example, the experimental scientist resolutely committed to the outcomes of painstaking observation and contrived interventions in natural processes is, despite clear precursors in intellectual history, a distinctively modern figure. So, too, the conscientious worshipper, emboldened by *individual* conscience to stand before institutional authorities and proclaim, "Here I stand; I cannot do otherwise",[8] is to a great extent a modern figure. Finally, the rational self, presuming to possess within its own inner resources the epistemic authority to discover the truth, *on its own*, is a decidedly modern persona (cf. Alexis de Tocqueville[9]). And here modernity refers to the West and, more narrowly, Europe (including significant parts of the Americas).

From the pragmatist perspective on human habituation, however, the only adequate account of human consciousness is one that formally links the role and forms of consciousness to the operation, dissolution, and (in the wake of such dissolution), transformation of habits. In brief, the emergence and operation of consciousness is inseparable from the operation and modification of habits (West 2014, 2015a, b). Nothing is more central to Peirce's accounts of mind, consciousness, subjectivity, and agency than this understanding of consciousness vis-à-vis habits. Indeed, nothing is more central to what might be justly identified as the pragmatist account of these interrelated phenomena than this understanding (even if the thought of William James at times cuts against the grain). That is, these accounts are wider than Peirce; they encompass most of those whom we identify as pragmatists.

[8]Of course, this refers to Martin Luther's assertion in 1521 at Diet of Worms.

[9]In the 1830s, Alexis de Tocqueville judged this to be an especially manifest tendency of citizens of the United States. Of all the countries in the world, "America is," he observed, "the one in which the precepts of Descartes are least studied and best followed" (1969[1840]: 429). The main reason for this is that, "each man is narrow shut up in himself, and from that basis makes the pretension to judge the world" (430). At the heart of the pragmatist consensus there is the rejection of Cartesianism not only in the technical philosophical sense but also a much broader cultural sense. This encompasses a rejection of the authority of the isolated or insular individual. Put positively, it entails the acceptance of a *community* of inquirers as the principal locus of epistemic authority.

The disruption of our habitual modes of comportment is an occasion for the more heightened (or vivid) forms of consciousness. Such disruptions might take place in the most trivial way (e.g., the door handle might unexpectedly resist what on countless occasions beforehand had proven to be a fluid, effective exertion of the hand in grasping and turning the handle, one of the most commonplace of our somatically rooted habits). But such habits might also take place in quite momentous ways (e.g., the emerging sense of no longer being a child might prompt an individual to refuse to grasp the hand of a parent, as both child and adult are about to cross the street, thereby signaling the desire to be independent, especially the desire not to be seen in public *as a child*; or the disruptive sense of no longer being able to walk into one's habitual place of religious worship and participate in good faith with the customary rituals). Whether such disruptions are trivial or significant, they signal the eruption of awareness. Actors tend to become aware of some aspects of their world and indeed of themselves when the unimpeded flow of their habitual exertions encounter unexpected or significant resistance.

A world in which our efforts were never impeded or frustrated would be one in which our awareness of both the determinate features of the world and the defining facets of the self would be, at best, indistinct. The world in which we actually live, one in which our efforts frequently meet resistance, in which our habits not infrequently prove themselves to be inadequate to the demands of the situation into which they themselves have thrust us, is a world in which both the world as an arena of action and the self as an agent are destined to become distinct. By virtue of such resistance to our exertions and challenges to our habits, we cannot help but become *aware* of the world in its radical otherness and of the self in its ineradicable fallibility (i.e., an encompassing sphere in which obstacles and affordances abound, a situated self for whom endeavor always carries the risk of frustration and failure). Thus, the fluency of action, inseparably tied to the efficacy of habits, dispenses with the need for vivid or heightened awareness, while the frustration of our exertions shifts the gears of our conscious agency, making our heightened awareness of our agential entanglements truly critical. For it is, *as agents*, that our awareness ordinarily comes into play and, on such occasions, that human awareness fulfills what appears to be its evolutionary function (the function of enabling agents to do effectively what they are habitually doing or, if this proves impossible, to do something else).

The Integration of Habits

Our use of the noun *consciousness* is misleading insofar as it designates an agent. The adjectival and adverbial forms of the word are more basic than the nominal form (cf. Dewey 1925, *LW* 1: 215–221; *Unmodern Philosophy and Modern Philosophy* [posthumously published in 2012]: 203, 206–209, 219–223). The agency in question is not consciousness, but the organism or a group of organisms acting in concert (for there is such a thing as collective agency). Agents are conscious; consciousness is not agential. Actions may be more or less *consciously* performed; in other words, actions may be more or less conscious. The overwhelming bulk of human exertions are, from the perspective of Peirce and other pragmatists such as Dewey and Mead, unconscious, though not necessarily in the dynamic or psychoanalytic sense of this ambiguous word. In effect (if not in consciousness and indebtedness) following Samuel Butler,[10] these pragmatists hold that the more perfect our knowledge is, the *less* conscious it is. Their emphasis upon habits implies not so much a slighting of consciousness as a refusal to privilege or exaggerate the role of consciousness. The ideal of knowledge here is not a set of explicitly articulated propositions, ultimately derived from intuitively known principles, but a network of more or less integrated habits, growing out of innate dispositions and continually modified by the unexpected challenges of novel situations into which somatic agents are thrust by the dynamic propulsion of their defining dispositions. "Knowledge … lives", Dewey pointedly asserts, "in the muscles, not in consciousness" (1922, *MW* 22: 124). He adds elsewhere, the "level of action fixed by *embodied* intelligence is always the important thing" (1927, *LW* 2: 166). And embodiment here signifies, for the most part, incarnated in habits, especially integrated, flexible, and nuanced ones. Verbal formalization, articulations, and codifications operate on a relatively superficial level, whereas agential dispositions, crises,[11] and transformations work on a much deeper one.

For the pragmatists at least, explicit, formal knowledge is rooted in tacit, "unconscious" knowing, operating at levels and in a manner (characteristically, a nuanced, flexible, mobile, and transferable manner) hardly ever appreciated by epistemologists and other theorists. Such tacit, unconscious processes and abilities are, at once, somatic, affective, and "practical" (i.e., they concern the exercise of agency, for the most part, in the everyday contexts of human engagement). Again, the more effective are these to the tasks at hand, the less aware the agent tends to be. This frees the consciousness of such agents to attend to either other facets of the undertaking or even matters unrelated to this undertaking.

[10]In a fuller treatment of this complex topic, the details of Butler's direct influence on Peirce and, as a result of this influence, indirect influence on subsequent pragmatists would need to be detailed. The sole reference in volume 7 of Peirce's *Collected Papers* to Butler's *Unconscious Memory* is misleading: The influence is far more extensive and profound than this suggests.

[11]Crises might be quite specific, such as not knowing how to resolve a particular doubt, or more radical and encompassing, such as not knowing how to go on as one has up to this point. At one point Alice asks the Footman, once again though in a louder voice, "How am I to get in?" He responds by stressing, "*Are* you to get in at all?," pointing out "That's the first question, you know." "How are we to *go on*?" In moments of crisis where our very identity is at stake, we come to realize that there is no guarantee that we will be able to find a way to go on (Saul becomes Paul or, in the case of Nietzsche, a traditionally trained scholar becomes a radically innovative thinker for whom traditional scholarship comes more and more to seem like a living death). Doubt is always a moment of crisis, hence one demanding a decision. But decisions typically do not concern a crisis of identity. But they can and perhaps more frequently than we imagine do concern such crises.

For example, the skillful driver effectively ignores any number of vibrations and noises in the car, ones often capturing and even arresting the attention of the novice; as a result, this driver can attend to a host of other factors bearing upon the activity of driving. But, then, skillful walkers might direct their attention to the various scenes (or sounds) brought within their purview by their ambulatory activity, while paying virtually no attention to the dynamics of the activity itself. The skilled point guard can dribble a ball as though it is part of her body and, as a consequence, keep her head up and attend to what her teammates and opponents are doing. Obliviousness at one level opens the possibility of attention at another level, often what is not improperly identified as a "higher" level (the level of the elementary skill of dribbling is "lower" than that of the more complex set of skills involved in evading a defender, faking out the defender of one's teammate, and then hitting that teammate with an exquisite pass). The skills involved in dribbling, "juking" one's own defender, feinting the defender of one's teammate, and passing the ball are integrated into a single graceful act.

As a result of habituation, the consciousness required initially to acquire these distinct skills and, then, to integrate them in variable patterns gives way to what has been called the cognitive unconscious. Far from being a locus—in a sense, a source —of repressed desires and fears, the cognitive unconscious is a resource of nuanced abilities and skills. Rather than being that which resists being known, it is what manifests itself positively and creatively in virtually every exertion of any skillful actor. Of course, the Freudian unconscious has manifests *itself*, but it characteristically does so in a stultifying and disabling way.[12] The cognitive unconscious is, in contrast, enabling.

There is nothing mysterious about it. It is simply the vast array of our unconscious habits as they are more or less integrated which provides an abiding resource for effective action. Just as we ordinarily think *with* signs, rather than *about* them, we typically act on habits, rather than attend to them. Like signs, habits enable us to attend to the most salient features of a complex situation in which so often seemingly insignificant and even negligible details are the very pivots on which the success of an endeavor turns (cf. Dewey 1922, *MW* 22: 67).[13]

The hand is, for example, aware of the smoothness of the door handle without being *consciously* aware of this feature of the very thing it is grasping.[14] The human organism is continually making differential responses to a more or less expansive

[12]The matter is in fact more complex than this, since the unconscious in the Freudian or psychoanalytic sense does not play merely a stultifying or negative role in human life. Our creativity is bound up with the unconscious in this sense, though neither Freud's nor anyone else's account of this role is entirely satisfactory.

[13]The metaphor of the pivot is also central to Peirce's thought. "The idea of other, or not, becomes", he memorably wrote, "the very pivot of thought" (1903, CP 1, 324).

[14]This is a misleading way of stating the matter. The hand is not aware of the qualities of the door handle, the organism by virtue of the totality of its organs is aware. The brain does not think; the organism does, though of course it does so (in part) by virtue of the brain and the central nervous system. Cf. Dewey, *LW* 1: 222.

range of discernible differences. Most of these responses are below the threshold of consciousness (or what might be called conscious awareness). Even so, these responses themselves might be taken as instances of awareness: they clearly indicate that the organism, at some level, is taking into account what is going on around it and within itself.

It might seem as though *conscious awareness* is a pleonastic expression, but it need not be. For example, one can be aware of the buzzing of an insect nearby and then become *consciously* aware of this sound. The awareness just below the threshold of conscious awareness is, in certain respects, able to be experimentally established. Indeed, Peirce, with his student Joseph Jastrow, devised an experiment to test just this; and they published their findings ("On Small Differences of Sensation" [1884] CP 7.21–46).[15] While *awareness* and *consciousness* are often used as virtually synonyms, they are occasionally used differentially, often with awareness designating the more basic form of those states and processes by which organisms are responsive to objects and events (somatic as well as extra-somatic). We might think of these states and processes as those in which a sentient being differentially responds to what is *going on* (e.g., the increasing gnawing of hunger pains, an expected explosion in very close proximity, or the quite muted sound of very distant thunder).

A difference somewhere makes a difference elsewhere, but the locus where it makes a difference is bound up with vital processes, including temporal patterns of active exertion, typically ones of complex entanglement with potentially salient features of the actual environment. For instance, the sound of a twig breaking prompts the deer to freeze. That of another sound emanating from the same place, immediately following the initial auditory sign, prompts the deer to take flight, running swiftly in the direction opposite of that from which these sounds have just emanated. Even before hearing the first sound, it is likely that just below the threshold of its auditory awareness, the deer is picking up clues regarding the possibility of a predator nearby. Research on plants suggests a degree of sensitivity not previously suspected by most scientists, though arguably some gardeners and horticulturalists have long been attuned to the more subtle and less dramatic modes by which plants are attuned to a host of factors and forces. Whether we are entitled to ascribe awareness, even in the most rudimentary form, to plants, is however not a question that we need to resolve here. Whatever term turns out to be apt here, we need to take seriously sensitivity to intra- and extra-somatic goings on, as they are exhibited by every living being.

There is a continuum here, where the organism is first subliminally and, then, becomes vividly or consciously aware of a phenomenon. By conscious awareness, I do not mean self-consciousness, the awareness that one is aware of some aspect of one's comportment or feature of one's ambience. I mean simply heightened awareness, such that one might recall that one was aware of such an aspect or

---

[15]This paper was originally published in *Memoirs of the National Academy of Sciences* 3, Part 1 (1884), 73–83.

feature. One might pick up on the clues from a friend's facial expressions and other signs that this person is sad or even depressed, without explicitly or consciously drawing this conclusion (it is in fact an abduction, hence only a conjecture, but it might for all that be true). And here too there are gradations: an unconsciously drawn conclusion might be one we resist acknowledging, even though it is one we have ourselves drawn, or it might be one we are disposed readily to acknowledge. Between one extreme (the disposition to resist strenuously the acknowledgment of what we at some level know) and the other (the disposition readily or "immediately" to acknowledged an unconsciously drawn conclusion or framed hypothesis), there are, without exaggeration an infinitesimal array of possibilities. To claim that there is a continuum here implies, at least for Peirce, that there is such an array of possibilities. If we want to avoid the apparent redundancy or paradox of an expression such as conscious awareness, we might draw the requisite distinction in terms of implicit and explicit awareness. (Until now, I have always resisted the use of this expression, taking it to be pleonastic. I have obviously changed my view of this.) However we mark the distinction, it is important to do so. And it is imperative to keep distinct explicit (or conscious) awareness from reflexive consciousness (or self-consciousness).

The Acquisition of Autonomy

In the Sartrean tradition, consciousness is always implicitly reflexive (i.e., consciousness always implies a reference to the self). In the Peircean, it does not. Jean-Paul Sartre effectively continues the Cartesian approach, making the *I* (or ego) in its awareness of itself the pivot around which everything ultimately turns, while Peirce thoroughly rejects this approach. Both thinkers however take alterity or otherness to be central to the acquisition or (in Sartre's case) the actualization of self-consciousness. (It is hard to know what the right word is here; I am far from happy with "actualization".) The confrontation with the other can in certain circumstances throw the self back upon itself such that the self is forced to become aware of itself as distinct or possibly separate. For Peirce, this occurs originally and ever after in our *experience* of error (1873, CP 7.345). In the experience of error, the otherness of what is other than the self is brought home to the self in an inescapable way. From the beginning of its existence, the human organism is entangled with what is other than itself, but in the earliest stages of its development it apparently lacks the capacity to distinguish between itself and what is other than itself. Very early in its development, however, such an organism acquires an awareness of itself as distinct from others—in a word, it attains self-consciousness. As it turns out, this being is hardly separable from others. In its innermost nature, this organism is a continuum, moreover, one whose very being is interwoven with the being of countless other continua. It is accordingly anything but separable from others; indeed, others are integral to the being of the finite, personal or individual self as a distinct identity. At the very least, this identity is definable (or identifiable) in terms of a cluster of habits, the more or less integrated dispositions characteristic of this *type* of organism and so singularly modified by *this* token (or instance) of the type. Humans are not only expressive animals but also their modes of expression fall into extremely broad, even if irreducibly vague, categories, such as speech, gesture, action, ritual, and various other ones. Theoretically inclined anthropologists are arguably the most reliable guides for the identification of these categories. Whoever proves to be most adept at this important task, the diverse modes of human expression demand attention, not simply linguistic utterance or expression.

There is however a thorny question here. For a cluster of habits does not secure the degree of unity characteristic of the identity of a self. Only feeling does. In a suggestive but undeveloped hint, Peirce asserts: "Of course, each personality is based upon a 'bundle of habits, as the saying is that a man is a bundle of habits. But a bundle of habits would not have the unity of self-consciousness. That *unity* must be given as a centre for the habits" (1898, CP 6.228).[16] The emphasis here might have fallen differently: That unity must be *given*. It is given by the very nature of feeling, but not feeling understood reductively as a physiological event or state. So Peirce continues: "The brain shows no central cell" (1898, CP 6.229), implying that whatever unity there is has its source in something other than in such a cell or even cluster of cells. From this Peirce draws a conclusion likely to be judged as fallacious by many, if not most, theorists today: "The unity of consciousness is therefore not of physiological origin. It can only be metaphysical. So far as feelings have any continuity, it is the metaphysical nature of feeling to have *unity*" (ibid.).

One facet of Peirce's claim is, on the surface, far more plausible than the other. The connection of unity with continuity seems to me, at least, just right: whatever unity there is here, it is an instance of continuity, not least of all because it is a unity discernible or discoverable over the course of time. While it is given at any instant, what is thus given is stretched over the complex course of an indefinite duration. It essentially a temporal and historical unity (in other words, a continuity, such that the present self is continuous with what it has been what but also what it is destined to be). While Peirce's understanding of *destiny* is hard to get at, it is as important as it is elusive. For our purpose, however, the principal point concerns what *would be* the case, if certain conditions obtained. Whether these conditions obtain is a matter of contingent historical events. But Peirce's stress on what *would* happen suggests that contingency and "destiny" work conjunctively.

While the suggestion that the unity in question is an instance of continuity seems plausible, the assertion that the unity or continuity is metaphysical, not physiological, is not one tailored to the fashions of our time. But before rejecting this claim out of hand, it is imperative simply to understand what Peirce is contending. As I interpret this claim, the unity of the self is more fundamental than anything secured by purely physiological means, though it is secured through such means.

[16]There is no contraction between this claim regarding a centre for habits being given, in feeling, and the claim that there is the need for an integration of habits (this integration being not an antecedent fact but an eventual accomplishment).

Let us develop an argument by analogy. The meaning of a word is not primarily a function of its inscriptions or embodiments, even if the actual conveyance of the meaning of such a sign depends upon such actualities.

In terms of Peirce's categories, meaning is an instance of thirdness, but is as such inseparable from secondness and indeed firstness. What this means in this context is that the word as a habit (for that is what a word at bottom is) actualizes itself through its enactments (or "uses"), but is not reducible to those actualizations. Thirdness in itself (i.e., thirdness in its firstness) should not be confused with thirdness in its secondness and, for that matter, in its thirdness. Thirdness in its secondness is thirdness in its actual operation and such an operation is anything but ethereal or "spiritual": it operates in and through concrete physical means, though variability of means is characteristic of its operation.

The Peircean conception of evolutionary processes is indeed one in which one observes evolving ends in effect "experimenting" with alterable means. From his perspective, any instance of evolution is, on the one hand, a vivid instance of what he calls "developmental teleology". The ends themselves emerge in the course of history and, moreover, even the most entrenched, consolidated ends are continually altered, sometimes in the most dramatic fashion. Such an instance is, on the other hand, one in which variability of means is no less salient a feature than the development of the ends themselves. Biological evolution "answers" the complex question of visual perception in myriad ways and, more fundamentally, the general question of perceptual sensitivity in almost certainly more multitudinous fashion.

Peirce's analogy to explain the relationship between the thirdness of laws and the secondness of their embodiment is the decrees of a court and the enforcement of these decrees by a sheriff (cf. Potter). The decrees of the court apart from the arm of the sheriff are impotent, while the exertion of force apart from the ends being made possible by such exertion are utterly meaningless.

Peirce claims, "the majority of logicians [and presumably many others] are in the habit of conceiving of a universe of absolutely distinct individual objects" (1906, CP 6.176). This in effect, if not also in intent, reduces reality to existence (or actuality). It is the position of nominalism. Peirce however rather slyly is disposed to point out that this is truly a disposition, a deeply rooted habit to conceive reality as a meaningless collection of separate existents. One difficulty with this position is that it exemplifies what it denies—the presence and importance of habits! The generality of habits, hence, the reality of generality, cannot be gainsaid, even by those whose tendency (or disposition) is to deny just this.

Whether or not our philosophical theories have the resources to allow us to recognize ourselves as such, we are bundles of habit (1898, CP 6.228). But we are not *mere* bundles of innate and acquired dispositions. These habits are more than merely a miscellany. They are to some extent unified, though likely far less integrated than we commonly suppose. Unity is, in one sense, given and, in another, attained or won.

Paul Ricoeur overstates the case when he asserts: "Everything that can be said about consciousness after Freud seems to be contained in the following formula: consciousness is not a given but a *task*" (1974[1969]: 108). Consciousness is indeed a task: it is something to be won and, moreover, it can be won only be perseverance, patience, courage, and ingenuity. Our habits in some respects facilitate this task (Indeed, it would be utterly impossible without their contribution) but in other respects divert and frustrate this undertaking. Even after Freud (I am inclined to say, *especially* after Freud), however, consciousness is not simply a task or achievement. It is always to some extent a gift, something made available to one by the grace of forces and events over which one frequently has no control, often factors about which one is largely, if not wholly, unaware.

Our habits are, at once, given and won. They are given, since they are part and parcel of the very constitution of the human organism *and* the inevitable result of our fateful initiation into a distinctive form of human life. By the grace of evolution, we possess a distinctively plastic organism possessing an utterly remarkable capacity to acquire, modify, and lose habits. By the grace of experience, our innate dispositions evolve into an incredibly complex network of more or less integrated habits. That is, the operation of experience encompasses the generation of habits. But habits are not simply given; they are also won. In part, they are won because any gift itself enjoins, in this context, the work of appropriation, though very often this work can hardly be shunned or avoided (*given* its imperative drives and affective attachments, the infant—the non-talker—*is* ineluctably initiated into this linguistic community, the community of speakers with whom its own life is so intimately and persistently bound up).

As it turns out, the task of becoming conscious and that of transforming our habits are not two separable or even completely distinct tasks. They are, in truth, two sides of the same coin. The self-cultivation of consciousness and the self-control of our actions, in the service of ideals and with the hope of becoming virtuous (that is—acquiring good, presumably effective, habits), point to a process in which the replication of self-control upon self-control generates the vir (an agent possessing the capacity to maintain for an indefinite duration the sustained, intense attention or awareness requisite for the execution of any worthwhile task) (1906, CP 5.402n.3). The very nature of habit is such that disruption and breakdown of our habits generate conscious awareness.

The very nature of the ideals and norms governing the acquisition and alteration of habits is, however, such that the efficacious integration of otherwise disparate habits into a functional unity becomes, to some extent, an inevitable result of the actual operation of these norms and ideals, but to a far greater extent a deliberately espoused ideal. That is, our ideals and norms willy-nilly tend toward the integration of our habits. In the course of our development, however, our captivation by certain ideals (e.g., the figure of the skeptic or the experimentalist or the *bon vivant* or the intellectual iconoclast) drives us to identify conscientiously with this figure. "As a scientist, I am," as one might pronounce, "committed to rejecting everything that is not underwritten by observation and experimentation". The functional unity requisite for effective agency (and what is agency stripped of efficacy?) is, in its more rudimentary forms, simply the result of the more or less inevitable integration of a vast number of formally distinct dispositions. But such unity can only be, in its more sophisticated forms, a personal accomplishment, albeit one ordinarily resting upon communal support. In these forms, the unity of the self is not the blind result of mechanical processes, but the envisioned goal of deliberative agents, committed to the ongoing task of becoming virtuous individuals.

The consciousness of the disparity between one's actual and aspirational self is at the heart of this task. Here one discerns the interplay between consciousness and habituation, in particular, the deliberately cultivated consciousness of moral agents and, in light of such consciousness, deliberately cultivated habits (presumed by the agents to be strengths of character, in a word, to be virtues). We are often acutely conscious of just how far our actual selves fall short of our aspirational selves. It may even be the case that we catch a glimpse of our tendency—that is, our disposition—to portray ourselves in an idealized light, thereby thrusting from view the extent to which we significantly diverge from what we actually profess. Our awareness of this disposition might drive us to frame a more honest self-portrait. More generally, reflecting on our habits and beliefs is an integral part of deliberating, especially when deliberation moves back and forth between considering what ought to be in a specific circumstance and what we are making ourselves by acting in this way rather than that.

For Peirce, at any rate, deliberative agents alone are autonomous actors. But their autonomy is a function of the interplay between awareness and habits, at various levels. The moral, political, pedagogical, and psychological implications of this seemingly simple yet truly profound account of the relationship between awareness and habits are numerous. Physiological mechanisms are of unquestionable importance. In particular, great significance attaches to those mechanisms by which habits are formed and, of greatest importance, to those by which our disposition to jettison old dispositions and acquire new ones throughout the entirely of our lives. But the deliberative processes by which the cultivation of habits becomes a self-conscious, self-critical, and self-controlled task are certainly no less important to understand, thus to investigate. What connects questions concerning these physiological mechanisms and ones concerning these deliberative processes is a simple yet pervasive and indeed profound point: our awareness or consciousness is a function of the breakdown of our habits. While habits do indeed operate behind and beyond conscious awareness, such awareness itself is comprehensible only in reference to the operation of habits, including the arrest or disruption of that operation.

Conclusion

My objective here has been simply to bring the innermost core of this Peircean account into sharpest focus. In the foreground of this account, there is the *experience* of doubt. But in the background, though not too far in the background, there is the self-imposed task of self-cultivated virtues. Doubts befall us willy-nilly, though we come to realize that they do so because of our beliefs, other habits, and particular exertions. Apart from background habits and energetic engagements with environing circumstances, no doubt would ever befall us, willy-nilly or otherwise.

In contrast, virtues are deliberately cultivated. Humans find they cannot be satisfied without an effective government over their spontaneous impulses; but in order to be effective this must be a "self-government, instituted by himself to suit himself – copied, it is true, largely from the government his parents wielded when he was a child, but only continued because he find it answers HIS OWN purposes" (1909, EP 2: 459; cf. Krolikowski 1964; Colapietro 1997). To repeat, my aim has been to bring this account into focus. If I have accomplished merely this (!), I have done something far from insignificant.

References

Allport, Gordon W. 1939. Dewey's individual and social psychology. In *The philosophy of John Dewey*, ed. Paul Arthur Schilpp, 263–290. New York: Tudor Publishing Company.

Backe, Andrew. 1999. Dewey and the reflex arc: The limits of James's influence. *The Transactions of the Charles S. Peirce Society* 35(2): 312–326.

Butler, Samuel. 1880. *Unconscious memory*. London: David Bogue.

Colapietro, Vincent. 1989. *Peirce's approach to the self*. Albany: State University of New York Press.

Colapietro, Vincent. 1997. The dynamical object and the deliberative subject. In *The rule of reason: The philosophy of Charles Sanders Peirce*, ed. Jacqueline Brunning, and Paul Forster, 262–288. Toronto: University of Toronto Press.

Colapietro, Vincent. 2004. Toward a truly pragmatic theory of signs: Reading Peirce's semeiotic in light of Dewey's gloss. In *Dewey, pragmatism, and economic methodology*, ed. Elias L. Kahlil, 102–129. London: Routledge.

Colapietro, Vincent. 2006. Reflective acknowledgment and practical identity. In *Semiotics and philosophy in Charles Sanders Peirce*, ed. Rossella Fabbrichesi and Susanna Marietti, 128ff. Newcastle, UK: Cambridge Scholars Press.

Dewey, John. 1925. *Experience and nature. The later works of John Dewey*, vol. 1. Carbondale: Southern Illinois University Press. Cited as *LW* 1.

Dewey, John. 1934. *Art as experience. Later works of John Dewey*, vol. 10, ed. Jo Ann Boydston. Carbondale: Southern Illinois University Press.

Dewey, John. 1972[1896]. The reflex arc concept in psychology. In *The early works of John Dewey*, vol. 5, 96–109. Carbondale: Southern Illinois University Press. Cited as *EW* 5.

Dewey, John. 1984[1927]. *The public and its problems. The later works of John Dewey*, vol. 2, ed. Jo Ann Boydston. Carbondale: Southern Illinois University Press. Cited as *LW* 2.

Dewey, John. 1988[1922]. *Human nature and conduct. The middle works of John Dewey*, vol. 14, ed. Jo Ann Boydston. Carbondale: Southern Illinois University Press. Cited as *MW* 14.

Dewey, John. 2012. *Unmodern philosophy and modern philosophy*, ed. Phillip Dean. Carbondale: Southern Illinois University Press.

Freud, Sigmund. 1989[1915]. The unconscious. In *The Freud Reader*, ed. Peter Gay, 572–584. New York: W.W. Norton and Company. Originally published in 1915 in two issues of *Zeitschrift*.

Hegel, G.W.F. 1977[1807]. *Phenomenology of spirit*. (Translated from the German by A.V. Miller.) Oxford: Oxford University Press.

James, William. 1976[1912]. *Essays in radical empiricism*. Cambridge: Harvard University Press.

James, William. 1981[1898]. *Principles of psychology*. Cambridge: Harvard University Press.

Krolikowski, Walter P. 1964. The Peircean *Vir*. In *Studies in the philosophy of Charles Sanders Peirce*, ed. Edward C. Moore, and Richard S. Robin, 257–270. Amherst: University of Massachusetts Press.

Nöth, Winfried. (this volume). Habits, human and nonhuman, and habit change according to Peirce. In *Consensus on Peirce's concept of habit: Before and beyond consciousness*, ed. Donna E. West and Myrdene Anderson. (Studies in Applied Philosophy, Epistemology and Rational Ethics [SAPERE].) New York: Springer.

Peirce, Charles Sanders. i. 1867–1913. *Collected papers of Charles Sanders Peirce*. Vols. 1–6, ed. Charles Hartshorne and Paul Weiss. Cambridge: Harvard University Press, 1931–1935. Vols. 7–8, ed. Arthur W. Burks. Cambridge: Harvard University Press, 1958. [References to Peirce's papers will be designated by CP, followed by volume, period, paragraph number.].

Potter, Vincent G. 1997. *Charles S. Peirce on Norms and Ideals*. New York: Fordham University Press.

Ricoeur, Paul. 1974. *The conflict of interpretations: Essays in hermeneutics*. (Translated from the French by Kathleen McLaughlin et al.) Evanston, Illinois: Northwestern University Press.

Smith, John E. 1992. *America's philosophical vision*. Chicago: University of Chicago Press.

Tocqueville, Alexis de. 1969. *Democracy in America*, ed. J.P. Mayer. (Translated from the French by George Lawrence.) Garden City, NY: Doubleday.

West, Donna. 2014. From habit to habituescence: Peirce's continuum of ideas. In *Semiotics 2013*, ed. Jamin Pelkey and Leonard G. Sbrocchi, 117–126. Ottawa, Canada: Legas Press.

West, Donna. 2015a. Dialogue as habit-taking in Peirce's continuum: The call to absolute chance. *Dialogue* 54(4): 685–702.

West, Donna. 2015b. The work of secondness as habit in the development of early schemes. *Public Journal of Semiotics* 6(2): 1–13.

Wittgenstein, Ludwig. 1972[1969]. *Über Gewissheit/On certainty*, ed. G.H. von Wright. (Translated from the German by Denis Paul and G.E.M. Anscombe.) Bilingual edition. New York: Harper and Row.

Wittgenstein, Ludwig. 1984[1980]. *Culture and value*. (Translated from the German by Peter Winch.) Chicago: University of Chicago Press.

Chapter 18
Culture as Habit, Habit as Culture: Instinct, Habituescence, Addiction

Sara Cannizzaro and Myrdene Anderson

Abstract We consider Charles Sanders Peirce's insights regarding the dynamics he associated with the concept of habit, so that we might periscope into some realms he left under-explicit: first, culture itself, and then, addiction, the forms of which are necessarily relative to particular cultures at particular times. Peirce's groundwork on habit includes deliberations on instinct, habituescence (the taking of habits), the habit of habit-taking, and the changing of habits, enabling us to think through individual habits that are both marked and unmarked (that is, noticed or not), and how these feed into contemporary cultural practices whether deemed to be innocuous or extreme. With respect to extreme habits, we use the term "addiction" as a suitable gloss for behaviors marked by actual or perceived dysfunction, regardless of any involvement of use or abuse of substances. Finally, we propose that Peirce's reflections on habits (perhaps colored by his own habits-unto-addictions), and particularly his phanaeroscopy (phenomenology) of thirds—moving from vagueness to generality, from belief to doubt, from habit-taking to habit-breaking—suggest paths for exploring the debate surrounding the "reversibility" or "irreversibility" of addictions, including implications for self-control, and in turn, for our increasingly domesticated 21st-century society.

Keywords Markedness · Habituescence · Addiction · Habit-change · Abduction · Belief · Doubt · Self-control

---

S. Cannizzaro (✉)
Middlesex University, London, UK
e-mail: s.cannizzaro@mdx.ac.uk

M. Anderson
Purdue University, West Lafayette, IN, USA
e-mail: myanders@purdue.edu

D.E. West and M. Anderson (eds.), *Consensus on Peirce's Concept of Habit*,
Studies in Applied Philosophy, Epistemology and Rational Ethics 31,
DOI 10.1007/978-3-319-45920-2_18

Sneaking up on Habit, via Culture, Communication, Cognition

In much of contemporary cultural analysis outside of anthropology, culture has been assumed as plain regularity, tradition, or even a kind of "code". Via the mediary notion of "code", culture has often been coupled with communication and cognition in the contemporary West. The synergistic triplet of culture, communication, and cognition, will point us to the dynamics inherent in habit, while the notion of code will not. That's because true to its theoretical origins in early computing technology, a code assumes tame linear associations, closed systems, rigidity, and fixedness—in short, communicative codability. Yet none of these notions—habit, culture, communication, cognition—reveals unproblematic transparency; as such, they demand continued probing, while codes promise solution and resolution.

Antecedent scholars of culture, however, already intuited a less orderly assemblage of notions behind the term, with other notions coming along for the ride. In 1871, Sir Edmund Burnett Tylor ventured an overlap between "culture" and "civilization", focusing on social institutions rather than material culture, but otherwise sounding quite astute—culture being a "complex whole", inclusive thoughts and actions, be they learned, shared, or social, even using the expression, "capabilities and habits". Eighty years later, many dozens of formulations had been published in the emerging discipline of anthropology in English alone—164 of these summarized by Alfred L. Kroeber and Clyde Kluckhohn (1952). A tendency to associate culture with high culture persisted, despite the title of Tylor's treatise (1871:1), but culture also pertained to a quality associated with the entire species, finally becoming, also or instead, a label for the unique configuration of beliefs and practices of a single society, this referenced to Franz Boas and the emergence of anthropology in North America (cf. Benedict 1934).

More relevant for the diachrony of this term, "culture", however, was published after the Second World War by Raymond Williams (1958, 1976, 1981). Williams reflects on how his absence from academe during that war allowed him to notice how nuances had shifted to explicit formulation in releasing culture from older associations with high culture and civilization. Throughout the 20th century, scholars continued to document the vital signs of keywords such as culture (cf. Bennett et al. 2005; Hintikka 1976; Knuuttila 1980; Lovejoy 1936), tending to resist any closed definition, or even plural definitions, of culture, to embrace more open notions that include generative and semiosic systems. As an aside, by the end of the 20th century, the word "primitive" became inappropriate in any context, being ethnocentrically fused with either negative judgment or romantic nostalgia.

Hence, there's been a robust movement away from any simplified conception of culture, while at the same time clinging to this inscrutable term itself. In general, its utility has tended to rest on its inferred role in a deeper structure, as an analogue to language-qua-language, in contrast with a derived empirical manifestation in a surface structure as society, or behavior, or, in the case of language, spoken and signed languages. The movement from culture as a paradigmatic structure to society and behavior as synergistic processes is patterned but not strictly predictable—analogous in fact to the movement from genetic and epigenetic potential to phenotypic actual, shaped though not determined, at every instant, by a significant surround, an Umwelt.

If there now appears one more paradigm consonant with Peirce semiotics generally, and biosemiotics in particular, it just might resemble that simmering and building up steam roughly since Andy Clark and David J. Chalmers published "The extended mind" in 1998 (cf. Bennett 2010; Rowlands 2010; Sparrow 2014), such that now even more "rhymes" (cf. Humphrey 1973) grace academic pages and conversations: beyond "extended", also "embodied", "embedded", "enacted". Peirce, and other pragmatists—albeit drawing on Aristotle, Kant, Hegel, and Darwin—often appear to have anticipated much of later thought without being directly implicated, even when it comes to postmodernism (cf. Deely 2001).

Here we pick up on this discourse, regarding culture as both patterned and incidental, to consider the idea of culture as habit, including those habits amenable to addictions, and conversely, the idea of habit as culture too, to put forth the idea that habits and addictions, generate in turn, their own forms of culture. Unlike codes, habits—bearing the same nature as Peircean interpretants—grow, both linearly and nonlinearly, to allow for both suspense and surprise (cf. Salthe 1993). As such, the notion of habit could not be farther away from code. The idea of culture as habit, then, would be inclusive of codes, rules, and laws, yet it is not fully reducible to any of them, regardless of the nuances of such laws and rules as described by Winfried Nöth (this volume). In Peircean terms, habits are flexible, plastic, organic, malleable; they can be predictable in their regularity, as well as unpredictable, as they can and *do* and *must* change when one least expects them to, and occasionally, also when we hope or intend them to change. For Peirce, habits are, at root, *always* fully changeable—an appealing model for approaching culture, and addiction as well.

Either culture or language, if not a synthetic linguïculture (Anderson and Gorlée 2011), are all potential frames for studying the psyco-ecological relations of a society, and of culture as a whole at some moment in time, allowing some diagnoses—if not of societies in their entireties, then of some of their practices. An earlier student of culture, Jules Henry, well appreciated this in his volumes, *Culture Against Man* (1963) and *Pathways to Madness* (1965). However, current social science steers clear of presuming to analyze society holistically, under pain of accusations of reductive stereotyping; yet some of the lay literature addressing social problems now steps up to the plate, particularly as it interrogates, as it should, society, and culture, as addiction.

Charles Sanders Peirce's Groundwork on Habit

While habit for Peirce prominently references regularities, generalities, and laws of thirdness, he allows that habit manifests also in firstness, with chance, as sensation and instinct (EP 1: 220), and in secondness, with sensation and volition (CP 6.146). Furthermore, regularities in empirically-accessed outward behavior have antecedents and analogues (though never causes) in inner behavior and experience, and the reverse, as these processes are all emphatically nonlinear. In contemporary semiotic discourse, as with Peirce a century ago, linear thinking is eschewed, along with any assumptions of closed loops of cause and effect. In fact, for Peirce, habit-taking at any level, like life itself, grows (cf. MS 404; cf. Merrell 1996). To growth, regularity, and generality, Peirce captures the "consciousness of taking a habit" (c. 1913: MS 930: 31) in his suggestive term, "habituescence". Moving to consciousness brings habit back to thirdness, the unmarked realm in ordinary discourse. Hence, the engagement of cognition through habit-taking at various levels—instinct, volition, habituescence—spawns culturally-shaped beliefs (assumptions), magic, medicinal efficacy, obsessive-compulsive disorders inclusive of gambling and hoarding, attention-deficit disorders, Rorschachs, addictions generally, all thriving on feedback within the neuro-unto-cognitive system, variously engaging individual psychology and collective culture and language. At the same time, in the literature, the most serious addictions embedded in our culture and personalities go unidentified as such. Such addictions, as habitual conditions, include some of our social relations and all of our culturally-shaped excesses: codependencies (so-called relationship addictions), the seven lively sins, as well as extractive industries, consumerism and consumption, convenient beliefs, gossip, and even war, a recent human invention, and overpopulation, a recent human accomplishment.

Marked and Unmarked Habits

As we intuit, nascent habits must change even to be the same, and in that process they may grow, amplify, intensify, and spread, to become full-blown habits in thirdness, just as they may lurk pregnantly or dim unto oblivion. As such, whether incidental or deliberate and whether congenial or disowned, any noticeable, more regular habits may be sufficiently marked to name and perhaps presume to control, in the realms of secondness and thirdness—energetic interpretants in the realm of effort and contiguity, and logical interpretants in the realm of cognition.

With respect to humans, perhaps the most ubiquitous potential categories of habit may be either transparent or opaque to analysis. Transparent habits comport with givens in a system; they are difficult to dredge from their un(re)markedness. They manifest in patterns that would ordinarily remain mundane, submerged in firstness or vagueness, inferrable as emotional interpretants with their links to feeling and resemblance, without being singled out for notice or naming. Opaque habits may well be labeled as such, but in their general, domesticated state as labeled, they may resist analysis until pulled into other frames, for example, when discussing them in relation to addiction and markedness theory (Jakobson 1970; Waugh 1982).

The most entrained and lawful of these habits, if overdetermined, may become opaque to the system harboring them. Like the infinity of potential transparent habits, the infinity of opaque habits may be invisible to awareness, to be literally as well as figuratively unmarked (cf. Jakobson 1970; Waugh 1982). Given that systems at every level are themselves in process of constant modification (by necessity and accident [cf. Monod 1971], both in degree and in kind, digitally and analogically, intentionally and incidentally), what will be noticed—as a habit, or as addiction (with or without labeling)—will perforce leave even more processes and conditions un(re)marked under the radar.

Routine Versus Repetition

Culture and cultural behavior is replete with, if not consisting in, repetition, routinization, ritual. The term "ritual" may apply both to compulsive repetitive behaviors requiring attention and to culturally-shaped ceremonies that rely on internalized procedures; these two notions contrast sharply regarding any involvement of conscious concentration and inculcation of habit. Contemporary neurocognitive research suggests that repetition of a behavior engages cognition in individuals, while ritualization liberates cognition especially when socioculturally instantiated (E. Kandel 2009, 2012, 2015; Boyer and Liénard 2006; Liénard and Boyer 2006). Following Nobel Laureate Eric Kandel, explicit declarative learning relies on conscious attention, while cognition is liberated from ritual's implicit procedural learning, these two habits engaging distinct brain processes. Peirce, unlike some of his pragmatist contemporaries, intuited this very distinction, insisting that repetition was not what contributes to habit and habituation, because culture (as habit) cannot be reduced to any one of its components.

In a sense, the possibility that mere repetition could be internalized as habit seems akin to Lamarckian notions of the inheritance of acquired characteristics sans interludes of selection. As an aside, however, we note that contemporary evidence of powerful extra-genetic, that is epigenetic, influence on all biological processes (and beyond, into culture) does loosely recall Lamarck, but in this case with specific explanatory models involving methylization in ontogeny of the phylogenetic DNA information, allowing of the inheritance of the entire fused record.

Framed individually, especially when conscious awareness overtakes sensation and perception, habit can shanghai cognition with intermittent feedback (cf. Bateson 1972) shaping belief and habit-taking. This is the process by which magical beliefs, for instance, become indelible in personal experience and in a society as a whole. "Feedback", as a more recent operational term in cybernetics from the last mid-century, is routinely specified as either positive or negative—the positive ampliative and the negative dampening. Intermittent feedback with respect to habit must be taken as positive; predictability bores, unpredictability alerts. The recent term "feedback" now aligns and overlaps with earlier and continuing terms in psychology of "reinforcement" and in sociology of "sanction". Today the unmarked aspect of reinforcement is positive and the unmarked aspect of sanction is negative, while for feedback, neither value is unmarked as yet in contemporary discourse; these trends in semantic fields show habit in more abstract realms.

A Contemporary Menu of Habits, and Addictions

Habits, then, can be opaque or transparent, marked or un(re)marked, good or bad, extreme or moderate. Framed historically and prehistorically, humans have primarily trafficked in habits, even addictions. The most ubiquitous of trade items have been and still are luxuries, transmogrified into necessities. That is, a society could conceivably survive and even thrive without spice, salt, sugar, chocolate, alcohol, other drugs, fancy commodities, and anything the Madison Avenues of the period conjure. Material culture, like ritual—that in its routines skips mindfulness—operates below the threshold of consciousness of habit. Duhigg (2012) uses "big box" retailer Target to show how America's most influential businesses capitalize on unrecognized habits. These phenomena may be givens, and opaque to the habituated.

But the motivation to indulge in these habits, whether compulsive luxuries or spurious experiences—a desire, a want—renders them, together with the social relations of the overarching ecology inclusive of economy, into imperative must-haves, or needs. In contemporary addiction parlance, "needs" translate as "cravings", these potentially growing to compulsions, and compulsions to addictions, taken literally or figuratively.

The term "addiction" had little currency until a century ago (Hart 2015), toward the end of Peirce's life. Perhaps that is why there is no direct reference in Peirce's philosophical writings to his own possible addictions, other than descriptions of his reliance on pain-relieving substances (that up to 1914 were just sold over the counter), and inferences we make of his compulsive scholarly productivity (cf. Brent 1993: 149). But as we know, Peirce's works reveal a sophisticated thinking-through of habit—one might say a compulsive habit in itself—that would scarcely have been achieved without some reflexive intimacy with ideas, actions, and laws that seemed to have minds of their own, possessing Peirce himself, and now us as well.

Addictions, as other habits, may be judged good or bad—or just better or worse, within some context and interpretation that will always be inflected by socioculture. Compulsive eating can potentially make a person obese and the habit severely sanctioned, observers inferring a lack of self-control, while compulsive writing, drawing, or painting can be a prerequisite for becoming a renowned artist (whether talented or not), who will paradoxically be admired for the assumed self-control.

It is interesting in fact how it is mainly the opaque and marked habits that have historically been recognized as addictions.

Most empirical and applied research centering on habit, extreme habits, and so-called addictions, has been carried out without foregrounding that all such processes, whether or not substances be involved, will be profoundly nuanced by general contexts and even more by the specific culture and historical period. Peele (1975, 2014) would remind us that stimulants—caffeine, nicotine, cocaine—suit our western culture with its stress on individuality and self-reliance, while depressants—alcohol, narcotics, barbituates—in suppressing sensitivity to surroundings, come to be severely judged, as are hallucinogens (cf. Raikhel and Garriott 2013). Interestingly, the opposite attitude towards depressants was adopted in the eighteenth century, when there was little concern about opium use since people who used opium were not considered to be troublesome, in contrast with many alcohol-users (McMurran 1994: 5). Looking farther back in time for western culture, Nicholas K. Humphrey (1999) muses whether the Enlightenment would have been launched at all without the energized atmospheres of the coffee-houses of Europe (cf. Humphrey 2011; Makari 2015).

For societies that presently refer to the Diagnostic and Statistical Manual of Mental Disorders (DSM) (American Psychiatric Association 2013), it bears mentioning that in the most recent edition of 2013 (DSM-5), the former section on "substance-related and addictive disorders" avoids the term "addiction", instead using the label, "substance use disorder". In this emerging medical paradigm, there is still wiggle-room between "use", "overuse", "abuse", "dependence", and "dysfunction", as these may fall into a "disorder", and this is just with respect to substances. Some scholars and scientists had never regarded behavioral disorders as addictions, yet the DSM-5 retains the term "addiction" for just one condition—that pertaining to compulsive gambling (cf. Schull 2014).

Addiction then appears to be a concept legitimized by scientific, and especially medical and interventionist, discourse. The discourse as such must rely on, and at the same time, constitutes, rhetorical strategies for its own longevity. We have identified two of these strategies: the dismissal of extreme behavioral habits as opposed to substance-habits as forms of addictions, and also, arguments as to whether addiction of any sort is actually treatable, that is, whether there are both "reversible" and "irreversible" habits—these unfortunate notions rampant in the literature. To us it is clear that a Peircean perspective on addiction may ease or even erase these dichotomies.

Substance Versus Behavioral Addiction

Much of the well-funded neurological and cognitive sciences' empirical research has focused on substance-induced addictions, often resorting to non-human subjects and artificial settings at that. Such neurocognitive research often seeks to tie behavior to a finite number of structural substrates in the brain, a "scientific habit" we sense is too reductive despite its sometimes being interesting. This approach actually provides a suitable context for the emergence of the dichotomy questioning extreme behavioral habits, as opposed to substance-habits, as forms of addictions.

Failing to recognize culture as habit means failing to recognize instances of culture, and behavior, as addictions too. It is no surprise then, that in conventional approaches to addiction, compulsive practices such as gambling, sex, and video-game addictions, binge-eating, pet-hoarding, but also work, email, internet, and mobile phone addictions, are often referred to as "behavioral" or even "natural" addictions, to be placed in opposition to substance addictions, these restricted to physically-ingested substances—stimulants, narcotics, hallucinogens. Peele (1975) was early to point out the addictive elements in pathological interpersonal relationships, later called "co-dependent", posing under the cover of "love". While society is aware of problems of self-medication, some genres of self-medication may be opaque to both society and the agent/patient, for instance when women turn to pregnancy for relief from depression.

For some, behavior is simply more "habitual" than addictive. The reason for sharply separating habit from addiction for the dichotomy-supporters, is that the latter implies something altogether different and far more serious than habitual behavioral routines—a mental disorder that makes self-destructive behavior nearly impossible to stop (Watkins 2009: 133–134). An exponent of this view is Goldberg (interviewed by Wallis in 1997), who argues that, "if you expand the concept of addiction to include everything people can do, then you must talk about people being addicted to books, addicted to jogging, addicted to other people." However, by arguing that deleterious behaviors are "just" habits, this view is guilty of a superficial understanding of habit itself, one which dismisses the semiotic approach that addiction is, above all else, a very general habit to take extreme habits.

To wit, the theme of the 51st annual Nobel Conference was "Addiction: Exploring the Science and Experience of an Equal Opportunity Condition". For this theme, only one of the six plenary speakers was a Nobel Laureate, Eric Kandel. Besides the other five and diverse experts (Owen Flanagan, Carl Hart, Denise Kandel, Marc Lewis, Sheigla Murphy), there was a panel of four practitioners on the frontlines of intervention. Each group included individuals afflicted or once afflicted with "addiction", but the lines of often acrimonious disagreement within each group were not strictly consistent with personal experience. In fact, both groups—addiction-free and addiction-familiar—fractured along another dimension, that of whether or not addictions are "diseases" and, somehow therefore, indelible, only to be controlled through avoidance of substance or context, with the mantra: abstinence is not cure. Otherwise, outside of addiction, we would ordinarily speak of "curing" diseases, not simply denying or avoiding them (cf. Fletcher 2013; Levy 2013).

Supporters of the "behavioral" addiction argument contend that both behavioral and substance addiction can be merged on the grounds of their capacity to cause unfortunate "effects" such as "failure to manage time, loss of sleep, skipped meals, social isolation, poor performance at school or work" (Watkins 2009), the experiences of "withdrawal, tolerance, interpersonal, and/or health problems" (Rosen 2012a, b: 62)—all indicating that brain-reward pathways can indeed be chemically and structurally altered by excessive exposure to "natural" awards (Watkins 2009: 139). For example, in reference to technology addiction, Rosen (2012a, b: 49) remarks how overuse of wireless mobile devices leads to great anxiety, while Greenfield (2011: 150) points out how the "internet and other digital devices alter mood and consciousness and therefore should be respected and limited." Hence, Porter and Kakabadse (2006: 536) defend the choice of referring to behavioral addiction by stating that these can dominate people's lives, leading to (initial) tolerance, to strong withdrawal symptoms if and when the behavior is stopped, and thereafter to decreased tolerance, ironically itself an explanation behind a significant number of deaths from overdose.

However, even an argument in favor of addiction based on "effects" may be stronger when complemented by an argument based on the common theoretical generalities that both behavioral and substance addictions exhibit, for example, their being malleable to *change*, as outlined below. The substance versus behavioral dichotomy that has so long characterized conventional understanding of addiction could be rendered, when approached from semiotics, into a complementary "substance ~ behavioral" pair united by the symbol denoting transversal communication, the tilde "~", as proposed by Semetsky (2013).

Irreversible Versus Reversible Addictions

As if the substance versus behavioral dichotomy was not enough, there also exists a polarization of views on whether addiction is permanent or not—in the words of specialists, distinguishing "reversible" from "irreversible" habits—or what in a Peircean perspective we would be more comfortable in describing under the rubric of habit-change in general (since even "irreversible" habits involve change in order to be maintained). The argument in favor of an "irreversible" view of an addiction stems from an understanding of addiction as a disease. In that model, addiction is a straightforward case of physiological malfunction—emotion, pain, and reward pathways in the brain being irreversibly modified, endogenous opioid receptors (responsible for the smooth functioning of the reward system) being fooled by psychoactive substances, even deleterious genes being inherited from prone-to-addiction parents or epigenetic DNA functioning (although we now know that alcohol addiction may be the only disposition having any significant genetic or epigenetic links, when compared with the sociocultural and economic ones) (Nieratschker et al. 2013; Schaler 2002).

Where no obvious physical malfunction can explain a person's diagnosis of addiction, the possibility of mental illness is invoked (McMurran 1994: 2). The root for the disease model of addiction can be traced to the late 1800s, when addiction was understood as both a moral vice and actual disease (McMurran 1994: 12); the then-president of the Society for the Study of Inebriety (formed in Britain in 1884), theorized that inebriety was a disease of the nervous system, allied to insanity (McMurran 1994: 11). One must also note that on both sides of the Atlantic, theories of addiction were highly racialized, as were patterns of presentation and practices of intervention (Flanagan 2013, 2015; Hart 2013, 2015; Lewis 2013, 2015a, b).

The disease model assumes that addicts are different from other people in that they have some constitutive biological or psychological abnormality. The obvious implication to the idea of "constitutive addiction" is that it forces a conception of addiction as a necessarily unavoidable condition and/or "irreversible" habit. For example, Alcoholic Anonymous's now famous view on the workings of addiction is that once an individual is addicted to alcohol, the person remains addicted for life. In this view, substance addiction is a progressive, irreversible, and fundamentally incurable process that assumes a recovering alcoholic will never be able to drink in moderate quantities, leaving the only path that of totally avoiding the substance (Alcoholics Anonymous 2015). Consequently, the mantra of the AA and more generally of supporters of the (irreversible) disease model, including those in the aforementioned Nobel conference is, simply: "abstinence is not recovery"; there can be no "cure". Interestingly, at AA meetings, even as portrayed in the media, one witnesses group members bonding over the ingestion of other substances in great quantities: nicotine, caffeine, and sugar, over and beyond the codependencies.

However, the other side of the debate poses that addiction is indeed "reversible" or at least modifiable. McMurran challenges addiction's assumed irreversibility as she reports that there are now many studies (e.g., Davies 1962) that report regular, unproblematic drinking by people previously diagnosed as alcoholics once they have received abstinence-oriented treatment (McMurran 1994: 23). Also, she considers the findings of a research study on opiate addiction (McMurran 1994: 24) showing that, after the Vietnam War, drug use amongst soldiers (that had increased during the war from 2 % to roughly 50 %) decreased to pre-service levels, thus indicating that the opiate addicted are not necessarily addicted permanently. As McMurran concludes,

> … there is sufficient evidence to suggest that the disease model of addiction does not fit the facts. Loss of control is not inevitable and invariable; tolerance, withdrawal, and craving vary; "addicts" are not different (except that they drink or use drugs excessively); and the problem is not necessarily irreversible or progressive. (1994: 29)

It then appears conclusive that the disease model that conceives of addiction as an irreversible condition, or a regularly fixed habit, is no longer a viable assumption (Flanagan 2013, 2015; Hart 2013, 2015; Lewis 2013, 2015a, b; Peele 2014; Schaler 2002; Schaller 2006). Going along with the philosophy of indelibility of phenomena in addiction, succumbs to simplifying what, in semiotics, we know is a dynamical system rooted in a more plastic conception of habit, down to one without any room for conscious or non-conscious change.

Habit-Change and the Plasticity of Addiction

For Peirce, the intelligent laws of habit, as Nöth explains (this volume), do not act through cause-effect in secondness as do the blind laws of physics, but through final causation in thirdness (cf. Salthe, this volume). According to Peirce, the law of the mind is strikingly different from the laws of physics in that it is statistical and probabilistic rather than mechanic/deterministic:

> A physical law is absolute. What it requires is an exact relation. Thus a physical force introduces into a motion a component motion to be combined with the rest by the parallelogram of forces; but the component motion must actually take place exactly as required by the law of force. On the other hand, no exact conformity is required by the mental law. Nay, exact conformity would be in downright conflict with the law; since it would instantly crystallize thought and prevent all further formation of habit. The law of mind only makes a given feeling more likely to arise. It thus resembles the 'non-conservative' forces of physics, such as viscosity and the like, which are due to statistical uniformities in the chance encounters of trillions of molecules. (EP 1: 292, 1891)

Habit is not something fixed once for all, but, on the contrary, a flexible rule of procedure adopted for the practical purpose of successfully interpreting the sign (Gorlée 2004: 63). Similarly to habit, the workings of addiction could also be "accounted for by the principles of probability" (EP 1: 223, 1883–1884). Considering addiction as a habit, that is, a flexible rule or law of mind, would imply that a reaction to particular circumstances is only more or less likely to occur, not that it will necessarily occur. In this sense, the perpetuation of an addictive habit within a triggering context becomes neither strictly unavoidable, necessary, nor naively irreversible. As such, addiction would be subject to the probabilistic law of habit-change, leaving room for habit-taking (substituting one addiction for another) and for the higher order change of habit-breaking (turning addiction into moderation, or moderation into addiction).

Peirce most often spoke of habit-taking and the habit of habit-taking, and also of habit-breaking, as being accessible to logical analysis. These processes are particularly pertinent to any unpacking of addiction—its emergence, its maintenance, and its possible moderation or erasure. There is never a moment in one's life that is without habits, whether judged "good" or "bad", and/or, orthogonally, extreme or moderate. Because of the malleability of habit, each new habit can then be considered as a more or less radical modification of a pre-existing habit. As Peirce put it, "when a feeling emerges into immediate consciousness, it always appears as a modification of a more or less general object already in the mind" (CP 6: 142, 1891–1892). Habit-taking often implies habit-breaking, hence even to remain the same, the habit must change. In fact, in virtue of their plasticity, habits have varying degrees of strength and can evolve in two directions: increase or decrease, or be maintained in some stable or unstable equilibrium. When plasticity decreases, they become fixed laws (paraphrasing Nöth, this volume). The opposite tendency, when plasticity increases, is the weakening of habit, which is fertile ground for forming new habits.

Taking up an Addiction as a Habit

In a Peircean perspective, the emergence of addiction as a general habit to take extreme habits can be traced to the emergence of lower order, single addictive habits (thirdness in firstness), which are in turn based on the emergence of individual feelings in abduction (more on abduction below). This view is not dissimilar to that assumed by Denise Kandel (1975), who, in her recent research (2015), concedes that the only legal substance that acts as a gateway, or rather, intensifier, for illegal substance use, is nicotine, and that because nicotine lays down pathways in the brain, through epidemiological learning, that anticipate and powerfully enhance the effects of cocaine; in fact, without recent exposure to nicotine, cocaine's effects are neither dramatic nor dangerous.

As an instantiation of habit-taking, taking up an addictive habit can be seen, at a basic level, as a journey from first to third, and at a more complex level, as a journey across triads, so from first to first of thirds and of seconds, and to third of thirds; in other words, from the sheer qualia of simple feelings to habits of feelings, of actions, and finally of thought. Thellefsen (2000: 102) explains why, in habit-change, one ought to pay particular attention to the interaction between firstness and thirdness.

> Peirce stresses that Thirdness is a category of habits and habits tend to become subconscious. So, the evolutionary way of Thirdness is that semiosis through Thirdness forms a habit. This habit gradually becomes more and more subconscious, and Thirdness begins its regress to Firstness. Not the monadic Firstness in nature but the Firstness of Thirdness—the Rheme.

This "firstness of thirdness" as described by Thellefsen would belong to a higher order than the original firstness. Hence, the consolidation of a habit, from its origination in feelings, to its initial suggestion of a general idea (thirdness in firstness, cf. Nöth, this volume), and then its sinking back into the unconscious as a habit of feelings (firstness in thirdness, cf. Thellefsen) hence as a habit of a higher order, is a key passage when one considers habit-taking (See Fig. 18.1).

An addiction can form as an ordinary habit, through an intersection of abduction, via induction, with deduction. Abduction explains the journey from the emergence of isolated feelings to that of a new habit of feelings, playing a key part in interacting with old habits and in creating the conditions for the emergence of new habits, some of which may be addictions. It is abduction that proposes a new habit by forming and bringing to awareness a perceptual judgment—in addiction parlance, a *craving*. Cravings feed forward, finding affordances between internal states and external situations (cf. Gibson 1979, Uexküll 1957[1934]). Deduction then, confirms the new habit by influencing the feelings that will arise in a certain situation. In fact, the habit one already has will influence the kind of sensation one can afford to perceive in the future. "It is habit, ultimately, which partially determines the nature of the sensory cue" (Rosenthal 1982: 232). In other words, deduction will consolidate the habits of feelings into thirdness and originate the regular sense of

FIRSTNESS
First of First
Feelings

SECONDNESS
Second of First
Actions

THIRDNESS → regresses to → FIRSTNESS
Third of First **First of Third**
General Idea *Habits of Feelings*

SECONDNESS
Second of Third
Habits of Actions

THIRDNESS
Third of Third
Habits of Thought

Fig. 18.1 A Peircean trajectory of habit-generation. The initial stage of habit-taking as described by the passage from thirdness of firstness (a general idea-unto-habit) to firstness of thirdness (habit of feelings) in a triad of a higher order

craving that will accompany the more complex layers of the addictive habit, those of actions and thought, hence compulsion and regularity.

In Peircean terms, habit-taking and habit-breaking can be triggered by doubt. According to Magnani et al. (this volume), "Doubt is the common ground between the repetition of a habit […] its critical break, and the attempt to replace it with a better one". Emerging through abduction, it is doubt that prepares the ground for old habit-breaking and for further habit-taking, by promising a moment of relief. This sense of relief is what the addicted person desperately seeks. As Maté (2010: 107) put it, "The addict craves the absence of the craving state. For a brief moment he's liberated from emptiness, from boredom, from lack of meaning, from yearning, from being driven or from pain. He is free."

The creation of a new belief relies on the loss of confidence the individual has about his own beliefs (Magnani et al., this volume). The person adopting an addictive behavior as a new habit has typically lost trust or belief in something important in the immediate environment (relationships, work, objects, security). In the irritation of belief through abduction, the lost trust is replaced by a new habit, which could be an addiction, thus creating a new dependence. Doubt can emerge from pain or boredom. "A hurt is at the centre of all addictive behaviors", states Maté (2008: 36), who also reports how he asked a patient why he kept using a substance, and obtained the answer: "Emptiness in my life. Boredom. Lack of direction." (Maté 2008: 36) Also, Rosen (2012a, b: 48) describes how people seem to use technologies compulsively [...] because they are worried about missing out important information. In this sense, doubting boredom and pain is a strategy to bring novelty in someone's life.

> Human beings want not only to survive, but also to live. We long to experience life in all its vividness, with full, untrammeled emotion. Adults envy the open-hearted and open-minded explorations of children; seeing their joy and curiosity, we pine for our own lost capacity for wide-eyed wonder. Boredom, rooted in a fundamental discomfort with the self, is one of the least tolerable mental states. (Maté 2008: 37)

That is why, contrary to an understanding of addiction as static regularity, it may not be an overstatement to say that addictions of various types are responses to an emotional, compulsive, and even intellectual quests for *novelty*. In this sense, "novelty" is a more neutral and less laden term than the almost mechanical concept of "reward".

The habit-change initiated by doubt in abduction is not deliberate, in that, similarly to originary qualia of feelings, habits of feelings are also unreflectively established in abduction. It is interesting in Peirce's conception, how abduction as form of hypothetical reasoning would ideally lead semiotic systems towards the "truth" more often than not. As Peirce put it, "It is the simpler hypothesis in the sense of the more facile and natural, the one that instinct suggests, that must be preferred; for the reason that, unless man have a natural bent in accordance with nature's, he has no chance of understanding nature at all" (6.477: 1908). However, and paradoxically, it is in the context of strong doubt of existing beliefs, through abduction, that addiction can instinctively feel and even act, like a plausible solution. In fact, West (2014: 125) states that abductions may become habits when they represent the best explanation possible. But, when examining the possibility that addictive habits can and do establish themselves in abduction, the expression "best explanation" has to be considered solely in neutral logical terms. An addictive substance-use can be the "best explanation" only logically, but doubtless it is not in an actual life context because in such context, the surplus value of "bad" habit becomes attached to it through the normative sciences.

The abductions that will concur to establish a new addictive habit are based on the general idea (belief) that the new habit will relieve anxiety and pain. As Maté (2008: 37) explains, "For the addict, the drug provides a route to feeling alive again, if only temporarily." When this general idea, effectively a perceptual judgment, is taken as utterly true, then there are strong chances that it will consolidate into a stable habit, of feeling, of action, and of thought. The problem is that all abductions are plausible, even the fallible ones. As Magnani et al. (this volume) remind us, the emotional condition brought about by the irritation of doubt does not guarantee that the individual so affected will necessarily progress on to enhanced wisdom.

Hence the "gut belief"—that substance or behavior will calm the state of anxiety brought about by pain or boredom—is plausible to the addict-to-be, but not necessarily true, since due to the mingling of unreflective feelings (firstness) with a compulsive component (secondness), there are strong possibilities that engaging in the addictive behavior not only will *not* reduce anxiety, but in the long term will amplify it by orders of magnitude not initially foreseen or even foreseeable. Hence the grounding of a habit of thought into a fallible abduction, or, as Gorlée (2014: 408) put it, in the "informal degeneracy of signs and purposes", can explain how, after encountering strong doubt in one's own beliefs, one may end up taking a destructive addictive habit rather than a moderate habit or perhaps may not change at all.

Habit-Change, or Breaking an Addiction

Like habit-taking, habit-breaking was part of the very fabric of habit for Peirce. Habit-change may occur through a spontaneous insight triggering the spark of change, or, through deliberate effort at creating a new habit, which offers a similar reward but different routine to the "bad" habit to be overcome or replaced. As Gorlée (2004: 153), explains, "Semiosis…changes with time and space; it entertains successively new doubts, new beliefs, and new persuasions. Under duress of new circumstances a habit formation cycle is re-generated."

Neither quotidian habit nor addiction at any level is indelible, although their modification or cessation does not strictly speaking indicate reversibility, but rather change. One should recall Heraclitus' observation that one cannot step into the same river twice. Very much as rivers are not reversible, the flow of semiosis is continuous, never static to allow any identical second step. Addiction cannot be considered reversible either. As Magnani et al. (this volume) states, "Once the abductive process is achieved, the agent cannot come back to the initial stage of belief/knowledge-based habit", but can only progress from the point of view of one's own experience, though not predictably (due to the final causation aspect of habits), and not necessarily to a better or worse state.

Regarding substance addiction, Moyers (2012: 98) states, "Addicts and alcoholics are never the same after they've been exposed to treatment, because they're smarter about their problems and wiser about the solution and what's at stake." Even in the case of a relapse occurring "as it often does, you haven't really gone back to square one", and "the key, for me, and an essential tool for anyone who wants to ward off the possibility of relapse, is to never stop learning" (Moyers 2012: 149–150). That is why it may be wise to resist "habitual" mechanical descriptions, and refer to "mutable" and "persisting" habits rather than "reversible" or "irreversible" ones.

Besides, the literature acknowledges that addicts and abusers typically outgrow their conditions, or "mature out", sometimes within a finite period of time, even a decade (cf. Flanagan 2013; Hart 2013; Pantalon 2015; Steinberg 2014). This jibes well with emerging evidence that responsible "maturity", evidenced in self-control, is still developing well into the 20s (Johnson et al. 2009; Schwartz and Begley 2003). Demographically documented in the West, while puberty has come earlier, maturity continues to be delayed, introducing an entirely new stage of development, not adequately described as "teenage", nudging "adolescence" closer and closer to the age of 30 (Siegel 2014), and reminding us that "childhood" itself was a recent invention (cf. Aries 1962[1960]).

Incidentally or deliberately broken habits ripple with further potential habit. Both habit and more intense addiction could be positive, negative, or even neutral, but, despite the fallible potential of habits gained through abductions, the positively evaluated conditions are less apt to be extinguished, via either natural or artificial selection. Broken habits and addictions can pose their own positive and/or negative consequences, even becoming positively marked in emergent "cultivated" forms in craft, art, and science, where abduction plays powerful roles. In fact, the surfacing of feelings to consciousness, seen as the moment in which doubt emerges in abduction and irritates an established habit, is a moment of vulnerability in which an old habit is broken whilst a new habit is taken. Psychiatric treatment, psychotherapy, cognitive behavioral therapy, counseling, group therapy, friendship, family support, sport, artistic production, or even a good nap, can be seen as ways to trigger doubt into existing addictions, hence creating the conditions for habit-breaking, via serendipitous "accident" and/or deliberate "necessity". Revealingly, singer Amy Winehouse, notoriously known for her musical talent as well as for her substance abuse-unto-dependency, was reported during a "really difficult time", to be "channelling [the substance abuse] into her music" (Rubin 2007).

This phenomenon is referred to as a "habit replacement" strategy, amounting to keeping the trigger and reward of a habit consistent, while changing the routine, or the habit of actions (Duhigg 2012: 62). This strategy results in substituting one habit's routine with another—for example, switching from heroin to methadone, or from alcohol or an obsessive compulsive disorder to religion. Of course it may be easier to switch from one addictive habit to another addictive habit than to a moderate habit. Habit change in this latter direction is testified by the founder of Alcoholic Anonymous, often quoted as saying how he turned to God to stop his addiction. In general, organized support groups such as Alcoholics Anonymous also provide the ideal grounding for the habit-replacement strategy by also creating a community of belief that substitutes the addict's "trust" in the rewarding effects of the addictive habit for the trust in oneself, aided and abetted by nicotine, caffeine, and sugar. Here the belief in the addictive habit may be substituted, in habituescence, by the belief in each other's effort.

Broken habits and overcome addictions can themselves become obsessive habits at another level, when dissonant firstnesses of incompleteness succumb to repetition in ritual, mantras, litanies, all compelling harmony, symmetry, consistency, completion, in cognitive thirdness aware of this loop of anticipation. That is why and how habit-breaking brings the chance of re-elapsing into the old addictive habit. Relapse, also known as "euphoric recall" or a "selective memory of the pleasure of the high" (Moyers 2012: 152), is a strikingly fallible abductive process whereby the sudden sensuous experience of the addictive trigger strongly resonates with a sense of pleasure buried in some recesses of the mind and body, at the expenses of the strong sense of displeasure precipitated by the very same trigger later on, also similarly buried in memory. Moyers explains how euphoric recall intended as craving "grows exponentially when we don't maintain a healthy respect for our disease … we don't keep in mind, foremost in our consciousness, the inability to control our disease" (Moyers 2012: 152; cf. Moyers 2007). How such a selection of the sensual component of information from memory takes place is a fuzzy business, to the point that "selection" is no longer seen as an appropriate term to describe the phenomenon (cf. Brier 2008), however such a "selection" fundamentally concerns the problem at stake in figuring out when a potential fact becomes information, that is, in Bateson's terms (1972b: 459), when we come across "a difference which makes a difference."

Switching from abuse to moderation is without doubt a very complex change that happens both within and across Peircean triads, and encompasses all types of interpretants (energetic, logical, emotional, cf. Fernández 2011), with or without requiring a forced break, or abstinence, as a mid-step. Changes of this kind are not logically impossible, just much harder to obtain either by chance or via deliberate intervention, or a mixture of both, because they rely on several changes of habits at the level of the lower triads (energetic and logical interpretants, or triads of firstness and of secondness).

Changes taking place in habit may include re-creating the conditions for genuine surprise, and spontaneously precipitating habit-breaking. Attempts at recreating the conditions for surprise may take the form of creative activities, including research. As Glanville (1998: 61) states,

> … learning involves accepting states outside of those we already have, which are, therefore, in a sense unmanageable […] if we want to enhance our creativity, then we need to gain options (variety). We can gain variety where ever we find in our environment more variety than we find in ourselves, so long as we are open to that variety and we do not try to control (i.e., restrict) it. When we are faced with unmanageability we have opportunity: opportunity to find more variety than we currently have.

It is possible, then, that a constant flux of novelty through different types of research, be it artistic, scientific, institutional, or just amateur treasure-hunting, may satisfy the addict's cravings for novelty and divert attention to perhaps similarly obsessive but certainly less harmful novelty-seeking routines. The rewards of intermittent reinforcement recur in this process.

Glanville argues that through *not* trying to exercise control, excessive variety becomes a "source of renewal and creativity" (Glanville 2004: 91). It is interesting how the Alcoholics Anonymous view on "hitting bottom" encapsulates this idea of unmanageability, as it involves releasing control and giving way to chance, hence in Peircean terms would be a trigger for change, through chance, in firstness (EP1: 220). Bateson (1999a[1972]: 330) explains how hitting "…'bottom' is a spell of panic which provides a favorable moment for change, but [in line with the probabilistic aspect of habit as a law of mind] not a moment at which change is inevitable". Through unmanageability the system in question can learn, change, and grow. In fact, when the addict has "hit bottom" but self-destruction hasn't occurred, then an individual may decide to self-help, or either seek or accept help.

Whether habit-change is deliberate, in habituescence, or unreflective, through chance alone, or whether such change is just a mixture of deliberate effort and chance —that corresponding to perceived conjoined "necessity" and "accident", recalling Monod's phrasing regarding biological evolution (1971). Addicts variously report both avenues at play—revulsion accumulating after repeated relapses (particularly when there are withdrawal symptoms to deal with), as well as spontaneous abductions precipitating "cold turkey" cessation of behavior (independent of substance). The 1947 Mickey Spillane novel, *I, the Jury*, introduced this rather opaque expression. Many afflicted with unwanted habits, even addictions, claim that abruptly quitting the behavior, including the behavior of substance use, has been their salvation. For them, the effort in maintaining moderation was not an option. However, "cold turkey" is not an option with respect to some depressant substances.

High-order learning and habit-change through either habituescence in thirdness, or chance in Firstness and Secondness—once one has put oneself, consciously or unconsciously, in a context that is favorable to the emergence of feelings—outcomes exist that cannot be fully predicted or guaranteed, duly recognized in all dynamical systems. This fact becomes obvious when considering addiction in a Peircean framework, because deliberate effort at habit-breaking can never be grounded in complete self-control, as it contains uncontrolled habits of feelings (cf. Noth, this volume); at the same time in hitting-bottom and unmanageability, there is always the inevitability of taking habits, and a tension towards thirdness, built into living itself.

Hence in light of Peirce's theory of habits, one may conceive of addiction as an extreme, but open and *evolutionary*, habit. But note that following Salthe, evolution consists in the "irreversible accumulation of historical information", which is unpredictable, allowing for surprise, and is set apart from "predictable irreversible change" or development, that can engender suspense (Salthe 1993: 29–32). Peirce was ahead of the times in many fields, but did not anticipate contemporary movements in biology and systems sciences. It seems that his references to "evolution" may have that process fused in some respects with "development", although that observation does not concern us. The habits of habits, coming into and out of addictions and other habits, reflect both trajectories, evolutionary and developmental.

Self-control: A Cultural Caveat

Addiction might appear to be a habit totally inaccessible to self-control, while habit as thirdness, and, in the case of moderation, a third of a third, is necessarily imbued with a high degree of self-control. Certain addicts do choose to be so. To consume or to otherwise act, or not, and to stop or ramp up or down any behavior, all entail some level of effort. However, categorizing addiction in thirdness responsive to self-control is a possibility that should not be excluded, as it could illustrate thirdness with less generality than an ultimate, rational, and fully self-controlled third, which anyway, being akin to an ultimate final interpretant, can never be fully achieved (for discussion of the ultimate final interpretant, see Santaella, this volume).

For example, the famous longitudinal Dunedin study in New Zealand links poor childhood self-control with a higher chance of developing substance-addiction in adult life (1972 to present; cf. Moffit et al. 2013). Quite varied research in the "individualistic" and "Calvinistic" West, is finding lesser self-control to bear strong associations with a variety of deleterious behaviors, from developmental issues in children and young adults (cf. Steinberg 2014) to health-impacting practices (cf. Schaller 2006), to overspending at individual and also national levels (Chen 2013). These issues bear clear linkages with each other, and with emerging neurological research pointing to brain maturation continuing well into the 20s, with its concomitant natural emergence of self-control (cf. Flanagan 2013; Hart 2013; Pantalon 2015; Steinberg 2014), altogether explaining the how and why so many addicts, and others encumbered by unsavory habits (to self and other), actually "age out" of their conditions. The majority of addicts outgrow their afflictions, and do so without exposure to intervention.

Much of Peirce's thinking assumed that self-control would be a viable modality for taking and shedding habits. This links with the persistent discussion about free-will in western thought. In North America, the inherited cosmologies further emphasize independence, individuality, and self-reliance—summing up our assets and our liabilities. These features of culture and personality, and even language (cf. Durst-Andersen 2011), go beyond stereotypes, but even as stereotypes these patterns can be compelling, even sobering, for reflection.

Indeed, some treatment practitioners assert that addictions are fostered and evaluated within very particular sociocultural crucibles (Peele 1975, 2014; Steinberg 2014). It seems suspiciously over determined that North American avenues out from addiction foreground self-control, reflecting a culture inclined to assume all is possible, even the "American dream". Immigrants animated the new world in waves of dogged if not also optimistic explorers, epitomizing the "juvenile systems" described by Salthe (1993) for biological systems generally. Those were the independent, individualistic, self-reliant generations able to actualize a modicum of self-control along the way, even literally moving on lest stumbling, at least until recently.

Since the last mid-century, society has changed to shape (and be shaped by) a less mobile and adventuresome, a more cautious and dependent, citizenry (Peele 1975)—a domesticated public, homogenized by market forces and imploding economies. The North American society that could be described a generation (or two or three) ago as evolutionarily open, or "juvenile", replete with possibilities, has proceeded toward a more developmentally closed, or "mature" or even "senescent" shape, using Salthe's (1993) metaphors. The potential for surprise has been replaced with the probability for suspense, as many view a future more predictably unpredictable, in an ominous way. At least in the United States, the public exhibits little self-control when it concerns matching purchasing to budgets, savings to period until retirement, meals to waistlines, or how many clothes, cars, and homes anyone actually needs or even might use. The confusion between needs and wants is fertile ground for advertising; and overnight shipping from internet marketplaces leaves no time for deferred gratification. Self-control, in the past as much imposed by circumstances as voluntary, has become a quaint notion.

The United States is still self-ascribed as a young nation, but in terms of its temperament by the end of the recent century, the nation was already showing its age in terms of cultural habits. Mature and post-mature systems, in biology and generally, as in culture and as in personality, enjoy fewer degrees of freedom, and may move from flexibility to fragility, from elastic to plastic, as their rigidities can lead to interruption and collapse (Salthe 1993). Contemporary conditions, also in at least some other parts of the West, continue to foster fresh opportunities for habits-unto-addictions, with new perspectives on them. Whether deliberative self-control will continue as a generator as well as a moderator of extreme habits is playing itself out in this early 21st century.

References

Alcoholics Anonymous. 2015. *The twelve steps of alcoholics anonymous*. http://www.alcoholics-anonymous.org.uk/About-AA/The-12-Steps-of-AA. Accessed 8 Oct 2015.

American Psychiatric Association. 2013. *Diagnostic and statistical manual of mental disorders, Fifth Edition* (DSM-5). Arlington, Virginia: American Psychiatric Association.

Anderson, Myrdene, and Dinda L. Gorlée. 2011. Duologue in the familiar and the strange: Translatability, translating, translation. In *Semiotics 2010*, ed. Karen Haworth, Jason Hogue, and Leonard G. Sbrocchi, 221–232. Ottawa: Legas Press.

Aries, Pierre. 1962[1960]. Centuries of childhood: A social history of family life. (Translated from the French [L'Enfant et la vie familiale soud l'Ancien Régime] by Robert Baldick.) New York: Alfred A. Knopf.

Bateson, Gregory. 1999a[1972a]. The cybernetics of "self": A theory of alcoholism. In *Steps to an ecology of mind: Collective essays on anthropology, psychiatry, evolution, and epistemology*, 309–337. San Francisco: Chandler Publications.

Bateson, Gregory. 1999b[1972b]. Form, substance, and difference. In *Steps to an ecology of mind: Collective essays on anthropology, psychiatry, evolution, and epistemology*, 454–471. New York: Ballantine.

Benedict, Ruth. 1934. *Patterns of culture*. Boston: Houghton Mifflin.

Bennett, Jane. 2010. *Vibrant matter: A political ecology of things*. Durham: Duke University Press.

Bennett, Tony, Lawrence Grossberg, and Meaghan Morris. 2005. *New keywords: A revised vocabulary of culture and society*. New York: Wiley-Blackwell.

Bernacer, Javier, and Jose Ignacio Murillo. 2014. The Aristotelian conception of habit and its contribution to human neuroscience. *Frontiers in Human Neuroscience* 8: 883. doi:10.3389/fnhum.2014.00883.

Boyer, Pascal, and Pierre Liénard. 2006. Why ritualized behavior? Precaution systems and action parsing in developmental, pathological, and cultural rituals. *Behavioral and Brain Sciences* 29 (6): 613–650.

Brent, Joseph. 1998[1993]. *Charles Sanders Peirce: A life.* Bloomington: Indiana University Press.
Brier, Søren. 2008. *Cybersemiotics: Why information is not enough!* Toronto: University of Toronto Press.
Chen, M. Keith. 2013. The effect of language on economic behavior: Evidence from savings rates, health behaviors, and retirement assets. *American Economic Review* 103(2): 690–731.
Clark, Andy, and David J. Chalmers. 1998. The extended mind. *Analysis* 56: 7–19.
Deely, John N. 2001. *Four ages of understanding: The first postmodern survey of philosophy from ancient times to the turn of the twenty-first century.* Toronto: University of Toronto Press.
Dewey, John. 1930. *Human nature and conduct.* New York: The Modern Library, Inc.
Duhigg, Charles. 2012. *The power of habit: Why we do what we do and how to change.* London: William Heinemann.
Dunedin Multidisciplinary Health and Development Research Unit (DMHDRU). 1972–present. Publications in numerous venues from New Zealand longitudinal study, cf. Moffit et al., 2013.
Durst-Andersen, Per. 2011. *Linguistic supertypes: A cognitive-semiotic theory of human communication.* Berlin: Mouton.
Fernández, Eliseo. 2011. Peircean habits and the life of symbols. In *Semiotics 2010*, ed. Karen Haworth, J. Hogue, Leonard Sbrocchi, 98–109. Ottawa: Legas Press.
Flanagan, Owen. 2013. The shame of addiction. *Frontiers in Psychiatry* 4: 120. doi:10.3389/fpsyt.2013.00120.
Flanagan, Owen. 2015. Willing addicts? Drinkers, dandies, druggies, and other Dionysians. Plenary lecture at the 51st Annual Nobel Conference, Addiction: Exploring the Science and Experience of an Equal Opportunity Condition, Gustavus Adolphus College, St. Peter, Minnesota, 6–7 October 2015.
Fletcher, Anne M. 2013. *Inside rehab: The surprising truth about addiction treatment and how to get help that works.* New York: Penguin.
Gibson, James J. 1979. *The ecological approach to visual perception.* Boston: Houghton Mifflin.
Glanville, Ranulph. 1998. A (cybernetic) musing: Variety and creativity. *Cybernetics and Human Knowing* 5(3):63–70.
Glanville, Ranulph. 2004. A (cybernetic) musing: Control, variety and addiction. *Cybernetics and Human Knowing* 11(4):85–92.
Gorlée, Dinda L. 2004. *Translating signs: Exploring text and semio-translation.* Amsterdam: Rodopi.
Gorlée, Dinda L. 2014. Peirce's Logotheca. In *Charles Sanders Peirce in his own words: 100 years of semiotics, communication, and cognition*, (Semiotics, Communication and Cognition, 14), ed. Torkild Thellefson and Bent Sørensen, 405–409. Berlin: De Gruyter Mouton.
Greenfield, David. 2011. The addictive properties of internet usage. In *Internet addiction. A handbook and guide to evaluation and treatment.* Hoboken, New Jersey: Wiley.
Hart, Carl. 2013. *High price: Drugs, neuroscience, and discovering myself.* New York: Harper.
Hart, Carl. 2015. Why drug-related research is biased: Who benefits and who pays. Plenary lecture at the 51st Annual Nobel Conference, Addiction: Exploring the Science and Experience of an Equal Opportunity Condition, Gustavus Adolphus College, St. Peter, Minnesota, 6–7 October 2015.
Henry, Jules. 1963. *Culture against man.* New York: Random House.
Henry, Jules. 1965. *Pathways to madness.* New York: Random House.
Hintikka, Jaakko. 1976. Gaps in the great chain of being: An exercise in the methodology of the history of ideas. *Proceedings and Addresses of the American Philosophical Association* 49: 22–38.
Humphrey, Nicholas K. 1973. The illusion of beauty. *Perception* 2: 429–439.
Humphrey, Nicholas K. 1999. *A history of the mind: Evolution of the birth of consciousness.* New York: Copernicus.

Humphrey, Nicholas K. 2011. *Soul dust: The magic of consciousness*. Princeton: Princeton University Press.
Jakobson, Roman. 1970. Linguistics. *Main trends of research in the social and human sciences; Part I: Social Sciences*, 419–463. Paris and Berlin: Mouton.
Johnson, Sara B., Robert W. Blum, and Jay N. Gledd. 2009. Adolescent maturity and the brain: The promise and pitfalls of neuroscience research in adolescent health policy. *Journal of Adolescent Health* 45(3):216–221.
Kandel, Denise B. 1975. Stages in adolescent involvement in drug use. *Science* 190: 912–914.
Kandel, Denise B. 2015. Molecular basis for the gateway hypothesis. Plenary lecture at the 51st Annual Nobel Conference, Addiction: Exploring the Science and Experience of an Equal Opportunity Condition, Gustavus Adolphus College, St. Peter, Minnesota, 6–7 October 2015.
Kandel, Eric R. 2015. We are what we remember: Memory and age-related memory disorders. Plenary lecture at the 51st Annual Nobel Conference, Addiction: Exploring the Science and Experience of an Equal Opportunity Condition, Gustavus Adolphus College, St. Peter, Minnesota, 6–7 October 2015.
Kandel, Eric R. 2009. The biology of memory: A 40-year perspective. *The Journal of Neuroscience* 29(41): 12748–12756.
Kandel, Eric R. 2012. *The age of insight: The quest to understand the unconscious in art, mind, and brain, from Vienna 1900 to the present*. New York: Random House.
Kessler, David A. 2016. *Capture: Unraveling the mystery of mental suffering*. New York: Harper/Collins Publishers.
Knuuttila, Simo, ed. 1980. *Reforging the great chain of being: Studies of the history of modal theories*. New York: Springer.
Kroeber, Alfred L., and Clyde Kluckhohn. 1952. Culture: A critical review of concepts and definitions. In *Papers of the Peabody Museum of archaeology and ethnology*, Harvard University, vol. 47(1). Cambridge: Peabody Museum.
Levy, Neil. 2013. Addiction is not a disease (and it matters). *Frontiers in Psychiatry* 4: 24. doi:10.3389/fpsyt.2013.00024.
Lewis, Marc. 2013. *Memoirs of an addicted brain: A neuroscientist examines his former life on drugs*. New York: Public Affairs.
Lewis, Marc. 2015a. *The biology of desire: Why addiction is not a disease*. New York: Public Affairs.
Lewis, Marc. 2015b. Willing addicts? Drinkers, dandies, druggies, and other dionysians. Plenary lecture at the 51st Annual Nobel Conference, Addiction: Exploring the Science and Experience of an Equal Opportunity Condition, Gustavus Adolphus College, St. Peter, Minnesota, 6–7 October 2015.
Liénard, Pierre, and Pascal Boyer. 2006. Whence collective rituals? A cultural selection model of ritualized behavior. *American Anthropologist*, 108(4): 814–27.
Lovejoy, Arthur O. 1936. *The great chain of being: A study of the history of an idea*. Cambridge: Harvard University Press.
Magnani, Lorenzo, Selene Arfini, and Tommaso Bertolotti. this volume. Of habit and abduction: Preserving ignorance or attaining knowledge? *Consensus on Peirce's Concept of Habit: Before and Beyond Consciousness*, ed. Donna E. West and Myrdene Anderson. (Studies in Applied Philosophy, Epistemology and Rational Ethics [SAPERE]) New York: Springer.
Makari, George. 2015. *Soul machine: The invention of the modern mind*. New York: W.W. Norton.
Margulis, Lynn. 1971. Whittaker's five kingdoms of organisms: Minor revisions suggested by considerations of the origin of mitosis. *Evolution* 25(1): 242–245.
Maté, Gabor. 2008. *In the realm of Hungry ghosts: Close encounters with addiction*. Toronto: Knopf Canada.
McMurran, Mary. 1994. *The psychology of addiction*. London: Taylor and Francis.
Merrell, Floyd. 1996. *Signs grow: Semiosis and life processes*. Toronto: Toronto University Press.

Moffitt, Terrie E., Richie Poulton, and Avshalom Caspi. 2013. Lifelong impact of early self-control: Childhood self-discipline predicts adult quality of life. *American Scientist* 101 (September–October 2013): 352–359.

Monod, Jacques. 1971. *Chance and necessity: An essay on the natural philosophy of modern biology*. New York: Alfred A. Knopf.

Moyers, William Cope. 2007. *Broken, my story of addiction and redemption*. New York: Penguin.

Moyers, William Cope. 2012. *Now what? An insider's guide to addiction and recovery*. Center City: Hazelden.

Nieratschker, Vanessa, Anil Batra, and Andreas J. Fallgatter. (2013). Genetics and epigenetics of alcohol dependence. *Journal of Molecular Psychiatry* 1: 11. doi:10.1186/2049-9256-1-11.

Nöth, Winfried. this volume. Habits, human and nonhuman, and habit change according to Peirce. In *Consensus on Peirce's concept of habit: Before and beyond consciousness*, ed. Donna E. West and Myrdene Anderson. (Studies in Applied Philosophy, Epistemology and Rational Ethics [SAPERE]) New York: Springer.

Pantalon, Michael V. 2015. Panelist, at the 51st Annual Nobel Conference, Addiction: Exploring the Science and Experience of an Equal Opportunity Condition, Gustavus Adolphus College, St. Peter, Minnesota, 6–7 October 2015.

Peele, Stanton, with Archie Brodsky. 1975. *Love and addiction*. New York: New American Library.

Peele, Stanton, with Ilse Thompson. 2014. *Recover! Stop thinking like an addict*. Boston: Da Capo Lifelong Books Press.

Peirce, Charles Sanders. i. 1867–1913. *Collected papers of Charles Sanders Peirce*. Vols. 1–6, ed. Charles Hartshorne and Paul Weiss. Cambridge: Harvard University Press, 1931–1935. Vols. 7–8, ed. Arthur W. Burks. Cambridge: Harvard University Press, 1958. [References to Peirce's papers will be designated by CP, followed by volume, period, paragraph number].

Peirce, Charles Sanders. i. 1867–1893. *The essential peirce: Selected philosophical writing*. Vol. 1 (1867–1893), ed. Nathan Houser and Christian Kloesel. Bloomington: Indiana University Press, 1992. [References to this volume will be designated by EP 1, followed by colon, page number].

Peirce, Charles Sanders. i. 1893–1913. *The essential peirce: Selected philosophical writing*. Vol. 2 (1893–1913), ed. the Peirce Edition Project. Bloomington: Indiana University Press, 1998. [References to this volume will be designated by EP 2, followed by colon, page number].

Peirce manuscripts in Texas Tech University Library at Texas Tech University, Institute of Studies of Pragmaticism, beginning with MS—or L for letter—and followed by a number, refer to the system of identification established by Richard R. Robin in Annotated Catalogue of the Papers of Charles S. Peirce (Amherst: University of Massachusetts Press, 1967), or in Richard R. Robin, "The Peirce Papers: A Supplementary Catalogue," Transactions of the Charles S. Peirce Society.

Peirce, Charles Sanders. i. 1839–1914. *The Charles S. Peirce papers. Manuscript collection in the Houghton Library*. 930: 31–33, 1913. Cambridge, MA: Harvard University.

Porter, Gayle, and Nada K. Kakabadse. 2006. HRM perspectives on addiction to technology and work *Journal of Management Development* 25(6):535–560.

Prigogine, Ilya, and Isabelle Stengers. 1984[1979]. *Order out of Chaos. Man's new dialogue with nature*. Translated from the French. Boulder, Colorado: New Science Library.

Raikhel, Eugene, and William Garriott, eds. 2013. *Addiction trajectories*. Durham, North Carolina: Duke University Press.

Rosen, Larry. 2012a. [*i*] *Disorder: Understanding our obsession with technology and overcoming its hold on us*. New York: Palgrave MacMillan.

Rosen, Robert. 2012b[1985]. *Anticipatory systems: Philosophical, mathematical, and methodological foundations*. New York: Pergamon.

Rosenthal, Sandra B. 1982. Meaning as habit: Some systematic implications of Peirce's pragmatism. *The Monist* 65(2):230–245.

Rowlands, Mark. 2010. *The new science of the mind: From extended mind to embodied phenomenology*. Cambridge: MIT Press.
Rubin, Courtney. 2007. “Amy Winehouse ‘Determined’ to Attend Grammys”, *Wayback Machine* (12 October 2007), https://web.archive.org/web/20130920153720/, http://www.people.com/people/article/0,,20165333,00.html. Accessed 5 Sept 2015.
Salthe, Stanley N. 1993. *Development and evolution: Complexity and change in biology*. Cambridge: MIT Press.
Salthe, Stanley N. this volume. Habit-taking, final causation, and the big bang theory. In *Consensus on Peirce's concept of habit: Before and beyond consciousness*, ed. Donna E. West and Myrdene Anderson. (Studies in Applied Philosophy, Epistemology and Rational Ethics [SAPERE]) New York: Springer.
Santaella, Lucia. this volume. The originality and relevance of Peirce's concept of habit. In *Consensus on Peirce's concept of habit: Before and beyond consciousness*, ed. Donna E. West and Myrdene Anderson. (Studies in Applied Philosophy, Epistemology and Rational Ethics [SAPERE]) New York: Springer.
Semetsky, Inna. 2013. *The edusemiotics of images: Essays on the art science of tarots*. Rotterdam: Sense Publishers.
Schaler, Jeffrey A. 2002. *Addiction is a choice*. Chicago: Open Court Publishing.
Schaller, James. 2006. *Suboxone: Take back your life from pain medications*. New York: Hope Academic Press.
Schneider, Samuel. 1998. *An identification, analysis, and critique of Thorstein B. Veblen's philosophy of higher education*. Lewiston, New York, and Lampeter, Wales: Edwin Mellen Press.
Schüll, Natasha Dow. 2014. *Addiction by design: Machine gambling in Las Vegas*. Princeton: Princeton University Press.
Schwartz, Jeffrey M., and Sharon Begley. 2003. *The mind and the brain: Neuroplasticity and the power of mental force*. New York: Harper.
Sebeok, Thomas A., and Jean Umiker- Sebeok. 1981. ‘You know my method’: A juxtaposition of Charles S. Peirce and Sherlock Holmes. In *The Play of Musement*, ed. Thomas A. Sebeok, 17–52. Bloomington: Indiana University Press.
Siegal, Daniel J. 2014. *The power and purpose of the teenage brain*. New York: TarcherPerigee/Penguin.
Sparrow, Tom. 2014. *The end of phenomenology: Metaphysics of the new realism*. Edinburgh: Edinburgh University Press.
Spillane, Frank Morrison (Mickey). 1947. *I, The Jury*. New York: E.P. Dutton.
Steinberg, Laurence. 2014. *Age of opportunity: Lessons from the new science of adolescence*. Boston: Houghton Mifflin.
Thelleffsen, Torkild L. 2000. Firstness and thirdness displacement: Epistemology of Peirce's trichotomies. *Applied Semiotics* 4(10): 91–103.
Thom, René. 1968. Topologie et signification. *L'iige de la science* 4: 219–242.
Tylor, Sir. Edward Burnett. 1871. *Primitive culture*, vols. 1 and 2. London: John Murray.
Uexküll, Jakob von. 1957[1934]. A stroll through the worlds of animals and men: A picture book of invisible worlds. In *Instinctive behavior: The development of a modern concept*, edited and translated from the German by Claire H. Schiller, 5–80. New York: International Universities Press, Inc.
Wallis, David. 1997. Just Click No, *The New Yorker*, 13 January 1997: 28. http://www.newyorker.com/magazine/1997/01/13/just-click-no. Accessed 20 Oct 2015.
Watkins, S. Craig. 2009. *The young and the digital. What the migration to social-network sites, games and anytime, anywhere media means for our future*. Beacon Press: Boston.
Waugh, Linda R. 1982. Marked and unmarked—A choice between unequals in semiotic structure. *Semiotica* 38(3/4): 299-318.

West, Donna E. 2014. From habit to habituescence: Peirce's continuum of ideas. In *Semiotics 2013*, ed. Jamin Pelkey and Leonard G. Sbrocchi, 117–126. Ottawa: Legas Press.
Williams, Raymond. 1983[1958]. *Culture and society: 1780–1950*. New York: Columbia University Press.
Williams, Raymond. 1985[1981]. *The sociology of culture*. Chicago: University of Chicago Press.
Williams, Raymond. 2014[1985, 1976]. *Keywords: A vocabulary of culture and society*. New York: Oxford University Press.

Chapter 19
The Habit-Taking Journey of the Self: Between Freewheeling Orience and the Inveterate Habits of Effete Mind

Fernando Andacht

Abstract How to account for the interplay of change and permanence in human identity? I discuss Peirce's contribution to solve a paradox: the certainty of humans of always being themselves, always the same person, despite the overtly evolving nature of the self. Intriguingly, what epitomizes the regular and predictable nature of habit and of habit-taking centrally involves the incidence of the most volatile element in Peirce's theory, namely, Orience ("free originality"), spontaneity. To describe the emergence of change in identity construed as a process of habit-taking, this chapter examines three decades of Peircean writings on habit (1878–1908). A conclusion of this study is how fundamental in his account is the role of the imagination when it comes to the shaping of habits. That elusive element in humanity bears tangible consequences, since imaginary considerations "will affect my real action should those circumstances be realized" (CP 2.148). Phaneroscopy provides an essential support to Peirce's mature account of habit.

Keywords Identity · Habit-taking · Orience · Spontaneity · Imagination · Change

From Baking Apple Pies to Coming to Terms with Our Evolving Self

In the following, I ponder on the origin and the operation of change in the realm of human identity construed as habit-taking, a semiosic process that comes about through fleeting flashes of originality, the manifestations of what Peirce called "Orience", namely, the genesis of meaning in "irresponsible, free originality" (CP 2.85, 1902). Interestingly, in the development of what seems to be the epitome of regularity and predictability, namely habit, Peirce's account attaches a decisive weight to a wholly unpredictable element, to wit, freewheeling spontaneity, thus positing a tension between the tendential proper of habits and beliefs—as the

F. Andacht (✉)
Universidad de la República, Montevideo, Uruguay
e-mail: fernando.andacht@fic.edu.uy

D.E. West and M. Anderson (eds.), *Consensus on Peirce's Concept of Habit*, Studies in Applied Philosophy, Epistemology and Rational Ethics 31,
DOI 10.1007/978-3-319-45920-2_19

closure of doubts—and the light imaginary flights that bring about slight or appreciable departures from rule-like behavior. To accomplish my goal, I will consider some relevant passages drawn from the logician's work during the three decades from 1878–1908. I will complement the perusal of texts with an illustration from a fictional film whose plot enacts almost didactically the onset of a crisis of the self, a striking instance of habit-change. This momentous alteration in a character's life is brought about by habit-change; most of all it exemplifies how "a belief-habit formed in the imagination simply, as when I consider how I ought to act under imaginary circumstances, will equally affect my real action should those circumstances be realized" (CP 2.148, c. 1902).[1]

The example will enable me to describe Peirce's unique contribution to solve an age-old paradox: the indispensable permanence that is inseparable from the unavoidable, constant change in human identity. The imperceptible but fateful journey of the self sets off in dream-like, immaterial images, namely the sheer qualities of Firstness as Orience, the coming into being or emergence of absolute "feeling", a notion that is distinct from the embodied, physiological "emotion". I will now refer to a passage in which Peirce explains his distinct kind of phenomenology in uncharacteristic domestic, down-to-earth terms.

To introduce the phaneroscopic categories, Peirce chooses to portray an everyday situation that makes it easy to grasp the triple distinction between the original, the singular, and the general as these "valencies" (1.291, 1905) manifest themselves in our experience, what is formally known as Firstness, Secondness, and Thirdness, respectively. Such is the phenomenological basis of the triadic process of sign generation or semiosis. In a text that ostensibly describes the mundane passage from somebody's experiencing the desire for an apple pie to his actual consumption of that pastry, and which includes its detailed preparation, Peirce describes the functioning of the three phaneroscopic categories. In that context, the logician claims that human desire is always of a general kind, so "to speak of a single individual pleasure is to use words without meaning" (CP 1.341, c. 1875). This is a central aspect of his account of the working of generality in life.

I will focus on one remarkable element from that fragment titled "Thirdness". At the beginning of his description of the steps that must be followed to consume and prepare the apple pie, the example with which Peirce aims "to examine the idea of generality", he writes: "An apple pie is desired" (CP 1.341). This is a syntactically marked form—it would have been simpler, that is, unmarked, to write, "I desire an apple pie". This alternative corresponds to the mode of being of habit as the paradigmatic instance of Thirdness. It does not refer to a single, isolated action, but

---

[1]This quote also comes from the 1902 *Minute Logic*, specifically from Chap. 2, "Why study logic?" This is relevant for my approach since in that text, Peirce discusses the notion of *logica utens*, namely, a non-systematic but still basically reliable manner of making inferences in everyday life: "Every reasoner, then, has some general idea of what good reasoning is. This constitutes a theory of logic: the scholastics called it the reasoner's *logica utens*. Every reasoner whose attention has been considerably drawn to his inner life must soon become aware of this" (CP 2.186, 1902).

to "a conduct", which insofar as it is "order and legislation" is third (CP 1.338, c. 1875). Both the grammar of the sentence and its subject matter denote a general, collective taste or preference, something nicely illustrated by the proverb "as American as apple pie", which signifies a typical, collective way of satisfying a desire, a widespread belief that "involves the establishment in our nature of a rule of action, or, for short, a *habit*" (CP 5.397, 1878). It is as if Peirce had written, I, as an American, like many more people, probably the majority of those who live in this nation, we all tend to satisfy the craving for something edible and sweet for dessert by consuming apple pies.

For the same reason, to talk about a definite possibility or of a well delimited quality lacks any sense. The indeterminate possibility mode is the way our imagination works according to the phaneroscopic category of Firstness. Neither it nor Thirdness has any concrete, embodied singularity as its defining feature; that is left to the operation of Secondness. In Peirce's illustration, this corresponds to the act of picking "any apples that are handy and seem good" (CP 1.341, c. 1895), the actual fruits that the cook must eventually take to prepare the desired apple pie. In the next section, I will discuss another illustration that also places great importance on the Orience or nascence of habit-change.

Saving Peirce's Mother: An Imaginative Feat and a Key Ingredient of Pragmatism

A Peircean take on the self as an evolving sign shows that one of our most cherished "belongings", that which makes us what we believe we are for ourselves and for others, may turn out to be not truly our own, not something that we possess as our DNA, but a developing sign among signs in the world, both external and internal: "Every sane person lives in a double world, the outer and the inner world, the world of percepts and the world of fancies" (CP 5.487, c. 1906). For three decades, Peirce reflects upon the genesis and the evolution of habits, which are inseparable from our beliefs, and which in his mature writings on the semiotic become the "pragmatic interpretants" (Alston 1956: 87). In "How to make our ideas clear", the classic text where Peirce introduced the pragmatic maxim in 1878, he uses an everyday situation to illustrate the relevance of the imagination in the taking and changing of habits and of the beliefs that are inseparable from them. In contrast with the epistemic function of genuine doubts—those that Peirce conceives as the irritating, necessary stimulus of all scientific inquiry, in that article he argues for the contribution of "feigned hesitancy" (CP 5.394, 1878) to that serious endeavor. While waiting for a train at the railway station, as a way to kill time, Peirce tells us that he begins to observe ads for trips that he knows well he will never take. Through his engaging in a state of imaginary doubt, the logician fancies many possible itineraries, and he reaches a surprising conclusion about the virtues of this kind of exercise:

> Most frequently doubts arise from some indecision, however momentary, in our action. Sometimes it is not so. I have, for example, to wait in a railway-station, and to pass the time I read the advertisements on the walls. I compare the advantages of different trains and different routes which I never expect to take, merely fancying myself to be in a state of hesitancy, because I am bored with having nothing to trouble me. Feigned hesitancy, whether feigned for mere amusement or with a lofty purpose, plays a great part in the production of scientific inquiry. (CP 5.394, 1878)

The quotation brings out the importance of the imagination not just for our dealings with our *logica utens* (CP 2.186, 1902), the implicit inferential process that "the reasoner" uses automatically in everyday life, but also for the far more elaborate application of *logica docens*, the formal, explicit reasoning that is at the heart of the (fourth) method to attain belief in the realm of science. Here I consider the consequences of this kind of "feigned hesitancy" concerning the kind of behavior we are to adopt in life, either in a trivial pursuit, such as the "choice between paying for the price of the fare with five coins of a penny or with one of five cents" (CP 5.394, 1878), or in far more serious dilemmas such as those which human beings face in a life-changing crisis, as my fictional example shows. Directly, Peirce goes on to underline the importance of this imaginative activity in the formation of a habit, that is, in the subsequent consolidation of a conduct or belief that may be used, much later, and with full efficacy, for the attainment of palpable and even admirable effects, in ordinary life as well as in science. To show the importance of this process for Peirce, I will now refer to a family anecdote that he evokes twice with a slight variation, to drive home the conclusion of his mature formulation of the pragmaticist semiotic, namely the claim that "A man can be durably affected by his percepts and by his fancies" (CP 5.487, c. 1907).

In the earlier (1902), more detailed version of this family story, which Peirce takes up five years later in *A survey of pragmaticism*, but in which he introduces a different casting, and which he presents as only a footnote,[2] it is Peirce's mother who is presented as the victim of a negligence that could have been fatal:

> I remember that one day at my father's table, my mother spilled some burning spirits on her skirt. Instantly, before the rest of us had had time in order to think what to do, my brother, Herbert, who was a small boy, had snatched up the rug and smothered the fire. We were astonished at his promptitude, which, as he grew up, proved be characteristic. I asked him how he came to think of it so quickly. He said: "I had considered on a previous day what I would do in case such an accident should occur". (CP 5.538, c. 1902)

The conclusion that Peirce draws from this domestic incident has far-reaching implications: his account posits a powerful bond between imagination and habit-change. Such is the kind of logical mechanism whose main upshot is to increase the generality and reasonableness of the world. In Herbert Peirce's answer to his brother's query about the remarkable speed and readiness to react so

[2]In the 1907 version, it is a lady who is a visitor to the Peirce household and not his mother who is the victim of the accident that could have been far more harmful: "one day, as the whole family were at table, some spirit from a "blazer," or "chafing-dish," dropped on the muslin dress of one of the ladies and was kindled" (CP 5.487, fn1).

admirably in the face of danger, the logician (CP 5.538, c. 1902) believes to have found an instance "of the act of stamping with approval, [of] 'endorsing' as one's own, an imaginary line of conduct, so that that it shall give a general shape to our actual future conduct." Peirce describes such an act as "*a resolve*". Those "airy nothings" that are our dreams, as Peirce (CP 6.455, 1908) writes elsewhere, echoing Shakespeare, bear real fruits in the external world. Therefore the explanation of the boy's remarkable act of saving his mother on occasion of that domestic accident involves the imaginary as a key component of that most suitable reaction at a time of emergency.

A slight domestic incident that nowadays would have likely ended up in a photo or in a video due to its surprising outcome to be then uploaded on social media, led me to choose a fictional film narrative that depicts the development of a similar *resolve*, another instance of a "striking example of a real habit produced by exercises in the imagination" (CP 5.487, c. 1907). Thus, in the next section, I present a film scene portraying a fictional character's endorsing as his own an imaginary line of conduct that will effectively shape his future actions and the fate of others. Far from being an exceptional occurrence, these examples serve to demonstrate how "it is not the muscular action but the accompanying inward efforts, the acts of imagination that produce the habit" (CP 5.479, 1907).

Such is the theoretical endeavor regarding the origin and growth of habits that began in 1878, when the logician formulated the pragmatic maxim, and which concluded with the positing of the final or ultimate interpretant conceived as a habit (CP 5.476, c. 1907). During this period, there is a recurrent, still insufficiently studied emphasis on the relevance of the imagination in habit-taking, and most definitely in habit-change:

> a belief habit formed in the imagination simply, as when I consider how I ought to act under imaginary circumstances, will equally affect my real action should those circumstances be realized. Thus, when you say that you have faith in reasoning, what you mean is that the belief-habit formed in the imagination will determine your actions in the real case. (CP 2.148, 1902)

It is this oscillation between the world of brute, resistant, outwardly real and the internal working of our fancies, our dreams, those "airy nothings" (CP 6.455, 1908), that Alexander (1990) claims to be a distinctive feature of "the pragmatic imagination", contrasting with positivism—namely "a recognition of the importance of a mode of understanding whereby the actual was reinterpreted and reconstructed in the light of the possible" (1990: 325). A habit as "a general law of action" (CP 2.148, 1902) is the shuttle that weaves and momentarily stabilizes the relationship between "the world of percepts and the world of fancies" (CP 5.487, c. 1907). As Peirce realized, both ordinary apple pies and life-changing crises are good examples of the working of dreams, together with concrete circumstances and with enduring tendencies to act in a certain way in order to attain some specific goals.

Orience and the Self, or the Subtle Art of Slipping into Iconicity

In order to analyze a paradigmatic instance of habit change, the onset of a new belief in someone who nevertheless looks wholly unchanged outwardly—the etymological purport of Latin *habitus*[3]—Peirce's self-confident account of the emergence of novelty in the universe comes in handy:

> By thus admitting pure spontaneity or life as a character of the universe, acting always and everywhere though restrained within narrow bounds by law, producing infinitesimal departures from law continually, and great ones with infinite infrequency, I account for all the variety and diversity of the universe, in the only sense in which the really sui generis and new can be said to be accounted for. The ordinary view has to admit the inexhaustible multitudinous variety of the world, has to admit that its mechanical law cannot account for this in the least, that variety can spring only from spontaneity, and yet denies without any evidence or reason the existence of this spontaneity, or else shoves it back to the beginning of time and supposes it dead ever since. The superior logic of my view appears to me not easily controverted. (CP 6.59, 1892)

Elsewhere, with a co-author, we (Andacht and Michel 2007) consider the notion of the cinema as an "iconic-symbolic essay" on complex questions such as that of subjectivity and the changing self-definition of a person. Through a brief reflection on the German film *The Lives of Others* (*Das Leben der Anderen*, von Donnersmarck, 2006, set decades earlier in East German times—henceforth *DLA*), I intend to answer a question that has to do with change and permanence in identity matters: why and how are we capable of endlessly changing and of still claiming to be the same person that we say, feel, and believe that we (always) are? For instance, in the social realm that this film portrays, how does change take place in the self, despite the persistence of merciless laws imposed by an authoritarian, cruel regime? According to Peirce's view of meaning and of the universe, reality entails constant change. Be it microscopic and imperceptible, or overwhelming and dramatic, there is an unstoppable flow in which we live, and which transforms us, even when we are convinced that nothing is altered in us, that like robust trees we are firmly planted in the midst of our existence, with the same kind of natural unyielding stubbornness, thanks to the obstinacy of our will. But what happens with the uncontrollable emergence of spontaneity, of fleeting qualities, of the suggestion of visions, whether external or internal, which may result in the understanding that comes with an unexpected insight, one which may have unpredictable consequences for human subjectivity? In *DLA*, these qualities of feeling must seep through thick layers of habit, of deadening obedience and coercive identification with a ruthless system of politics and of a highly controlled daily existence, with

[3]I have drawn this notion from Miller's (1995) well-argued paper on "Peirce's conception of habit".

absolutely no room for humor.[4] Thus we contemplate how an utterly different self slowly but irresistibly grows out of the frozen shell of self-discipline produced by that inhuman way of life, one that is based on scrutinizing and punishing any deviance, no matter how slight, from the only authorized way of behaving and thinking, one in which State, society, and human subjectivity must merge flawlessly.

If we take into account the totalitarian oppression of every aspect of life, this German film may be construed as an allegorical reflection of an institution that sets out to overpower mind construed as "lively matter" (Hausman 1993: 148), as being susceptible to spontaneity, freewheeling variation, and perversely transforms it into quasi-effete mind, a rigid, fixed entity that is felt to be the most suitable for sheer survival. The upshot is a mode of life in which the imagination can lead to a deviation from iron-like norms, and which will invariably be brutally punished. At the heart of this fictional recreation of a historical dystopia, we get to observe the irrational, inhuman project for a nation to live in a realm where "there will be no indeterminacy or chance but a complete reign of law" (CP 1.409, 1887–1888). Were such a "complete reign of law" ever be attained in the life-world, it would resemble far more Franz Kafka's nightmarish fables than an orderly, idyllic human existence. What Peirce foresees as a hypothetical long term, an absolutely distant point in time, is sought after as a collective goal when a society seeks complete obedience and uniformity of thought. In that case, the dysphoric ideal is effete, not "lively" mind (Hausman 1993: 148).

So what happens when the embodiment of the ruthless political system that pervades every single aspect of life, an exemplary agent of the State secret police, succumbs to this "*musement*" (CP 6.458, 1908), through a "slippage into iconicity"? What takes over the film's protagonist is a contagion by the light but irresistible suggestion of qualities of feeling that are utterly alien to the rigid habits, the wooden-like pattern of existence that makes of this agent the epitome of "the shield and sword of the Party"?[5] Something momentous develops before our eyes, as we witness a habit change, the emergence of a new belief that goes together with a different disposition to act in a way that was never even entertained before by the character. This is the preamble of a kind of trial of his new habit, which takes place in a scene in an elevator that I describe below. At the heart of this fictional setting, we observe a remarkable consequence of "the tendency to take habits" (CP 6.32, 1891). Thus it seems particularly fitting to describe this scene in Peircean terms as the inception of an *ultimate logical interpretant*, which is the only one, writes Peirce (MS 318, 33–34, 1907, cited by Miller 1998: 73), that does not lead us to yet

[4]In one of its most sinister moments, *DLA* shows a young bureaucrat at the cafeteria of the headquarters of the East German secret police telling a joke about the Secretary of the ruling Party, Erich Honecker. When he realizes that there is a high-ranking officer nearby, the bureaucrat stops dead in his tracks, but it is too late. In an eerily cheerful way he is enticed to finish it, and then his career is over.

[5]That was the motto of the dreaded *Ministerium für Staatssicherheit,* the East Germany secret police popularly known as *Stasi.*

another sign, to another concept; such is the working of a "habit change" in the semiotic: "It can be proved that the only kind of mental effect that can be so produced and that is not a sign but is of general application is a habit-change" (CP 5.476 c. 1906).

Captain Gerd Wiesler is a master spy. The dour middle-aged man is not only an implacable Stasi interrogator, but he also imparts those skills to young apprentices in a formal educational setting. Wiesler is a lone wolf: as he lives alone, he has no confidant, no one with whom he could talk about his momentous change of heart, about the newly acquired habit to gaze at the world with benevolent, humane eyes. So the transformation that we observe in him manifests itself not conceptually, not as something expressed in so many words, but as a different way of organizing his actions, his very being in the world: "It can be proved that the only kind of mental effect that can be so produced and that is not a sign but is of general application is a habit-change" (CP 5.476, c. 1906). That is an instance of the "general application of a belief-habit", an expectation to act in a certain way to attain a specific result; it involves "a modification of a person's tendencies towards action, resulting from previous experiences or previous exertions of his will or acts, or from a complexus of both kinds of cause" (CP 5.476, c. 1906).

In that context, Peirce introduces a key distinction: "the real and living logical conclusion is that habit; the verbal formulation merely expresses it" (MS 318, 751, 1907). No other logical interpretant can be "final", warns Peirce, except for this living habit; it is a tendency that allows for both the tough resistance of otherness and the freewheeling spontaneity, the play of chance in the universe. This is the antithesis of matter construed as "effete mind", this is "lively mind" (Hausman 1993), which consists in patterns that are sufficiently organized so as to be reliable modes of conduct, such are the ways that people keep, because they are trusted to obtain some expected (and satisfactory) results. But they are also flexible enough to allow for spontaneity, for sheer chance to erupt and bring about change, adaptation to new circumstances, no matter how startling or challenging.

The Suggestive Power of Orience in an Elevator: Habit-Change and the Self

Little did Captain Wiesler know that life would be ambushing him with "a *petite bouchée* with the universes" (CP 6.458, 1908), an encounter with the unexpected not only outside but also within him. The real insofar as it functions as a dynamical object comes his way in the shape of an innocent person; only a child could be naïve and curious enough to ask the stern-looking Wiesler point blank a question about his job in very unfavorable terms. Up to that moment, the spy had been dedicated to the State cause to a fault, that is, to prying into people's lives and to hunting mercilessly any deviation from the Law of the land.

What takes place in this lonesome character's existence may be seen as an anticipation of the massive change that would come about only a few years later—the film takes place in 1984—with the fall of the Berlin Wall, and then of the entire socialist system, in a vast portion of the world. Wiesler's imaginative experience can be described as an irresistible "slippage into iconicity", his succumbing to the power of inner signs that take him far afield, and transform him. In his case, the habit change consists in "raising the strength" (MS 318, 361, 1907) of the new habit. He stops being a dedicated spy, a man who had almost literally turned into the "shield and sword" of an authoritarian political system dedicated to perfect an omnipresent human machinery of surveillance and punishment. This is the best possible clarification of the concept of being a different person that we can obtain by observing the character's reaction in a trivial but revealing encounter. In it, we witness a conduct that this Stasi agent would have considered irrefutable evidence of treason, and sufficient cause for imprisonment. As Hookway (2009: 14) writes, "We clarify a concept fully when we know in detail what is involved in having beliefs that contain it; and, since a belief is described as a 'habit of action', clarifying a concept involves identifying the habits involved in such beliefs". So it is time to get to know more about the new habits acquired by this most rigid man, who at first is shown in the film as being impervious to any form of habit-change. It is what we will find out during a brief but eventful trip on an elevator.

Before the spy's actual encounter with the child, we see a ball getting inside the elevator; the irruption of the toy is followed by the appearance of its very young owner. After a day's work, Wiesler is returning to his bleak, drab apartment—an accurate iconic sign of his personality. This chance meeting will reveal a new identity, a different person. It is as if the callous doorkeeper of Kafka's *Before the Law* thawed, and gave place to a sympathetic human being. As the boy starts to question him, inverting thus Wiesler's work routine, we notice a subtle but undeniable transformation in him. After staring at the adult as only children are wont to do, the boy asks Wiesler, who up to that point in time was utterly lost in his own thoughts and oblivious to the other's close presence in the elevator: "Is it true that you work in the Stasi?" After a few seconds that seem to last much longer, the official replies with another question: "Do you know what the Stasi is?" By then, viewers are already familiar with his poised monotone: it is deceivingly bland, lifeless but ominous; one expects something sinister to happen anytime now. The boy does not hesitate to give him a full explanation, one that he got at home, he replies: "Yes, they are bad men who put people in prison, my dad says". The words are uttered with child-like candor, and while the boy talks, he looks Wiesler in the eye. What the boy has said is tantamount to a signed confession, albeit an involuntary one. The man's reply has an even duller, muted tone, if that were possible: "I see". It is easy to predict his next lethal move: to find out the identity of the adult who conveyed the seditious ideas. And the opening of his utterance confirms our worst expectations: "What is the name of…" but then Wiesler interrupts himself. He does not finish his question in the way it seemed certain he would, by behaving as the Stasi high-ranking officer that he has been for so long. Again the pause seems to drag on for a very long time, but the child is curious, and goes at it again: "Of my what?" he asks the stern-looking man, as if unwittingly the child was in a rush to arrive at the woeful denouement. At last, Wiesler finishes his question: "The ball, how is your ball called?" After the fashion of the mythical child who once upon a time exclaimed in front of the vain Emperor that he was not elegant but stark naked, the boy in the elevator can't help but remark, still looking at him in the eyes: "You're funny! Balls don't have names!"

And he is right, of course, soccer balls do not bear names in the way people or even pets do. But that is precisely the work of Orience, of Firstness, of sheer spontaneity in the phenomenology that underlies the semiotic and habit-taking. Faced with this first challenge, after his having slipped into an unexplored realm of uncharted qualities of feeling, Wiesler must come up with unheard-of signs, he finds a verbal novelty that matches his habit-change, his new beliefs. I will now briefly describe the process that leads to the habit-change depicted in *DLA*.

In his 1878 reflection on the suggestive power of the imagination, Peirce explains that what we fancy or envision about ourselves is not only or mostly a mere escape to the realm of fantasy, but a genuine component of our adaptive readiness to act in certain ways. Just like hard, concrete objects of the world—be they oak chairs or theories—have a bearing on our being in the world, and on our continuous endeavor of signifying it correctly, so does the imaginary component of reality. Possibilism is that element that, even in the most unfavorable conditions for the survival of creativity, still enables us to imagine and then express spontaneously our beliefs, doubts, and longings, to conceive of new meanings and experience them as a result of habit change. It is an intangible influence that manifests itself as a light suggestion in the life-world, but becomes an often imperceptible beginning of momentous alterations of it. But how can absolute qualities, which pertain to the "category the First, the idea of that which is such as it is regardless of anything else" (CP 5.66, 1903), have such world-changing effects? Since Firstness neither works forcefully as the "outward clash" (Secondness), nor as an ideal, teleological influence (Thirdness), its mode of action may be construed as a way of enabling our musement (CP 6.458, 1908), which helps generate new meanings that cannot be entirely predicted from those qualities. Thus it comes about through our "considering some wonder in one of the Universes" (CP 6.458, 1908).

Slipping into Iconicity as a Possible Path to Habit-Change

The prevalence of Firstness, even if only fleeting, is what makes those moments in our lives deserve to be described as "aesthetic", albeit their being unrelated with works of art of any kind. In what follows, I intend to show how the imagination is a key component of the self as a sign, a determining influence on its constant growth in complexity. Without it there is little hope to understand oneself and the other, in the almost literal sense of being able to be empathetic with another source of meaning-generation. Regardless of their particular ideological definition, there are historical forms of social organization that make of dualism their existential axis.

Such a reductionistic mode of being, understanding, and existing in the life-world affects the self considerably. In tyrannical political systems, the subversive, creative reintegrative power of synechism, Peirce's logic of continuity, may re-establish imaginatively a state of affairs wherein reasonableness in the life-world may grow once again.

I call the manifestation of the imaginative influence on the self a "slippage into iconicity". This semiotic mechanism allows for a powerful experience of qualitative immediacy, one that is not too different from the joyful, ecstatic feelings that art brings about, but which takes place in relationship with others.[6] It implies the categoreal prevalence of Firstness, of a liberating, spontaneous feeling, an imaginary journey which maximizes the aesthetic, qualitative component of subjectivity, and which thus increases the reasonableness of the self as semiosis, and that of the world that the self inhabits. Superficially it resembles the act of emulating the other, someone who is felt to be admirable. But emulation presupposes both a distinct other, someone who is separate from me, and also a conscious, intentional act to be or to become similar to the inspiring other.[7] "Slipping into iconicity" refers to the free, spontaneous play of qualities that neither belong to anyone nor "involve a second" (CP 1.358, 1887–1888). What gradually and almost insensibly overtakes Captain Wiesler in *DLA* bears no name or definition. As the image of pure *play* that Peirce proposes, it acts as the wind, which "bloweth where it listeth" (CP 6.458, 1908), without any conscious aim. An instance of this freewheeling dimension of experience occurs when, up in the attic of the building where the couple of East German artists Christa-Maria Sieland and Georg Dreyman live, a full surveillance 24/7 secret operation is set up. Wiesler himself is in charge; he must eavesdrop and write the report of one of two daily shifts. One evening, while the implacable sentinel is engrossed in his surveillance, he is gradually drawn into the quality of intimacy of the scene that is taking place some floors below. The actress has not been able to resist any longer the advances of the powerful functionary who blackmails her into consenting to him brutally having his way with her.[8] As she returns to her apartment, the desolate woman begs her lover to embrace her, without asking any questions, and to shelter her from her anguish. While we watch this moment of silent tenderness, vulnerability, and empathy, upstairs we witness a stunning scene: the body of the spy is no longer rigid and focused on his scrutiny; it involuntarily iconizes the posture of the woman who is being comforted. It is important to mention that he cannot see them, as there are only hidden microphones, besides there is barely any word exchanged at that point. It is an experience of sheer Firstness, of a pervasive *tone*—in the acoustic and technical sense of a

[6]This is one of the three-dimensional modes of experience about which Ransdell (2002) gives an enlightening account in his article on the artwork considered from a semiotic perspective.

[7]Emulation is the "ambition or endeavor to equal or excel others (as in achievement)", *Merriam-Webster*, 2008.

[8]The surveillance has been ordered by this powerful figure, in order to get rid of his rival, the playwright Dreyman.

qualisign (CP 4. 537, 1906)[9]—that irresistibly overtakes Wiesler. Thus what began as a tenacious hunt for incriminatory evidence ends up being a liberating, life-changing crisis set off by irresponsible qualities of feeling. Viewers ignore at that point of the narrative what the outcome of this slippage into iconicity may be, since this is a nondeterministic component of semiosis. There is also an experience of listening to music being played below, and of literature—a book of poems that Wiesler steals from the apartment. Such moments are all part of a semiotic process that brings to mind Peirce's own account of his brother Herbert's internal rehearsal of "an imaginary line of conduct, so that that it shall give a general shape to (his) actual future conduct" (CP 5.538, c. 1902). When the right time comes, this character is able to react in a strikingly different manner; there has been a habit change that alters the purport of Wiesler's entire existence.

Despite the man's unchanged appearance—his old *habit* in the sense of a kind of presentation (Miller 1995: 72),[10] "a condition, appearance, dress"—there is a new disposition to act in stark contrast with what appeared to be mineral-like in its steadfastness, but which turned out to be alive and teeming with original possibilities. As Barrena (2007: 270) asserts, creativity is not the exclusive attribute of a few chosen individuals; it can emerge as new meanings in anyone's life. Nevertheless, we should not forget that the creative endeavors that make us free are far from being inexorable; they are a goal to be attained at our own risk because …

> both our consciousness and our freedom are at least in their higher manifestations, not so much gifts as achievements, not so much things conferred upon us from without but things won by dint of our own courage. (…) It takes courage to open ourselves to the possibility of becoming aware of who we are. (Colapietro 1989: 41)

To describe this fictional situation in pragmatic terms, we could say that viewers observe the upshot of what it means to muster the kind of courage needed to fallibly come to terms with the true meaning of the evolving self, insofar as it is as a process of endless, hard-won self-inquiry. Undaunted by what he ever so imperceptibly begins to imaginatively contemplate as wholly unexplored paths of his being, the Stasi agent is increasingly and perilously diverted from the official road of absolute obedience and complete vocational fusion with the ways of the law of a dreary, ever-vigilant State.

Despite its seeming permanence, the self turns out to be "the union of the hazardous and the stable" (Dewey 1925: 62). As any sign, the self bears the liberating and suggestive influence of the Present upon its development. Being a sign, the self partakes of a qualitative nature, as a quality among qualities, but it will momentarily harden when it becomes embodied and caught up in concrete situated actions—such is any particular conduct, which will be then judged to be good or bad, but which will always bear intimations of that subtle instigation without prompting which is the

---

[9]"An indefinite significant character such as a tone of voice can neither be called a Type nor a Token. I propose to call such a Sign a *Tone*" (CP 4.537, 1906).

[10]As it appears in the title of Erving Goffman's classic 1959 monograph: *The Presentation of the Self in Everyday Life*.

human imagination. The working of the Law, be it that of human desire or the motion of the planets, cannot function without the intervention of free-wheeling chance, of spontaneity, which brings about change in spite of or together with all forms of regularity, in any living process. This is the sense of Santaella's (1999: 505) critique of the mechanistic account of causality in the universe, when she describes an alternative, triadic account of the working of determination in Peirce: "In clear opposition to this deterministic notion of causation (=mechanical) based on a conception of law as absolute, Peirce conceived of law as a living power founded on a peculiar conception of habit giving room for chance, growth, and evolution." Indeed, to quote the logician: "the uncertainty of the mental law is no mere defect of it, but is on the contrary of its essence" (CP 6.148, 1892). But what happens when such natural growth is jeopardized by a lethal man-made threat? This is what the story told by *DLA* enables us to observe at close range: how human beings react to such danger, and also what relevance the imagination has in those situations, what kind of effect it has on the evolving self.

As in previous works, I rely on an opposition between human "identity", construed as a concrete interpretative outcome—an effect that is embodied at a specific moment in the history of a person or of a community (such as the role of father or that of civil servant)—and the "self" conceived as a semiotic process that evolves through a recurrent interpretative tendency, whether it be conscious or not, and which produces those concrete socio-historical nodes that are our situated identities, a series of actual forms of being in the world.[11] What takes place in the climax of this fictional narrative is a consequence of "the tendency to take habits" (CP 6.32, 1891).

It seems particularly apt to describe the change observed in that film scene in Peircean terms as an ultimate logical interpretant, the only one at that will not lead us to yet another sign, to a further concept is a "habit change", as Peirce asserts (MS 318, pp. 33–34, 1907, cited by Miller 1998: 73)[12]. Such a change involves a different manner of organizing one's actions: "the real and living logical conclusion is that habit; the verbal formulation merely expresses it" (MS 318, 751, 1907).

In a similar way to the family anecdote twice narrated by Peirce that has young Herbert Peirce as the unexpected man of the hour, through his having previously imagined that kind of emergency together with the most appropriate response to it, the film example portrays inadvertently a central claim of Peirce regarding the evolving nature of habits in our lives. It is a change that involves "a modification of a person's tendencies towards action, resulting from previous experiences or previous exertions of his will or acts, or from a complexus of both kinds of cause" (CP 5.476, c. 1906). To construe mind as "lively matter" (Hausman 1993: 148) entails to think of it as consisting in patterns that are sufficiently organized so as to be efficient modes of conduct, but also flexible enough to allow for spontaneity, for sheer chance to come into the picture and bring about change, the adaptation to new circumstances, no matter how surprising they may be.

---

[11]This theoretical distinction I drew from Wiley's (1994) study of the semiotic self.

[12]My quotes from MS 318 (1907) come from Miller's (1995) discussion of habit change.

Chance, Spontaneity and Variety Active in the Cosmos as It Is in Everyday Life

> Most of men betray endlessly the self that is waiting to be, and to tell the whole truth, our personal individuality is a character that is never entirely realized, an enticing utopia, a secret legend… (Ortega y Gasset 1980[1957]: 32–33)

In this section, I will look at other theoretical aspects of Orience and the evolving self, so as to close this reflection upon the development of the imagination within the theory of habit-taking and habit-change in Peirce's thought, during this selected period of time (1878–1908).

A not too well known influence of Peirce's analysis of experience is that which it had on the theory of the creator of Psychodrama, Jacob-Levy Moreno (1889–1974).[13] A keen critic of Freud, of a model of the psyche he considered too deterministic, the also Viennese Moreno relied on the idea of spontaneity to develop his own theory of the functioning of human identity. When he sets down the principles of psychodrama, Moreno (1946: 9) writes admiringly about "the astonishing references (of Peirce) to spontaneity which remained unpublished long after his death". He then goes on to quote one of the semiotician's accounts of this notion: "Now what is spontaneity? It is the character of not resulting by law from something antecedent" (CP 1.161, c. 1905). In Peirce's architectonic theory of the cosmos, indebted to Darwin's theory of evolution, the original indetermination of spontaneity plays a central role as part of the principle of growth of complexity in the universe:

> By thus admitting pure spontaneity or life as a character of the universe, acting always and everywhere though restrained within narrow bounds by law, producing infinitesimal departures from law continually, and great ones with infinite infrequency, I account for all the variety and diversity of the universe, in the only sense in which the really sui generis and new can be said to be accounted for. The ordinary view has to admit the inexhaustible multitudinous variety of the world, has to admit that its mechanical law cannot account for this in the least, that variety can spring only from spontaneity, and yet denies without any evidence or reason the existence of this spontaneity, or else shoves it back to the beginning of time and supposes it dead ever since. The superior logic of my view appears to me not easily controverted. (CP 6.59, 1892)

For both theories, semiotic and psychodrama, spontaneous change is a constitutive feature of the functioning of the universe and of humankind, something that is valid for human processes as well as for cosmic ones, and which is fully compatible with the regularity that is the upshot of all kinds of law. Firstness, which is presupposed by spontaneity, accounts for the introduction of the new in a way that is inseparable from the working out of tendencies, of the law-like behavior that determines the operation of habits, but does so in a non-mechanical fashion (CP 1.174, c. 1905). This serves to correct Moreno's (1946: 9) misconception that Peirce reduced spontaneity to sheer randomness. For the psychodramatist (1946:

[13] I owe this reference to Michel's (2006) research on the semiotic study of subjectivity.

81), to be spontaneous implied the capacity to face new situations or unexpected challenges in an adequate manner.

What was missing from Moreno's critique regarding the notion of spontaneity in Peirce is the way in which chance combines with the ordering tendencies in the semiotic; it does not posit a dualistic and thus reductionistic opposition between the two components. The following quote provides a fitting reply to Moreno's objection about spontaneity in Peirce, which is inseparable from the notion of habit-taking, because there is the seed of regularity in the former:

> To undertake to account for anything by saying baldly that it is due to chance would, indeed, be futile. But this I do not do. I make use of chance chiefly to make room for *a principle of generalization, or tendency to form habits, which I hold has produced all regularities*. The mechanical philosopher leaves the whole specification of the world utterly unaccounted for, which is pretty nearly as bad as to baldly attribute it to chance. I attribute it altogether to chance, it is true, but to chance in the form of a spontaneity which is to some degree regular. (CP 6.63, 1892—emphasis added)

Miller's (1995) article, "Peirce's conception of habit", is akin to my own take on the emergence or Orience of habits. She accounts for "emergent meaning" (1995: 75) in relation with Peirce's attempt to define the "ultimate logical interpretant" as a sign whose nature cannot be "of an intellectual kind" (MS 318: 33–34, 1907). In that context, Peirce develops the idea of "a habit change, a modification of a person's tendencies towards action, resulting from previous experiences, or from previous exertions of his will or acts, or from a complexus of both kinds of cause" (CP 5.476, c. 1907). In the 1907 manuscript (MS 318), Peirce claims that such a habit change "often consists in raising or lowering the strength of a habit", as I wrote above in reference to the film example. To account for a change that is not fathomable if we were to identify habits exclusively with laws, Miller (1995: 74) differentiates habit from the far more firmly entrenched notion of law, by asserting that we should characterize habit "as a tendency to act, (which) is generality-in-the making, the mediation between first and second which is the instituting of a third". On the other hand, law, she claims (ibid.), "appears to be the deepest degree of habit".

However, the proposed distinction between the two related concepts of Thirdness—habit and law—is not so clear in a text written a few years before. In the *Minute Logic* (1902), Peirce gave a thorough description of the process whereby new habits come about in human beings. It is an inferential process that combines perception, imagination, and desire, and whose outcome is law-like and inseparable from future, potential action:

> An expectation is a habit of imagining. A habit is not an affection of consciousness; it is a general law of action, such that on a certain general kind of occasion a man will be more or less apt to act in a certain general way. An imagination is an affection of consciousness which can be directly compared with a percept in some special feature, and be pronounced to accord or disaccord with it. (…) Of course, every expectation is a matter of inference. What an inference is we shall soon see more exactly than we need just now to consider. For our present purpose it is sufficient to say that the inferential process involves the formation of a habit. For it produces a belief, or opinion; and a genuine belief, or opinion, is

> something on which a man is prepared to act, and is therefore, in a general sense, a habit. A belief need not be conscious. (…) For every habit has, or is, a general law. Whatever is truly general refers to the indefinite future; for the past contains only a certain collection of such cases that have occurred. The past is actual fact. But a general (fact) cannot be fully realized. It is a potentiality; and its mode of being is *esse in futuro*. The future is potential, not actual. What particularly distinguishes a general belief, or opinion, such as is an inferential conclusion, from other habits, is that it is active in the imagination. (CP 2.148, 1902)

What comes out most vividly in this passage is the pivotal role of the imagination in habit-change, in the evolving process whereby what seems most regular and predictable has in its womb, as it were, the light, suggestive seeds of the new that will produce "infinitesimal departures from law continually" (CP 6.59, 1892). In Peirce's recollection of the family incident that has his very young brother acting in a surprising way and in the film scene that depicts a character whose identity has hardened to the point of becoming a quasi "effete mind" due to "inveterate habits" (CP 6.25, 1891), we witness the impact of the imagination on habit-change. Between these two instances there is a wide range of habit-taking processes: Herbert Peirce's conduct embodies the unfettered plasticity of childhood, the power of the imagination at its height; the German film is an allegory of the forlorn fate of humanity when it is forced to lead a shackled existence, when life is a realm where spontaneity is under constant threat. But the boy's rehearsal of "an imaginary line of conduct" as a preparation for his "actual future behavior" (CP 5.538, c. 1902) in Peirce's text is echoed in the slippage into iconicity of the Stasi agent, as the self of the latter undergoes a momentous change, which ends up in the acquisition of a new habit. The social context in which the fictional example takes place favors the degradation of lively matter into quasi-effete mind, a situation in which "objects become material by losing the plasticity and openness of mind, taking on the fixity of materiality by virtue of developing fixed habits" (Miller 1995: 74). The plasticity at the origin of the heroic deed in the Peirce family anecdote, and the fictional transformation of a most rigid disposition of the self into an open one through an immersion in possibilism, both point out the centrality of the imagination in the endless human struggle against "the slavery of inveterate habit", because "the highest quality of mind involves a great readiness to take habits, and a great readiness to lose them" (CP 6.613, 1893).

I would like to highlight Miller's (1995: 74) notion of "emergent meaning" in relationship with habits as dispositions to act, not yet solidly entrenched as genuine laws, inasmuch as habit-change takes place against an older, previous "pattern of expectations" (1995: 76). Both Herbert Peirce and the character depicted in the film allegory of a historical event reach their momentous decisions to act in an unheard of manner after having tried imaginatively a different identity. In both cases, it is that of being a savior of the other in a life-endangering situation. Likewise, we may try in the imagination a different style of clothes, a new appearance or look, to prepare or rehearse mentally for a momentous change in our life. In my examples, the courageous act of rescuing the Other is the ultimate logical interpretant of the evolving self. Be it a real person in life-endangering distress or a fictional innocent human being who unwittingly exposes his family to incarceration in the dungeons of the secret police for betrayal, the habit-change implies an increase of concrete reasonableness, of a growth in self-control through an imaginary exercise.

Concluding Remarks: Habit-Taking Without the Imagination Is Unimaginable

There is an often-quoted image that Peirce uses to describe the teleological functioning of the laws in the cosmos and in society. Just as any court of law needs the physical enforcement of a sheriff, so does final causation need to be implemented by the brute haecceity of efficient causation (CP 1.213, 1902). But then the logician decides to raise the stakes, and he gives us an extreme account of what a lawless universe (one not guided by teleology) would be like: "Efficient causation without final causation, however, is worse than helpless, by far; it is mere chaos; and chaos is not even so much as chaos, without final causation; it is blank nothing" (CP 1.220, 1902). In a similar fashion, we could claim that without the vague, freewheeling but powerful activity of the imagination, our habit-taking would inevitably turn into a deadening mechanism, a process that instead of helping us to adapt to the changing circumstances of the world would prevent us from coming up with new ways of attaining our purposes. The self would be chained to a single, unchanging, and unchangeable identity. This dreadful, inhuman outcome is something about which Peirce is fully aware, when he criticizes "the old psychology which identified the soul with the ego", and proposes an alternative account that allows for the work of the imagination in the dialectical relationship between the self and the successive identities that emerge as localized interpretations: "the soul may contain several personalities and is as complex as the brain itself, and that the faculties, *while not exactly definable and not absolutely fixed*, are as real as are the different convolutions of the cortex" (CP 1.112, c. 1896—emphasis added). The lack of a fixed identity, one that is defined in absolute terms forever is essential for habit-taking to serve as an adaptation mechanism.

The purpose of the chapter that now concludes was to bring to the reader's attention the vital role of the imagination, the Orience or Firstness of habit-taking, without which people—as well as the universe that we inhabit—would be condemned to the rigidity of matter, of mind enslaved by "inveterate habits". Just as important as final causation for the "working out of results in this world" (CP 1.220, c. 1902) is the imagination for habit-change in the inferential process of habit-taking that creates regularity in nature and in culture. It is through this blending of the imaginary, the actual, and the general that the semiotic with its phenomenological basis, phaneroscopy, and pragmaticism, all appear as the inseparable components of the elaborate theory that Peirce built throughout his life. Concerning the specific role of habit, of the beliefs that are inseparable from them and from our conduct in life, I hope to have shown that that this particular period in his life, the one that goes from 1878 to 1908, is a most fruitful one. In his mature elaboration of the semiotic, the centrality of Peirce's definition of the ultimate logical interpretant as a habit is nicely brought out by Alston (1956), when he claims that the "production of this pragmatic interpretant" is "a condition for meaningfulness" (1956: 87).

My approach emphasized the relevance of that flight into the imagination, which Peirce called "musement" at the end of the period under study (CP 6.458, 1908), when he was trying to formulate the most enduring upshot of semiosis, namely, our conduct, the ways in which we try to adapt to the environment, whether it be natural or sociocultural. I think that the validity of the Peircean account of meaning-generation as being inseparable from the fruits that this logical process bears in our experience, in a realm that we share with others, depends to an equally important degree on the emergent, varying meaning that is affected by the imagination, the Orience or genesis of habit-change. It is what makes us capable of adapting to the most challenging circumstances through a rehearsal with inner signs in continuity with the outer signs, that is, with percepts, and also with the regularities of the world in a synechistic fashion.

I would like to leave the closing statement on this still not fully acknowledged component to Peirce. According to the semiotician, what enables us to become reasonable and reasoning beings is "the development of concrete reasonableness" (CP 5.3, 1902). The following quote also belongs to the period in which Peirce works on these ideas on habit-taking and habit-change:

> People who build castles in the air do not, for the most part, accomplish much, it is true; but every man who does accomplish great things is given to building elaborate castles in the air and then painfully copying them on solid ground. Indeed, the whole business of ratiocination, and all that makes us intellectual beings, is performed in imagination. Vigorous men are wont to hold mere imagination in contempt; and in that they would be quite right if there were such a thing. (…) Mere imagination would indeed be mere trifling; only no imagination is *mere*. "More than all that is in thy custody, watch over thy phantasy," said Solomon, "For out of it are the issues of life." (CP 6.286, c. 1893—emphasis in the original).

References

Alexander, Thomas A. 1990. Pragmatic imagination. *Transactions of The C. S. Peirce Society* 26 (3): 325–348.

Alston, William P. 1956. Pragmatism and the theory of signs in Peirce. *Philosophy and Phenomenological Research* 17(1): 79–88.

Andacht, Fernando. 1996. El lugar de la imaginación en la semiótica de C.S. Peirce. *Anuario Filosófico* 29(3): 1265–1289.

Andacht, Fernando. 1998. On the relevance of the imagination in the semiotic of C.S. Peirce. *Versus. Quaderni di studi semiotici* 80–81: 201–228.

Andacht, Fernando. 2000. the other as our interpretant. *S. European Journal for Semiotic Studies* 12(4): 631–655.

Andacht, Fernando, and Mariela Michel. 2005. A semiotic reflection on self-interpretation and identity. *Theory and Psychology* 15(1): 51–76.

Andacht, F., & Mariela Michel. 2007. El turista accidental: el cine como ensayo icónico-simbólico sobre la identidad humana. *Colección de Semiótica Latinoamericana, 'Semióticas del cine'*, 23–40. Maracaibo: Asociación Venezolana de Semiótica/Laboratorio de Investigaciones Semióticas/Universidad del Zulia.

Andacht, Fernando, and Mariela Michel. 2009. The predictable and accidental journey of the self as semiosis. In *Semiotics 2008—specialization, semiosis, semiotics,* ed. John Deely and Leonard G. Sbrocchi, 347–362. Ottawa: Legas Press.

Barrena, Sara. 2007. *La razón creativa*. Madrid: Rialp.

Colapietro, Vincent. 1989. *Peirce's approach to the self. A semiotic perspective on human subjectivity*. Albany: State University of New York Press.

Colapietro, Vincent. 1990. The integral self: Systematic illusion or inescapable task? *Listening* 25 (3): 192–210.

Dewey, John. 1958[1925]. *Experience and nature*. New York: Dover Publications.

Hardwick, Charles S., ed. 1977. *Semiotics and significs: The correspondence between Charles S. Peirce and Victoria Lady Welby*. Bloomington: Indiana University Press.

Hausman, Carl R. 1979. Value and the Peircean categories. *Transactions of the C.S. Peirce Society* 15(3): 203–223.

Hausman, Carl R. 1993. *Charles S. Peirce's evolutionary philosophy*. New York: Cambridge University Press.

Hookway, Christopher. 2009. Habits and interpretation: Defending the pragmatist maxim. Paper presented at the *Peirce and Early Analytic Philosophy Symposium*, Helsinki Peirce Research Center. http://www.helsinki.fi/peirce/PEA/. Accessed 5 May 2014.

Michel, Mariela. 2006. O self semiótico: desenvolvimento interpretativo da identidade como processo dramático. (PhD Thesis unpublished). Universidade Federal do Rio Grande do Sul, Brasil. Available in: http://hdl.handle.net/10183/7783 . Accessed on July 10th, 2014.

Miller, Marjorie. 1995. Peirce's conception of habit. In *Peirce's Doctrine of Signs: Theory, Applications, and Connections*, ed. Vincent Colapietro and Thomas M. Olshewsky, 71–78. Berlin: Mouton de Gruyter.

Moreno, Jacob L. 1946. *Psychodrama*, vol. 1. New York: Beacon House.

Ortega y Gasset, José. 1980[1957]. *El hombre y la gente*. Madrid: Revista de Occidente.

Peirce, Charles Sanders. i. 1867–1913. *Collected papers of Charles Sanders Peirce*. Vols. 1–6, ed. Charles Hartshorne and Paul Weiss. Cambridge: Harvard University Press, 1931–1935. Vols. 7–8, ed. Arthur W. Burks. Cambridge: Harvard University Press, 1958. [References to Peirce's papers will be designated by CP, followed by volume, period, paragraph number.].

Ransdell, Joseph. 1989. Teleology and the autonomy of the semiosis process. *Arisbe: The Peirce Gateway*. http://members.door.net/arisbe/. Accessed 10 Nov 2005.

Ransdell, Joseph. 2002. The semiotical conception of the artwork. In Paper presented in *Advanced Seminar on Peirce's Philosophy and Semiotics*, Graduate Program of Communication PUC-SP, Sao Paulo.

Robin, Richard. 1967. *Annotated catalogue of the papers of C. S. Peirce*. Worcester: University of Massachusetts Press.

Santaella, Lucia. 1999. A new causality for understanding the living. *Semiotica* 127(1/4): 497–519.

Wiley, Norbert. 1994. *The semiotic self*. Chicago: University of Chicago Press.

Chapter 20
Of Habit and Abduction: Preserving Ignorance or Attaining Knowledge?

Lorenzo Magnani, Selene Arfini and Tommaso Bertolotti

Abstract "Habit" is not an easy term in Peirce's epistemology: on the one hand it often signifies the rule of action that is attained with the fixation of belief (1877) [EP 1: 109–123]; on the another hand, it is also described as an almost instinctual process that determines further reasonings, the element "by virtue of which an idea gives rise to another" (1873) [CP 7.354]. Stressing the apparently wide separation between these two traits of habit in the epistemic continuum between doubt and belief, we will be able to illustrate (a) a knowledge-based kind of habit, for the analysis of which we will also exploit Gibson's concept of "affordance" (1950), which also plays a pivotal role in the justification of the agent's own beliefs; and (b) an ignorance-based kind of habit, which will be *proved as necessary* for the beginning of thought, and which is at the base of the ampliative reasoning, condensed in another Peircean key topic (often qualified as "instinctual" in his writings): abduction.

Keywords Epistemology · Abduction · Doubt · Belief · Gibson · Affordance

Introduction: The Double Meaning of "Habit"

Among the terms composing Peirce's theoretical lexicon, the "habit" is endowed with an undoubtedly strong logical, epistemological, and cognitive value which justifies its various applications. Usually it expresses the idea of "some general

---

L. Magnani (✉) · T. Bertolotti
University of Pavia, Pavia, Italy
e-mail: lmagnani@unipv.it

T. Bertolotti
e-mail: bertolotti@unipv.it

S. Arfini
University of Chieti and Pescara, Chieti-Pescara, Italy
e-mail: selene.arfini@unich.it

D.E. West and M. Anderson (eds.), *Consensus on Peirce's Concept of Habit*,
Studies in Applied Philosophy, Epistemology and Rational Ethics 31,
DOI 10.1007/978-3-319-45920-2_20

principle working in a man's nature to determine how he will act" (1902) [CP 2.170].[1]

This is also the idea of habit received by William James, informing the pragmatist tradition. It is a *behavioural* notion of habit, related to how a person is able to react in a habitual way to a given array of stimuli because of some culturally-impressed, or self-impressed, inscription at the neural level.[2] James also relates habit to an issue that will prove crucial for our present inquiry, that is its ability *to avoid being entangled in situations of perennial indecision*, so that we need to concentrate less thought on trivial aspects of life.

> There is no more miserable human being than one in whom nothing is habitual but indecision, and for whom the lighting of every cigar, the drinking of every cup, the time of rising and going to bed every day, and the beginning of every bit of work, are subjects of express volitional deliberation. Full half the time of such a man goes to the deciding, or regretting, of matters which ought to be so ingrained in him as practically not to exist for his consciousness at all. [James (1892/1920) 1983:145]

When reworked by Peirce, "habit" becomes a richer and more intricate concept that both bears the trustworthiness of reasoning and comes to represent the actual meaning of the fixation of belief. To be more clear, if on the one hand habit signifies the rule of action that the *fixation of belief* implies (1877) [EP 1: 109–123], on the other hand it is also described as an almost *instinctual process* that determines a more or less correct reasoning, "by virtue of which an idea gives rise to another" (1873) [CP 7.354].

This hiatus within the notion of habit begs for an analysis. It is placed at the core of the complex interplay between the irritation of doubt and the fixation of *belief*, the latter described by Peirce as the aim of thought. We can see the habit both as the *final outcome* of the inferential activity of the mind (achieving, thanks to belief, a rule for action), and the *initial push* that drives the agent to the irritation of doubt—and hence to the attainment of belief itself. In both these cases, the habit is seen as a necessary element that either completes or drives the mechanism of the agent's reasoning.

Stressing such double-nature of the concept of habit, it must be said that we can follow two paths of reasoning. We can see it as just one phenomenon of which

[1]To be more accurate, this general definition only refers to Peirce's epistemological and psychological analysis of habit, as our study is structured within it. Indeed, it would be difficult to encompass Peirce's many uses of the concept of habit in a single definition, no matter how broad. As many contributors in this volume highlighted, habit is "by no means exclusively a mental act" [Coletta, this volume], nor a notion that belongs just to the analysis of human or animal cognition (even if it amply regards emotion, experience, and understanding [Gorlee, this volume]); indeed, it is a concept used by Peirce and following researchers also in the philosophical study of physics and biology to comprehend natural disposition [Stjernfelt, this volume], physical laws [Pickering, this volume], and regularities as energy dispersal and biological system propagation [West, this volume]. Moreover, it also appears to be a relevant concept in the Peircean semiotic triad of *Firstness*, *Secondness* and *Thirdness*: for a thorough analysis of this topic, we refer the reader to West [2014].

[2]Indeed James' psychological treatment stresses the neural grounding of habit.

Peirce highlights two different roles, with respect to the different subjects about human reasoning he is analyzing (that is as the way the fixation of belief is externalized and the conditions that affect the irritation of doubt). Or, we can think of it as two different states of human cognitive activity that share the same broad definition but that must be studied separately. Despite the fact that they are both legitimate analyses, could we ask which one of these is the most correct and fair? Is it possible to choose one ignoring the other for the sake of straight argumentation?

This question, which may cause ontological or philological issues, becomes much more interesting from an epistemological perspective. In our opinion, indeed, we can usefully put together the possible answers and see the Peircean process of thought in this new light. We can effectively say that the complex state of mind that habit implies presents two important roles in the creation and modification of beliefs. The distinction between these two roles is so neat that it leads us into thinking that they are totally different stages of the reasoning process. If indeed we stress this apparently wide separation and arrange the results in the epistemic continuum between doubt and belief, we can take advantage of a rare opportunity to shed more light on the mechanism of belief creation and change in the Peircean framework.

Thus, in order to strike a balance between the two aforementioned views, we will focus on the separation between a pre-inferential and a post-inferential kind of habit and then we will see how they interact in Peirce's model of thought. Using this strategy, and focusing on the duplicity of habit in an epistemic dynamics, we can individuate:

- a **knowledge-based kind of habit** (for the explanation of which we will also use Gibson's concept of "affordance" (1950)) which plays a pivotal role in the justification of the agent's own beliefs;
- an **ignorance-based kind of habit**, which will be demonstrated as necessary for the beginning of thought, and which is at the base of the ampliative reasoning, condensed in the concept of abduction.

Ampliative and Non-Ampliative Reasoning: From Peircean A-B Reasonings to A-B Habits

We can start by considering Peirce's simplest description of habit—that is a "general principle working on a man's nature to determine how he will act" (1902) [CP 2.170]. In other words a habit, of which the agent can be more or less aware, is what *drives reason* to follow a certain path of action. Obviously, the entanglement between reasoning and action is a headline of the pragmatistic view, from which our analysis starts and where the double nature of habit can be productively investigated. So, when the habit is described as a rule of action, often it also stands for "habit of reasoning" and vice versa: it is something that, affecting our thought, also makes us choose a pattern for action and, affecting our behavior, it modifies also our way of thinking. As recently reminded by Ippoliti, there are just "two main roots of logic and reasoning: ampliative reasoning, heuristics, and methods for discovering on one hand, and non-ampliative reasoning, deduction, and methods for justifying and grounding our findings on the other" [Ippoliti 2015: 1]. This dichotomous perspective should help us to understand the useful separation that can be found in the Peircean structure of thought between the two kinds of habit (of reasoning and action) illustrated above.

Following this dichotomy, we can present again the distinction between types of habit we are interested in. Indeed, what we call the knowledge-based kind of habit refers to a justificatory, non-ampliative reasoning, which characteristically supports actions that can reinforce the believing system of the agent. Alternatively, what we call the ignorance-based kind of habit forces the agent to adopt an ampliative reasoning and so eventually to expand knowledge through actions that let the agent attain new information and create new beliefs. Interestingly, the two kinds of habit are different with respect to the level of knowledge they attain and use. On this issue, an interesting definition given by Peirce is worth quoting:

> Your reasonings are determined by certain general habits of reasoning, each of which has been, in some sense, approved by you. But you may recognize that your habit of reasoning are of two distinct kinds, producing two kinds of reasoning which we may call A-reasoning and B-reasonings. You may think that of the A-reasonings very few are seriously in error, but that none of them much advance your knowledge of the truth. Of your B-reasonings, you may think that a large majority are worthless their error being known by their being subsequently found to come in conflict with A-reasoning. It will be perceived by this description that the B-reasonings are a little more than guesses. (1902) [CP 2.189]

The division between A-reasonings and B-reasonings reinforces the ampliative and non-ampliative dichotomy with a further specification: the exploitation of the idea of different "habits of reasoning" as the condition of a certain kind of thinking is what influences the *knowledge attained* and the *knowledge used* in the cognitive processes. The B-Reasonings—characterized by the expression "little more than guesses"—do not always lead to certainty and truth, but certainly contribute to an extension of knowledge that is impossible to obtain with the correct and certain A-reasonings.

Nevertheless, it is *more likely* that reasonings of A-type are valid inferences compared with a B-type of reasoning, the latter being justified by *logica utens*[3] (a general and individual theory of what good reasoning is) but cannot be employed in a scientific argument, for instance.[4]

---

[3]On Peirce's concept of habit in the distinction between his definition of *logica utens* and *logica docens*, cf. Pietarinen (2005).

[4]Even if it is very natural to identify "valid inferences" with "deductions", we should say that A-type reasonings do not have to be necessarily considered as valid deductions. Indeed, the agent can apply justificatory A-reasoning even while generating fallacies that *in specific cases* lead from correct (true) premises to correct (true) answers. In these controversial cases, and in a practical sense, some fallacies can effectively justify some beliefs and so they can be correctly included in

Now, in order to further clarify how the knowledge-based habit effectively produces A-reasonings, and how the ignorance-based one causes B-reasonings, we will analyze in the following subsection why Peirce defines the *formulation of a habit in a universal proposition* as a "guiding principle of inference".

The Guiding Principle of Inference and Its Possible Problematic Results

According to Peirce, the "guiding principle of inference" is the logical base of a habit, consisting in a universal proposition that is the formulation of the same habit (1877) [EP 1: 112–113]. Such formulations "guide" the inferences in modalities that depend on the type of habit. The non-ampliative and justificatory reasonings (A-reasonings) are driven by a guiding principle of inference that is entirely derived from the knowledge that the agent already has. On the contrary, ampliative reasonings (B-reasonings) have to start from a guiding principle that has only its premises in the knowledge of the agent, but that formulates its conclusion as "little more than guesses". In other words, an A-reasoning is focused on the agent's system of beliefs: it simply applies notions that the agent *already* thinks are true, and from which the guiding principles of inference are derived.

B-reasonings, instead, should only start as a kind of uncertain *analogical correlation* supported by a "habitual" model of thought and action. Surprise or a physical effort[5]—according to Peirce—may break the correlation, and the principle of inference becomes an ampliative reasoning (as we will see, an abduction), which produces eventually causal knowledge for a given phenomenon.[6] At the end of

(Footnote 4 continued)

the reasonings of A-type. This point of view about fallacies follows not only the Peircean perspective towards "valid inferences" and their related habits (1877) [EP 1: 112], but also the recent theory against the EAUI conception of fallacies originating in the "Agent-Based Logic"—also called "New Logic"—developed by Gabbay and Woods (2001) and Woods (2007). According to the classical EAUI conception, fallacies are negatively considered to be "Erroneous", "Attractive", "Universal", and "Incorrigible". In the agent-based perspective advocated by Gabbay and Woods, instead, fallacies are still attractive, universal, and incorrigible also because they can often drive the agent to adopt intelligent and practical solutions (cf. also Woods 2013). We will further analyze this aspect in the next subsection, illustrating the consequences of a fallacious A-reasoning in terms of habit creation: indeed, if a useful habit emerges from an efficacious, even if fallacious, justificatory reasoning, it will certainly contribute to make the initial A-reasoning more "attractive" and "incorrigible".

[5]Peirce's stress on physical effort is related to an embodied perception of surprise that may not be totally available to consciousness, as a "full" surprise would be: the result of a mismatch between one's beliefs and the external world can be tacitly revealed by the increased physical effort required to carry out the planned action, when it is driven by a habit that is not valid anymore.

[6]We will further analyze the mechanism that drives the ignorance-based habit to the creation of an abduction in Sect. 4.1.

these processes, the agent has either attained the justification of some of his beliefs by habits and reasonings of the *A-type* or he has changed some of them (no longer adequate) through *B-type* habits and reasonings.

Summing up, the knowledge-based habit is what we have first described as the rule of action implied by the fixation of belief: it is the final result of thought, it structures the subsequent actions of the agent on the basis of his beliefs. Ignorance-based habit, instead, allows the agent to produce an abduction (the creative inference most extensively analyzed by Peirce), in order to solve a problematic "breakthrough" of a doubt (caused by surprise, physical effort, or the like).

Obviously, both kinds of reasoning (emerging from the two types of habit) can be incorrect, yet there is a difference about what an error of reasoning is if it occurs because of the first kind of habit or of the second one. Specifically, the justificatory habit can force the agent to adopt a "fallacious" behavior defending an only apparently right belief. For example, making him buy products using a "bandwagon effect": he believes, incorrectly, that the more a product is sold, the more it is worth buying. So, he developed a habit to buy products based on how many people have already bought them. It is, of course, a known fallacy, but as long as that habit does not fail (i.e. making him buy an unsatisfactory product), he is going to defend its guiding principle ("the more it is sold, the more you should buy it") even if it is based on a fallacious belief.

Instead, the ampliative habit can always lead to an abduction that maintains the agent ignorant about some aspects of the topic investigated. For example we can speak about the performance of manipulative abductions, that is, when the agent is *thinking through doing* and not only, in a pragmatic sense, *about doing* (cf. Magnani 2009: 233). Let us say that the agent believes himself to be an advanced-level user of a computer writing program, but suddenly faces a problem never encountered. Using what he knows about that computer program, he has to figure out the solution "trying" different combinations of keys and using heuristics of unknown effect. In the end, after solving the problem, he will react to a similar situation with the solution made up during the manipulative process, but without knowing why that method was effective and not others. The agent probably could not have tried all possible methods, but only stopped when reaching the first effective one. The habit that guaranteed the adoption of a manipulative abduction is effective when it contrasts the agent's incorrect certainty, but does not guarantee a complete ignorance-free solution.[7]

The separation between the two species of error can be better understood if we think about the contrasting meanings of the Latin word "errare" (the etymological root of error), as recently highlighted by Boumans and Hon (2014): indeed, it could stand for "to go this way and that, to walk at random" or "to go off the track, to go astray". In this sense, when the principle of inference of a knowledge-based habit gives birth to erroneous reasoning, we have to analyze the belief from which it

[7]On the ignorance-preserving trait of abduction, cf. Woods (2013), Magnani (2013), Aliseda (2005).

started, where the agent has committed the error of "going off a specified track", that is believing a fallacious notion instead of a logically legitimate one. The error, instead, implicit in an ampliative B-reasoning, is condensed in the ignorance-preserving trait of the inference itself. The knowledge added thanks to the ampliative reasoning has not eliminated the ignorance that the agent still possesses in a given field, but it has just filled a gap to answer a specific problem. The error here consists in the fallibility of the inference itself: we have to remember that B-reasonings are also defined as "little more than guesses" and one's going wrong comes from "walking at random" through the more or less valuable possibilities. This aspect, which represents the major limit of abductive reasoning, is also the key to comprehend its broad application to various fields of knowledge.

Thus, in the next sections, in order to better understand the specific differences between the two kinds of habit (and their related reasoning), we will introduce the interplay between mental states of doubt and belief in Peircean epistemology to see, specifically, how the two kinds of habit allow such interplay. Then, we will deal with the epistemological value of belief as a knowledge container (from which the knowledge-based kind of habit emerges) and the role of doubt as the first step out of a problematic situation (guided by the ampliative and ignorance-based type of habit).

A Knowledge-Based Habit: Beliefs and Affordances

The role of the dynamics between doubt and belief is pivotal to understand human inferential reasoning, as highlighted by Peirce particularly in the articles published in 1877 and 1878, "The Fixation of Belief" and "How to Make Our Ideas Clear".

> Doubt and Belief, as the words are commonly employed, relate to religious or other grave discussions. But here I use them to designate the starting of any question, no matter how small or how great, and the resolution of it. (1878) [EP 1: 127–128]

Indeed, doubt and belief represent the starting point and the wanted conclusion of thought, its dwelling extremities and its elementary parts. In particular, Peirce focused his attention on the modalities through which the psychological and emotional conditions connected to the states of doubt and belief profoundly affect the rational dynamic itself. Indeed, the mechanism that drives our thoughts is mostly guided by the fact that we *repel* the *irritating* state of doubt and *wish for* attaining a *quiet* state of belief. It mostly annoys us to remain in a state of doubt, even if it is necessary, and want a functional answer, even if it is not the most correct one.

As Peirce wrote: "It is true that we do generally reason correctly by nature. But that is an accident [. . .] We are, doubtless, in the main logical animals, but we are not perfectly so. Most of us, for example, are naturally more sanguine and hopeful than logic would justify" (1877) [EP 1: 112]. Instead of explaining the tendency of the human mind to produce beliefs that are often less than logically correct as a mere side-effect of limited human nature, Peirce tried to understand why we rather prefer having an unwarranted belief than a rationally extended state of doubt.

In order to describe this (almost instinctual) problematic trait, he claimed that our epistemic dynamic is driven more by its *psychological and emotional consequences*, than by the will to formulate always correct inferences. He focused on the fact that moving from the state of doubt to belief is satisfactory and relieving per se, as passing from a very stressful situation to a moment of rest. Hence, in order to see how knowledge-based habit works, we are going to examine the situation of momentary rest for thought that is belief, with all its emotional and cognitive consequences.

Belief: Habit as a Rule for Action

As already mentioned, the mental state of belief is a peaceful one. Thought has reached its aim: it has found an answer to a problem, and the agent can proceed to act *as if* that answer were completely correct and functional (while, obviously, it might not be the case). In fact, this mental state gives the agent a rule for action and this is what makes a state of belief so pleasant: the agent thinks that he has the knowledge to perform an action, and so feels confident to execute it.

The rule for action attained in a state of belief is, indeed, the knowledge-based habit we mentioned before. It is the outcome of the belief that actually *verifies* its feasibility. When Peirce calls the mental state of habit "a rule for action", he is not just connecting thought and action in a strict causal relation, but he is also indicating the extension of the mental state in the empirical world as a natural, hence satisfactory, component of the former. A belief proves functional when it reaches a practical meaning and gives the agent a rule for action, a habit that affects behavior and future choices. It is simple: since we think our beliefs to be certain (if not, we would have no reasons to believe them), we have all the reasons to apply the principles they suggest.

This thought-action connection looks functional, but there is a catch: while the whole of our knowledge is set up by our beliefs, our beliefs are not always grounded on knowledge. In this sense, we call this a *knowledge-based* trait of habit since it relies on the structure of *confidence* the agent possesses about his own knowledge. Indeed, when a habit is the result of a "static affirmation of a principle" (represented by the state of belief in Peirce's writings), it makes the agent confident about believing. There is no place for doubt in this kind of habit: the agent is driven by trust in his own beliefs to repeatedly apply them practically.

Obviously, though, what permits us to use the "knowledge-based" term in relation to a believing state, is ultimately the analysis of the duplicity that habit has in Peirce's writings.[8] Here, knowledge refers to the dynamics between the

[8]In the past few decades, many philosophers and logicians focused on the relationship between the amount of knowledge the agent thinks he possesses and the amount he actually has; in a case, these studies have led to the creation of the so-called "New Logic" and the research on the epistemic dimension of a realistic agent [Gabbay and Woods 2001]. This research eventually reached the

emotional states that compound the inferential process of the agent, and not to *episteme* as scientific, or inter-subjectively valid, "knowledge". Hence, this perspective allows us to see the satisfactory state of belief as justifiable and legitimate from the agent's point of view and, at the same time, permits us to focus on the process that creates habit based on belief. If the agent has no need to doubt his own beliefs, he can apply the believed principle without fear of being wrong about it. So, he puts himself in a habitual repetition of the same belief. As Peirce pointed out, this is the essence of the believing state itself:

> And what, then, is belief? […] We have seen that it has just three properties: first, it is something that we are aware of; second, it appeases the irritation of doubt; and, third, it involves the establishment in our nature of a rule of action, or, say for short a *habit*. […] The essence of belief is the establishment of a habit, and different beliefs are distinguished by the different modes of action to which they give rise. (1878) [CP 5.397]

At this point, in order to show how this kind of habit is supported by an amount of knowledge, we are going to introduce the concept of affordance introduced by Gibson (1950, 1977) and the capacity of some knowledge-based habits to frame the agent's perspective on the grounding of his beliefs.

Learning Affordances as Knowledge-Based Habits

First of all, we have to set up the conceptual environment where the theory of affordance finds its place and value. In the most general definition, affordances are specific clusters of information that an agent can find in particular objects merely through perception: affordances inform the agent about the adaptive value of the object or the event being observed or manipulated. The information is not received and elaborated by one specific sense but perceived in an embodied way (cf. Mace 1977). The eco-cognitive epistemological analysis of the inferential processes of the mind—as claimed by current epistemologists (cf. Magnani 2009) and promoters of the extended cognition paradigm (as Clark 2008)—has permitted the recognition of the affordance theory as a conceptual tool, not only for investigating (mostly, but not only) human perception, but also (mostly, but not only) human manipulation and distribution of cognitive meanings in suitable environmental supports.

(Footnote 8 continued)

concept of "epistemic bubble" to explain the complex interplay between knowledge and belief [Woods 2005, 2013; Magnani and Bertolotti 2011]. Indeed, the epistemic bubble can be described as the automatic *ascription of knowledge* to the agent's belief system. The result is that the embubbled agent is unable to perfectly distinguish what *is known* from what *is merely believed*; the difference can be spotted only by a third-person perspective.

Gibson, indeed, defined the "affordance" as what the environment offers, provides, or furnishes. For instance, a chair affords (for a human) an opportunity for sitting, air-breathing, water-swimming, stair-climbing, and so on. It appears to the agent every bit as clear as the piece of confectionery saying "Eat me!" in Carroll's *Alice in Wonderland* (1865) [2010: 10]. One of the main tenets of this theory is that agents do not retain in their memory an explicit and complete model of the environment, including its variables and applications, but they actively manipulate it by picking up information and resources ad hoc. Information and resources are not only given, but they are actively sought for and even manufactured. In this framework it is possible to understand the other, more specific, definitions of affordance that Gibson provided, considering them as: (1) opportunities for action; (2) values and meanings of things which can be directly perceived; (3) ecological facts; (4) elements that imply the mutuality of perceiver and environment.

> An important fact about the affordances of the environment is that they are in a sense objective, real, and physical, unlike values and meanings, which are often supposed to be subjective, phenomenal, and mental. But actually, an affordance is neither an objective property nor a subjective property; or it is both, if you like. An affordance cuts across the dichotomy of subjective-objective and helps us to understand its inadequacy. It is equally a fact of the environment and a fact of behavior. It is both physical and psychical, yet neither. An affordance points both ways, to the environment and to the observer. [Gibson 1979: 129]

His hypothesis is highly stimulating: "[. . .] the perceiving of an affordance is not a process of perceiving a value-free physical object [. . .] it is a process of perceiving a value-rich ecological object", and then, "physics may be value free, but ecology is not" [Gibson 1979: 140].

Gibson stressed the adaptive value of affordances, relying on the idea that they are primarily an "opportunity for action": something that triggers the agent's mind to perform a given activity, just as the knowledge-based habit in Peirce. As he pointed out, "a habit arises when, having had the sensation of performing a certain act, *m*, on several occasions *a*, *b*, *c*, we come to do it upon every occurrence of the general event, *l*, of which *a*, *b* and *c* are special cases" (1868) [EP 1: 28–55]. As we mentioned, the dimension of repetition of the habit is a necessary consequence of the attainment of belief: if we believe that a particular word in another language as "*spingere*" in Italian means "*push*" in English, we will push the door when we find that word written upon it in order to open it. In the same way, this habit is the representation of an affordance too: after knowing the meaning of the word "*spingere*", a door with the word upon it will be perceived as a door that can be opened just by pushing it, not by pulling it. It will say "push me" to the agent as the chair says "sit on me" and the cake "eat me" to Alice.

So, in exactly the same way, we can say that agents can "modify" or "create" affordances by manipulating their environment just as they modify and create habits by manipulating their believing system. That is to say, the reinforcement of habit may acquire a more epistemic and less behavioral nature than originally hypothesized by William James. Moreover, the manipulation of the environment affects the manipulation of the believing system, and the believing system affects the manipulation of the environment. The knowledge behind this process relies on different kinds of sources: from the automatic, almost instinctive, response to an object with a strong ecological value—for instance, a huge stone with a flat surface can afford sitting—to the learning process that permits us to believe that the word "spingere" upon a door means "push" and allows us to open it immediately. The pragmatic application of a belief through a knowledge-based habit comes to be a spontaneous outcome of ontogenesis and phylogenesis: we can obviously guess that even the more basic and wired perceptual affordances available to our ancestors were very different from the present ones, as the habits that can be detected in children are different from those that can be seen in adults.

Affordances can then be helpful to illuminate what we defined as knowledge-based habits: they are a form of ecological-and-psychological-at-once knowledge, that is either pre-wired in the agent's cognition (for instance, implicit knowledge that the ground affords walking), or learned through the acquisition of knowledge, for instance in the trivial example of the door sign—but also in less trivial examples of learned affordances (and therefore habits) concerning the use of artifacts.[9]

So far, we explored the development of habit as a consequence of some knowledge acquisition on the agent's part. By resorting to the notion of affordance, we suggest that the habit, that is the ability to react in a given (habitual) way to a given stimulus, can be associated more or less directly with the recognition and enactment of a given affordance.[10] However, the argument we structured, assimilating affordances and knowledge-based habits, begs the question about the problematic access and development of the state of doubt. If our beliefs shape our knowledge and perception, how can we be able to ignore our certainties and formulate doubt? How can we go from a knowledge-based habit to the doubt over the belief which caused it? This is where, raising the stakes, we must introduce and analyze the second kind of habit, the ignorance-based one.

[9]Though this is not the appropriate lieu of discussion, we clearly side along those maintaining that to define affordances as immediate, direct perception of possibilities does not imply the necessary impossibility to learn and develop new affordances apart from those that are naturally available to our cognitive system—chiefly because of phylogenesis. Gibson himself seemed to be quite clear in assimilating the artifactual dimension to the natural one, in contending (right after the definition of affordance quoted above) that the artifactual environment "is not a *new* environment—an artificial environment distinct from the natural environment—but the same old environment modified by man. It is a mistake to separate the natural from the artificial as if there were two environments; artifacts have to be manufactured from natural substances. It is also a mistake to separate the cultural environment from the natural environment, as if there were a world of mental products distinct from the world of material products. There is only one world, however diverse, and all animals live in it, alright we human animals have altered it to suit ourselves" [Gibson 1979: 130].

[10]Another interesting interpretation of the connection between habit and affordance is given by West [2014: 119].

The Ignorance-Based Habit: The Rising of Doubt

First of all, we should remember that the specific difference between the states of doubt and belief in Peirce is pragmatic: that is, the distinction between them is grounded on the different behaviors they are connected to. The quiet and satisfactory state of belief prompts the agent to act—through the adoption and the defense of a principle; instead, the irritation of doubt freezes the agent's attempt to act since it pushes his reason to find an explanation to an unanswered question, and try to compensate the lack of certainty with the creation of another belief. As we can see, the habit is the center of the interplay between doubt and belief; it actually plays both the role of the pleasant result of belief attainment, and of the ground where the irritation of doubt can start harassing the agent.

We saw earlier how habit can be the application of a principle that is thought to be utterly true, and how habit alters and affects our point of view and decision-making strategies (its non-ampliative role if we refer to the first distinction we mentioned). Now we are going to show the way the habit can induce a state of doubt, so that it becomes the trigger of an ampliative reasoning. Essentially, we are going to see how a knowledge-based habit can turn into an ignorance-based one. Thus, we should begin with the study of the habit as a starting-point for the research.

> But since belief is a rule for action, the application of which involves further doubt and further thought, at the same time that it is a stopping-place, it is also a new starting-place for thought. That is why I have permitted myself to call it *thought at rest*, although thought is essentially an action. (1878) [EP 1: 129]

The initial stage of the process of reasoning is not, as it can be imagined, a substantial *void*. Even the "genuine doubt", which should emerge from surprise, does not begin in the absence of beliefs on the matter.[11] Something in doubt (and in the surprise at its origin) implies, though, the *breaking of certainties that surround our beliefs in a given field* or about a determined situation. It starts with the realization that the agent is in a condition of ignorance about something, whether suspected or not. Surely, doubt is not a form of abyssal negation, but just a lack of cognitive confidence over some particular matter.

Thus, also the laborious work of investigation sparked by doubt could be described as the implementation of a *very specific question* born out of the negation of the preceding, abandoned, belief.[12] But what is actually negated is not just the belief, but also the habit that it implied: if the habit is the application of a certain

---

[11]Peirce himself stressed that "genuine doubt always has an external origin, usually from surprise", all the more because it is not possible to give oneself a "genuine surprise" by an "act of the will" (1905) [CP 5.443]. The agent's misrepresentation of the emergence of surprise, connected with habit formation, is interestingly analyzed by Colapietro (this volume). Indeed, he advocates the possibility that the agent can effectively play an active part in the stimulation of his own state of surprise, while being cognitively prevented from fully recognizing his role in the process.

[12]Peirce himself employed the word *struggle* to stress the violent trait of this condition (1877) [EP 1: 114].

principle that we are no longer certain to be true, should it not be reasonable that we stop doing it, and instead reflect?

Hence, precisely, doubt is the common ground between the repetition of a habit (as the rule of action created by the attainment of a belief), its critical break, and the attempt to replace it with a better one. Obviously, the emotional condition of the *irritation of doubt* does not warrant that the agent is going to necessarily improve his knowledge from the external and scientific perspective that we mentioned above.[13] That is because doubt is primarily an unwanted state from which the individual wants to escape no matter what: it originates from a loss of confidence about a belief, which as a consequence cannot sustain any more practical decision-making. At the same time, this process obviously does not warrant that the replacement will be a correct (or a more correct) one. Surely, that would be convenient, but it does not often happen. What does actually occur is the emergence of a specific inferential reasoning that matches the repetitional structure of the habit: a hopeful breaking of an unsatisfactory analogy: an instance of abductive reasoning.

Abductive Reasoning as the Starter of the New Belief Creation

In the previous section we examined how non-ampliative reasoning (A-reasoning, in Peirce's terms) can be produced from the attainment of a belief and the establishment of a relative habit. The analogical disposition, which makes the derived habit just an application of the believed proposition, is the reason for the "knowledge-based" attribute: what makes the agent confident in his own belief is the fact that he thinks it is correct and actually counts as knowledge. The connected habit becomes a sort of deductive mechanism that infers the right behavior from the belief indication.

> That which determines us, from given premises, to draw one inference rather than another is some habit of mind, whether it be constitutional or acquired. The habit is good or otherwise, according to whether it produces true conclusions from true premises or not; and an inference is regarded as valid or not, without reference to the truth or falsity of its conclusion specially, but according to whether the habit that determines it is such as to produce true conclusions in general or not. (1877) [EP 1: 112]

As already said—and as Peirce's quotation reminds us—the specification of each habit is determined by the outcome it produces. It is a knowledge-based habit if it contributes to justify the agent's knowledge, and it is an ignorance-based one if it rose from ignorance. The agent, in order to move from a type of habit to another and generate an ampliative reasoning from a justificatory one, must change the very structure of the habit. Of course, the new habit must be derived from the attainment of a belief, but it is not any more just a part of the deductive process that has its

[13]cf. subsection "Belief: Habit as a Rule for Action" above.

premise in the believed proposition and its conclusion in its external application through a rule for action. It must become the "test" of the belief, not its mere consequence. As Peirce wrote, "the habit-change often consists in raising or lowering the strength of a habit" (1907) [CP 5.477]; in this analysis we can say that instead of a habit that is the model of the agent's knowledge, the lowered strength of the habit makes it more eager to admit a doubt on the proposition it applies.[14] Such movement allows the habit to change out of "surprise": the will to test the believed proposition is the first step to the creation of the aforementioned abduction.[15]

In brief, abduction is a process of *inferring* certain facts and/or laws and hypotheses that render some sentences plausible, that *explain* or *discover* some (eventually new) phenomenon or observation.[16] The surprising event, once it breaks the knowledge-based habit, has to be explained by the agent, who is facing a new significance into the mass of facts that cannot be solved with the content of beliefs possessed before.[17] The inferential process that starts from this habit-breaking mechanism is driven by the irritation of doubt, by the cognitive rush to find a new explanation. The reasoning that is performed by the agent is what allows him to individuate a new pattern out of a cluster of data. Thus, the first consequence of the abductive process is the production of a permanent change in the agent's point of view: reading a new pattern consists in a new way of using the collected information, and it implies the possibility of individuating the same pattern again should it occur again. Even if the problematic condition of doubt is unpleasant, once the abductive process is achieved, the agent cannot come back to the initial stage of belief/knowledge-based habit.

---

[14]In Peirce's unfinished essay *Pragmatism* (1907) [EP 2: 398–433], analyzed in this volume by Bergman (this volume), the complex dynamic of habit-change is considered in terms of the "ultimate logical interpretant". While addressing the reader to Bergman's paper for further enlightenment, we should mention that the ultimate logical interpretant, defined as the "concluding goal of cognitive sign action", refers to many topics we already discussed: for instance the clarification of a habit in terms of the actions it would produce, the establishment of such a habit of action in our nature, and the revision of existing habits.

[15]Aliseda (this volume) also richly analyzes abductive reasoning as the process that guides the transition between doubt and belief in Peirce's epistemology.

[16]The classical schematic representation of abduction is expressed by what Gabbay and Woods (2005) call the *AKM-schema*, as contrasted with their own GW-schema (Gabbay-Woods). In the AKM, *A* refers to Aliseda (1997, 2006), *K* to Kowalski (1979), Kuipers (1999), and Kakas et al. (1993), *M* to Magnani (2009) and Meheus et al. (2002). A detailed illustration of the AKM schema is given in Magnani (2009), together with the recent EC-Model (Eco-Cognitive Model) of abduction.

[17]A more vivid explanation of the creation of belief through surprise that we described can be found in the Peirce's famous first definition of abduction: "A mass of facts is before us. We go through them. We examine them. We find them a confused snarl, an impenetrable jungle. We are unable to hold them in our minds. [. . .] But suddenly, while we are poring over our digest of the facts and are endeavouring to set them into order, it occurs to us that if we were to assume something to be true that we do not know to be true, these facts would arrange themselves luminously. That is *abduction* [. . .]" (1903) [EP 2: 226–241].

Hence, abduction concerns the passage from what is known to what is not known yet: it is indeed an inferential process aiming at finding out explanatory information starting from a cluster of data that breaks out of a habitual recurrence. Instead of just a form of knowledge, we should remark that what leads the creation of the new belief and its rule of action is a type of *ignorance* that relies on the loss of confidence the agent has about his own beliefs. At this point, we can call over our epistemological exploration of Peirce's habit: as the passage from doubt to belief is a transition from uncertainty to confidence in the agent's resolution, the passage from the knowledge-based habit to the ignorance-based one is the opposite mechanism. At first, the agent is capable of justifying his beliefs, letting them draw his course of action, confident that his beliefs are well-grounded. Then, the habit enables the agent to enhance his knowledge after the irritation of doubt occurs, as it triggers the research of a better solution for what could not be previously explained.

Conclusion

In the end, we can say that the solution we have presented to the initial question about the duplicity of habit in the Peircean model of thought solves, if not the ontological and philological issues around it, at least the epistemological and cognitive foundation of the mechanism of belief creation and change. The interplay between knowledge-based and ignorance-based habit allows the agent to discover new information, and move out of an unconfirmed fixation of belief, thus possibly expanding his knowledge. The first type of habit justifies the agent's knowledge and improves its grounding. As we saw, it also modifies perception and contributes to discovering affordances that distribute the agent's information on the environment. Thus, the adoption of a knowledge-based habit is cognitively comforting, and —from an epistemological perspective—it is very effective for the maintenance of a well-composed knowledge. The ignorance-based habit has instead the role of effectively testing the effectiveness of the agent's beliefs. It is a trigger of doubt and, within it, abductive reasoning can emerge and increase the possibility of improving the agent's knowledge.

References

Aliseda, Atocha. 1997. *Seeking explanations: Abduction in logic, philosophy of science and artificial intelligence*. Amsterdam: Institute for Logic, Language and Computation.

Aliseda, Atocha. 2005. The logic of abduction in the light of Peirce's pragmatism. *Semiotica* 153 (1/4): 363–374.

Aliseda, Atocha. 2006. *Abductive reasoning. Logical investigations into discovery and explanation*. Berlin: Springer.

Aliseda, Atocha. this volume. Belief as habit. In *Consensus on Peirce's concept of habit: Before and beyond consciousness*, eds. Donna E. West and Myrdene Anderson. (Studies in Applied Philosophy, Epistemology and Rational Ethics [SAPERE].) New York: Springer.

Bergman, Mats. this volume. Habit-change as ultimate interpretant. In *Consensus on Peirce's concept of habit: Before and beyond consciousness*, eds. Donna E. West and Myrdene Anderson. (Studies in Applied Philosophy, Epistemology and Rational Ethics [SAPERE].) New York: Springer.

Boumans, Marcel, and Giora Hon. 2014. Introduction. In *Error and uncertainty in scientific practice, of history and philosophy of technoscience*, ed. M. Boumans, G. Hon, and A.C. Petersen, 1–13. London: Pickering and Chatto Publishers.

Carroll, Lewis (Charles Lutwidge Dodgson). 2010 [1865]. *Alice's Adventures in Wonderland*. London: Bibliopolis.

Clark, Andy. 2008. *Supersizing the mind. Embodiment, action, and cognitive extension*. Oxford/New York: Oxford University Press.

Colapietro, Vincent. this volume. Habits, awareness, and autonomy. In *Consensus on Peirce's concept of habit: Before and beyond consciousness*, eds. Donna E. West and Myrdene Anderson. (Studies in Applied Philosophy, Epistemology and Rational Ethics [SAPERE].) New York: Springer.

Coletta, W. John. this volume. The "Irrealevance" of habit formation: Peirce, Hofstadter, and the rocky paradoxes of physiosemiosis. In *Consensus on Peirce's concept of habit: Before and beyond consciousness*, eds. Donna E. West and Myrdene Anderson. (Studies in Applied Philosophy, Epistemology and Rational Ethics [SAPERE].) New York: Springer.

Coletta, W. John. 2005. *The reach of abduction*. Volume 2 of *A practical logic of cognitive systems*. North Holland, Amsterdam: Elsevier.

Gibson, James J. 1950. *The perception of the visual world*. Boston: Houghton-Mifflin.

Gibson, James J. 1977. The theory of affordances. In *Perceiving, acting, and knowing*, ed. R.E. Shaw, and J. Bransford, 127–143. Hillsdale, NJ: Lawrence Erlbaum Associates.

Gibson, James J. 1979. *The ecological approach to visual perception*. Boston, MA: Houghton Mifflin.

Gorlée, Dinda L. this volume. On habit: Peirce's story and history. In *Consensus on Peirce's concept of habit: Before and beyond consciousness*, eds. Donna E. West and Myrdene Anderson. (Studies in Applied Philosophy, Epistemology and Rational Ethics [SAPERE].) New York: Springer.

Ippoliti, Emiliano. 2015. Reasoning at the frontier of knowledge: Introductory essay. In *Heuristic reasoning* 16 (forthcoming), ed. E. Ippoliti (Studies in Applied Philosophy, Epistemology and Rational Ethics [SAPERE]), 1–10. London: Springer.

James, William. 1983 [1892]. *Psychology. The briefer course*. New York: Holton. (First publication 1920; last publication 1983).

Kakas, Antonis, Robert A. Kowalski, and F. Toni. 1993. Abductive logic programming. *Journal of Logic and Computation* 2(6): 719–770.

Kowalski, Robert A. 1979. *Logic for problem solving*. New York: Elsevier.

Kuipers, Theo A.F. 1999. Abduction aiming at empirical progress of even truth approximation leading to a challenge for computational modelling. *Foundations of Science* 4: 307–323.

Mace, William M. 1977. James J. Gibson's strategy for perceiving: Ask not what's inside your head but what your head's inside of. In *Perceiving, acting and knowing: Toward an ecological psychology*, eds. R. Shaw and J. Bransford, 43–65. Hillsdale: Lawrence Erlbaum Associates.

Magnani, Lorenzo, and Tommaso Bertolotti. 2011. Cognitive bubbles and firewalls: Epistemic immunizations in human reasoning. In *CogSci 2011, XXXIII annual conference of the cognitive science society*, ed. L. Carlson, C. Hölscher, and T. Shipley, 3370–3375. Boston, MA: Cognitive Science Society.

Magnani, Lorenzo. 2009. *Abductive cognition. The epistemological and eco-cognitive dimensions of hypothetical reasoning*. Berlin/Heidelberg: Springer.

Magnani, Lorenzo. 2013. Is abduction ignorance-preserving? Conventions, models, and fictions in science. *Logic Journal of the IGPL* 21: 882–914.

Meheus, Joke, Liza Verhoeven, Maarten Van Dyck, and Dagmar Provijn. 2002. Ampliative adaptive logics and the foundation of logic-based approaches to abduction. In *Logical and computational aspects of model-based reasoning*, ed. L. Magnani, N.J. Nersessian, and C. Pizzi, 39–71. Dordrecht: Kluwer Academic Publishers.

Peirce, Charles Sanders. i. 1867–1913. *Collected papers of Charles Sanders Peirce*. Vols. 1–6, eds. Charles Hartshorne and Paul Weiss. Cambridge: Harvard University Press, 1931–1935. Vols. 7–8, ed. Arthur W. Burks. Cambridge: Harvard University Press, 1958. [References to Peirce's papers will be designated by CP, followed by volume, period, paragraph number.].

Peirce, Charles Sanders. i. 1867–1893. *The essential Peirce: Selected philosophical writing*. Volume 1 (1867–1893), eds. Nathan Houser and Christian Kloesel. Bloomington: Indiana University Press, 1992. [References to this volume will be designated by EP 1, followed by colon, page number.].

Peirce, Charles Sanders. i. 1893–1913. *The essential Peirce: Selected philosophical writing*. Volume 2 (1893–1913), ed. the Peirce Edition Project. Bloomington: Indiana University Press, 1998. [References to this volume will be designated by EP 2, followed by colon, page number.].

Pickering, John. this volume. Is nature habit-forming? *Consensus on Peirce's concept of habit: Before and beyond consciousness*, eds. Donna E. West and Myrdene Anderson. (Studies in Applied Philosophy, Epistemology and Rational Ethics [SAPERE].) New York: Springer.

Pietarinen, Ahti-Veikko. 2005. Cultivating habits of reasoning: Peirce and the *Logica Utens* versus *Logica Docens* distinction. *History of Philosophy Quarterly* 22(4): 369–373.

Stjernfelt, Frederik. this volume. Dicisigns and habits: Implicit propositions and habit-taking in Peirce's pragmatism. *Consensus on Peirce's concept of habit: Before and beyond consciousness*, eds. Donna E. West and Myrdene Anderson. (Studies in Applied Philosophy, Epistemology and Rational Ethics [SAPERE].) New York: Springer.

West, Donna E. 2014. From habit to habituescence: Peirce's continuum of ideas. In *Semiotics 2013,* eds. Jamin. Pelkey and Leonard G. Sbrocchi, 117–126. Toronto: Legas Press.

West, Donna E. this volume. Indexical scaffolds to habit-formation. In *Consensus on Peirce's concept of habit: Before and beyond consciousness*, eds. Donna E. West and Myrdene Anderson. (Studies in Applied Philosophy, Epistemology and Rational Ethics [SAPERE].) New York: Springer.

Woods, John. 2005. Epistemic bubbles. In *We will show them: Essay in honour of Dov Gabbay (Volume II),* eds. S. Artemov, H. Barringer, A. Garcez, L. Lamb, and J. Woods, 731–774. London: College Publications.

Woods, John. 2007. The concept of fallacy is empty: A resource-bound approach to error. In *Model-based reasoning in science, technology and medicine*, vol. 64, eds. L. Magnani and L. Ping, (Studies in Computational intelligence), 60–90. Amsterdam: Springer.

Woods, John. 2013. *Errors of reasoning. Naturalizing the logic of inference*, vol. 45 of *Studies in Logic: Logic and Cognitive Systems*. London: College Publications.

Chapter 21
Social Minds and the Fixation of Belief

Nathan Houser

Abstract To survive and thrive in our always evolving world, humans must learn from experience and conserve vital intelligence for immediate access when needed for the resolution of existential predicaments. Peirce described this process as one of inquiry, a trial and error proceeding that resolves quandaries (doubts) and forges habits (beliefs) that accumulate into stores of practical intelligence that are programs for survival. These stores of practical intelligence are the integrated systems of belief that constitute the cores of our minds. But humans aspire to intelligence of a more theoretical sort that has no immediate practical or vital importance but which satisfies intellectual yearnings and advances knowledge in general. The quest for knowledge in general, scientific inquiry, does not yield a body of fixed beliefs (as with practical intelligence) but only a body of provisional beliefs never quite accepted as final. Our minds also encompass theoretical intelligence of this sort. But individual human experience can never achieve the comprehensive practical intelligence necessary for the survival of civilization and the human brain is not adequate for the storage of the theoretical intelligence necessary for the advancement of science. The survival and advancement of civilization depends on the extension of mind beyond individual biological organisms into social groups and institutions. This accords with Peirce's generalized conception of mind and his idea that minded organisms function within mind that is, at least in part, external to them. Not only can institutions develop minds of their own, some institutions evolve to become the crucial reservoirs of intelligence that define cultures and perpetuate civilization. We may speculate that religion is the human institution that embodies the intelligence that addresses the matters of vital importance for civilizations, while it is the institution of science that embodies the theoretical intelligence that addresses the human aspiration to find things out and our only hope to advance toward the truth.

Keywords Inquiry · Doubt · Habit · Belief · Instinct · Intelligence · External mind · Distributed semiosis · Religion · Science

N. Houser (✉)
Indiana University, Indianapolis, USA
e-mail: nhouser@iupui.edu

D.E. West and M. Anderson (eds.), *Consensus on Peirce's Concept of Habit*,
Studies in Applied Philosophy, Epistemology and Rational Ethics 31,
DOI 10.1007/978-3-319-45920-2_21

The Role of Habit

Peirce's writings are replete with distinctive key ideas and doctrines that range over his philosophy and give it unity. His categories in various forms come to mind, as well as his objective idealism, and his fallibilism. Peirce's pragmaticism, too, is certainly a signature component of his philosophy, as is his metaphysical triad—agapism, tychism, and synechism—the basis of his evolutionary metaphysics. One would be hard-pressed to single out one most paramount conception running throughout, and tying together, Peirce's systematic thought. His life-long self-identification as a logician might lead one to suppose that if any conception deserves to be recognized as of key importance, it may well be *reasoning,* or perhaps, *inference*. There is ample support in Peirce's writings for supposing that nothing was of more constant concern to him than the growth of knowledge through inferential processes and how different kinds of inferences preserve or contribute to the advancement of truth. However, one might just as rightly suppose that semiosis is the most crucial conception given the fundamental role of signs in Peirce's philosophy and the inferential nature of all sign action. But as the contributors to this volume reveal, Peirce's conception of habit, which was a continual focus of concern throughout his life from his early development of pragmatism to his most advanced conception of semiosis, must be ranked of very high importance. In three areas in particular, the role of habit is especially prominent: (1) in the fixation of belief, where habit, as belief, is the end of inquiry; (2) in semiotics, where habit, as the final interpretant, is the end of semiosis; and (3) in cosmology, where habit as law is the end of evolution. It is some of the ramifications of Peirce's views of the role of habit in mental development and the formation of belief that will be dealt with in this essay.

Doubt-Belief Theory of Inquiry

Peirce's theory of belief formation from his classic presentation of pragmatism in his *Popular Science Monthly* papers of 1877–1878 provides the general framework for the considerations that follow.[1] A convert to the evolutionary paradigm, though not exclusively to Darwin's theory, Peirce supposed that human thought is a natural process generated by the need to resolve difficulties or quandaries that arise in the course of experience—Peirce calls these quandaries doubts and their resolutions beliefs. This process of resolving doubts and acquiring beliefs (habits, or stored

---

[1]Peirce introduced his pragmatism in a series of six papers published in 1877 and 1878 in the *Popular Science Monthly.* These six papers, as a group, were planned as a book to be titled "Illustrations of the Logic of Science", and have been reprinted in the main editions of Peirce's writings. The first two papers are regarded as the classic presentation of pragmatism: "The Fixation of Belief" (W3: 242–257) "How to Make Our Ideas Clear" (W3: 257–276).

programs, to expedite future experiential interactions) is essentially a matter of practical reasoning, and the body of beliefs that is built up over time is a store of practical intelligence.

This is the gist of Peirce's doubt-belief theory of inquiry. Doubt arises in the course of life as a state of irritation or anxiety due to a felt need to respond to an object or event that does not comport well with the usual run of experience—no ready behavioral reaction is forthcoming. The natural response to doubt is a state of cerebral excitement and trial-and-error physical activity that Peirce called inquiry. Inquiry ceases when successful action is taken and the irritation of doubt subsides. This process of inquiry leads to the formation of cognitive and behavioral habits that Peirce called beliefs. Thought, or inquiry, is no longer requisite after satisfactory habits are formed—after beliefs become fixed. So the natural purpose of thought is to produce belief: "thought in action has for its only possible motive the attainment of thought at rest" (from "How to Make Our Ideas Clear", 1878: W3: 263; see also EP 1: 129 and CP 5.396). In "The Fixation of Belief", Peirce discussed four methods for fixing belief: tenacity, authority, the a priori method, and the method of science. Although the first three methods facilitate habit formation (behavioral programming) preparatory for future experience, only under the method of science are beliefs formed in direct interaction with experience and without prejudice. A special attraction of the method of science, for Peirce, is that it can be used to advance knowledge and not simply to adjust behavior for practical purposes.

Two features of Peirce's theory of belief are especially noteworthy. First, instead of regarding belief as an attitude toward a proposition, a view common among philosophers of mind, Peirce's behavioral account focused on dispositions instead of propositions. Quine regarded Peirce's behavioral (dispositional) theory of belief as a major advance toward a naturalized philosophy (Quine 1981: 36–37). The second feature of note is Peirce's explicit interest in neuroscience, such as it was in his time. He wrote of thinking as cerebration (from a draft of the first chapter of a work on logic, 1880: W4: 45; see also EP 1: 200 and CP 3.155) and of habit formation as a "property of the nervous system" and the physiological basis of learning (from "A Guess at the Riddle", 1887–1888: W6: 190; see also CP 1.390). But Peirce was not a scientistic reductionist as the naturalized philosophers today, under the sway of nominalism, tend to be. Peirce's realism enabled him to recognize the governing power of beliefs (habits) as *rules* of action and to understand belief in the context of final causation. Beliefs are not simply brain states that act efficiently, but are states of mind that act semiotically. So Peirce's account of belief begins with interaction between a subject, a sentient being, and what is external to that subject, along with the presumption that the well-being of the subject requires satisfactory interactions with what is external to it. Experiential cruxes generate doubt and initiate inquiry, or thought, the function of which is to eliminate the irritation of doubt and produce belief. Thus, on the one hand, the upshot of thought goes beyond thought in evoking action. Beliefs as physiological states serve this function. Yet, in their generality as mental states, beliefs function not only as embedded behavioral programs but as *rules* of action—like conditional propositions. In "How to Make Our Ideas Clear", Peirce stressed that the purpose of thought in producing habits of action was not to conclude in individual acts, but to produce habits as *rules* of action. This led him to the key insight of his pragmatism, namely, that every real distinction of thought and, therefore, every distinction of meaning consists in a *possible* difference in practice, with emphasis on "possible" (from "How to Make Our Ideas Clear", 1878: W3: 265; see also EP 1: 131 and CP 5.400).

Two Kinds of Belief

In 1898, in a series of lectures in Cambridge, Peirce reconsidered the question of belief, largely in response to the appearance of William James's essay, "The Will to Believe" (James 1896; see Houser 2016 for further discussion of Peirce's reaction to James). James was concerned with the spread of agnosticism among his students and colleagues and wanted to lend support for the choice of religious life. In his famous essay, James offered a "justification of faith" and defended "the right to believe". James, trained in science, accepted that beliefs should be based on evidence whenever possible, but he championed one's right to follow the guidance of "our passional nature" in cases where adopting a belief is unavoidable and the intellect is stymied by an absence of evidence. "Our passional nature", he wrote, "not only lawfully may, but must, decide an option between propositions, whenever it is a genuine option that cannot by its nature be decided on intellectual grounds".[2] It is not clear that Peirce disagreed with James about the value of religious life, nor, for that matter, about the influence of our passional nature on our cognitive development, but he objected strongly to James's attempt to justify the *deliberate* adoption of beliefs without evidence—that blocked the road of inquiry and thwarted the will to learn. James, Peirce thought, had succumbed to "the Hellenic tendency to mingle philosophy and practice", whereas, in Peirce's opinion, were philosophy to influence religion and morality it should do so "with secular slowness and the most conservative caution" (from "Philosophy and the Conduct of Life", 1898a: EP 2: 29; see also CP 1.620). Peirce emphasized the importance of the distinction between practice and theory focusing especially on the nature and function of belief in ordinary life in contrast to its role in theoretical pursuits.

A key idea that Peirce developed in his 1898 Cambridge lectures was that there are two distinct types of beliefs: the kind we depend on in matters of vital importance and the kind we hold provisionally in our intellectual pursuits. Beliefs proper are cerebral habits, or states of mind, that develop in the regular course of life as nature's way of adapting the species to its given environment. According to Peirce, belief is demonstrated by a readiness to act—and, therefore, we believe a proposition if we are prepared to act on it. "Full belief", Peirce said, "is willingness

[2]A genuine option is one which is live, forced, and momentous. For details see James (1896: 11).

to act upon the proposition in vital crises" (from "Philosophy and the Conduct of Life", 1898a: EP 2: 33; see also CP 1. 635). These are natural beliefs and they are of practical, or as Peirce put it, of "vital" importance.

The second type of belief, one Peirce hesitated even to designate by that name because it involved mental states that are more properly regarded as hypotheses and are of little if any practical importance, are theoretical beliefs that serve as stepping stones on a path toward knowledge and truth. Even though action must be taken in the pursuit of pure science, and notwithstanding that the fruits of science may be put to important, even vital, practical service, the aim of pure science is not really practical at all, but only to find things out—"to learn the lesson that the universe has to teach" (from "The First Rule of Logic", 1898b: EP 2: 54; see also CP 5.589). Scientific propositions are only held provisionally and any true person of science "stands ready to abandon one or all as soon as experience opposes them" (from "Philosophy and the Conduct of Life", 1898a: EP 2: 33; see also CP 1.635. The quotation that follows is also from EP 2: 33). So Peirce maintained that there is "no proposition at all in science which answers to the conception of belief"—not the "full belief" that can be relied on without hesitation.

What Peirce calls inquiry in "The Fixation of Belief", cerebral activity triggered by experiential cruxes that concludes in actions that resolve the "irritation of doubt" and build up the habits of action that he regarded as beliefs proper, is, in effect, *natural reasoning*. A peculiarity of natural reasoning and the beliefs it generates and contributes to practical intelligence is that truth and validity, the core values of deliberative reasoning, are largely irrelevant. Although the pragmatists may have been right in holding that in the give and take of experience the beliefs that prove to be dependable for the long run, the beliefs that work, may be expected to have an increasing congruency with the truth, these natural beliefs are not embraced because they are true but because they solve problems or dissolve difficulties (as Peirce says, they successfully address matters of vital importance). There is no doubt that human intelligence extends well beyond the body of natural beliefs that address vital concerns and includes a great deal of theoretical belief that expands human thought into the realms of science and art where truth and beauty are paramount. Nevertheless, it is natural reasoning and the practical intelligence it fosters that underwrites the continuation of human life and that testifies to our kinship with all the other life forms in the universe.

As Peirce continued distinguishing the two types of belief in his Cambridge lectures, he emphasized the limitations of reasoning in the conduct of life. In really vital situations, on-the-spot action is essential; reasoning is not an option. What is required is something like instinct, a disposition to act immediately and effectively in such circumstances. Such dispositions, which can be confidently activated at a moment's notice without any deliberation, are what Peirce conceived of as beliefs proper. It is not too difficult to imagine why such beliefs would undermine science, which relies on doubt and deliberation.

However, Peirce's rigid insistence that no proposition in science answers to the conception of belief may have been a little hyperbolic intended to emphasize the distinction between theory and practice (see CP 5.538 ff. where Peirce discusses further the distinction between practical and theoretical beliefs). Nevertheless, scientific beliefs seem clearly to fall in on the side of theory and Peirce was resolute in distinguishing theoretical inquiry from practical inquiry. "It is notoriously true", Peirce insisted, "that into whatever you do not put your whole heart and soul in that you will not have much success. Now, the two masters, ***theory*** and ***practice,*** you cannot serve. That perfect balance of attention which is requisite for observing the system of things is utterly lost if human desires intervene, and all the more so the higher and holier those desires may be" (from "Philosophy and the Conduct of Life", 1898a: EP 2: 34; see also CP 1.642).

External Mind

Assuming, with Peirce and the early pragmatists, that mind is an integral part of the natural world and that an essential feature of nature is continuous growth (evolution), and assuming too that the natural growth of mind more or less follows Peirce's doubt-belief model, then the natural tendency of mental development for minded organisms would be to accumulate intelligence, in the form of beliefs or habits of thought, that prepares them for future experience. Mind, on this view, is essentially a system of beliefs (cognitive habits) built up over the course of experience and retained because they have helped settle past quandaries and doubts and can be called on in the future when environmental conditions resemble those that were resolved by the responses those beliefs actuated in the past. The mental operations involved in this process of resolving experiential cruxes by locating them within a pattern of previous experience and associated beliefs is a rational process although it is usually not deliberative but is more or less automatic, even including instinctive responses: "our logically controlled thoughts compose a small part of the mind, the mere blossom of a vast complexus which we may call the instinctive mind…" (from "Pragmatism as the Logic of Abduction", 1903a, b, c, d: EP 2: 241; see also CP 5.212). This natural reasoning is practical reasoning and it fulfils the natural functions of mind to resolve experiential cruxes and to program minded organisms for safe and flourishing passage through life. Surely this is correct for the most part, although Peirce rejected the presumption that mind resides only in minded organisms and held a more generalized conception of mind making it more accurate to say that minded organisms function within mind which is, at least in part, external to them. And, as will briefly be taken up in what follows, there may also be minded institutions which are not organisms, properly speaking.

It was Peirce's contention that the proper subject of study for psychology is mind and he repeatedly remarked on the mistake psychologists so often make by focusing almost exclusively on consciousness. Consciousness, Peirce maintained, "is really in itself nothing but feeling… a mere property of protoplasm, perhaps only of nerve matter" (this and the following quotation are from Peirce's "Minute Logic", 1901–1902: CP 7.364). Peirce recognized that "biological organisms" and, especially, nervous systems, "are favorably conditioned for exhibiting the phenomena of mind" so "it is not surprising that mind and feeling should be confounded". According to Peirce, "modern philosophy has never been able quite to shake off the Cartesian idea of the mind, as something that 'resides'... in the pineal gland. Everybody laughs at this nowadays, and yet everybody continues to think of mind in this same general way, as something within..." (from "The Three Normative Sciences", 1903b: EP 2:199; see also CP 5.128). But Peirce regarded this as a very narrow view of mind. Psychology would not "be set to rights", he said, until it is understood that "feeling is nothing but the inward aspect of things, while mind on the contrary is essentially an external phenomenon.... In my opinion it is much more true that the thoughts of a living writer are in any printed copy of his book than that they are in his brain" (this and the following quotation are from Peirce's "Minute Logic", 1901–1902: CP 7.364).

Peirce concluded that consciousness is a relatively simple matter, while mind, "when you once grasp the truth that it is not consciousness nor proportionate in any way to consciousness, is a very difficult thing to analyze" (CP 7.365). Writing to William James in 1904 about James's article, "Does 'Consciousness' Exist", Peirce pointed out that conscious experience, or what he sometimes called the "contents of consciousness", fell into three classes: feelings, two-sided consciousness, and thought (CP 8.281–283).[3] With reference to Peirce's ubiquitous categories, these three classes are consciousness of firstness, consciousness of secondness, and consciousness of thirdness. Peirce told James that he included thought, the world of triadic relations, in consciousness only because "we are aware of it", but that it is not in consciousness in the way firstness and secondness are. It is this world of triadic relations, Peirce told James, that constitutes the main contents of the mind. In his fourth Harvard Lecture of 1903, Peirce said that "Thirdness, as I use the term, is only a synonym for Representation" (from "The Seven Systems of Metaphysics", 1903c: EP 2: 184; see also CP 5.105). Later that year, in the syllabus for his Lowell Lectures, he said that he was using mind as a synonym of representation and he remarked playfully that his auditors should bear in "mind that this mind is not the mind that the psychologists mind if they mind any mind" (R 478).

This world of triadic relations, or representation, is Peirce's universe of signs. One of Peirce's earliest key insights was that all thought is in signs and by "thought" he meant to include all conscious mental events: "whenever we think, we have present to the consciousness some feeling, image, conception, or other representation, which serves as a sign" (from "Some Consequences of Four Incapacities", 1868: W2: 223; see also EP 1: 38 and CP 5.283).

But Peirce recognized that there are also unconscious mental events that function semiotically and should, therefore, be regarded as thought in a generalized sense. So we can reframe Peirce's doubt-belief theory of inquiry in semiotic terms. Without going deeply into technicalities, we know that all signs mediate between their objects and interpretants: "A sign is anything, of whatsoever mode of being,

[3]Peirce used various names for the three forms of consciousness; in CP 7.551 he calls feeling "primisense," consciousness of otherness "altersense," and thought "medisense."

which mediates between an object and an interpretant; since it is both determined by the object *relatively to the interpretant,* and determines the interpretant *in reference to the object,* in such wise as to cause the interpretant to be determined by the object through the mediation of this 'sign'" (from Peirce's famous manuscript, R 318, 1907: EP 2: 410; this manuscript, a substantial portion of which is published in EP 2: 398–433, is the source for the rest of the discussion of semiotics in this paragraph). We know, further, that there are three kinds of interpretants: emotional, energetic, and logical. The logical interpretant is essentially in a relatively future tense, what Peirce called a "would-be", and constitutes "the intellectual apprehension of the meaning of a sign". Peirce maintained that while concepts, propositions, or arguments may be logical interpretants, only habits can be final logical interpretants. The habit alone, though it may be a sign in some other way, does not call for further interpretation. It calls for action. Based on this sketch of Peirce's theory of signs, it would seem that the natural process of inquiry, initiated by the irritation of doubt, would consist of semiosis beginning at the level of feeling and generating emotional interpretants but quickly transitioning to the active level with energetic interpretants. But only when the semiosis (inquiry) concludes in a habit, the final interpretant, does the "irritation" subside and the inquiry come to an end. Peirce's identification of thought with semiosis, and his recognition that signs are external transactions, led him to what he called a logical conception of mind according to which minds are sign systems and thought is sign action or semiosis.

So Peirce concluded that thought is intrinsically an external process and that mind is a relational network of signs (consisting most fundamentally of interrelated general conceptions and associated interpretative habits) that we participate in and operate within but which is not really *ours.* Peirce did not believe that thought is necessarily connected with brains, but is operative in semiosic processes at work within groups and in the external physical world (see "Prolegomena to an Apology for Pragmaticism", 1906: CP 4.551; cf. n. 6, below). Peirce extended the scope of mind to encompass all sign action—any process governed by purpose or final causation. Mind is the ground of all semiosis.

Social Mind

According to Peirce's generalized conception, mind began to develop in the universe with the origination of sign action or, in other words, with the onset of habit formation (see Houser 2014a: 9–32). This was the beginning of what would, over the long course of evolution, develop into the relational networks of signs and interpretative habits that constitute full-fledged minds. Peirce knew that mental states have to be embodied somehow, but, like Hilary Putnam more than half a century later, Peirce understood that the essence of a mental state is logical or functional, not physical, and that mind is independent of its embodiment, just as a computer program is independent of its embodiment (see CP 7.364, from Peirce's 1901–1902 "Minute Logic"; Putnam introduced his computer analogy in Putnam

1960). Peirce also understood that the development of mind, the aggregation and interweaving of conceptions and interpretative habits that constitutes mind proper, depends on a sentient and goal-driven (loosely speaking) substratum but that the required sentience may well be distributed across countless individual sub-systems and even over extended periods of time. Furthermore, just as he realized that a writer's thoughts are more literally in her books than in her brain, Peirce knew that operative minds can be embodied in institutions and cultural practices and traditions, and even in artifacts.[4] In fact, Peirce believed that it was these extended minds that serve as the cognitive base, or ground, for human thought (semiosis).

It is often supposed that because Peirce was a self-proclaimed objective idealist, holding that "matter is effete mind, inveterate habits becoming physical laws" (from "The Architecture of Theories," 1890: W8: 106; see also EP 1: 293 and CP 6.25), he must have assumed that mind is primordial, but that view is mistaken. Although the details of Peirce's cosmology are unsettled, even disputed (e.g. Houser 2014a: 17–18, including footnotes), it is evident that in the pre-historic chaos that Peirce posited as the "ovum of the universe", neither mind nor matter were present. Peirce was a thoroughgoing evolutionist and his cosmology, which he offered as a logical narrative of the origin of the universe, depicts the primal chaos as a state of possibility and indeterminacy out of which, purely by chance, some random happening acquired an "incipient staying quality" and a tendency to take habits was started (CP 6.204; see also RLT: 262). This signaled the birth of semiosis and mind (thus establishing an arena for objective idealism).

Peirce was convinced that mind arose in the "inorganic world" and that thought (semiosis) first developed there (CP 4.551).[5] If a sentient, or quasi-sentient, substrate was necessary even for the original primitive beginnings of habit-taking, then it must have been what Peirce supposed was an unconscious "background psychic life", perhaps best understood as an attractive force that somehow imbued events with a tendency to repeat or reproduce themselves—a nascent will to survive.[6] But however Peirce conceived of this background psychic life, he was careful to distinguish it from original mind, which only emerged from the primeval chaos with the onset of habit formation and flourished as relational networks of habits expanded and evolved into functioning semiotic systems. When mind developed in the organic world it did not develop exclusively within the brains of sentient creatures but as instinctive programs of group behavior such as those exhibited by bees in their hives and ants in their colonies (e.g., CP 4.551). The central point is that consciousness, or feeling, is not an essential attribute of mind: "feeling is nothing but the inward aspect of things, while mind on the contrary is essentially an external phenomenon" (from "Minute Logic", 1901–1902: CP 7.364). So, just as

[4]This distinction is similar to that made in contemporary philosophy of mind and cognitive science between engrams and exograms—see Adams and Aizawa (2010: 144).

[5]Note that "organic" was mistakenly set in type by *The Monist* and was reproduced in *The Collected Papers.* In Peirce's original manuscript (his printer's copy) he clearly wrote "inorganic".

[6]Arthur Burks, editor of volumes 7 and 8 of the Harvard edition of Peirce's papers, rejected Peirce's panpsychism but believed it to be ingenious for times. See Houser (2014a: 29).

Peirce said that "we are in thought and not that thoughts are in us" (from "Some Consequences of Four Incapacities," 1868: W2: 227, n. 4; see also Peirce's first footnote on EP 1: 42 and CP 5.289, n. 1), he might as well have said that we are in mind and mind is not essentially in us. Peirce made this point forcefully in remarking on how knowledge is gained: "all knowledge comes to us by observation, part of it forced upon us from without from Nature's mind and part coming from the depths of that inward aspect of mind, which we egotistically call **ours;** though in truth it is we who float upon its surface and belong to it more than it belongs to us" (from "Of Reasoning in General", 1895a, b: CP 7.558; see also EP 2: 24).

Nature's mind was a key conception for Peirce. He understood that the system of natural laws that had evolved to direct the physical operation of the universe constituted a generalized mind that was in crucial respects recapitulated in the sub-minds that developed under its influence. As simple biological organisms evolved, natural selection must have favored species that as a whole embodied adaptive intelligence, the kind of distributed or swarm intelligence that is exhibited in the mass behavior of insects and schools of fish and flocks of birds. Survival of a species depends on its adequate attunement to nature's mind, which Peirce believed was accomplished by instinctive programming. As the intellectual capacity of living organisms increased, more behavioral programming could be rooted in individual minds, but the key for survival was the deep-seated instinctive species-wide programming that routinely attuned minded-organisms to nature's mind. Even humans, who, as far as we know, have greater individual mental capacity than any other living organisms, and therefore are able by means of the doubt-belief process of inquiry to amass environmentally and culturally sensitive habits of behavior (natural beliefs), remain enormously dependent on instinctive insight, a natural endowment consequent on our minds (the minds of members of our species) having evolved under the influence of nature's laws (from "How to Theorize", 1903d: CP 5.604). In fact, as noted above, Peirce supposed that the portion of an individual's mind developed through inquiry is but a "mere blossom" of the "vast complexus" that constitutes the instinctive mind (from "Pragmatism as the Logic of Abduction", 1903: EP 2: 241; see also CP 5.212). Sometimes Peirce referred to this instinctive mind as the "light of nature" or "an inward light of reason," which, because it attunes us with nature, provides us with an adequately successful capacity for abductive inference—enough to give us the edge necessary for surviving when confronted with dangerous experiential cruxes.[7] But to direct any process of inquiry, instinctive mind must operate semiotically, for all thought is in signs, and Peirce tells us that all signs address future thought and thus, in effect, other minds (or subsequent states of the signifying mind).

[7]See CP 2.24 where Peirce speculated that the idea of the light of reason, or nearly equivalent conceptions, can probably be found in most cultures. He mentions, for example "the 'old philosopher' of China, Lao-Tze" and "the old Babylonian philosophy of the first chapter of Genesis."

Dialogue is Peirce's paradigm for semiosis (see Colapietro 1989: 22), which implies that all mind is structured to function in a social context. In order for an utterer and an interpreter to communicate, Peirce thought that parts of their minds had to fuse into a single mind, which he called "the commens" (from draft letter to Victoria Lady Welby, 1906: EP 2:478; see also Peirce/Welby 1977: 197). So he believed that what we might think of as shared mind can operate across separate individuals, and even in extended group processes, but there can also be a more autonomous mind produced by social inquiry (communal thought) and widely distributed across groups in the sub-minds of the members of those groups. Human thought is far more dependent on this social mind than is generally understood. Peirce held unambiguously that "man is essentially a social animal" (CP 1.11), "mere cells of the social organism" (from "Philosophy and the Conduct of Life," 1898a: EP 2: 40; see also CP 1.647), and "that everything that bears an important meaning to him must receive its interpretation from social considerations" (from an unidentified fragment: R 1573.273).[8]

According to this way of thinking, our social communities and cultural associations can be minded institutions, reservoirs of *social habits*—our culture's stores of social beliefs.[9] The intelligence acquired over time by active social minds, although partly instantiated in the distributed nervous systems (the sub-minds) of the individual members of the social institutions in question, may be stored more enduringly in cultural practices and traditions, in written documents and in language itself, in artworks and musical compositions, and in artifacts of all sorts—in whatever supports and maintains a civilization.[10] But social minds are not merely reservoirs of social beliefs; they are operational programs and relational networks for ongoing distributed semiosis—animated minds of the social groups whose behavior they regulate. Not only did Peirce attribute minds to institutions and social groups, he believed there is something like "personal consciousness" in groups of persons who "are in intimate and intensely sympathetic communion" (from "Man's Glassy Essence", 1892a, b: W8: 182; see also EP 2: 313 and CP 6.271). Peirce called these kinds of social groups "greater persons"—living communities with collective personalities.[11] Josiah Royce, Peirce's erstwhile disciple, expressed well the view he shared with Peirce:

> A community is not a mere collection of individuals. It is a sort of live unit... [that] grows or decays, is healthy or diseased, is young or aged, much as any individual member of the community possesses such characters.... Not only does the community live, it has a mind of its own,—a mind whose psychology is not the same as the psychology of an individual human being. The social mind displays its psychological traits in its characteristic products,

[8]For a discussion of this quotation, see Houser (2014b).

[9]The idea of social beliefs has been discussed and defended by Émile Durkheim and Margaret Gilbert, among others.

[10]See n. 5, above, about the distinction made between engrams and exograms.

[11]See W8: 183 (also EP 1:350 and CP 6: 271), from "Man's Glassy Essence", (1892a, b), for Peirce's use of the expression "greater persons" and W8: 196 (also EP 1: 364 and CP 6.307), from "Evolutionary Love", 1892, for his use of "collective personality".

—in languages, in customs, in religions,—products which an individual human mind, or even a collection of such minds, when they are not somehow organized into a genuine community, cannot produce. Yet language, custom, religion are all of them genuinely mental products (Josiah Royce 1918: 80–81).

Current research into mind from this perspective is generally conducted by social psychologists (who study collective behavior)[12] or cognitive scientists (who study distributed cognition),[13] and to evaluate Peirce's ideas and do them justice considerable effort would be required to compare and connect Peirce's work with ongoing research. Requisite though that effort is, the aim of this essay is far more modest and is mainly conjectural, namely, that if we take seriously the idea of social minds, not merely as reservoirs of intelligence but as operational programs and relational networks for ongoing distributed semiosis, then we may suppose that these social minds are subject, at least indirectly, to Peirce's theory of belief formation and fixation that we usually associate with individual human minds. It is further supposed that Peirce's controversial insistence on the crucial difference between natural beliefs that concern matters of vital importance and theoretical beliefs that concern pure science has instructive relevance for understanding social institutions.

Although the idea of social mind and its essential role in the emergence and functioning of individual human minds and the development of self-conceptions may be traced to many sources (see n.13), usually George Herbert Mead is recognized as pivotal in the historical development of this approach. Working within the American pragmatist tradition, especially under the influence of his colleague, John Dewey, and his university teacher, Josiah Royce, Mead formulated a general theory of mind, heedful of its neurological base in sentient organisms but predicated on the broader foundation for mind in the interaction of organisms with each other —in their social environment (see, especially, Mead 1934).[14] Mead maintained that mind develops in these social interactions, especially in interactions involving language, and that individual self-consciousness is a by-product of these interactions: "The self is not so much a substance as a process in which the conversation of gestures has been internalized within an organic form. This process does not exist for itself, but is simply a phase of the whole social organization of which the individual is a part. The organization of the social act has been imported into the organism and becomes then the mind of the individual" (Mead 1934: 178).

The similarities between Peirce's and Mead's views on mind are striking, enough to make one wonder why Peirce's work did not figure centrally in the

[12]For a useful overview of the social mind from the standpoint of social psychology and of the history of social psychology see Forgas et al. (2001). Also see Valsiner and van der Veer (2000).

[13]For a helpful overview see Adams and Aizawa (2010).

[14]Mead's social behaviorism was the foundation for the school of sociology known as symbolic interactionism.

development of social psychology.[15] It would be especially interesting to compare and contrast Peirce's and Mead's views on self-consciousness and the implications for personal identity but that is a separate study.[16] The reason for mentioning Mead's theory of social mind here is to illustrate that Peirce's prior theory coheres with a pragmatic train of thought running through Royce and Dewey to Mead and then blossoming into the social psychology movement. But, for present purposes, what is important is that Peirce advanced a social theory of mind, holding not only that there is social mind external to and separate from individual human mentality (which is, however, an outgrowth of social mind), but that social mind constitutes its own network of beliefs and may even exemplify a unique character or personality.

The Fixation of Social Belief

In the remainder of this essay, I will briefly expand on my conjecture that social minds are subject to Peirce's theory of belief formation and fixation and that his controversial insistence on the crucial difference between natural beliefs and theoretical beliefs has instructive relevance for understanding social institutions. A more in-depth study of Peirce's theory of social belief would require taking a close look at different ways beliefs become fixed in social minds. Probably the four methods Peirce described in "The Fixation of Belief" for how beliefs become fixed in individual human minds have complements in social psychology, but with some modifications the method of authority may be most relevant: this method, Peirce said, addresses the problem of "how to fix belief, not in the individual merely, but in the community" (from "The Fixation of Belief", 1877: W3: 250; see also EP 1: 117, CP 5.378, and W3: 25 from an earlier variant). And perhaps the method of science finds its complement in the research practices of the scientific community, at least ideally. It would also be necessary to carefully examine Peirce's more explicit considerations of the evolution of mind, or what he once called "the historical development of human thought" (from "Evolutionary Love", 1892: W8: 196; see also EP 1: 363 and CP 6.307). Guided by his categories, Peirce identified three principal modes of mental evolution—tychastic, anancastic, and agapastic—and claimed that "the agapastic development of thought" depends on sympathy and the continuity of mind.[17] He said this kind of development "may affect a whole people or community in its collective personality, and be thence communicated to such individuals as are in powerfully sympathetic connection with the collective people, although they may be intellectually incapable of attaining the idea by their

[15]For discussion of Peirce's influence on the development of sociology, see Wiley (2006).

[16]For some contributions to this study, see Nguyen (2011); Singer (1991); Wiley (1994); Gelpi (2008).

[17]A good treatment of Peirce's evolutionary philosophy can be found in Hausman (1993).

private understandings or even perhaps of consciously apprehending it" (from "Evolutionary Love", 1892: W8: 196; see also EP 1: 364 and CP 6.307).[18] But an in-depth examination of Peirce's accounts of the evolution of mind and the formation of social beliefs is beyond the purview of this essay. For this precursory consideration, Peirce's account of the functional distinction between natural belief and theoretical belief provides a thought-provoking schema for making some important preliminary distinctions.

How do social minds evolve is the crucial question. Presumably, in the evolutionary struggle for species survival, living organisms (principally considering humans) have acquired social sympathies and sensibilities that have drawn them together into accordant groups where, first of all, they have strength in numbers. Of special importance is the symbiosis that develops between individuals and social institutions. Although individuals depend in part on instinctive programming and on direct learning from experience to guide them safely through life, they especially depend on the social minds of the organizations and institutions they belong to for behavioral programming that pertains to social practices and cultural identity. Presumably, the natural affinity humans have for others of their kind, and their penchant for social cohesion, fosters the formation of social institutions and systems which, somewhat like instincts, infix, or at least advance, patterns of behavior (including habits and patterns of thought) that are generally conducive to individual well-being but that also enhance the long-term viability of the social institutions themselves. The more successful institutions, those that survive and prosper over the ages, accumulate vast systems of useful information as indicated in the previous section. These systems of useful information, the beneficial habits distilled from the trials and errors, and *successes,* of generations of human experience, constitute the social minds of the institutions that define cultures and perpetuate civilizations. Through these social minds, individuals can draw on consequential beliefs, and systems of beliefs, they could never develop (program) on their own, thus greatly increasing their access to advantageous intelligence.

At the same time, minded institutions depend on the separate individuals they sustain to sustain them in return by animating them by virtue of the collective sub-minds networked together into extended semiotic systems. Civilizations are rife with institutions and social organizations of various kinds serving a multitude of purposes—including security, communication, education, aesthetic and spiritual elevation, entertainment, comfort, and companionship. The prevailing function of social institutions is to program individual behavior to propagate the endeavors they promote. To fulfill this function, institutions accumulate and provide access to whatever specialized intelligence (programming) is required. These institutional stores of intelligence are the extended social minds, which, like nature's mind, we constantly depend on and egotistically suppose to be our own (see CP 7.558, quoted above in the section on social mind). These minded institutions, in order to fill their evolutionary role in human civilization, themselves become greater persons, to use

[18]As is typical with Peirce, he identified three kinds of agapastic development.

Peirce's term, or genuine communities, to use Royce's, with evolutionary needs of their own. This is a key point. The animated social minds of the institutions individuals depend on most are flourishing semiotic systems which are subject to evolutionary pressures, along with everything else that grows and evolves by interacting with environmental forces that have the power to eliminate or select.

This apparent anthropomorphizing of social institutions will not sit well with some readers. It is one thing to attribute an extension of mind, as a cognitive system, from sentient humans to non-sentient systems or institutions. This is of a kind with supposing that the books in our libraries or our iPhones are part of our extended cognitive systems—or, as Peirce said admiringly of Lavoisier, he carried "his mind into his laboratory, and literally [made] his alembics and cucurbits instruments of thought… manipulating real things instead of words and fancies" (from "The Fixation of Belief", 1877: W3: 243–244; see also EP 1: 111 and CP 5.363). But it is quite something else to suppose that social institutions have functioning minds of their own. Maybe minds as stores of networked information can extend into non-sentient structures but surely, it is commonly supposed, actual cognitive processing has to take place in living brains.[19] I believe this view links mind too closely with consciousness, as Peirce reproved psychologists for tending to do. The systematic unity of a mind derives from the kinds of experiences that have programmed it and how the network of interpretants (the contents of that mind) cohere with respect to a purpose and perhaps a character. Whatever consciousness is necessary for an institution to engage actively with ongoing affairs in the here and now is provided by the community of humans dependent on and animating that institution. That gives the institution a distributed sentience—a distributed consciousness—even though, obviously, institutions do not have individual nervous systems of their own.

I believe institutions do, however, develop minds of their own that over time become far more indispensable for the preservation and advancement of culture, of civilization itself, than the individual minds of humans could ever be. Of particular importance is that social institutions crucial for human flourishing attract individuals to carry on their work and to spread their influence. It is this dependence of individual humans, and of society more generally, on the social intelligence programmed in institutional minds that guarantees a dominant role for social organizations and institutions in the order of things. It is my conjecture that the social minds of institutions develop and evolve more or less in accordance with the doubt-belief process Peirce described for individual minds. Presumably, social minds remain at rest, simply directing behavior according to settled beliefs (behavioral habits) already programmed, until disturbed by doubt. The irritation of doubt for a social mind would presumably arise within the community of sentient individuals who are members of, or dependent on, the institution under

[19]This is more or less the view put forward by Adams and Aizawa (2010), who argue that "the only cognitive processes [even in extended cognitive systems] are those found in the brain" (2010: 146).

consideration, and would likely stem from an experiential conflict or disturbance for which usual community- or institution-sanctioned responses were not forthcoming or seemed inadequate. This irritation of doubt, at least if distributed widely enough throughout the community, would generate the analogue of inquiry, responsive actions, more or less trial and error, until a satisfactory state of affairs was restored —thus adapting the social mind to changing environmental realities. The core beliefs that are selected and assimilated into social minds are habits of behavior that have helped to resolve existential predicaments that have threatened the survival or well-being of the group in question or that have interfered with the pursuit of some common purpose. If we consider extended groups of humanity that constitute entire civilizations, then the most urgent concerns would likely be to resolve the experiential predicaments (the irritations of doubt) that threaten to disrupt or destabilize the social order and, in so doing, to foster or strengthen practices and systems of belief that fortify the social fabric and sustain established social mores. The strongest institutions evolve over the long course of social development and presumably fill the greatest needs and serve the most crucial ends or human aims.

I think it would be difficult to overstate the importance of institutions for humanity—civilization, itself, might justly be said to consist primarily of a great interplay of institutions. I suppose it is up to sociologists and anthropologists to map out the complex network of institutions that structures society in any given age and up to social psychologists to work out how the intelligence accumulated and disseminated by these institutions constitutes the minds of the different ages. These studies would be needed to support any authoritative classification of institutions or ranking of their relative importance. But an insightful, if provisional, philosophical investigation of this topic was conducted by Fisch who concluded that the basic human institutions, the ones all others depend on, are family and speech, and he identified several others as of particular importance for social life—including agriculture, industry, commerce (so "we can work and… supply ourselves with the necessaries and commodities of life"); games (so "we can play"); the arts (so "we can create and enjoy objects of beauty"); religion (so "we can worship and pray and give solemnity to the great occasions of life"); schools, laboratories, libraries, museums, observatories, the sciences (so "we can engage in research"); and government (so "we can give some sort of working harmony to the other institutions") (Fisch 1955–1956: 45).

According to Fisch, Aristotle held that "the basic institutions come into being in order that men may live, and they continue in being, and others are added, in order that men may live well" (Fisch 1955–1956: 46). This conforms to Peirce's basic division of the undertakings of life into those that pertain to matters of vital importance (concerning the practical needs of life) and those pertaining to human aspirations (in particular, the pursuit of knowledge and truth). As Peirce contended in his 1898 Cambridge lectures, these two great classes of human endeavor rely on different kinds of beliefs, the firm beliefs developed in the course of experience that program individuals for survival (natural beliefs) and the provisional beliefs generated by abductive reasoning with the aim of finding things out (theoretical beliefs). It is surely the case that, in the regular course of life, substantial bodies of practical beliefs develop naturally as stores of behavioral habits programmed into individual brains, and something like that, although less automatically, must also occur with theoretical beliefs for individuals whose lives are devoted to intellectual or scientific pursuits. But if what I have said about the role of social minds and durable institutions in accumulating specialized knowledge and augmenting individuals' intellectual capacity is true, then we would expect that powerful institutions have evolved to support these two great classes of human endeavor.

Is there a powerful natural human institution that contributes in an immediate and vital way to the survival of societies and cultures in the way instinctive and fixed behavior contributes to the survival and thriving of lower life forms and species? The function of such institutions would be to inculcate and sustain practices and values that are life enhancing at a fundamental level, in effect, programming individuals to embrace systems of belief that give an evolutionary advantage to societies that foster them. Probably many long-standing institutions serve this function to some extent, but I believe religion, above all, plays this role in human life and civilization. Just as truth and validity are largely irrelevant to natural reasoning and the practical beliefs that have proven themselves in the vital give and take of real life experience, so with religion it is largely beside the point to question the truth or validity of the systems of values and beliefs that it propagates. The relevant question is whether those values and beliefs are life enhancing at the human level and whether they are effective in programming individuals to cooperate readily and selflessly to promote their cultural communities.

Peirce appears to have supposed that religion played this role in human life and culture. "The ***raison d'être*** of a church", he wrote, "is to confer upon men a life broader than their narrow personalities" (from "Religion and Politics", c. 1895a, b: CP 6.451). "[R]eligion is a great, perhaps the greatest, factor of that social life which extends beyond one's own circle of personal friends. That life is everything for elevated, and humane, and democratic civilization; and if one renounces the Church, in what other way can one as satisfactorily exercise the faculty of fraternizing with all one's neighbours?" (from "Religion and Politics", c. 1895a, b: CP 6.449).[20] He did not think that religion in its truest sense was a set of theological doctrines or beliefs but something far more vital. "Religion is a life", he said, "and can be identified with a belief only provided that belief be a living belief—a thing to be lived rather than said or thought" (from "What is Christian Faith", 1893a, b: CP 6.439). Peirce supposed that his theory of greater persons ("corporate personality") was especially applicable to religion: "If such a fact is capable of being made out anywhere, it should be in the church" (from "Man's Glassy Essence", 1892a, b: W8: 183; see also EP 1: 350 and CP 6.271).

Obviously the conception of religion presented here is limited to what I regard as its fundamental natural function. Religion in its fullness, like most of the institutions that make up human cultures, is a complex mix of practical and theoretical

[20]Perhaps today, in some cultures, professional sports serve this function as well, or better, than the institution of religion, but that would not have been the case in Peirce's day.

intelligence. But I think it is the crucial role religion plays in promoting social stability and human welfare by programming individuals for the vital affairs of life by inculcating attitudes, values, and ideas that strengthen civilization, that helps explain the ubiquity of religion within human cultures.

Is there another powerful human institution that supports the other great class of human endeavor, an institution that is disinterested in the vital affairs of human life but that supports the human aspiration to advance knowledge—to find things out? Such an institution would have been born not of human existential concerns but of intellectual yearnings; not of a need to believe but a desire to know. Many intellectual pursuits and disciplines contribute to this endeavor, including theology in its philosophical guise, but it is science, as Peirce conceived of it, that is most directly and fully devoted to this purpose. In reflecting on the scientific advances of the 19th century, Peirce remarked on how privileged he felt to have experienced, often in his own home, "the steadily burning enthusiasm of the scientific generation of Darwin", and he reflected on the changed meaning of the word "science". It no longer meant "systematized knowledge", as it once had, "nor anything set down in a book", but "science" had come to mean "a mode of life; not knowledge, but the devoted, well-considered life-pursuit of knowledge: devotion to Truth—not 'devotion to truth as one sees it,' for that is not devotion to truth at all, but only to party —no, far from that, devotion to the truth that the man is not yet able to see but is striving to obtain" (from "The Century's Great Men in Science", 1901a, b: HP 1: 490–491). Science, in this sense, is the quintessential communal enterprise; it aims for a goal that individuals can only hope to contribute to but can never expect to achieve on their own: "all of the greatest achievements of mind have been beyond the powers of unaided individuals" (from "Evolutionary Love", 1892: W8: 203; see also EP 1:369 and CP 6.315). But even though it is the practice of science and devotion to the pursuit of knowledge that is the lifeblood of science, it is the prodigious accumulation of knowledge and methods, the vast network of intelligence distilled from generations of trials and errors, of successes and failures, and preserved in the extended mind of science the institution, that makes it possible for a mere individual, with a brain so small and short-lived, to participate in such a grand enterprise.

Unlike religion, which cultivates a practical intelligence to address and resolve matters of existential import and immediate vital concern, science cultivates a theoretical intelligence, concerned not with practical matters, except incidentally, but with finding out the truth about things. Science is never urgent. The practical beliefs that religion inculcates have resolved doubts and program believers for automatic action without resort to deliberation. The theoretical beliefs that science sanctions are always only quasi-beliefs and are never doubt free. Fallibilism is at the core of the scientific attitude; even when following out the consequences of the most established scientific truths, the genuine scientific investigator will be prepared to admit error when confronted with disconfirming experience. As Peirce understood from his early acquaintance with the leading scientists of the mid-19th century, the advancement of science can only result from a community endeavor—and there will be many setbacks along the way. The contribution of any individual investigator will at most be small.[21]

Place of the Individual

In conclusion, it seems appropriate to question the significance of the human individual in light of Peirce's theory of the externality of mind and the critical dependence of human life and thought on instinctive and social mind. It may seem as if individuals are merely means for animating minded institutions and for furthering institutional ends, whether those ends are to fix beliefs that promote beneficial behavior and stabilize society or to embrace conjecture and experiment in the hope of contributing to the body of "established truths". Many of Peirce's remarks support this view. He said that we are "[m]ere cells of the social organism" and that "[o]ur deepest sentiment pronounces the verdict of our own insignificance" (from a variant of "Philosophy and the Conduct of Life", 1898a: CP 1. 673; see also EP 2: 40 and CP 1.647). Our separate existence as individuals "is manifested only by ignorance and error" and, insofar as we are anything apart from our fellows, we are "only a negation" (from "Some Consequences of Four Incapacities", 1868: W2: 241–242; see also EP 1:55 and CP 5.317). Yet we must not take lightly the fact that it is living individuals who have the power to really make things happen in the world. Peirce criticized Comte for turning his heroes into biased abstractions and neglecting their "living reality and passion" and their "concrete souls" (from "The Comtist Calendar", 1892a, b: W8: 267–268). The importance of institutions and social minds for human survival and civilized life is profound and certainly individual achievements must always be understood in the context of communal life and thought, yet it is the individual, however rare, who has the power to step out of the mainstream and introduce something novel and unexpected to confront old ideas. It is then, Peirce said, that the future of civilization trembles in the balance.[22]

[21]To avoid misunderstanding, it is worth repeating that I make no pretense of making definitive pronouncements about the nature of religion or science. My aim here is to consider how Peirce's conception of belief, in particular as it applies to matters of vital importance and to theoretical concerns, may help illuminate the development of external minds in institutions and society. The institutions of religion and science are of key importance for this inquiry—which clearly is still preliminary.

[22]For more extended discussion of Peirce's views on the importance of individuals see Houser (2013, 2014b).

References

Adams, Fred, and Kenneth Aizawa. 2010. *The bounds of cognition*. Hoboken, NJ: Wiley Blackwell.

Colapietro, Vincent. 1989. *Peirce's Approach to the Self.* Albany: State University of New York Press.

Fisch, Max H. 1956. The critic of institutions. *Proceedings and Addresses of the American Philosophical Association* 29: 42–56.

Forgas, J. P., K. D. Williams, and L. Wheeler, eds. 2001. *The social mind: cognitive and motivational aspects of interpersonal behavior.* New York: Cambridge University Press.

Gelpi, Donald L. 2008. *The gracing of human experience: Rethinking the relationship between nature and grace*. Eugene, Oregon: Wipf and Stock Publishers.

Hausman, Carl. 1993. *Charles S. Peirce's evolutionary philosophy*. New York, : Cambridge University Press.

Houser, Nathan. 2013. Peirce's neglected views on the importance of the individual for the advancement of civilization. *Cognitio; Revista de Filosofia* 14.2: 163–77.

Houser, Nathan. 2014a. The intelligible universe. *Peirce and biosemiotics: A guess at the riddle of life,* eds. Vinicius Romanini and Eliseo Fernández, 9–32. Heidelberg: Springer.

Houser, Nathan. 2014b. Bohemians like me. *Charles Sanders Peirce in his own Words: 100 years of semiotics, communication and cognition,* eds. Torkild Thellefsen and Bent Sorensen, 137–144. Boston/Berlin: Walter de Gruyter.

Houser, Nathan. 2016. The imperative for non-rational belief. Forthcoming in *Cognitio*.

James, William. 1896. The will to believe. *The New World* 5: 327–347. Reprinted in *The Will to Believe and Other Essays in Popular Philosophy,* 1–31. New York: Longmans, Green and Company.

James, William. 1920[1904]. Does "consciousness" exist? *The Journal of Philosophy, Psychology and Scientific Methods* 1.18: 477–491. Reprinted in *Essays in Radical Empiricism.* New York: Longmans, Green and Company, 1920).

Mead, George Herbert. 1934. *Mind, self and society from the standpoint of a social behaviorist,* ed. Charles W. Morris. Chicago: University of Chicago Press.

Nguyen, Nam T. 2011. *Nature's primal self: Peirce, jaspers, and corrington.* Lanham: Maryland: Lexington Books.

Peirce, Charles S. 1839–1914. The Charles S. Peirce Papers, The Houghton Library, Harvard University (referenced using the Robin numbering system—e.g. the 273rd sheet from the manuscript folder numbered 1573 in the Robin Catalogue is cited as R 1573.273).

Peirce, Charles S. 1868. "Some consequences of four incapacities". *Journal of Speculative Philosophy* 2: 140–157; reprinted in CP 5.264–5.317, W2: 211–242, and EP 1: 28–55.

Peirce, Charles S. 1877–1878. "Illustrations of the Logic of Science". *Popular Science Monthly* [*PSM*] (consisting of: "The fixation of belief," *PSM* 12: 1–15, November 1877; "How to Make Our Ideas Clear", *PSM* 12: 286–302, January 1878; "The Doctrine of Chances", *PSM* 12: 604–615, March 1878; "The Probability of Induction", *PSM* 12: 705–718, April 1878; "The Order of Nature", *PSM* 13: 203–217, June 1878; and "Deduction, induction, and hypothesis", *PSM* 13:470–482, August 1878); these papers are reprinted, with variations, in CP 5.358–5.410, 2.619–2.693, 6.395–6.427, W3: 242–338, and EP 1: 109–199.

Peirce, Charles S. 1887–88. "A guess at the riddle". Unfinished draft of work on Peirce's system of philosophy built on his categories. W6: 165–210; see also EP 1: 245–279 and, for disconnected selections, CP 1.354, 1.1–2, 1.355–1.368, 1.373–1.375, 379–383, and 1.385–1.416.

Peirce, Charles S. 1891. "The architecture of theories". *The Monist* 1: 161–176; reprinted in EP 1: 284–297 and CP 6.7–6.34. Critically edited printer's copy manuscript of 1890 is published in W8: 98–110.

Peirce, Charles S. 1892a. "Man's Glassy Essence". *The Monist* 3: 1–22; reprinted in EP 1: 334–351 and CP 6.238–6.271. *Critically edited printer's copy manuscript is published in* W8: 165–183.

Peirce, Charles S. 1892b. "The Comtist Calendar". *The Nation* 54.1386: 54–55. *Reprinted in* W8: 267–270.

Peirce, Charles S. 1893. "What is christian faith?" *The Open Court* 7: 3743–3745. Reprinted in CP 6.435–6.448 and in forthcoming W9.

Peirce, Charles S. 1894. "How to reason: A critick of arguments" (known as "The Grand Logic"). Peirce was unable to find a publisher for this book. Parts of this work appear throughout the *Collected Papers*—see Burks' bibliography (CP 8, pp. 278–280) for details. It will be published as W11 of the Indianapolis critical edition.

Peirce, Charles S. 1895. "Of reasoning in general". First and apparently only completed chapter of "Short Logic," a book on logic that Peirce began after his manuscript of "How to Reason" was rejected by Ginn & Co. Published in EP 2: 11–26. See Burks' bibliography (CP 8, p. 280) for where parts appear in the *Collected Papers*.

Peirce, Charles S. c.1895. "Religion and politics". Apparently written for a newspaper (see Burks' bibliography, CP 8, p. 286). Published in part in CP 6.449–451.

Peirce, Charles S. 1898a. "Philosophy and the conduct of life". First Cambridge Conferences Lecture. Published in EP 2: 27–41, CP 1.616–648 (and 1.649–1.677 for variants), and RLT: 105–122.

Peirce, Charles S. 1898b. "The first rule of logic". Fourth Cambridge Conferences Lecture. Published in EP 2: 42–56, CP 5.574–589, and RLT: 165–180.

Peirce, Charles S. 1901a. "The century's great men in science". The New York Evening Post, 12 January 1901. Reprinted in Annual Report of the Board of Regents of the Smithsonian Institution, 693–700;Washington, D.C.: Government Printing Office. Also reprinted in The 19th Century: A Review of Progress during the Past One Hundred Years in the Chief Departments of Human Activity, 312–322; New York and London: G. P. Putnam's Sons: the Knickerbocker Press). *Published in HP* 1: 489–496.

Peirce, Charles S. 1901–1902. "Minute logic". Manuscript (R 425–434), unpublished in Peirce's lifetime. Selections published piecemeal in the *Collected Papers* (see Burks' bibliography, CP 8, pp. 293–294 for details). "Minute Logic" will be published whole as a volume of the Indianapolis critical edition.

Peirce, Charles S. 1903a. "Pragmatism as the logic of abduction". Seventh Harvard Lecture. Published in EP 2: 226–241, CP 5.180–212 (in part), and HL: 241–256.

Peirce, Charles S. 1903b. "The three normative sciences". Fifth Harvard Lecture. Published in EP 2: 196–207, CP 5.120–150, and HL: 205–220.

Peirce, Charles S. 1903c. "The seven systems of metaphysics". Fourth Harvard Lecture. Published in EP 2: 179–195, HL: 189–203, and scattered in CP 5 (see EP 2: 179 headnote for details).

Peirce, Charles S. 1903d. "How to theorize". Eighth Lowell Lecture. Published in part in CP 5.590–5.604 (for more detail see Burks' bibliography, CP 8, p. 295).

Peirce, Charles S. 1906. "Prolegomena to an apology for pragmaticism." *The Monist*, 16.4: 492–546; reprinted in CP 4.530–572.

Peirce, Charles S. 1907. Manuscript of letter to the editor of *The Nation* (R 318). A substantial selection from this manuscript is published in EP 2: 298–433 (entitled "Pragmatism").

Peirce, Charles S. i. 1867–1913. *Collected papers of Charles Sanders Peirce*. Vols. 1–6, eds. Charles Hartshorne and Paul Weiss. Cambridge: Harvard University Press, 1931–1935. Vols. 7–8, ed. Arthur W. Burks. Cambridge: Harvard University Press, 1958. [References to Peirce's papers will be designated by CP, followed by volume, period, paragraph number.].

Peirce, Charles S. i.1867–1913. *Writings of Charles S. Peirce: A chronological edition*. Vols. 1–6 to date, ed. the Peirce Edition Project. Bloomington: Indiana University Press. [References to these volumes will be designated by W, followed by volume number, colon, page number.].

Peirce, Charles S. 1985. *Historical perspective on peirce's logic of science,* 2 vols. Ed. Carolyn Eisele (New York: Mouton). Example reference to vol. 1, p. 491: HP 1: 491.

Peirce, Charles S. i. 1867–1893. *The essential peirce: Selected philosophical writing.* Volume 1 (1867–1893), eds. Nathan Houser and Christian Kloesel. Bloomington: Indiana University Press, 1992. [References to this volume will be designated by EP 1, followed by colon, page number.].

Peirce, Charles S. i. 1898. *Reasoning and the logic of things: The cambridge conferences lectures of 1898,* ed. Kenneth Laine Ketner. Cambridge: Harvard University Press, 1992. [References to this volume will be designated by RLT, followed by lecture number, colon, page number.] Introduction, and comments, by Kenneth Laine Ketner and Hilary Putnam: 1992: 1–102.S.

Peirce, Charles S. Pragmatism as a principle and method of right thinking: The 1903 Harvard Lectures On Pragmatism. edited by Patricia Ann Turrisi. (Albany: State University of New York Press, 1997).

Peirce, Charles S. i. 1893–1913. *The essential Peirce: Selected philosophical writing.* Volume 2 (1893–1913), ed. the Peirce Edition Project. Bloomington: Indiana University Press, 1998. [References to this volume will be designated by EP 2, followed by colon, page number.].

Peirce, Charles Sanders and Victoria Lady Welby. Semiotic and significs: The correspondence between Charles S. Peirce and Victoria Lady Welby, Charles S. Hardwick, ed. with the assistance of James Cook. (The Press of Arisbe Associates, Second Edition, 2001).

Putnam, Hillary. 1975[1960]. Minds and machines. *Dimensions of mind,* ed. Sidney Hook, 148–180. New York: New York University Press. Reprinted in *Mind, Language and Reality*, 362–385.

Quine, W. V. 1981. The pragmatists' place in empiricism. *Pragmatism, its Sources andProspects,* eds. R. J. Mulvaney and P. M. Zeltner, 36–37. Columbia: University of South Carolina Press.

Royce, Josiah. 1918. *The problem of christianity*. New York: The MacMillan Company.

Singer, Milton. 1991. *Semiotics of cities, selves, and cultures: Explorations in semiotic anthropology*. Berlin: Walter de Gruyter.

Valsiner, J., and R. van der Veer. 2000. *The social mind; Construction of the idea*. New York: Cambridge University Press.

Wiley, Norbert. 1994. *The semiotic self.* Chicago: University of Chicago Press.

Wiley, Norbert. 2006. Peirce and the founding of American sociology. *Journal of Classical Sociology* 6(1): 23–50.

Chapter 22
Habit as a Law of Mind: A Peircean Approach to Habit in Cultural and Mental Phenomena

Elize Bisanz and Scott Cunningham

> *A habit is not an affection of consciousness; it is a general law of action.*
>
> Charles S. Peirce

Abstract In *The Architecture of Theories* and *The Law of Mind,* Charles S. Peirce declares the categories of *Chance* and *Continuity* as determinant for the emergence and evolution of ideas on their way from individuality to generality. Ideas, as the emergence of mental activity, spread continuously affecting other ideas and establish patterns of activities. In this process of spreading they *lose energy intensity* and by merging into other ideas *gain generality.* Hence, ideas embody both individuality in the sense of occurring once, and continuity as the bonding law of cultural unity, through the force of habit. For Peirce, all thought, including the manifestation of ideas, is performed by means of signs (Peirce 1868: 103–114) anchored in sign-systems. Accordingly, general ideas are performed by general signs called symbols. Symbols are the most fundamental sign-category used to develop and establish cultural "evolution" (Cassirer 1953). Being general ideas, symbols also follow the laws of individuality and continuity. This paper will compare the function of patterns and regularities in mind activity with the function of symbol activity in the establishment of cultural patterns and will argue that both phenomena can be understood as the result of the law of habit.

Keywords Symbolic form · Bonding (weld) · Energy · Individuality · Reason (reasoning)

E. Bisanz (✉)
Leuphana Universität, Luneburg, Germany
e-mail: bisanz@uni-lueneburg.de

S. Cunningham
Texas Tech University, Lubbock, TX, USA
e-mail: scott.cunningham@ttu.edu

D.E. West and M. Anderson (eds.), *Consensus on Peirce's Concept of Habit,*
Studies in Applied Philosophy, Epistemology and Rational Ethics 31,
DOI 10.1007/978-3-319-45920-2_22

Habit as Mind Activity

There is no doubt that habit, as a behavior-pattern, should be exclusive to humans. Nevertheless, there are fundamental differences in both the structure and the purpose of habit.

In order to understand the necessity underlying habit activity we need first to identify its purpose or function as it applies to human activity. Is there a law of habit? If there is a law of habit, is it part of the process of habit activity itself, or part of the larger logically necessary context within which habits develop? Peirce offers some characteristic features of the law of habit that instantiate its role in human activity. Peirce's general ideas or propositions concerning habit can be summarized as follows:

- as a process of reasoning;
- as a working process, an ongoing conscious activity; and
- as the result of intelligent behavior.

Peirce's thoughts concerning habit are offered mostly within the contexts of reasoning and consciousness. Reasoning is understood as, "… the process in which the reasoner is conscious that a judgment or judgments, the premises, according to a general habit of thought, which he may not be able precisely to formulate, but which he approves as conducive to true knowledge" (Baldwin 1902: 426), with consciousness being Habit, an act of reasoning understood as the process of conscious action according to a "general habit of thought" (Baldwin 1902: 426). Peirce explores habit most frequently within the specific context of reasoning-processes. Accordingly, here we will consider the laws contributing to mind-processes as well as laws involved in habit-processes.

Reasoning, for Peirce, underlies a general law of mind activity. One of Peirce's main concerns, especially in his late thought, centered around establishing a scientific approach to the observation and analysis of thought-processes. Many of his writings seek to establish the necessity of a close bond between reasoning and signs. Throughout *The Monist Series* one can observe the unfolding of several basic concepts concerning how mind-activity and sign-activity are intertwined and submerged within universal meaning-processes. The focus will now turn to three core categories and to what extent these determine habit formation as they appear in "The Law of Mind". These core categories are *Chance*, *Continuity*, and *Individuality*.

In his article entitled "The Architecture of Theories", of these three core categories, Peirce declares *Chance* to be the fundamental category pertaining to evolution. In order to see this he suggests the examination of the *General Law of Mental Action*. Peirce understands *mind*—the process of brain activity—as always being subject to spontaneity; although the law of mind as reasoning and as the activity of brain process follow the three main classes of logical inference: Abduction, Deduction, and Induction. The bottom line, for Peirce, is that the essence of the law of mental action, when mind is in fact both affected and effected, is for the most part induced by *Chance* or uncertainty.

In a subsequent paper, "The Law of Mind", Peirce introduces *Continuity* as a genuine component of the law of mind. With *Continuity* Peirce explains how, in the emergence of mind activity, ideas spread continuously and then affect further ideas, "Thus, by induction, a number of sensations followed by one reaction become united under one general idea followed by the same reaction; while by the hypothetic process, a number of reactions called for by one occasion get united in a general idea which is called out by the same occasion. By deduction, the habit fulfills its function of calling out certain reactions on certain occasions" (Peirce [1892]2009: 97). By continuously spreading, habit, depletes its energy and therefore its ability to affect and bond, while simoultaneously *gaining generality*.

Nevertheless, ideas are individual in the sense of occurring once—for one cannot have the same idea twice in one's mind—and yet have it remain continuous. Furthermore, the law of *Continuity* is closely related to the role of *time-flow*. Time-flow in thought-processes follow a specific vector: along the chain of the past-present-future time axis, each potential sequence of mind activity is induced by a previous sequence. As present can only be affected by the past, any future sequence is always entailed from and induced by present mind-activity. This implies that both past and future ideas can be present and observable by direct perception without being physically present.[1] Peirce argues that since immediate consciousness is embodied by infinite intervals of time, the object of consciousness is perceived as a continuous entity through immediate feeling. Therefore, infinity and *Continuity* determine our consciousness of time, and ideas evolve in this time-flow, whereas time interval in mind is understood as a series of *comparative perceptions* where, "The last moment of the instant is the point of connection of the past to the present" (Peirce [1892]2009: 91).

Furthermore, the reasoning process or mind-activity is strongly affected by the basic element of *feeling* and accordingly its laws. Feeling as a purely sensational category also underlies the law of continuity. It spreads when brain activity is initiated through the process of excitation. Here again, Peirce explains the spreading as an additive process of independent states of excitations within the brain. "When a particular kind of feeling is present, an infinitesimal continuum of all feelings differing infinitesimally from that is present" (Peirce [1892]2009: 91). This unified feeling, "… is a mere amorphous continuum of protoplasm, with feeling passing from one part to another" (Peirce [1892]2009: 91). Just as we cannot be aware of when and how an idea reaches its "conclusion", in the same manner we cannot define a specific kind of feeling or even a specific time or context. Instead, there is an *immediate continuity* of feeling between proximate parts of activated brain that in turn determine the pattern of the coordinating structure established within the accompanying nerve-matter activity.

[1]For a nice discussion of the real/existent distinction see Samway (1995: 77–79).

Like ideas, feelings evolve in a flow where, "… every state of feeling is affectable by every earlier state" (Peirce [1892]2009: 91–92).[2] This suggests that a feature of the *law of mind* is that it has its own connectivity pattern which is not of the same nature as the *law of physical force*. This is due to the fact that the time relation involved, moving from past to future, is different from the time relation moving from future to past.

To summarize, the laws of bonding for the chain of brain excitations, that hold and bring brain activity together as a singular conscious sensation are induced by:

- the tendency of an idea to evoke further ideas;
- the energy affecting other ideas; and
- intrinsic qualities as feelings.

The process of reasoning includes all of the above. Peirce explains, "Whenever ideas come together they tend to weld into general ideas; and wherever they are generally connected, general ideas govern the connection; and these general ideas are living feelings spread out" (Peirce [1892]2009: 96). Inquiry into these states of conscious feeling unity can be performed using objective scientific observation as they are feelings *in presence*. Any changes and or resemblances in the process of affecting are both perceivable and persistent.

How do these laws affect habit? We begin with the role of Sensation. About this role Peirce writes, "When we think, we are conscious that a connection between feelings is being determined by a general rule, we are aware of being governed by habit. Intellectual power is nothing but facility in taking habits and in following them in cases essentially analogous to, but in non-essentials possibly widely remote from, the normal cases of connections of feelings under which those habits were formed" (Peirce [1891]2009: 63).

This analogous situation implies that feeling or sensation not only induces habit-processes but are also induced by them. Peirce continues by substantiating sensation at the most basic level and ascribes to habit the ability to evoke neural sensations: "Feeling may be supposed to exist, wherever a nerve-cell is in an excited condition. The disturbance of feeling, or sense of reaction, accompanies the transmission of disturbance between nerve-cells or from a nerve-cell to a muscle-cell or the external stimulation of a nerve-cell. General conceptions arise upon the formation of habits in the nerve-matter, which are molecular changes consequent upon its activity and probably connected with its nutrition" (Peirce [1891]2009: 64).

The law of habit differs substantially from other laws. Unlike all physical laws it is never absolute but is instead determined by the laws of *Chance*, *Continuity* and *Individuality*. As the law of habit is closely connected to mind-activity, no exactitude can be expected in it. Thoughts, for Peirce, unfold within the flow of signs that he called Semiosis. Ideas within this sign flow are unique, individual,

[2]Possibly from hearing Cajal's lectures given at the Decennial Celebration at Clark University in 1899. For a fine account of the details surrounding Cajal's visit see Haines (2007).

non-static, mind-activities that are not susceptible to absolute predictability. Peirce declares this to be a basic requirement for the transformation of habits within the evolution of human behavior. Similar to sign-activity, habits grow and spread in both time and cultural space. They are also uniquely hypothetical in that the probability of acquiring a habit is deeply interrelated with the basic probability of the occurrence of a conscious behavior.

Habit as a relational process, in spite of its unique hypothetical character, must be conceived within logic. It promotes the process of mediation between feeling and reaction, contributing to the fixation of the hypothetical character and enhances evolution. Peirce writes: "Three conceptions are perpetually turning up at every point in every theory of logic, and in the most rounded systems they occur in connection with one another. They are conceptions so very broad and consequently indefinite that they are hard to seize and may be easily overlooked. I call them the conceptions of First, Second, Third. First is the conception of being or existing independent of anything else. Second is the conception of being relative to, the conception of reaction with, something else. Third is the conception of mediation, whereby a first and second are brought into relation." Peirce further elaborates these by explaining: "Chance is First, Law is Second, the tendency to take habits is Third. Mind is First, Matter is Second, Evolution is Third" (Peirce [1891]2009: 68).[3]

Habit organizes the universe of knowledge in evolving entities by forming a bond between feelings and regularities of general concepts understood as cultural conventions. To be able to capture the evolving character of culture it is essential to understand habit as thirdness:

> But no mental action seems to be necessary or invariable in its character. In whatever manner the mind has reacted under a given sensation, in that manner it is the more likely to react again; were this, however, an absolute necessity, habits would become wooden and ineradicable, and no room being left for the formation of new habits, intellectual life would come to a speedy close. Thus, the uncertainty of the mental law is no mere defect of it, but is on the contrary of its essence. … There always remains a certain amount of arbitrary spontaneity in its action, without which it would be dead. (Peirce [1892]2009: 97)

Within the flow of sign-processes (Semiosis), habits inform, reinforce, and enhance memory as they induce activity-patterns in reasoning. As Peirce explains, in reasoning, the formation of habit is conscious activity, which first begins in individual mind and is passed over to a collectively culturally acquired process.

Habit contributes to the efficiency of energy-control within the brain and cultural activities. It is the ability to organize effort and develop general principles to transform them into activity patterns. The above introduced laws of mind-activity, *Chance*, *Continuity*, and *Individuality*, are also at work in habit-activity. Habits evoke and induce further habits, they determine both the transformation and the dynamics of the evolution of future habits. As Peirce states in correspondence with Lady Welby, "It appears to me that the essential function of a sign is to render

[3]Firstness, Secondness, and Thirdness will not be addressed here as they are discussed extensively in several chapters within this volume.

inefficient relations efficient,—not to set them into action, but to establish a habit or general rule whereby they will act on occasion. According to the physical doctrine, nothing ever happens but the continued rectilinear velocities with the accelerations that accompany different relative postions of the particles. All other relations, of which we know so many, are inefficient. Knowledge in some way renders them efficient; and a sign is something by knowing which we know something more" (Hardwick [1904]2001: 31–32).

Habit as Symbol-Processing Activity

It is true that Peirce discusses habit mostly in the context of individual consciousness and reasoning processes. In those contexts it is defined as an element of individual perception, nevertheless, we find a building of the intentional force of general principles in his semeiotic within the sphere of sign-activity, where habits are also involved in collective meaning processing along with knowledge acquisition, assimilation, and maintenance.

Symbols, more than any other sign type, exemplify the features evolving of universal meaning-processing in its full scope. In his "Prolegomena to an Apology of Pragmaticism", Peirce explains, "A Symbol incorporates a habit, and is indispensable to the application of any intellectual habit, at least" (Peirce [1906]2009: 309). Symbols are means of hypostatic thought. They exemplify experience and concepts of future possibilities, a step essential to activate memories and transform them into effective communication tools in *praesentia* in the form of habits. In fact, as Peirce points out, symbols only function in and through habits, they are highly dependent on the repetitive use of a given sign-relation. "But since symbols rest exclusively on habits already definitely formed but not furnishing any observation even of themselves, and since knowledge is habit, they do not enable us to add to our knowledge even so much as a necessary consequent, unless by means of a definite preformed habit" (Peirce[1906]2009: 309).

Strong bonds form between habits and signs. Their very reality depend upon one another: Habit formation is evoked, induced, processed, and established primarily through signs and their activity, and vice versa. The activity of meaning processing can only come about through the habit activities of the interpretant. Habit merges the subject into the cultural sphere where symbolic activity unfolds into multiple forms. Within the cultural sphere interactivity occurs between emergent, spreading new habits with the already present, established dominant habits. In this interactivity is found the very essence of cultural dynamic. This inner energy of cultural activity which, primarily implemented by habit, conserves while at the same time renews, replenishes and reenergizes cultural knowledge.

Peirce refers to the bond (between habits and signs) on several levels (including as the sign-habit relation). Signs can be established and maintained as signs through continuity following habitual action with an inherent potential for further growth. Peirce makes two propositions in this respect:

> What are signs for, anyhow? They are to communicate ideas, are they not? … The pragmaticist insists that this is not all, and offers to back his assertion by proof. He grants that the continual increase of the embodiment of the idea-potentiality is the *summum bonum*. But he undertakes to prove by the minute examination of logic that signs which should be merely parts of an endless viaduct for the transmission of idea-potentiality, without any conveyance of it into anything but symbols, namely, into action or habit of action, would not be signs at all, since they would not, little or much, fulfill the function of signs; and further, that without embodiment in something else than symbols, the principles of logic show there never could be the least growth in idea-potentiality. (Peirce [1905–1906]2009: 276–277)

Habits, being deeply rooted in cultural space, develop into physical laws in culture by the promotion of symbolic processes. They, the symbolic processes, in turn, via habit, transform cultural information into triadic tool structures to inform, reinforce, and enhance the cultural sphere. Accordingly, as Symbols are embodiments of habitual reasoning, the concept of habit should be regarded also as both knowledge-preserving and knowledge-maintaining process.

The role of habit to inform, reinforce, enhance, and maintain the structure of cultural knowledge via symbols comprise the bedrock of contemporary cultural science. In his book, *The Logic of Cultural Sciences*, Ernst Cassirer elaborates the historical conflict between the two main opposing scientific perspectives (science and humanities) to explain mind and culture. This opposition in turn led to a redundant cultural conflict known as the "tragedy of culture". The term refers primarily to the situation of a constantly evolving cultural production in which the cultural objects or symbolic-systems are no longer able to serve their original purpose that would be to organize human thought and work, but instead, develop in intrinsic dynamic destabilizing habitual patterns. Cassirer explains this redundancy as a continual agitation or clash of the symbolic forms against cultural habits and proposes a new approach to understand its resulting impact. To reconcile the schism between the biological and cultural understanding of mind, Cassirer suggests we focus on the binding character of the symbolic forms, both on structural and semantic levels, arguing that as representations, symbolic forms or symbols, in general, unfold a strong binding force within the cultural body. Symbolic forms, for Cassirer, organize our thoughts and actions in patterns, and in doing so link the individual activity of sign-interpretation with the general structural pattern of symbolic evolution.

In this sense, symbols exemplify the law of habit processing. Accordingly then habits organize our universe of signs, meanings, and information in efficient practical schemes of cultural activity.

Habits are separate individual real entities, likened to abstract behavioral concepts, of commensurate power and while quite often find themselves reduced to abstract behavioral concepts, habits not only should not but can not be conceived as abstract behavioral concepts or reduced to such. To this situation Peirce comments,

> There is a well-known law of habit-taking which means that many kinds of behavior of many kinds of things such that if such a thing behaves in such a way more than once on the same one of many kinds of occasions, then there will be a slightly increased probability that that individual thing will behave on the next occasion of that kind in the same kind of way.

> This statement is, as the reader will not have failed to remark, as vague as it possibly could be in three distinct respects; first, as to the kinds of individual things to whose behavior it applies; secondly, as to those kinds of behavior; and thirdly, as to the kinds of occasions as to whose occurrence it relates. In this tridimensional infinity of vagueness, I suppose everybody would assent to its truth; and I am not prepared to make the statement definite in either of these respects. (Peirce 1911: 018–019)

Habit, as real mind-process, becomes submerged deep in our conscious reasoning activities in the form of memorizing skill. Habit then serves to inform, reinforce, enhance, and maintain the structure of cultural knowledge. Habit, like all mind-activity, is in essence diagrammatic. It unfolds in the processing of a set of information to become a meaningful scheme. Hence, "That which every sign does is determine its interpretant. The responsive interpretant, or signification, of one kind of signs is a vague presentation, of another kind is an action, while of a third is involved in a habit and is general in its nature. It is to this third class that a diagram belongs. It has to be interpreted according to conventions embodied in habits."[4]

The importance of this quote for Peirce is to be found in the last two sentences. It is clear in the next to last sentence that Peirce thought the nature of diagrams to be that of thirds, or triadic in nature. In the last sentence Peirce shows the connection between the structure of the diagram and the law of habit. The diagram will result from the playing out of the conventions of the law of habit.

Peirce and Cassirer both consider signs to be integral for the establishment and development of habits. The importance of the relation between signs and the development of habit can be the classical (popular) habit-pattern of a cue-routine-reward/punishment, that can also be verified by the triadic sign activity of Semiosis including a *sensation* involved in an *activity* and resulting in a fixation of *meaning*. This basic pattern of Semiosis occurs in learning, symbolic perception, and behavioral or habit-formations.

Thus, like *sensation* and *volition*, *habit* is one of the three forms of consciousness and is closely intertwined with them, as the beginning of every habitual activity starts by sensation and is endured by volition: "We now come to the third mode of consciousness that I consider to be utterly unlike either of the two already considered [sensation and volition]. I may add that I hold it to be the only other undecomposable[5] mode; but this will not be a matter of any particular moment to us in the course of this essay. If you ask me what this mode of consciousness is, I shall reply, in brief, that it is that being aware of acquiring a habit" (Peirce 1913: 23).

Ernst Cassirer in *The Philosophy of Symbolic Forms,* aptly declares the natural and cultural urge of humans to understand, communicate, and express themselves to be the core attribute of human behavior, which he summarizes by the term

[4]Signs, Thoughts, Reasoning (title given in forthcoming *Prolegomena to a Science of Reasoning, 2016*), c. 1906, Institute for Studies in Pragmaticism, Peirce Manuscripts (MS) 293, Texas Tech University.

[5]Indecomposable elements for Peirce were the stable, invariant elements that could be found in any phaneron.

"symbolic animal".[6] Consequently, to understand a person we have to understand his symbolic ability, "We cannot define man by any inherent principle which constitutes his metaphysical essence—nor can we define him by any inborn faculty or instinct that may be ascertained by empirical observation. Man's outstanding characteristic, his distinguishing mark, is not his metaphysical or physical nature—but his work. It is this work, it is the system of human activities, which defines and determines the circle of 'humanity'" (Cassirer 1944: 68).

Genuine examples of human work as symbolic form can be found in activities such as rites, prayers, and myths. To understand their underlying systems and structures, Cassirer proposes to approach them as embodiments of knowledge, that endure within a culture as both collective and singular habits. As such they, "show us what *happens* in certain human groups under certain conditions" (Cassirer 2000: 42). This would mean that what happens actually shows what it purports to show.

Cassirer describes Symbolic Systems as "languages"[7] with specific logical structure, accordingly culture results through the intertwinement of "languages" such as, "the language of science, the language of art, of religion, and so on—in its particularity" (Cassirer 2000: 42). As cultural objects they have a twofold representational relation, they represent an interpretant as well as a structure, "Religion, language, art … are tokens, memorials, and reminders in which alone we can grasp a religious, linguistic, or artistic meaning. And it is just in this reciprocal determination that we recognize a cultural object" (Cassirer 2000: 42).

The cultural object's symbolic value is immersed in a common knowledge sphere of culture that the physical objects embody. Symbols as cultural entities are the results of human *work*, "… the physical existence, the objectivity represented, and the personal expression—are determining and necessary for everything that is a 'work' and not merely an 'effect', and which in this sense belongs not only to 'nature' but also to 'culture'" (Cassirer 2000: 43).

In *The Philosophy of Symbolic Forms* Cassirer symbol is explained primarily as a transmitter of cultural meaning by its growth. Signs are explained as vehicles of experience along with conception, expression, and their use of signs as mental process are considered to be the, "beginning of intelligence" (Langer 1948: 23).

In Susanne K. Langer's seminal book, *Philosophy in a New Key: A Study in the Symbolism of Reason, Rite, and Art*, she declares the modern human mind to be, "an incredible complex of impressions and transformations; and its product is a fabric of meanings that would make the most elaborate dream of the most ambitious tapestry-weaver look like a mat. The warp of that fabric consists of what we call 'data' the signs to which experience has conditioned us to attend, and upon which

[6]Symbolic forms, for Cassirer, are embodiments of symbolic meaning. In symbolic meaning Cassirer distinguishes three meaning categories: a. the *expressive* meaning, b. *representational* meaning, and c. *signifying* (a literal translation would be signification) meaning. Together all three build the sign structure of symbolic forms. Therefore, according to Cassirer, the study of symbolic forms gives us access to the forms of human reasoning.

[7]Cassirer has a general understanding of "language" as the structure of symbolic forms (see note 22).

we act often without any conscious ideation. The woof is symbolism. Out of signs and symbols we weave our tissue of 'reality'" (Langer 1948: 227).

In relation to habit, Langer points out an additional feature of signs, "The only single habit involved in the whole process [sign-processes] is the habit of constantly obeying signs" (Langer 1948: 227–228). Humans navigate in culture by virtue of obeying signs to communicate meaning. Meaning for Langer has both, "a logical and a psychological aspect." … "Psychologically, any item that is to have meaning must be *employed* as a sign or a symbol; that is to say, it must be a sign or a symbol *to* someone. Logically, it must be *capable* of conveying a meaning, it must be the sort of item that can be thus employed" (Langer 1948: 42). Such is also the case for habit. Habit is simultaneously both logical and psychological activity. It has to be fixed by or in a *sign* and it must be carried out by *someone*. Symbol, of all signs, is the most profound and common forms of habit-vehicles.

> "The fundamental difference between signs and symbols is their difference of association, and consequently of their *use* by the third party to the meaning function, the subject; signs *announce* their objects to him, whereas symbols *lead him to conceive* their objects" (Langer 1948: 49). Langer, like Peirce, points to Sensation as necessary for building habits. For Langer *practical vision*[8] (Sensation), is the ability of spontaneous[9] abstraction from the stream of sense-experience. Spontaneous Abstraction, she argues, is an elementary sense of knowledge and could be regarded as the juxtaposition of symbolic thought (reasoning) and animal behavior (sign-perception). Summarizing, Langer says, "Here, in practical vision, which makes symbols for thought out of signs for behavior, we have the roots of *practical intelligence*. It is more than specialized reaction and more than free imagination; it is conception anchored in reality." (Langer 1948: 217)

Accordingly, Peirce's *habit* can be understood as closely comparable to Langer's *practical intelligence*, an idea which reinforces the notion of habit-acquisition as a skill for the efficient use of resources, not only intellectual, but also at a simple brain-activity level. In fact, in brain-science this theory has been elaborated in sufficient manner, where several empirical experiments have verified a decrease in neural synaptic activities during habit-processing.[10] This notion, that habit acquisition as a means for the optimization of resource use, also applies to the area of cultural activity and can be especially seen in its emphasis in the concept of symbolic formation. Cassirer also emphasizes that the logical and psychological aspects of cultural evolution are brought about through symbolic form. "In all spiritual growth we can discern a twofold trend. It is related to natural, purely

[8]What Langer called practical vision could be considered as equivalent in Peirce to diagrammatic thought.

[9]Spontaneous not intuitive, see Charles S. Peirce, "Questions Concerning Certain Faculties Claimed for Man", *Journal of Speculative Philosophy*, no. 2, (1868): 103–114.

[10]See Christian Hoppe and Jelena Stojanovic. "High-Aptitude Minds," *Scientific American Mind*, 19 (2008): 60–67.

organic growth insofar as both are subject to the law of continuity. The later phase is not absolutely alien to the earlier one but is only the fulfillment of what was intimated in the preceding phase. On the other hand this interlocking of phases does not exclude a sharp opposition between them" (Cassirer 1953: 448). Consequently, symbols embody cultural knowledge in space and time and grow in three "dimensions": representation, expression, and pure meaning. Symbols have a direct meaning that they represent through the expressive features, and third, they develop a meaning level beyond the iconic and expressive spheres. "It is through this transition. … that the form of scientific knowledge is first truly constituted, that its concept of truth and reality definitively breaks away from that of the naïve world view" (Cassirer 1953: 448–449).

Peirce also confirms that knowledge is based on experience as well as habit. In *Studies in Logic by Members of the Johns Hopkins University*,[11] edited by Peirce, he ends his own chapter entitled "A Theory of Probable Inference", with the following paragraph: "Side by side, with the well established proposition that all knowledge is based on experience, and that science is only advanced by the experimental verifications of theories, we have to place this other equally important truth, that all human knowledge, up to the highest flights of science, is but the development of our inborn animal instincts" (Peirce et al. 1883: 181).

The ultimate goal of a comprehensive theory of knowledge and the study of habits, Peirce's *semeiotik* presents multiple theoretical concepts in which three essential elements find their place, (1) the subject as the embodiment of experience, (2) the experiment as the conscious process of sign-activity, and (3) the interpretant immersed in a universal and instinct-guided experience. With these concepts habits comprise the leading role. Habits can be read as footprints of individual and cultural evolution. They can be applied as practical tools to navigate through the vast data of either cultural or natural laws, to build up our own sign-strategies, in order to cope with the unknown, and to communicate and defend our collective or singular sign-worlds. On the other hand, exclusive extensive habit-dependency might lead an instance above mentioned redundant continual agitation or clash of symbolic forms, in which symbols develop their own life of mechanical technocracy, beyond humanity. Peirce thought it up to civilization as to when and whether this happens, "Civilization has certain tendencies toward the strengthening of some habits and the weakening of others" (Peirce 1913: 008).

A profound study of the human habit to develop habits would help culture to find the appropriate doses.

---

[11]A book edited by Peirce containing papers written by both him and his students during his appointment at John Hopkins University as Lecturer in Logic.

Appendix 22.1: General Bibliography of Basic Reference Works on Peirce

Collected Papers of Charles Sanders Peirce, edited by Charles Hartshorne and Paul Weiss (Volumes 1–6) and Arthur Burks (Volumes 7–8) (Cambridge: Harvard University Press, 1931–58). References to Peirce's papers begin with CP and are followed by volume and paragraph numbers. Also available in the Past Masters series, InteLex Corporation.

A Comprehensive Bibliography of the Published Works of Charles Sanders Peirce with a Bibliography of Secondary Studies (second edition, revised), edited by Kenneth Laine Ketner with the assistance of Arthur Franklin Stewart and Claude V. Bridges (Bowling Green, Ohio: Philosophy Documentation Center, Bowling Green State University, 1986). A microfiche edition of Peirce's extensive lifetime publications is available from the same source. References to Peirce's publications begin with P, followed by a number from this bibliography.

Peirce manuscripts in Texas Tech University Library at Texas Tech University, Institute of Studies of Pragmaticism, beginning with MS for manuscript—or L for letter—and followed by a number, refer to the system of identification established by Richard R. Robin in the Annotated Catalogue of the Papers of Charles S. Peirce (Amherst: University of Massachusetts Press, 1967), or in Richard R. Robin, "The Peirce Papers: A Supplementary Catalogue," Transactions of the Charles S. Peirce Society. Vol. 7, No. 1, 1971, pp. 37–57.

Peirce, Charles Sanders. Studies in the Scientific and Mathematical Philosophy of Charles S. Peirce: Essays by Carolyn Eisele. R.M. Martin, ed. (The Hague: Mouton, 1979)

Peirce, Charles Sanders. Historical Perspectives on Peirce's Logic of Science (2 Vols.). Carolyn Eisele, ed. (Berlin: Mouton, 1985)

Peirce, Charles Sanders. The New Elements of Mathematics (4 Vols.). Carolyn Eisele, ed. (The Hague-Paris: Mouton, 1976)

Peirce, Charles Sanders and Victoria Lady Welby. Semiotic and Significs: The Correspondence between Charles S. Peirce and Victoria Lady Welby, Charles S. Hardwick, ed. with the assistance of James Cook. (The Press of Arisbe Associates, Second Edition, 2001)

Peirce, Charles Sanders. Pragmatism as a Principle and Method of Right Thinking: The 1903 Harvard Lectures On Pragmatism. edited by Patricia Ann Turrisi. (Albany: State University of New York Press, 1997)

Peirce, Charles Sanders. Reasoning and the Logic of Things: The Cambridge Conferences Lectures of 1898 by Charles Sanders Peirce, edited by Kenneth Laine Ketner with an introduction by Kenneth Laine Ketner and Hilary Putnam (Cambridge: Harvard University Press, 1991).

Peirce, Charles Sanders. Charles Sanders Peirce: Contributions to *The Nation*: Part One: 1869–1893, edited by Kenneth Laine Ketner and James Edward Cook (Lubbock: Texas Tech University Press, 1975).

Peirce, Charles Sanders. Charles Sanders Peirce: Contributions to *The Nation*: Part Two: 1894–1900, edited by Kenneth Laine Ketner and James Edward Cook (Lubbock: Texas Tech University Press, 1978).

Peirce, Charles Sanders. Charles Sanders Peirce: Contributions to *The Nation*: Part Three: 1901–1908, edited by Kenneth Laine Ketner and James Edward Cook (Lubbock: Texas Tech University Press, 1979).

Peirce, Charles Sanders. Charles Sanders Peirce: Contributions to *The Nation*: Part Four: Index, edited by Kenneth Laine Ketner and James Edward Cook (Lubbock: Texas Tech University Press, 1987)

Peirce, Charles Sanders. The Monist Series, Chicago: The Open Court Publishing Co. 1891–1892 "The Architecture of Theories," "The Doctrine Of Necessity Examined," "The Regenerated Logic," "The Law of Mind," "Man's Glassy Essence," and "Evolutionary Love" (original appearance)

Peirce, Charles Sanders. The Popular Science Monthly Series, New York: D. Appleton and Company. 1877–1878 "The Fixation of Belief," "How to Make Our Ideas Clear," "The Doctrine of Chances," "The Probability of Induction," "The Order of Nature," and "Deduction, Induction, and Hypothesis" (original appearance)

Peirce, Charles Sanders. The Journal of Speculative Philosophy Series, St. Louis: R.P. Studley & Co. 1868–1869 "Questions Concerning Certain Faculties Claimed for Man," "Some Consequences for Four Incapacities," and "Grounds of Validity of the Laws of Logic: Further Consequences of Four Incapacities" (original appearance)

Peirce, Charles Sanders. The Logic of Interdisciplinarity: The Monist-Series, edited by Elize Bisanz, (Berlin: Akademie Verlag GmbH, 2009)

Appendix 22.2: Glossary of Peircean Terminology

Peirce was scientifically meticulous in his use of language. In order to better grasp Peirce's meaning it is necessary to have his terminological understanding available. The definitions below are but a few of the many contributed by Peirce to The Century Dictionary and Cyclopaedia, and the Dictionary of Philosophy and Psychology (Baldwin's). Counts of the number of times the words appear in The Collected Papers are also included.

Bonding—Peirce often used the term "weld" to indicate that one item becomes continuous with a second item. While this term was not contributed by Peirce to either *The Century Dictionary* or *The Dictionary for Philosophy and Psychology* it does appear several times in *Reasoning and the Logic of Things* (Peirce) 1992 (1898), 49, 91–91, 95, 158–160.

The word "weld" appears in *The Collected Papers* 9 times.

Chance—Fall, falling. … A throw of dice; the number turned up by a dice. … Risk; hazard; a balanced possibility of gain or loss, particularly in gaming; uncertainty. … A contingent or unexpected event; an event which might or might not befall. … Vicissitude; contingent or unexpected events in a series or collectively. … Luck; fortune; that which happens to or befalls one. … Opportunity; a favorable contingency: as, now is your *chance*. … Probability; the proportion of events favorable to a hypothesis out of all those which may occur: as, the *chances* are against your succeeding. … Fortuity, especially, the absence of a cause necessitating an event, or the absence of any known reason why an event should turn out one way rather than another, spoken of as if it were a real agency; the variability of an event under given general conditions, viewed as real agency. *The Century Dictionary*, the complete entry for "chance" can be found on pp. 918.

While Peirce did not contribute an entry for the word "chance" to the *Dictionary of Philosophy and Psychology* the word does appear in several entries. "Idea" appears in the entries Peirce contributed for Matter and Form, and Necessity.

The word "chance" appears in *The Collected Papers* 207 times.

Conscious—attributing, or capable of attributing, one's sensations, cognitions, etc., to one's self; aware of the unity of self in knowledge; aware of one's self; self-conscious/present to consciousness; known or perceived as existing in one's self; felt; as *conscious* guilt/aware of an object; perceiving. (a) aware of an internal object; aware of a thought, feeling or volition.—*The Century Dictionary and Cyclopedia*, the complete entries for "conscious" and "consciousness" can be found on pp. 1202–1203.

While Peirce did not contribute a separate entry for either "conscious" or "consciousness" to the *Dictionary of Philosophy and Psychology* the word does appear in several entries. "Conscious" or "Consciousness' appears in the entries Peirce contributed for Inference, Knowledge, Leading Principle, Logic, and Necessity.

The word "conscious" appears in *The Collected Papers* 120 times and the word "consciousness" appears 389 times.

Continuity—Uninterrupted connection of parts in space or time; uninterruptedness. … In *math.* and *philos.*, a connection of points (or other elements) as intimate as that of the instants or points of an interval of time: thus, the continuity of space consists in this, that a point can move from any one position to any other so that at each instant it shall have a definite and distinct position in space. *The Century Dictionary and Cyclopedia*, the complete entries for "continual", "continue", "continued", "continuous", and "continuum" can be found on pp. 1228–1230.

While Peirce did not contribute an entry for the word "continuity" to the *Dictionary of Philosophy and Psychology* the word does appear in several entries. "Continuity" appears in the entries Peirce contributed for Individual, Involution, Logic, and Multitude.

The word "continuity" appears in *The Collected Papers* 180 times.

Determinate/Determine—In *logic,* to explain or limit by adding differences. *The Century Dictionary and Cyclopedia*, the complete entries for determinance, determinant, determinantal, determination, determinative and determinism can be found on pp. 1571–1573.

Peirce did not contribute an entry for either "determinate" or "determine" to the *Dictionary of Philosophy and Psychology*. However, in 1868 Peirce sent a letter to the editors of *The Journal of Speculative Philosophy* whose purpose was to clarify what he meant by "determined" (Vol 2, pp. 190–191). There are also helpful passages regarding the meaning of "determined" in *A Thief of Peirce: The Letters of Kenneth Laine Ketner and Walker Percy* (pp. 35–36).

The words "determinate" or "determine" appear in *The Collected Papers* 86 and 146 times respectively.

Efficient/Efficiency—Producing outward effects; of a nature to produce a result; active; causative. The quality of being efficient; effectual agency; competent power; the quality of power of producing desired or intended effects.

Peirce did not contribute an entry for the words "efficient" or "efficiency" to the *Dictionary of Philosophy and Psychology.*

The words "efficient" or "efficiency" appear in *The Collected Papers* and 48 and 13 times respectively.

Energy—In the *Aristotelian philos.*, actuality; realization; existence; the being of the being no longer in germ or in posse, but in life or in esse: opposed to *power, potency,* or *potentiality.* Thus *first energy* is the state of acquired habit; *second energy*, the exercise of a habit: one when has learned to sing is a singer *in first energy*; when he is singing, he is a singer in *second energy. The Century Dictionary and Cyclopedia*, the complete entry for "energy" can be found on pp. 1926–1927.

Peirce did not contribute an entry for the word "energy" to the *Dictionary of Philosophy and Psychology.*

The word "energy" appears in *The Collected Papers* 112 times.

Exact/Exactitude—precisely correct or right; real; actual; veritable: as, the *exact* sum or amount; the *exact* time. A statement is *exact* which does not differ from the true by any quantity, however small.

While Peirce did not contribute an entry for the word "exact" to the *Dictionary of Philosophy and Psychology* however the word appears in many of the entries Peirce contributed. It appears, for example, in the entries for Index, Individual, Involution, and Law of Thought.

The words "exact" and "exactitude" appear in *The Collected Papers* 220 and.

Generality—The state or condition of being general, in any of the senses of that word…. Something that is general, as a general statement or principle; especially, a saying of a general and vague nature. *The Century Dictionary and Cyclopedia*, complete entries for "general", "generale", "generalizable", "generalization", and "generalize" can be found on pp. 2482–2483.

While Peirce did not contribute an entry for the word "generalization" to the *Dictionary of Philosophy and Psychology* the word does however appear in many of the entries Peirce contributed. "Generalization" appears in the entries for Imagining and Logic.

"Generalization" appears in *The Collected Papers* 73 times and "generality" appears 102 times.

Habit—A usual or customary mode of action; particularly a mode of action so established by use as to be entirely natural, involuntary, instinctive, unconscious, uncontrollable, etc.; used especially of the action, whether physical, mental, or moral, of living beings, but also, by extension, of that of inanimate beings; hence, in general, custom; usage; also, a natural, or more generally an acquired proclivity, disposition, or tendency to act in a certain way.

Peirce did not contribute an entry for the word "energy" to the *Dictionary of Philosophy and Psychology*.

The word "habit" appears in *The Collected* Papers 223 times.

Hypostasis—In *metaph.*, a substantial mode by which the existence of a substantial nature is determined to subsist by itself and be incommunicable; subsistence,—… A hypothetical substance; a phenomenon or state of things spoken and thought of as if it were a substance. … Principle; a term applied by the alchemists to mercury, sulphur, and salt, in accordance with their doctrine that these were the three principles of all material bodies. *The Century Dictionary and Cyclopedia* the entire entries for "hypostatic", and "hypostatization" can be found on pp. 2957–2958.

The word "hypostatic" appears in *The Collected Papers* 9 times and the word "hypostasis" once.

Idea—In the Platonic philosophy, and in similar idealistic thought, an archtype, or pure immaterial pattern, of which the individual objects in any one natural class are but the imperfect copies, and by participation in which they have their being: in this sense the word is generally qualified by the adjective *Platonic*. … A mental image or picture. … In the language of Descartes and of English philosophers, an immediate object of thought—that is, what one feels when one feels, or fancies when one fancies, or thinks when one thinks, and, in short, whatever is in one's understanding and directly present to cognitive consciousness. … A conception of what is desirable or ought to be, different from what had been observed; a governing conception or principle; a teleological conception. … In the *Kantian philos.*, a conception of reason, the object of which transcends all possible experience, as God, Freedom of the Will, Immortality; in the *Hegelian philos.*, the absolute truth of which everything that exists is the expression—the ideal realized, the essence which includes its own existence: in the latter sense commonly used with the definite article; in other a priori philosophies, an a priori conception of a perfection to be aimed at, not corresponding to anything observed, nor even fully realized. An opinion; a thought, especially one not well established by evidence. … An abstract principle, of not much immediate practical consequence in existing circumstances. *The Century Dictionary and Cyclopedia*, the complete entry for "idea" can be found on pp. 2973.

While Peirce did not contribute an entry for the word "idea" to the *Dictionary of Philosophy and Psychology* the word does appear in several entries. "Idea" appears in the entries Peirce contributed for Genus, Index, Law of Thought, Limitative, Logic, Logomachy, Matter and Form, Method and Methodology or Methodeutic, Middle Term, Mixed, Modality, Monad, Name, Necessity, Negation, Negative and Nominal.

The word "idea" appears in *The Collected Papers* 724 times.

Individuality—The condition or mode of being individual. (a) The being individual in contradistinction of being general. (b) Existence Independent of other things; that which makes the possession of characters by the subject a distinct fact from their possession by another subject. (c) The unity of consciousness; the connection between all the different feelings and other modifications of consciousness which are present at any one instant of time. (d) The simplicity of the soul; the indivisible unity of the substance of the mind as it exists at any instant. (e) Personality; the essential characters of a person. [The use of this word, which has not a wide currency, tends to vagueness, owing to confusion with the meaning (b).]

Individual (in logic) [as a technical term of logic, *individuum* first appears in Boethius, in a translation from Victorinus, no doubt of atomon, a word used by Plato (*Sophistes*, 229 D) for an indivisible species, and by Aristotle, often in the same sense, but occasionally for an individual. Of course the physical and mathematical sense of the word were earlier. Aristotle's usual term for individuals is ta kaq ekasta, Lat. *singularia*, Eng. *singulars*.] Used in logic in two closely connected senses. (1) According to the more formal of these an individual is an object (or term) not only actually determinate in respect to having or wanting each general character and not both having and wanting any, but is necessitated by its mode of being to be so determinate. See PARTICULAR (in logic).

This definition does not prevent two distinct individuals from being precisely similar, since they may be distinguished by their heceeities (or determinations not of a generalizable nature); so that Leibnitz' principle of indiscernibles is not involved in this definition. Although the principles of contradiction and excluded middle may be regarded as together constituting the definition of the relation expressed by 'not,' yet they also imply that whatever exists consists of individuals. This, however, does not seem to be an identical proposition or necessity of thought; for Kant's Law of Specification (*Krit. d. reinen Vernunft*, 1st ed., 656; 2nd ed., 684; but it is requisite to read the whole section to understand his meaning), which has been widely accepted, treats logical quantity as a continuum in Kant's sense, i.e. that every part of which is composed of parts. Though this law is only regulative, it is supposed to be demanded by reason, and its wide acceptance as so demanded is a strong argument in favour of the conceivability of a world without individuals in the sense of the definition now considered. Besides, since it is not in the nature of concepts adequately to define individuals, it would seem that a world from which they were eliminated would only be the more intelligible. A new discussion of the matter, on a level with modern mathematical thought and with exact logic, is a desideratum. A highly important contribution is contained in Schröder's *Logik*, iii, Vorles. 10. What Scotus says (*Quaest. in Met.*, VII. 9, xiii and xv) is worth consideration.

(2) Another definition which avoids the above difficulties is that an individual is something which reacts. That is to say, it does react against some things, and is of such a nature that it might react, or have reacted, against my will.

This is the stoical definition of a reality; but since the Stoics were individualistic nominalists, this rather favours the satisfactoriness of the definition than otherwise.

It may be objected that it is unintelligible; but in the sense in which this is true, it is a merit, since an individual is unintelligible in that sense. It is a brute fact that the moon exists, and all explanations suppose the existence of that same matter. That existence is unintelligible in the sense in which the definition is so. That is to say, a reaction may be experienced, but it cannot be conceived in its character of a reaction; for that element evaporates from every general idea. According to this definition, that which alone immediately presents itself as an individual is a reaction against the will. But everything whose identity consists in a continuity of reactions will be a single logical individual. Thus any portion of space, so far as it can be regarded as reacting, is for logic a single individual; its spatial extension is no objection. With this definition there is no difficulty about the truth that whatever exists is individual, since existence (not reality) and individuality are essentially the same thing; and whatever fulfils the present definition equally fulfils the former definition by virtue of the principles of contradiction and excluded middle, regarded as mere definitions of the relation expressed by 'not.' As for the principle of indiscernibles, if two individual things are exactly alike in all other respects, they must, according to this definition, differ in their spatial relations, since space is nothing but the intuitional presentation of the conditions of reaction, or of some of them. But there will be no logical hindrance to two things being exactly alike in all other respects; and if they are never so, that is a physical law, not a necessity of logic.

This second definition, therefore, seems to be the preferable one. Cf. PARTICULAR (in logic). Dictionary of Philosophy and Psychology, this complete entry can be found on pp. 537–538. Peirce did not provide the entry for individuality in the *Dictionary of Philosophy and Psychology*.

The word "individuality" appears in *The Collected Papers* 28 times.

Reason—An idea acting as a cause to create or confirm a belief, or to induce a voluntary action; a judgment or belief going to determine a given belief or line of conduct. … A fact known or supposed, from which another fact follows logically, as in consequence of some known law of nature or the general course of things; an explanation. … An intellectual faculty, or such faculties collectively. Intelligence considered as having universal

A reasonable thing; a rational thing to do; an idea or a statement conformable to common sense. … The exercise of reason; reasoning; right reasoning; argumentation; discussion. … The intelligible essence of a thing or species; the quiddity. … In logic, the premise or premises if an argument, especially the minor premise. *The Century Dictionary and Cyclopedia*, complete entries for "reasonable", "reasoned", and "reasoner" can be found on pp. 4990–4991.

Reasoning—is a process in which the reasoned conscious that a judgment, the conclusion, is determined by other judgment or judgments, the premises, according to a general habit of thought, which he may not be able precisely to formulate, but which he approves as conducive to true knowledge. *Dictionary of Philosophy and Psychology*, the complete entry may be found on pp. 426–428.

The word "reasoning" occurs in *The Collected Papers* 776 times.

References

Baldwin, John M. 1902. *Dictionary of philosophy and psychology*. New York: Macmillan.
Cassirer, Ernst. 1944. *An essay on man*. New Haven: Yale University Press; London Milford.
Cassirer, Ernst. 1953. *The philosophy of symbolic forms. The phenomenology of knowledge*, vol. 3. Ralph Manheim, New Haven: Yale University Press.
Cassirer, Ernst. 2000[1942]. *The logic of the cultural sciences*. Translated and edited by Steve Lofts. New Haven: Yale University Press.
Haines, D.E. 2007. Santiago Ramón y Cajal at Clark University, 1899; his only visit to the United States. *Brain Research Reviews* 55: 463–480.
Hoppe, J., and J. Stojanovic. 2008. High-aptitude minds. *Scientific American Mind* 19: 60–67.
Langer, Susanne K. 1948. *Philosophy in a new key*. New York: New American Library.
Peirce, Charles S. 1868. Questions concerning certain faculties claimed for man. *Journal of Speculative Philosophy* 2: 103–114.
Peirce, Charles S. 1906. Signs, thoughts, reasoning (title given in forthcoming *Prolegomena to a Science of Reasoning, 2016*); Institute for Studies in Pragmaticism, Peirce Manuscripts (MS) 293, Texas Tech University.
Peirce, Charles S. 1911. Sketch of logical critics; Institute for Studies in Pragmaticism, Peirce Manuscripts (MS) 671:008, Texas Tech University.
Peirce, Charles S. 1913. A study of how to reason safely and efficiently; Institute for Studies in Pragmaticism, Peirce Manuscripts (MS) 681:023, Texas Tech University.
Peirce, Charles S. 2009. *Charles S. Peirce, the logic of interdisciplinarity: The monist-series*, ed. Elize Bisanz. Berlin: Akademie Verlag.
Peirce, Charles S., and Lady Victioria Welby. 2001. *Semiotic and significs: The correspondence between Charles. S. Peirce and Lady Welby*, ed. and comp. Charles S. Hardwick, with assistance of James Cook. Elsah, Illinois: The Press of Arisbe Associates.
Peirce, Charles S., Benjamin Gilman, Christine Ladd-Franklin, Allan Marquand, and O.H. Mitchell. 1883. *Studies in logic by members of the Johns Hopkins University*, ed. Charles S. Peirce. Boston: Little, Brown, and Company.
Samway, S.J., and H. Patrick, eds. 1995. *A thief of Peirce: The letters of Kenneth Laine Ketner and Walker Percy*. Jackson: University Press of Mississippi.
Story, William E. 1899. *Clark University, 1889–1899: Decennial celebration*. Worcester, Massachusetts: Clark University.

Chapter 23
Epilogue—Reflections on Complexions of Habit

Donna E. West

Abstract Peirce's entire semiotic rests upon his concept of habit. In supplying us with his categories, together with virtual habit as potential habit, Peirce opens up pregnant propositions, assertions, and arguments to arrive at novel inferences in the continuum of ideas. He makes emphatic his commitment to unify all members, by demonstrating the relevance of momentary, individual existents to the improvement of logic for all generations. Accordingly, habit welds members of the continuum—championing each accomplishment (physical, mental, conscious/unconscious), in light of their contribution to ontological and logical objectivity at large. In short, the compulsion to discover hidden patterns of truth for mankind provides a lofty purpose for the individual—to utilize signs (consciously and unconsciously) as tools of enlightenment beyond the self.

Keywords The categories · Continuum · Consciousness · Interpretants · Virtual habit

Introduction—Habit and the Continuum

Peirce's entire semiotic rests upon his concept of habit. It unifies individual existents, and integrates them with the fullness of the continuum. Peirce expresses this process as "concretion" or the all in the one (1.478, 5.107, 8.208), such that momentary existents/individuals can have sustained, living participation in the trajectory of events, in view of their eventual contribution to other co-existents and to the whole of creation. Boler likewise fastens on the issue of possibility in interpreting Peirce's notion of concretion: "the would-be is never contracted to the is" (Boler 1963: 141). In other words, individuals have a presence across space and time; they leave a trace which other members make a part of subsequent inquiry. Unlike Scotus' (i. 1290–1295: 80–81) and Leibniz' (c. 1686: 137) notions of

D.E. West (✉)
Department of Modern Languages, State University of New York at Cortland, Cortland, USA
e-mail: westsimon@twcny.rr.com

D.E. West and M. Anderson (eds.), *Consensus on Peirce's Concept of Habit*,
Studies in Applied Philosophy, Epistemology and Rational Ethics 31,
DOI 10.1007/978-3-319-45920-2_23

"contraction" in which each one ultimately represents the monad, (the single divine source), Peirce's concept of concretion celebrates the uniqueness and transcendence of each individual. In other words, for Peirce, because the individual makes its mark on the continuum when its unique patterns of belief/action (habits) influence the other members, discretely and as a whole, the continuum is consolidated and perpetuated. This is especially pronounced when the influence is taken up, and is implemented and modified by other faces, extending to future occasions. Accordingly, the individual as a distinct unit imprints its habits, and ultimately its identity upon the continuum—making immeasurable memories of its existence and contributions to the state of knowledge at large. In short, the import of the continuum is obviated through the lens that habit supplies.

In fact, habit serves an essential purpose—to supply continual sustenance for the continuum. It fuels novel inferences, which, in turn, form the inheritance to logically nourish its prodigy. In this way, habit-change has its ultimate expression—to act as the kernel for semiosis, such that when received by members (sentient and non-sentient alike) its interpretants remain open to further expansion. As such, a chain of sign interpretation is promoted. Peirce determines: "Man is not whole as long as he is single; he is essentially a possible member of society. Especially one man's experience is nothing, if it stands alone. If he sees what others cannot we call it hallucination. It is not my experience but our experience that has to be thought of; and this "us" has indefinite possibilities" (1893: 5.402). Here Peirce demonstrates how unique individuals amount to little, apart from their contribution to others. Peirce intimates here that the interpreter is the source for "indefinite possibilities" or expanded logical interpretants. But, later, (in 1906) when he introduces the emotional, energetic, and logical interpretants, Peirce makes explicit the critical function of the interpreter to semiosis, to enlarging or shrinking interpretants (whichever is more appropriate): "I have already noted that a sign has an Object and an Interpretant, the latter being that which the sign produces in the quasi-mind that is the interpreter *by determining the latter to a feeling, to an assertion, or to a sign,* which determination is the Interpretant" (4.536). Peirce points out that without the quasi-mind of the interpreter semiosis would fail, since determinations of feelings, exertions, and/or signs would be truncated; and the final interpretant could never even be an aspiration. Hence, interpreters stand at the core of Peirce's semiotic; the would-bes of their subsequent meanings/effects perpetuate new inferences and, in turn, introduce habit-change.

Effects upon the interpreter continue to occupy Peirce's focus. In a 1908 letter to Lady Welby he describes the interpretant as: "an effect upon a person" (SS: 80–81). The fact that in the same passage Peirce extends interpretership beyond humans lends further credence to his emphasis on the continuum [Short (1996: 496, 2007: 52) and Bergman (this volume) likewise observe this phenomenon]. The interpreter (humans and non-humans alike) opens up novel avenues of application for the sign, augmenting emotional and energetic interpretants to logical ones, and expanding/contracting the substance of logical interpretants. Peirce's existential graphs particularly illustrate the integral place for the interpreter in Peirce's semiotic. The existence of the interpreter impels liberal exchange of signs, and supplies the impetus for expressing subjunctive, imperative, and indicative assertions.

The salience of the interpreter underscores the function of dialogue, since it is dialogue which represents the prime means to make explicit implied propositions/assertions into arguments, likewise highlighting regularities of thought and action (cf. West 2015, this volume), and highlights the role of the interpreter to extract the habits of the message producer, and to re-situate them within his own interlanguage or (to coin a new term) "inter-semiotic." In short, habit welds individuals to the continuum (both synchronically and diachronically); in so doing, it advances the state of logic for all.

Application to Peirce's Categories

Without undermining the role of logic (and the need for objectivity in the reasoning process), Rosenthal (1982) supports the premise that pragmatic, action-based schemas form the basis for Peircean habit. She tacitly invokes the premise that absolute chance, not ordinary chance, is the primary force guiding belief and action habit. In support of her position, Rosenthal demonstrates the indispensability of Peirce's categories—how Peirce's phenomenological tenets inform his pragmatic ones. She remonstrates upon how Firstness, as Peirce's most basic category, flows in upon pragmatic, brute force impositions of Secondness, via sensorimotor pathways. She posits that foundational to habit are certain basic kinds of causative relations between events, determined not by means of intuition, but by a colonization or harnessing of Firstness to bear upon pragmatic elements in Secondness. The beauty of Rosenthal's approach is the fact that she features habit as the factor most responsible for translating insights from interior sources to the language of exterior action schemas. The inspiration of Peirce's categories to enlighten the process of habit's profound influence upon courses of action is invaluable; it reifies the primacy of Peirce's categories, while obviating the indispensability of habit as the gatekeeper to resolve logical dissonance. Although Rosenthal does not expressly articulate the latter claim, it is pregnant in the very fabric of her approach—to clarify habit's role as the agent for determining the plausibility and workability of predictions, together with habit's influence on establishing revisionary courses of action.

She demonstrates that without Firstness, Secondness would be virtually absent of inspiration; and without habit, Thirdness would amount to little more than conformity to precepts of law-like proportion—relegating Thirdness to a sterile set of invariant how-tos. Such would not do at all; it would undermine Peirce's primary injunction: not to "block the way to inquiry." Foundational to Rosenthal's message is that habit is the single, most critical factor monitoring the influence of Peirce's categories to real-world enactments. It is habit that makes possible the ingenuity from insight in Firstness into pragmatic genres, by discouraging fleeting and hasty assertions, and in turn, precluding practices destined to fail.

Rosenthal's treatment of causation especially illuminates the efficacy of habit as gatekeeper of abductions, promoting hunches which Peirce would claim are ultimately good guesses: "…unless a man had a tendency to guess right [habit], unless his guesses are better than tossing up a copper, no truth that he does not already virtually possess could ever be disclosed to him, while if he has any decided tendency to guess right, as he *may* have, then no matter how often he guesses wrong, he will get at the truth at last" (1903: EP2:250). Guessing right ultimately materializes in beliefs harvested as habit-change—as courses of action. Moreover, absent the implicit recognition of regularities (habit), many hypotheses would remain unrecognized; and getting at the truth would be truncated. Particularly useful is Rosenthal's characterization of the kinds of causation underlying event relations and their effects, especially Peirce's concept of Efficient Causation: "…it can be seen that efficient causation, in the sense of actualization of a possibility, requires the rational or 'ideal causality' of Thirdness to provide the positive potentialities, while Thirdness, apart from its relation to Secondness is not real" (Rosenthal 1982: 240). In other words, any actualization of novel predictions (possibilities) defy any potentiality of existence absent their groundedness in Thirdness—in the organism providing the essential structure for embodiment. In short, it is habit, as the agent of Thirdness, which validates the promise for and utility of guessing right via insights; likewise habit sanctions whether new approaches have sufficient merit (in light of past practices and the factor of probability) to recommend them as viable ontological operators.

Nonetheless, although the essence of habit is Thirdness, it, in itself, can never possess sufficient concreteness and status as an individual (in Peirce's sense) to materialize—it needs some remnant of Secondness to have momentary existence as an individual. Absent application to Peirce's other two categories, habit would lack relevance; and its status as a Third would serve little purpose. Peirce convinces us of the credibility of this line of reasoning as follows: "the third category—the category of thought, representation, triadic reality, mediation, genuine thirdness, thirdness as such—is an essential ingredient of reality, yet does not by itself constitute reality, since this category (which is that cosmology appears as the element of habit) can have no concrete being without acting, as a separate object on which to work its government, just as action cannot exist without the immediate being of feeling on which to act" (5.436: 1904).[1] Between 1904 and 1907, Peirce makes plain the indispensability of Secondness via index to ground the object and to suggest relations within and between events, culminating in more general tendencies of belief and action, namely, habit. Although in this way, Secondness illustrates relations, habit supersedes these actualities—it welds both feelings, and actions into tendencies, increasing still further the relevance of the combined effect of his categories. Habit beckons cohesiveness between feelings, actions, and concepts, which

[1]A portion written in *The Monist* in which he emphasizes his primary application of habit to pragmatic genres "…that in this way one can put in strong light a position which the pragmatist holds and must hold…" (5.436: 1904).

otherwise might remain unexpressed and consequently unrecognized as aggregate interpretants of signs.

In the previous passage, Peirce illustrates a pragmatic platform for habit. He asserts that habit is the governing force directing action, in its power to retrospectively and prospectively consider Efficient Causation. Peirce means that in integrating facts with possibility, habit determines how a particular kind of cause brings about certain consequences. As such, habit garners the potency necessary to govern decisions about how to act to produce a sought-after result, even when the outcome surfaces unexpectedly. Its capacity to govern internal and external happenings unifies phenomenological and perceptual operations, necessary to plausibly predict the relationship of one event/condition to the eventuality of another.

Energizing the place of Secondness is the elevated role of index. It serves as the basis upon which habit in Thirdness can operate—have expression, because without the forum of Secondness, spatial and temporal relations would garner little, if any, force to affect the state of knowledge (cf. Short 2007: 48–50 for a more foundational discussion). After 1885, Secondness via Index (for Peirce) constitutes the conduit by which sense data become relevant to what can be known—it updates the state of knowledge by acknowledging that particulars have ontological status; and thus they must be afforded a more central place in his semiotic. Peirce augments his notion of individuals as particulars existing for more than a single point in space and time. He does so by taking up Scotus' commitment to continuity within the physical world, in the haecceity of the moment (spatial and temporal features): "In truth, any fact is in one sense ultimate,—that is to say, in its isolated aggressive stubbornness and individual reality. What Scotus calls the haecceities of things, the hereness and nowness of them, are indeed ultimate.... Why IT, independently of its general characters, comes to have any definite place in the world, is not a question to be asked; it is simply an ultimate fact" (1887–1888: W6: 205). In providing for increased status of Secondness where Index reigns, Peirce encapsulates volitional and non-volitional action-habits of both non-sentient and sentient systems.[2] He expresses the primary purpose of Index to be attentional and momentary, but displays the relational continuity that Index in the forum of Secondness affords—in its "stubbornness," allowing the particular object of its focus a more permanent identity.

In 1885 and thereafter Peirce attributes to Index (and as such Secondness) with a riveting function, pointing out objects as individuals (but with some quality of continued existence) in time and space, thus rendering Index germane not merely to momentary attentional foci (which are often arbitrary and capricious), but to signs which convey direction for novel predictions about event relations (1885: 8.41). As such, viable predictions can be proffered regarding the processes contributing to

[2]"In fact, habits, from the mode of their formation, necessarily consist in the permanence of some relation...each law of nature would consist in some permanence such as the permanence of mass, momentum, and energy" (1887–1888: W6: 210).

particular geological effects and to other physical patterns of the natural world. It is obvious then, that shepherding habit is foundational as a path-maker for existent and potentially existent episodes (illustrating habit and virtual habit, respectively); and this becomes the primary responsibility of Index. It frames, locates, directionalizes; but most critically, it supplies the spatial and temporal axes in Secondness, so necessary to conceiving continuities/generalities in Thirdness. In short, the role of habit (facilitated by Index) to introduce and sustain new pathways of ontological knowledge is extraordinary. It enhances the import of the categories by implementing deictic features into deontic and epistemological genres. It secures the reality of objects even when they are momentarily absent, and highlights activity in possible events. Here hypotheses explaining causal relations conceived in Firstness into fields of Secondness are realized.

Habit has the potency to create a unified pathway drawn (inscribed upon the physical surround, and/or upon the mental canvas) from a host of actual or imagined venues. In fact, it is habit that rescues us from generating propositions and hypotheses solely upon observance of brute force features in the haecceity of the moment. Fastening upon patent, as well as latent factors is especially critical. Patent factors are particularly ripe given the likelihood of focusing on salient features over others. Nonetheless, while Index beckons observers to attend to relational components (often visual) within Secondness, the haecceity inherent in index, alone is insufficient to de-emphasize incidental factors, and promote instead propositions which rest upon logic alone. By its very nature as a visual sign referring to co-existent objects in Secondness, Index cannot be relied upon to either ignore other riveting influences of discrete features present in the here and now, nor can it naturally recognize similarities across uses. When constructing a plausible rendering of event relations in Secondness, index is hard pressed to ignore irrelevant features. Habit rescues Secondness from its momentariness, bringing it into communication with the whole of its instantiations and its logical interpretants. The lens of habit constitutes the single most critical operation conjoining all instantiations. Furthermore, habit affords consideration (consciously or unconsciously) of hidden contributory factors crucial to determine plausible explanations for abductive inferences.

Logical Interpretants and Virtual Habit

It is in habit that Peirce's categories gain their fullest expression. A forum for feelings in Firstness supplies an impetus to be recognized and to take flight; and facts in Secondness acquire new associations when performed in novel episodes and when informed by discoveries and motivated by affect. In fact, these new regularities constitute habits which insinuate novel law-based Thirdnesses to previous perceptual judgments, renovating regularities in Firstness and Secondness. Habit constitutes a re-interpretive tool for previous Firstnesses and Secondnesses—inscribing itself upon raw feelings and experiences (indirect and direct). In this way, the influence of the Logical Interpretant takes precedence—steering semiosis from Emotional/Energetic Interpretants toward Final/Ultimate Interpretants. This critical function for habit (of re-steering interpretants) illustrates the indexical quality of the Logical Interpretant, demonstrating further the indispensability of index in the business of habit formation. The indexical function measures strides toward reaching the Final interpretant. Indexical features direct semiosis of Logical interpretants to new heights—toward the realization of modality-based effects. In fact, without implications wrought from Dynamic Interpretants (Emotional or Energetic), the possibility of ascertaining Final Interpretants would be truncated. Ultimately, these three (Emotional, Energetic, Logical) as Immediate, Dynamical, or Final Interpretants [along the lines of Short's (1996: 509; 2007: 58) argument] can begin to incorporate the would-be properties characteristic of Final Interpretants.

Peirce alludes to this process in MS 620 (1909): "But Hell is paved with good resolutions; and therefore to this promise [of future conduct] must be attached good security, or…must be baked into the hard brickbat of a real determination of the habit-machinery of his organism, which shall have force to govern his actions. A determination is a virtual habit [not a habit]…but it works all the effects of habit…" For Peirce, a "resolution" does not qualify as either a habit or a virtual habit; it fails to possess the impetus or "force" to direct the action plan of the organism. Peirce implies resolutions lack the force to become determinations because of their absence of clarity. Conversely, determinations require clarity which is often supplied by the vividness of the image if the virtual habit is dream-like: "The effectiveness of the virtual habit relatively to that of a real habit is, I say, unquestionably than in proportion to the vividness of the imaginations that induce the former relatively to the vividness of the perception…" (MS 620: 24–25). In fact, the more vivid is the more likely the determination is to be implemented into action habits and/or to gain status as a permanent belief.

What Peirce's sense of determination imparts is constructed, truly directionalized path to implement a new approach, not merely a vague sense of wanting to orchestrate a change but a compulsion to do so consequent to a vivid imagination of the effect once the virtual habit is implemented. In contrast, a resolution is but a fleeting thought not involving a commitment to perform any planned action scheme. In fact, in a Nota Bene in the same manuscript, Peirce likens the distinction in the use of "virtual/el" between English and German, to the difference between resolutions and determinations. While resolutions articulate the need for a change in conduct/action (absent any impetus or plan), real determinations are driven by affect in Firstness to put into practice the sequence of preferred conduct. Peirce explains that determinations entail more than a possibility to perform the action, but a real investment—requiring intention and the will to perform. As such, Peirce considers resolutions to be weak predictors of a performance commitment, hence not embracing habit; he asserts in the same passage: "Hell is paved with good resolutions." Such implies that the interpretant of resolution-based assertions is nothing short of "fluff" or failed promises—effect being non-action, procrastination, and the like. Conversely, interpretants of determinations (their meaning and effects) are so reliable as to promote expectations that the conduct will be implemented without fail—in circumstances almost as if the promised action were already performed. To illustrate: later in the same passage, Peirce analogizes this point in a dialogue between Milton's Adam and Raphael.[3] The spiritual "touch" to which Adam alludes has the same effect as if a physical touch were extended: "…while it [virtual habit] is not the N [any common noun], it has nevertheless the characteristic, behavior and properties of an N…. By virtual touch Milton's Adam meant something that was not touch, but we might [*sic*] all the delight that touch can bring" (MS 620: 25: 1909).

Although Peirce's notion of virtual habit produces a similar effect to that of actual habit, it does not qualify as habit, since it is either but a fleeting experience, or a could-be/would-be one: "So a determination [a virtual habit] is not a habit, since it does not result in repeated performances on the same sort of occasions of the sort of action that it will cause to be again performed on the same sort of occasion; it works all the effects of habit and is therefore strictly speaking a virtual habit" (MS 620: 25: 1909). In other words, because virtual habits fail to demonstrate a regular pattern (either of action or mental patterns) they do not reach muster to qualify as habit. Despite the fact that they do not qualify as habit, Peirce particularly underscores the fact that they consist in a reliability far greater than the "*virtuel*" habits (German) which simply denote possibility (MS 620: 24: 1909). In short, virtual habits are equally real alongside habits; they merely lack consistency, consequent to their dependence upon the spontaneity characteristic of Firstness. Essentially, not even their identification with mere potentiality resigns instances of virtual habit to the periphery of habit proper.

Peirce does not allow the issue of "virtual habit" to fade. Two years later, in 1911, he rearticulates his conviction that determinations (as opposed to resolutions) represent a preferential precondition to transmit a simple awareness of the need to translate firstness into action. In MS (674: 1911) he applies this distinction to children's development of logical and moral skills: It seems to me that a command or act of will intended to cause the person who performed it to behave in a certain way on a future occasion or on all occasions of a certain kind is what the word resolution ought to be understood to denote, while that state of a person brought about by his own will that consists in its being true that he will behave in a certain way either on a certain occasion or on any other occasion of a certain kind that may take place ought to be called a determination; that is a "determination," or "settlement of his will." It is obvious that Peirce considers a resolution to be less reliably performed—not founded upon real belief, while a determination is a lasting state. "…if I am not mistaken may properly be called a habit; although it is not created, as most states to which this word is usually applied are, by frequent repetition" (MS 674: 14–15: c. 1911). It is critical to consider the context of Peirce's latter remarks; they are an outgrowth of a lengthy description of how to

[3]"Love not the Heav'nly spirits? And how their Love Express they? By looks only? Or do they mix Irradiance, virtual or immediate touch" (*Paradise Lost* VIII.615–617)?

advance meaningful self-control. Peirce refers to ascertaining this self-control as "the supreme art." In bringing into captivity one's feelings in Firstness and one's effort in Secondness, one can be liberated to give an imperative to one's self—ultimately to put into action what is dormant, slumbering and perhaps divorced from the continuum at large. As such, both icons in Firstness, and indexes as brute experiences in Secondness can be synthesized into modes of believing and behaving as habit.

In Peirce's Logic Notebook (October, 1909), he discusses the classes of interpretants which can support the existence of virtual habits. In fact, his proposal of the new interpretant classes may well be to directly anchor determinations as legitimate habits, despite their status as peripheral to habit. In ascribing discrete action-based properties to novel interpretants (1906: MS 339), virtual habits are raised to new heights; but, even these virtual habits must surface as potentialities, not as slight possibility or impossibility. Peirce constructs his argument for virtual habit by reiterating his continued conviction that inferences consequent to future but objective standards (without eliminating creative insights) must govern if the inferences are to be folded into habit proper. Peirce defines the boundaries of virtual interpretants as follows: "I do not think that the import of any word except perhaps a pronoun is limited to what is in the utter's mind actualiter...It is, on the contrary...what is in the mind, perhaps not even habitualiter, but only virtualiter, which constitutes the import" (c. 1905: 5.504). As such, Peirce solidifies the critical role of creatively imaginative propositions/assertions into the heart of triadic relations, especially energizing interpretants in Thirdness; but he does so privileging the incursion of subsequent potential meanings into the interpretant file. Although Peirce framed the influence of virtual habit with refinements in the logical interpretant (late in his semiotic), seeds of virtual habit were sown as early as 1868 (W2: 227): "Finally, no present actual thought (which is a mere feeling) has any meaning, any intellectual value (of a present thought) lies in what this thought may be connected with in representation by subsequent thoughts (which renders meaning) altogether something virtual."

Accordingly, effects which (although not actualized) possess a real potentiality of being so are subsumed into previous energetic/logical interpretants—expanding interpretants, and legitimizing virtual habits as habits. In other words, the interpretants of current thoughts are but a shadow of what they are intended to be in light of subsequent ones. Another critical factor afforded by virtual habit to semiosis is that these interpretants must be of a type to permit them to be actualized more universally, not merely to be germinated in a single mind, nor to fester in such mind, because absent sharing tenable suggestions with other minds, inquiry is narrowly defined. The kind of interpretant which Peirce insists qualifies as habit must possess the imperative to potentially affect others either by virtue of its future scientific benefits or its pragmatic advantages—recommending itself as a sound course of action (1909: MS 637: 12)—a way of believing and acting which results in broad benefits. Of course, the recommendations materialize in re-constructed logical interpretants which give rise to foreseen, particularized effects. Anything less, e.g., resolutions, vague renderings, unlikely possibilities, cannot hope to give rise to real, specific, or expected effects, in view of their indefinite character.

As articulated in this volume (Bergman, West), and as pointed out by Short (1996), Peirce's concept of the individual renovated his semiotic, as well as his metaphysic. His notion of the individual, a momentary state of affairs, was given a primary place in his continuum—allowing the individual to represent the whole. Accordingly, no longer need they (the individual and the continuum) oppose each other. In this way, Peirce intentionally integrated the individual into the continuum; and as such, the individual signs can enrich and perpetuate the continuum. The process is activated by momentary signs which penetrate already established meanings in the mind to confirm, disconfirm, or to lead the self or another to revisionary hypotheses and/or action-habits.

Conversely, virtual habits, given their already conceived of directed actions, can (if implemented) transform Dynamic Interpretants into Final ones. As such, habit as Logical Interpretant is permitted greater latitude when it takes advantage of potentialities pregnant in predictions of cause-effect structures, provided that determinations propel the actualization of the conduct. According to this taxonomy, potential propositions, assertions, and/or arguments can serve as potential habits, if they are sufficiently specified/vivid; but a proposal for generation of entire worlds is unlikely to result in a sufficiently definite objective to produce actual habit. Rationale lies in the fact that indefinite descriptions (as in possible worlds) are neither narrow nor specific, and cannot be expected to provide a clearly articulated energetic or logical interpretant. Worlds which are so indeterminate such that the likelihood of actualization is miniscule, ordinarily do not qualify as virtual habits, since real potentiality for their existence is de-legitimized.

The Role of Consciousness-Raising

For sentient beings, consciousness serves as a gatekeeper to discern the desirability of behavior schemas and/or the plausibility of hypotheses/remedial courses of action—a use of habit which Peirce considered to be central (MS 637:12: 1909). In fact, the presence of consciousness in habit-taking is paramount, particularly when constructing principles to govern the recommendation of viable courses of action. This is so given the indispensability of revising Logical Interpretants to mediate habit change, and perhaps reassigning them to different sign vehicles.

Consciousness takes many forms: from the unconscious, to awareness, to deliberate intervention, to the presence of intentionality, and finally to planning and employing metacognitive competencies. In 1905 Peirce makes plain that certain forms of consciousness need not underlie taking a habit; but "imagination consciousness" is ultimately paramount. It is evident that the import of Firstness in raising consciousness to control mental habits via construction of possible worlds in the imagination is a powerful tool in semiosis. Here habit-taking is obviated by the influence of affect to impel thoughts from a state of possibility (in which propositions are quasi- and nebulous) to actuality and mediated logical structures. But, because "immediate consciousness" pertains to indubitable assumptions which fail to give rise to defined expressible possibilities derived from genres of Firstness, Firstness-driven processes alone cannot hope to qualify as habit. Peirce uses meaning as "the conception [a word] conveys" (W 2:238) as the original point of departure for habit; and, contrary to Short, habit is not the "ultimate interpretant of [a concept]" (2007: 58). Instead, habit is an expression of the unconscious for sentient beings. In short, in its fullest sense, although habit affects and is affected by consciousness (especially when mental control is operational), it is not reducible to consciousness, since states of affairs and their outcomes are not propelled by living agents, and since even when they are, other forms of habit control, i.e., unconscious insight/automatic instinct.

Conclusion: The Ultimate Purpose for Habit

In supplying us with virtual habit as potential habit, Peirce opens up pregnant propositions, assertions, and arguments to arrive at novel inferences in the continuum of ideas. He makes emphatic his commitment to unify all members, by demonstrating the relevance of momentary, individual existents to the improvement of logic for all generations. In order to legitimately claim that habit cannot exist without habit-change, he had to include creative abductions as part and parcel of habit proper to be the quintessential form for diverse systems. Accordingly, habit welds members of the continuum—championing each accomplishment (physical, mental, conscious/unconscious), in light of their contribution to ontological and logical objectivity at large. In short, the compulsion to discover hidden patterns of truth for mankind provides a lofty purpose for the individual—to utilize signs (consciously and unconsciously) as tools of enlightenment beyond the self—for his fellow man; and it is only through assessing and reassessing interpretants that memories can obtain their full significance, and propositions/assertions/arguments their ultimate effect.

References

Bergman, Mats. this volume. Habit-change as ultimate interpretant. In *Consensus on Peirce's concept of habit: Before and beyond consciousness*, ed. Donna E. West and Myrdene Anderson. (Studies in Applied Philosophy, Epistemology and Rational Ethics [SAPERE].) New York: Springer.

Boler, John. 1963. *Charles Peirce and Scholastic realism: A study of Peirce's relation to John Duns Scotus*. Seattle: University of Washington Press.

Leibniz, Gottfried Wilhelm. 1988 [c. 1686]. Primary truths. In *Discourse on metaphysics and related writings*, ed. R. Niall, D. Martin and Stuart Brown, 131–138. Manchester: Manchester University Press.

Milton, John. 2005 [i. 1667–1674] *Paradise lost*, ed. Philip Pullman. Oxford: Oxford University Press.

Peirce, Charles Sanders. i. 1867–1913. *Collected papers of Charles Sanders Peirce*. Vols. 1–6, ed. Charles Hartshorne and Paul Weiss. Cambridge: Harvard University Press, 1931–1935. Vols. 7–8, ed. Arthur W. Burks. Cambridge: Harvard University Press, 1958. [References to Peirce's papers will be designated by CP, followed by volume, period, paragraph number.].

Peirce, Charles Sanders. i. 1867–1893. *The essential Peirce: Selected philosophical writing*. Volume 1 (1867–1893), ed. Nathan Houser and Christian Kloesel. Bloomington: Indiana University Press, 1992. [References to this volume will be designated by EP 1, followed by colon, page number.].

Peirce, Charles Sanders. i. 1893–1913. *The essential Peirce: Selected philosophical writing*. Volume 2 (1893–1913), ed. the Peirce Edition Project. Bloomington: Indiana University Press, 1998. [References to this volume will be designated by EP 2, followed by colon, page number.].

Peirce manuscripts in Texas Tech University Library at Texas Tech University, Institute of Studies of Pragmaticism, beginning with MS—or L for letter—and followed by a number, refer to the system of identification established by Richard R. Robin in Annotated Catalogue of the Papers of Charles S. Peirce (Amherst: University of Massachusetts Press, 1967), or in Richard R. Robin, "The Peirce Papers: A Supplementary Catalogue," Transactions of the Charles S. Peirce Society.

Peirce, Charles Sanders. i.1867–1913. *Writings of Charles S. Peirce: A chronological edition*. Vols. 1–6 to date, ed. the Peirce Edition Project. Bloomington: Indiana University Press. [References to these volumes will be designated by W, followed by volume number, colon, page number.].

Peirce, Charles Sanders and Victoria Lady Welby. Semiotic and Significs: The Correspondence between Charles S. Peirce and Victoria Lady Welby, Charles S. Hardwick, ed. with the assistance of James Cook. (The Press of Arisbe Associates, Second Edition, 2001).

Rosenthal, Sandra. 1982. Mean as habit: Some systematic implications of Peirce's pragmatism. *The Monist* 65(2): 230–245.

Scotus, John Duns. 2005 [i. 1290–1295]. *Early Oxford lecture on individuation*, trans. A. Wolter. St. Bonaventure: Franciscan Institute Press.

Short, Thomas L. 1996. Interpreting Peirce's interpretant: A response to Lalor, Liszka, and Meyers. *Transactions of the Charles S. Peirce Society* 32(4): 488–541.

Short, Thomas L. 2007. *Peirce's logic of signs*. Cambridge: Cambridge University Press.

West, Donna E. 2015. Dialogue as habit-taking in Peirce's continuum: The call to absolute chance. *Dialogue (Canadian Review of Philosophy)* 54(4): 685–702.

West, Donna E. this volume. Indexical scaffolds to habit-formation. In *Consensus on Peirce's concept of habit: Before and beyond consciousness*, ed. Donna E. West and Myrdene Anderson. (Studies in Applied Philosophy, Epistemology and Rational Ethics [SAPERE].) New York: Springer.

Index

D.E. West and M. Anderson (eds.), *Consensus on Peirce's Concept of Habit*,
Studies in Applied Philosophy, Epistemology and Rational Ethics 31,
DOI 10.1007/978-3-319-45920-2

CPSIA information can be obtained
at www.ICGtesting.com
Printed in the USA
LVOW13*2134101217
559299LV00008B/382/P